Jürgen H. Gross

Mass Spectrometry

Springer

Berlin
Heidelberg
New York
Hong Kong
London
Milan
Paris
Tokyo

Jürgen H. Gross

Mass Spectrometry

A Textbook

With 357 Illustrations

 Springer

Dr. Jürgen H. Gross
Institute of Organic Chemistry
University of Heidelberg
Im Neuenheimer Feld 270
69120 Heidelberg
Germany
author@ms-textbook.com

Problems and Solutions available via author's website

www.ms-textbook.com

ISBN 3-540-40739-1 Springer-Verlag Berlin Heidelberg New York

Bibliographic information published by Die Deutsche Bibliothek
Die Deutsche Bibliothek lists this publication in the Deutsche Nationalbibliografie;
detailed bibliographic data is available in the Internet at http://dnb.ddb.de.

Springer-Verlag is a part of Springer Science+Business Media

springeronline.com

© Springer-Verlag Berlin Heidelberg 2004
Printed in Germany

Typesetting: Data conversion by author
Cover-design: Künkel+Lopka, Heidelberg

Printed on acid-free paper   02/3020 ra – 5 4 3 2 1 0

Dedicated to my wife Michaela and my daughters Julia and Elena

Preface

When non-mass spectrometrists are talking about mass spectrometry it rather often sounds as if they were telling a story out of Poe's *Tales of Mystery and Imagination*. Indeed, mass spectrometry appears to be regarded as a mysterious method, just good enough to supply some molecular weight information. Unfortunately, this rumor about the dark side of analytical methods reaches students much earlier than their first contact with mass spectrometry. Possibly, some of this may have been bred by mass spectrometrists themselves who tended to celebrate each mass spectrum they obtained from the gigantic machines of the early days. Of course, there were also those who enthusiastically started in the 1950s to develop mass spectrometry out of the domain of physics to become a new analytical tool for chemistry.

Nonetheless, some oddities remain and the method which is to be introduced herein is not always straightforward and easy. If you had asked me, the author, just after having finished my introductory course whether mass spectrometry would become my preferred area of work, I surely would have strongly denied. On the other hand, J. J. Veith's mass spectrometry laboratory at Darmstadt University was bright and clean, had no noxious odors, and thus presented a nice contrast to a preparative organic chemistry laboratory. Numerous stainless steel flanges and electronics cabinets were tempting to be explored and – whoops – infected me with CMSD (chronic mass spectrometry disease). Staying with Veith's group slowly transformed me into a mass spectrometrist. Inspiring books such as *Fundamental Aspects of Organic Mass Spectrometry* or *Metastable Ions*, out of stock even in those days, did help me very much during my metamorphosis. Having completed my doctoral thesis on fragmentation pathways of isolated immonium ions in the gas phase, I assumed my current position. Since 1994, I have been head of the mass spectrometry laboratory at the Chemistry Department of Heidelberg University where I teach introductory courses and seminars on mass spectrometry.

When students ask what books to read on mass spectrometry, there are various excellent monographs, but the ideal textbook still seemed to be missing – at least in my opinion. Finally, encouraged by many people including P. Enders, Springer-Verlag Heidelberg, two years of writing began.

The present volume would not have its actual status without the critical reviews of the chapters by leading experts in the field. Their thorough corrections, remarks, and comments were essential. Therefore, P. Enders, Springer-Verlag Heidelberg (*Introduction*), J. Grotemeyer, University of Kiel (*Gas Phase Ion Chemistry*), S. Giesa, Bayer Industry Services, Leverkusen (*Isotopes*), J. Franzen,

Bruker Daltonik, Bremen (*Instrumentation*), J. O. Metzger, University of Oldenburg (*Electron Ionization* and *Fragmentation of Organic Ions and Interpretation of EI Mass Spectra*), J. R. Wesener, Bayer Industry Services, Leverkusen (*Chemical Ionization*), J. J. Veith, Technical University of Darmstadt (*Field Desorption*), R. M. Caprioli, Vanderbilt University, Nashville (*Fast Atom Bombardment*), M. Karas, University of Frankfurt (*Matrix-Assisted Laser Desorption/Ionization*), M. Wilm, European Molecular Biology Laboratory, Heidelberg (*Electrospray Ionization*) and M. W. Linscheid, Humboldt University, Berlin (*Hyphenated Methods*) deserve my deep gratitude.

Many manufacturers of mass spectrometers and mass spectrometry supply are gratefully acknowledged for sending large collections of schemes and photographs for use in this book. The author wishes to express his thanks to those scientists, many of them from the University of Heidelberg, who generously allowed to use material from their actual research as examples and to those publishers, who granted the numerous copyrights for use of figures from their publications. The generous permission of the National Institute of Standards and Technology (G. Mallard, J. Sauerwein) to use a large set of electron ionization mass spectra from the NIST/EPA/NIH Mass Spectral Library is also gratefully acknowledged. My thanks are extended to the staff of my facility (N. Nieth, A. Seith, B. Flock) for their efforts and to the staff of the local libraries for their friendly support. I am indebted to the former director of our institute (R. Gleiter) and to the former dean of our faculty (R. N. Lichtenthaler) for permission to write a book besides my official duties.

Despite all efforts, some errors or misleading passages will still have remained. Mistakes are an attribute that make us human, but unfortunately, they do not contribute to the scientific or educational value of a textbook. Therefore, please do not hesitate to report errors to me or to drop a line of comment in order to allow for corrections in a future edition.

Hopefully, *Mass Spectrometry – A Textbook* will introduce you to the many facets of mass spectrometry and will satisfy your expectations.

Jürgen H. Gross

University of Heidelberg
Institute of Organic Chemistry
Im Neuenheimer Feld 270
69120 Heidelberg
Germany
email: author@ms-textbook.com

Table of Contents

1 Introduction

Mass spectrometry is an indispensable analytical tool in chemistry, biochemistry, pharmacy, and medicine. No student, researcher or practitioner in these disciplines can really get along without a substantial knowledge of mass spectrometry. Mass spectrometry is employed to analyze combinatorial libraries [1,2] sequence bio-molecules, [3] and help explore single cells [4,5] or other planets. [6] Structure elucidation of unknowns, environmental and forensic analytics, quality control of drugs, flavors and polymers: they all rely to a great extent on mass spectrometry. [7-11]

From the 1950s to the present mass spectrometry has changed tremendously and still is changing. [12,13] The pioneering mass spectrometrist had a home-built rather than a commercial instrument. This machine, typically a magnetic sector instrument with electron ionization, delivered a few mass spectra per day, provided sufficient care was taken of this delicate device. If the mass spectrometrist knew this particular instrument and understood how to interpret EI spectra he or she had a substantial knowledge of mass spectrometry of that time. [14-18]

Nowadays, the output of mass spectra has reached an unprecedented level. Highly automated systems are able to produce even thousands of spectra per day when running a routine application where samples of the very same type are to be treated by an analytical protocol that has been carefully elaborated by an expert before. A large number of ionization methods and types of mass analyzers has been developed and combined in various ways. People bringing their samples to a mass spectrometry laboratory for analysis by any promising ionization method often feel overburdened by the task of merely having to select one out of about a dozen techniques offered. It is this variety, that makes a basic understanding of mass spectrometry more important than ever before. On the other extreme, there are mass spectrometry laboratories employing only one particular method – preferably matrix-assisted laser desorption/ionization (MALDI) or electrospray ionization (ESI). In contrast to some 40–50 years ago, the instrumentation is concealed in a "black box" actually, a nicely designed and beautifully colored unit resembling an espresso machine or tumble dryer. Let us take a look inside!

1.1 Aims and Scope

This book is tailored to be your guide to mass spectrometry – from the first steps to your daily work in research. Starting from the very principles of gas phase ion chemistry and isotopic properties, it leads through design of mass analyzers, mass spectral interpretation and ionization methods in use. Finally, the book closes with a chapter on chromatography–mass spectrometry coupling. In total, it comprises of twelve chapters that can be read independently from each other. However, for the novice it is recommended to work through from front to back, occasionally skipping over more advanced sections.

Step by step you will understand how mass spectrometry works and what it can do as a powerful tool in your hands that serves equally well for analytical applications as for basic research. A clear layout and many high-quality figures and schemes are included to assist your understanding. The correctness of scientific content has been examined by leading experts in a manner that has been adapted as *Sponsor Referee Procedure* by an established mass spectrometry journal. [19] Each chapter provides a list of carefully selected references, emphasizing tutorial and review articles, book chapters and monographs in the respective field. Titles are included with all citations to help with the evaluation of useful further reading. [20] References for general further reading on mass spectrometry are compiled at the end of this chapter.

The coverage of this book is restricted to the field of what is called "organic mass spectrometry" in a broad sense. It includes the ionization methods and mass analyzers currently in use, and in addition to classical organic compounds it covers applications to bio-organic samples such as peptides and oligonucleotides. Of course, transition metal complexes, synthetic polymers and fullerenes are discussed as well as environmental or forensic applications. The classical fields of inorganic mass spectrometry, i.e., elemental analysis by glow-discharge, thermal ionization or secondary ion mass spectrometry are omitted. Accelerator and isotope ratio mass spectrometry are also beyond the scope of this volume.

> **Note**: "Problems and solutions" sections are omitted from the printed book. These are offered free of charge at http://www.ms-textbook.com.

1.2 What Is Mass Spectrometry?

Well, mass spectrometry is somewhat different. The problems usually start with the simple fact that most *mass spectrometrists* do not like to be called *mass spectroscopists*.

> **Rule:** "First of all, never make the mistake of calling it 'mass spectroscopy'. Spectroscopy involves the absorption of electromagnetic radiation, and mass spectrometry is different, as we will see. The mass spectrometrists sometimes get upset if you confuse this issue." [21]

Indeed, there is almost no book using the term *mass spectroscopy* and all scientific journals in the field bear *mass spectrometry* in their titles. You will find such highlighted rules, notes and definitions throughout the book. This more amusing one – we might call it the "zeroth law of mass spectrometry" – has been taken from a standard organic chemistry textbook. The same author finishes his chapter on mass spectrometry with the conclusion that "despite occasional mysteries, mass spectrometry is still highly useful". [21]

Historical Remark: Another explanation for this terminology originates from the historical development of our instrumentation. [13] The device employed by Thomson to do the first of all mass-separating experiments was a type of *spectroscope* showing blurred signals on a flourescent screen. [22] Dempster constructed an instrument with a deflecting magnetic field with an angle of 180°. In order to detect different masses, it could either be equipped with a photographic plate – a so-called *mass spectrograph* – or it could have a variable magnetic field to detect different masses by focusing them successively onto an electric point detector. [23] Later, the term *mass spectrometer* was coined for the latter type of instruments with *scanning* magnetic field. [24]

To have a common platform to build on, we need to define mass spectrometry and several closely related issues, most of them being generalized or refined in later chapters. Then, we may gather the pieces of the puzzle to get a rough estimate of what needs to be known in order to understand the subject. Finally, it is indicated to agree on some conventions for naming and writing. [25-27]

1.2.1 Mass Spectrometry

"The basic principle of *mass spectrometry* (MS) is to generate ions from either inorganic or organic compounds by any suitable method, to separate these ions by their *mass-to-charge ratio* (*m/z*) and to detect them qualitatively and quantitatively by their respective *m/z* and abundance. The analyte may be ionized thermally, by electric fields or by impacting energetic electrons, ions or photons. The ... ions can be single ionized atoms, clusters, molecules or their fragments or associates. Ion separation is effected by static or dynamic electric or magnetic fields." Although this definition of mass spectrometry dates back to 1968 when organic mass spectrometry was in its infancy, [28] it is still valid. However, two additions should be made. First, besides electrons, (atomic) ions or photons, energetic neutral atoms and heavy cluster ions can also be used to effect ionization of the analyte. Second, as demonstrated with great success by the time-of-flight analyzer, ion separation by *m/z* can be effected in field free regions, too, provided the ions possess a well-defined kinetic energy at the entrance of the flight path.

1.2.2 Mass Spectrometer

Obviously, almost any technique to achieve the goals of ionization, separation and detection of ions in the gas phase can be applied – and actually has been applied – in mass spectrometry. This leads to a simple basic setup having all mass spectrometers in common. A mass spectrometer consists of an *ion source*, a *mass analyzer* and a *detector* which are operated under high vacuum conditions. A closer look at the front end of such a device might separate the steps of *sample introduction, evaporation* and successive *ionization* or *desorption/ionization*, respectively, but it is not always trivial to identify each of these steps clearly separated from the others. If the instrument is not a too old one, some data system will be added to the rear end which is used to collect and process data from the detector. Since the 1990s, data systems are also employed to control all functions of the instrument (Fig. 1.1).

The *consumption of analyte* by its examination in the mass spectrometer is an aspect deserving our attention: Whereas other spectroscopic methods such as nuclear magnetic resonance (NMR), infrared (IR) or Raman spectroscopy do allow for sample recovery, mass spectrometry does consume the analyte. This is the logical result of the sequence from ionization and translational motion through the mass analyzer to the detector during analysis. Although some sample is consumed for mass spectrometry, it may still be regarded as a practically non-destructive method because the amount of analyte needed is in the low microgram range and often by several orders of magnitude below. In turn, the extremely low sample consumption of mass spectrometry makes it the method of choice when most other analytical techniques fail because they are not able to yield analytical information from nanogram amounts of sample.

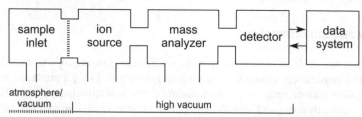

Fig. 1.1. General scheme of a mass spectrometer. Often, several types of sample inlets are attached to the ion source housing. Transfer of the sample from atmospheric pressure to the high vacuum of the ion source and mass analyzer is accomplished by use of a vacuum lock (Chap. 5.3).

1.2.3 Mass Spectrum

A *mass spectrum* is the two-dimensional representation of signal intensity (ordinate) versus *m/z* (abscissa). The intensity of a *peak*, as signals are usually called, directly reflects the abundance of ionic species of that respective *m/z* ratio which have been created from the analyte within the ion source.

The *mass-to-charge ratio, m/z*, (read "m over z") [29] is dimensionless by definition, because it calculates from the dimensionless *mass number, m*, of a given ion, and the number of its elementary charges, *z*. The number of elementary charges is often, but by far not necessarily, equal to one. As long as only singly charged ions are observed ($z = 1$) the *m/z* scale directly reflects the *m* scale. However, there can be conditions where doubly, triply or even highly charged ions are being created from the analyte depending on the ionization method employed. The location of a peak on the abscissa is reported as "at *m/z x*".

> **Note:** Some mass spectrometrists use the unit *thomson* [Th] (to honor J. J. Thomson) instead of the dimensionless quantity *m/z*. Although the thomson is accepted by some journals, it is not a SI unit.

The distance between peaks on that axis has the meaning of a neutral loss from the ion at higher *m/z* to produce the fragment ion at lower *m/z*. Therefore, the amount of this neutral loss is given as "*x* u", where the symbol u stands for *unified atomic mass*. It is important to notice that the mass of the neutral is only reflected by the difference between the corresponding *m/z* ratios. This is because the mass spectrometer detects only charged species, i.e., the charge-retaining group of a fragmenting ion. Since 1961 the *unified atomic mass* [u] is defined as $^1/_{12}$ of the mass of one atom of nuclide ^{12}C which has been assigned to 12 u exactly by convention.

> **Note:** In particular mass spectrometrists in the biomedical field of mass spectrometry tend to use the *dalton* [Da] (to honor J. Dalton) instead of the *unified atomic mass* [u]. The dalton also is not a SI unit.

Often but not necessarily, the peak at highest *m/z* results from the detection of the intact ionized molecule, the *molecular ion*, $M^{+\cdot}$. The *molecular ion peak* is usually accompanied by several peaks at lower *m/z* caused by fragmentation of the molecular ion to yield *fragment ions*. Consequently, the respective peaks in the mass spectrum may be referred to as *fragment ion peaks*.

The most intense peak of a mass spectrum is called *base peak*. In most representations of mass spectral data the intensity of the base peak is normalized to 100 % *relative intensity*. This largely helps to make mass spectra more easily comparable. The normalization can be done because the relative intensities are independent from the absolute ion abundances registered by the detector. However, within the ion source there are upper limits for the number of ions per volume where significant changes of the appearance of spectra will occur (Chap. 7). In the older literature, spectra were sometimes normalized relative to the sum of all intensities measured, e.g., denoted as $\%\sum_{ions}$, or the intensities were reported normalized to the sum of all intensities above a certain *m/z*, e.g., above *m/z* 40 ($\%\sum_{40}$).

Example: In the electron ionization mass spectrum of a hydrocarbon, the molecular ion peak and the base peak of the spectrum correspond to the same ionic species at *m/z* 16 (Fig. 1.2). The fragment ion peaks at *m/z* 12–15 are spaced at 1 u distance. Obviously, the molecular ion, $M^{+\cdot}$, fragments by loss of H$^\cdot$ which is the only possibility to explain the peak at *m/z* 15 by loss of a neutral of 1 u mass. Accordingly, the peaks at lower *m/z* might arise from loss of a H_2 molecule (2 u) and so forth. It does not take an expert to recognize that this spectrum belongs to methane, CH_4, showing its molecular ion peak at *m/z* 16 because the atomic mass number of carbon is 12 and that of hydrogen is 1, and thus $12 \text{ u} + 4 \times 1 \text{ u} = 16 \text{ u}$.

Removal of one electron from a 16 u neutral yields a singly-charged radical ion that is detected at *m/z* 16 by the mass spectrometer. Of course, most mass spectra are not that simple, but this is how it works.

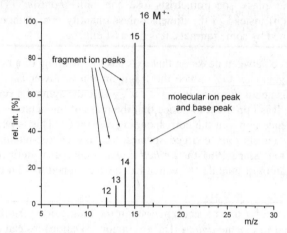

Fig. 1.2. Electron ionization mass spectrum of a hydrocarbon. Adapted with permission. © National Institute of Standards and Technology, NIST, 2002.

The above spectrum is represented as a *bar graph* or *histogram*. Such *data reduction* is common in mass spectrometry and useful as long as peaks are well resolved. The intensities of the peaks can be obtained either from measured peak heights or more correctly from peak areas. The position, i.e., the *m/z* ratio, of the signal is determined from its centroid. Noise below some user-defined cut level is usually subtracted from the bar graph spectrum. If peak shape and peak width become important, e.g., in case of high mass analytes or high resolution measurements, spectra should be represented as *profile data* as initially acquired by the mass spectrometer. *Tabular listings* of mass spectra are used to report mass and intensity data more accurately (Fig. 1.3).

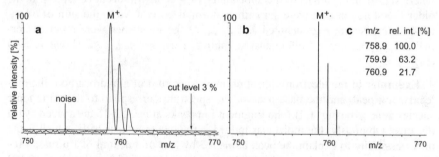

Fig. 1.3. Three representations of the molecular ion signal in the field desorption mass spectrum (Chap. 8) of tetrapentacontane, $C_{54}H_{110}$; (**a**) profile spectrum, (**b**) bar graph representation, and (**c**) tabular listing.

1.3 Filling the Black Box

There is no one-and-only approach to the wide field of mass spectrometry. At least, it can be concluded from the preceding pages that it is necessary to learn about the ways of sample introduction, generation of ions, their mass analysis and their detection as well as about registration and presentation of mass spectra. The still missing issue is not inherent to a mass spectrometer, but of key importance for the successful application of mass spectrometry. This is mass spectral interpretation. All these items are correlated to each other in many ways and contribute to what we call mass spectrometry (Fig. 1.4).

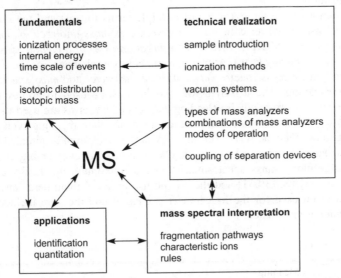

Fig. 1.4. The main contributions to what we call mass spectrometry. Each of the segments is correlated to the others in multiple ways.

1.4 Terminology

As indicated in the very first introductory paragraphs, terminology can be a delicate issue in mass spectrometry (shouldn't it be mass spectroscopy?). To effectively communicate about the subject we need to agree on some established terms, acronyms and symbols for use in mass spectrometry.

The current terminology is chiefly defined by three authoritative publications: i) a compilation by Price under the guidance of the *American Society for Mass Spectrometry* (ASMS), [25] ii) one by Todd representing the official recommendations of the *International Union of Pure and Applied Chemistry* (IUPAC), [26] and iii) one by Sparkman trying to bring the preceding and sometimes contradictory ones together. [27] IUPAC, for example, stays in opposition to the vast majority of practitioners, journals and books when talking about mass spectroscopy and defining terms such as *daughter ion* and *parent ion* as equivalent to *product ion* and *precursor ion*, respectively. Sparkman discourages the use of *daughter ion* and *parent ion* as these are archaic and gender-specific terms. On the other hand, Price and Sparkman keep using *mass spectrometry*. Unfortunately, none of these collections is fully comprehensive, e.g., only IUPAC offers terms related to vacuum technology and Sparkman does not give a definition of *ionization energy*. Nevertheless, there is about 95 % agreement between these guidelines to terminology in mass spectrometry and their overall coverage can be regarded highly sufficient making the application of any of these beneficial to oral and written communication.

One cannot ignore the existence of multiple terms for one and the same thing sometimes just coined for commercial reasons, e.g., *mass-analyzed ion kinetic energy spectrometry* (MIKES, correct) and *direct analysis of daughter ions* (DADI, incorrect *and* company term). Another prominent example concerns the use of MS as an acronym for *mass spectrometry*, *mass spectrometer* and *mass spectrum*, too. This is misleading. The acronym MS should only be used to abbreviate *mass spectrometry*. Unfortunately, misleading and redundant terms are used throughout the literature, and thus, we need at least to understand their meaning even if we are not going to use them actively. Terminology in this book avoids outdated or vague terms and special notes are given for clarification wherever ambiguities might arise. Furthermore, mass spectrometrist like to communicate their work using countless acronyms, [30,31] and there is no use to avoid them here. They are all explained when used for the first time in a chapter and they are included in the subject index for reference.

Table 1.1. Symbols

Symbol	Meaning
•	unpaired electron in radicals
+	positive even-electron ions
−	negative even-electron ions
+•	positive radical ions
−•	negative radical ions
⌒	arrow for transfer of an electron pair
⌒	single-barbed arrow for transfer of a single electron
⌇	to indicate position of cleaved bond
⟶	fragmentation or reaction
⟶o⟶	rearrangement fragmentation

1.5 Units, Physical Quantities, and Physical Constants

The consistent use of units for physical quantities is a prerequisite in science, because it simplifies the comparison of physical quantities, e.g., temperature, pressure or physical dimensions. Therefore, the *International System of Units* (SI) is used throughout the book. This system is based on seven units that can be combined to form any other unit needed. Nevertheless, mass spectrometers often have long lifetimes and 20 year-old instruments being scaled in inches and having pressure readings in Torr or psi may still be in use. The following tables provide collections of SI units together with some frequently needed conversion factors, a collection of physical constants and some quantities such as the charge of the electron and the mass of the proton (Tables 1.2–1.4). Finally, there is a collection of number prefixes (Table 1.5).

Table 1.2. SI base units

Physical Base Quantity	SI unit	Symbol
length	metre[a]	m
mass	kilogram[b]	kg
time	second	s
electric current	ampere	A
thermodynamic temperature	kelvin[c]	K
amount of substance	mole	mol
luminous intensity	candela	cd

[a] 1 m = 3.2808 ft = 39.3701 in; 1 in = 2.54 cm; 1 ft = 0.3048 m
[b] 1 kg = 2.2046 lb; 1 lb = 0.4536 kg
[c] $T[°C] = T[K] - 273.15$; $T[°F] = 1.8 \times T[°C] + 32$

Table 1.3. Derived SI units with special names

Quantity	Name	Symbol	Expression in terms of base units	Expression in terms of other SI units
frequency	hertz	Hz	s^{-1}	
force	newton	N	$m\,kg\,s^{-2}$	$J\,m^{-1}$
pressure	pascal[a]	Pa	$kg\,m^{-1}\,s^{-2}$	$N\,m^{-2}$
volume	liter	l	$10^{-3}\,m^3$	
energy	joule[b]	J	$m^2\,kg\,s^{-2}$	$N\,m$
power	watt	W	$m^2\,kg\,s^{-3}$	$J\,s^{-1}$
electric charge	coulomb	C	$A\,s$	
electric potential	volt	V	$m^2\,kg\,A^{-1}\,s^{-3}$	$W\,A^{-1}$
magnetic flux density	tesla	T	$kg\,A^{-1}\,s^{-2}$	

[a] 1 bar = 1000 mbar = 10^5 Pa; 1 Torr = 133 Pa; 1 psi = 6895 Pa = 68.95 mbar
[b] 1 cal = 4.1868 J; 1 eV = 1.60219×10^{-19} J = 96.485 kJ mol^{-1}

Table 1.4. Physical constants and frequently used quantities

Physical Constant/Quantity	Symbol	Quantity
charge of the electron	e	1.60219×10^{-19} C
mass of the electron	m_e	9.10953×10^{-31} kg
mass of the proton	m_p	1.67265×10^{-27} kg
mass of the neutron	m_n	1.67495×10^{-27} kg
unified atomic mass	u	1.66055×10^{-27} kg
speed of light in vacuum	c	2.99793×10^{-8} m s^{-1}
Avogadro's constant	N_A	6.02205×10^{23} mol^{-1}

Table 1.5. Number Prefixes

a	f	p	n	μ	m	c	d	k	M	G
atto	femto	pico	nano	micro	milli	centi	deci	kilo	mega	giga
10^{-18}	10^{-15}	10^{-12}	10^{-9}	10^{-6}	10^{-3}	10^{-2}	10^{-1}	10^3	10^6	10^9

Reference List

Direct Reference

1. Enjalbal, C.; Maux, D.; Martinez, J.; Combarieu, R.; Aubagnac, J.-L. Mass Spectrometry and Combinatorial Chemistry: New Approaches for Direct Support-Bound Compound Identification. *Comb. Chem. High Throughput Screening* **2001**, *4*, 363-373.
2. Enjalbal, C.; Maux, D.; Combarieu, R.; Martinez, J.; Aubagnac, J.-L. Imaging Combinatorial Libraries by Mass Spectrometry: From Peptide to Organic-Supported Syntheses. *J. Comb. Chem.* **2003**, *5*, 102-109.
3. Maux, D.; Enjalbal, C.; Martinez, J.; Aubagnac, J.-L.; Combarieu, R. Static Secondary Ion Mass Spectrometry to Monitor Solid-Phase Peptide Synthesis. *J. Am. Soc. Mass Spectrom.* **2001**, *12*, 1099-1105.
4. Beverly, M.B.; Voorhees, K.J.; Hadfield, T.L. Direct Mass Spectrometric Analysis of Bacillus Spores. *Rapid Commun. Mass Spectrom.* **1999**, *13*, 2320-2326.
5. Jones, J.J.; Stump, M.J.; Fleming, R.C.; Lay, J.O., Jr.; Wilkins, C.L. Investigation of MALDI-TOF and FT-MS Techniques for Analysis of Escherichia Coli Whole Cells. *Anal. Chem.* **2003**, *75*, 1340-1347.
6. Fenselau, C.; Caprioli, R. Mass Spectrometry in the Exploration of Mars. *J. Mass Spectrom.* **2003**, *38*, 1-10.
7. He, F.; Hendrickson, C.L.; Marshall, A.G. Baseline Mass Resolution of Peptide Isobars: A Record for Molecular Mass Resolution. *Anal. Chem.* **2001**, *73*, 647-650.
8. Cooper, H.J.; Marshall, A.G. ESI-FT-ICR Mass Spectrometric Analysis of Wine. *J. Agric. Food. Chem.* **2001**, *49*, 5710-5718.
9. Hughey, C.A.; Rodgers, R.P.; Marshall, A.G. Resolution of 11,000 Compositionally Distinct Components in a Single ESI-FT-ICR Mass Spectrum of Crude Oil. *Anal. Chem.* **2002**, *74*, 4145-4149.
10. Mühlberger, F.; Wieser, J.; Ulrich, A.; Zimmermann, R. Single Photon Ionization Via Incoherent VUV-Excimer Light: Robust and Compact TOF Mass Spectrometer for Real-Time Process Gas Analysis. *Anal. Chem.* **2002**, *74*, 3790-3801.
11. Glish, G.L.; Vachet, R.W. The Basics of Mass Spectrometry in the Twenty-First Century. *Nat. Rev. Drug Discovery* **2003**, *2*, 140-150.
12. Busch, K.L. Synergistic Developments in MS. A 50-Year Journey From "Art" to Science. *Spectroscopy* **2000**, *15*, 30-39.
13. *Measuring Mass - From Positive Rays to Proteins;* Grayson, M.A., editor; ASMS and CHF: Santa Fe and Philadelphia, 2002.
14. Meyerson, S. Reminiscences of the Early Days of MS in the Petroleum Industry. *Org. Mass Spectrom.* **1986**, *21*, 197-208.

15. Quayle, A. Recollections of MS of the Fifties in a UK Petroleum Laboratory. *Org. Mass Spectrom.* **1987**, *22*, 569-585.
16. Maccoll, A. Organic Mass Spectrometry - the Origins. *Org. Mass Spectrom.* **1993**, *28*, 1371-1372.
17. Meyerson, S. Mass Spectrometry in the News, 1949. *Org. Mass Spectrom.* **1993**, *28*, 1373-1374.
18. Meyerson, S. From Black Magic to Chemistry. The Metamorphosis of Organic MS. *Anal. Chem.* **1994**, *66*, 960A-964A.
19. Boyd, R.K. Editorial: The Sponsor Referee Procedure. *Rapid Commun. Mass Spectrom.* **2002**, *16*.
20. Gross, M.L.; Sparkman, O.D. The Importance of Titles in References. *J. Am. Soc. Mass Spectrom.* **1998**, *9*, 451.
21. Jones, M., Jr. Mass Spectrometry, in *Organic Chemistry*, 2nd ed.; W. W. Norton & Company: New York, 2000; 641-649.
22. Griffiths, I.W. J. J. Thomson - the Centenary of His Discovery of the Electron and of His Invention of Mass Spectrometry. *Rapid Commun. Mass Spectrom.* **1997**, *11*, 1-16.
23. Dempster, A.J. A New Method of Positive Ray Analysis. *Phys. Rev.* **1918**, *11*, 316-325.
24. Nier, A.O. Some Reflections on the Early Days of Mass Spectrometry at the University of Minnesota. *Int. J. Mass Spectrom. Ion Proc.* **1990**, *100*, 1-13.
25. Price, P. Standard Definitions of Terms Relating to Mass Spectrometry. A Report From the Committee on Measurements and Standards of the ASMS. *J. Am. Soc. Mass Spectrom.* **1991**, *2*, 336-348.
26. Todd, J.F.J. Recommendations for Nomenclature and Symbolism for Mass Spectroscopy Including an Appendix of Terms Used in Vacuum Technology. *Int. J. Mass Spectrom. Ion. Proc.* **1995**, *142*, 211-240.
27. Sparkman, O.D. *Mass Spec Desk Reference;* 1st ed.; Global View Publishing: Pittsburgh, 2000.
28. Kienitz, H. Einführung, in *Massenspektrometrie*, Kienitz, H., editor; Verlag Chemie: Weinheim, 1968.
29. Busch, K.L. Units in Mass Spectrometry. *Spectroscopy* **2001**, *16*, 28-31.
30. Busch, K.L. SAMS: Speaking With Acronyms in Mass Spectrometry. *Spectroscopy* **2002**, *17*, 54-62.
31. Busch, K.L. A Glossary for Mass Spectrometry. *Spectroscopy* **2002**, *17*, S26-S34.

Classical Mass Spectrometry Books

32. Field, F.H.; Franklin, J.L. *Electron Impact Phenomena and the Properties of Gaseous Ions;* Academic Press: New York, 1957.
33. Beynon, J.H. *Mass Spectrometry and Its Applications to Organic Chemistry;* 1st ed.; Elsevier: Amsterdam, 1960.
34. Biemann, K. *Mass Spectrometry;* 1st ed.; McCrawHill Book Co.: New York, 1962.
35. Biemann, K. *Mass Spectrometry - Organic Chemical Applications;* 1st ed.; McCraw-Hill: New York, 1962.
36. *Mass Spectrometry of Organic Ions;* 1st ed.; McLafferty, F.W., editor; Academic Press: London, 1963.
37. Budzikiewicz, H.; Djerassi, C.; Williams, D.H. *Mass Spectrometry of Organic Compounds;* 1st ed.; Holden-Day: San Francisco, 1967.
38. *Massenspektrometrie;* 1st ed.; Kienitz, H., editor; Verlag Chemie: Weinheim, 1968.
39. Cooks, R.G.; Beynon, J.H.; Caprioli, R.M. *Metastable Ions;* Elsevier: Amsterdam, 1973.
40. Levsen, K. *Fundamental Aspects of Organic Mass Spectrometry;* Verlag Chemie: Weinheim, 1978.

Introductory Texts

41. Duckworth, H.E.; Barber, R.C.; Venkatasubramanian, V.S. *Mass Spectroscopy;* 2nd ed.; Cambridge University Press: Cambridge, 1986.
42. McLafferty, F.W.; Turecek, F. *Interpretation of Mass Spectra;* 4th ed.; University Science Books: Mill Valley, 1993.
43. Watson, J.T. *Introduction to Mass Spectrometry;* 3rd ed.; Lippincott-Raven: New York, 1997.
44. Budzikiewicz, H. *Massenspektrometrie - eine Einführung;* 4th ed.; Wiley-VCH: Weinheim, 1998.
45. Barker, J. *Mass Spectrometry - Analytical Chemistry by Open Learning;* 2nd ed.; John Wiley & Sons: Chichester, 1999.
46. Smith, R.M. *Understanding Mass Spectra - A Basic Approach;* 1st ed.; John Wiley & Sons: New York, 1999.
47. Pretsch, E.; Bühlmann, P.; Affolter, C. *Structure Determination of Organic Compounds - Tables of Spectral Data;* 3rd ed.; Springer-Verlag: Heidelberg, 2000.
48. De Hoffmann, E.; Stroobant, V. *Mass Spectrometry - Principles and Applications;* 2nd ed.; John Wiley & Sons: Chichester, 2001.

49. Herbert, C.G.; Johnstone, R.A.W. *Mass Spectrometry Basics;* CRC press: Boca Raton, 2002.

Monographs

50. Porter, Q.N.; Baldas, J. *Mass Spectrometry of Heterocyclic Compounds;* 1st ed.; Wiley Interscience: New York, 1971.
51. Dawson, P.H. *Quadrupole Mass Spectrometry and Its Applications;* Elsevier: New York, 1976.
52. Beckey, H.D. *Principles of Field Desorption and Field Ionization Mass Spectrometry;* Pergamon Press: Oxford, 1977.
53. Meuzelaar, H.L.C.; Haverkamp, J.; Hileman, F.D. *Pyrolysis Mass Spectrometry of Recent and Fossil Biomaterials;* Elsevier: Amsterdam, 1982.
54. *Tandem Mass Spectrometry;* 1st ed.; McLafferty, F.W., editor; John Wiley & Sons: New York, 1983.
55. Message, G.M. *Practical Aspects of Gas Chromatography/Mass Spectrometry;* 1st ed.; John Wiley & Sons: New York, 1984.
56. Vogel, P. *Carbocation Chemistry;* Amsterdam, 1985.
57. *Gaseous Ion Chemistry and Mass Spectrometry;* Futrell, J.H., editor; John Wiley and Sons: New York, 1986.
58. *Secondary Ion Mass Spectrometry: Basic Concepts, Instrumental Aspects, Applications;* 1st ed.; Benninghoven, A.; Werner, H.W.; Rudenauer, F.G., editors; Wiley Interscience: New York, 1986.
59. Busch, K.L.; Glish, G.L.; McLuckey, S.A. *Mass Spectrometry/Mass Spectrometry;* 1st ed.; Wiley VCH: New York, 1988.
60. March, R.E.; Hughes, R.J. *Quadrupole Storage Mass Spectrometry;* John Wiley & Sons: Chichester, 1989.
61. Wilson, R.G.; Stevie, F.A.; Magee, C.W. *Secondary Ion Mass Spectrometry: A Practical Handbook for Depth Profiling and Bulk Impurity Analysis;* John Wiley & Sons: Chichester, 1989.
62. Prókai, L. *Field Desorption Mass Spectrometry;* Marcel Dekker: New York, 1990.
63. *Continuous-Flow Fast Atom Bombardment Mass Spectrometry;* Caprioli, R.M., editor; John Wiley & Sons: Chichester, 1990.
64. *Fourier Transform Ion Cyclotron Resonance Mass Spectrometry: Analytical Applications;* 1st ed.; Asamoto, B., editor; John Wiley & Sons: New York, 1991.
65. Harrison, A.G. *Chemical Ionization Mass Spectrometry;* 2nd ed.; CRC Press: Boca Raton, 1992.
66. Chapman, J.R. *Practical Organic Mass Spectrometry: A Guide for Chemical and Biochemical Analysis;* 2nd ed.; John Wiley & Sons: Chichester, 1993.
67. *Time of Flight Mass Spectrometry and Its Applications;* 1st ed.; Schlag, E.W., editor; Elsevier: Amsterdam, 1994.
68. *Forensic Applications of Mass Spectrometry;* Yinon, J., editor; CRC Press: Boca Raton, 1994.
69. *Applications of Mass Spectrometry to Organic Stereochemistry;* 1st ed.; Splitter, J.S.; Turecek, F., editors; Verlag Chemie: Weinheim, 1994.
70. *Practical Aspects of Ion Trap Mass Spectrometry;* March, R.E.; Todd, J.F.J., editors; CRC: Boca Raton, 1995; Vols. 1–3.
71. Lehmann W.D. *Massenspektrometrie in der Biochemie;* Spektrum Akademischer Verlag: Heidelberg, 1996.
72. *Electrospray Ionization Mass Spectrometry - Fundamentals, Instrumentation and Applications;* 1st ed.; Dole, R.B., editor; John Wiley & Sons: Chichester, 1997.
73. Cotter, R.J. *Time-of-Flight Mass Spectrometry: Instrumentation and Applications in Biological Research;* American Chemical Society: Washington, DC, 1997.
74. Platzner, I.T.; Habfast, K.; Walder, A.J.; Goetz, A. *Modern Isotope Ratio Mass Spectrometry;* John Wiley & Sons: Chichester, 1997.
75. *Mass Spectrometry of Proteins and Peptides;* Chapman, J.R., editor; Humana Press: 2000.
76. Kinter, M.; Sherman, N.E. *Protein Sequencing and Identification Using Tandem Mass Spectrometry;* John Wiley & Sons: Chichester, 2000.
77. Taylor, H.E. *Inductively Coupled Plasma Mass Spectroscopy;* Academic Press: London, 2000.
78. *Mass Spectrometry of Polymers;* 1st ed.; Montaudo, G.; Lattimer, R.P., editors; CRC Press: Boca Raton, 2001.
79. Budde, W.L. *Analytical Mass Spectrometry;* ACS and Oxford University Press: Washington, D.C. and Oxford, 2001.
80. Ardrey, R.E. *Liquid Chromatography-Mass Spectrometry - An Introduction;* John Wiley & Sons: Chichester, 2003.

2 Gas Phase Ion Chemistry

The mass spectrometer can be regarded as a kind of chemistry laboratory, especially designed to study ions in the gas phase. [1,2] In addition to the task it is usually employed for – creation of mass spectra for a generally analytical purpose – it allows for the examination of fragmentation pathways of selected ions, for the study of ion–neutral reactions and more. Understanding these fundamentals is prerequisite for the proper application of mass spectrometry with all technical facets available, and for the successful interpretation of mass spectra because "Analytical chemistry is the application of physical chemistry to the real world." [3]

In the first place, this chapter deals with the fundamentals of gas phase ion chemistry, i.e., with ionization, excitation, ion thermochemistry, ion lifetimes, and reaction rates of ion dissociation. The final sections are devoted to more practical aspects of gas phase ion chemistry such as the determination of ionization and appearance energies or of gas phase basicities and proton affinities.

Brief discussions of some topics of this chapter may also be found in physical chemistry textbooks; however, much better introductions are given in the specialized literature. [4-11] Detailed compound-specific fragmentation mechanisms, ion–molecule complexes, and more are dealt with later (Chap. 6.).

2.1 Quasi-Equilibrium Theory

The *quasi-equilibrium theory* (QET) of mass spectra is a theoretical approach to describe the unimolecular decompositions of ions and hence their mass spectra. [12-14,14] QET has been developed as an adaptation of Rice-Ramsperger-Marcus-Kassel (RRKM) theory to fit the conditions of mass spectrometry and it represents a landmark in the theory of mass spectra. [11] In the mass spectrometer almost all processes occur under high vacuum conditions, i.e., in the highly diluted gas phase, and one has to become aware of the differences to chemical reactions in the condensed phase as they are usually carried out in the laboratory. [15,16] Consequently, bimolecular reactions are rare and the chemistry in a mass spectrometer is rather the chemistry of *isolated ions in the gas phase*. Isolated ions are not in thermal equilibrium with their surroundings as assumed by RRKM theory. Instead, to be isolated in the gas phase means for an ion that it may only internally redistribute energy and that it may only undergo unimolecular reactions such as isomerization or dissociation. This is why the theory of unimolecular reactions plays an important role in mass spectrometry.

The QET is not the only theory in the field; indeed, several apparently competitive statistical theories to describe the rate constant of a unimolecular reaction have been formulated. [10,14] Unfortunately, none of these theories has been able to quantitatively describe all reactions of a given ion. Nonetheless, QET is well established and even the simplified form allows sufficient insight into the behavior of isolated ions. Thus, we start out the chapter from the basic assumptions of QET. Following this trail will lead us from the neutral molecule to ions, and over transition states and reaction rates to fragmentation products and thus, through the basic concepts and definitions of gas phase ion chemistry.

2.1.1 Basic Assumptions of QET

Due to QET the rate constant, k, of a unimolecular reaction is basically a function of excess energy, E_{ex}, of the reactants in the transition state and thus $k_{(E)}$ is strongly influenced by the internal energy distribution of the ions under study. The quasi-equilibrium theory makes the following essential assumptions: [12,17]

1. The initial ionization is "vertical", i.e., without change in position and kinetic energy of the nuclei while it takes place. With the usual electron energy any valence shell electron may be removed.
2. The molecular ion will be of low symmetry and have an odd electron. It will have as many low-lying excited electronic states as necessary to form essentially a continuum. Radiationless transitions then will result in transfer of electronic energy into vibrational energy at times comparable to the periods of nuclear vibrations.
3. These low-lying excited electronic states will in general not be repulsive; hence, the molecular ions will not dissociate immediately, but rather remain together for a time sufficient for the excess electronic energy to become randomly distributed as vibrational energy over the ion.
4. The rates of dissociation of the molecular ion are determined by the probabilities of the energy randomly distributed over the ion becoming concentrated in the particular fashions required to give several activated complex configurations yielding the dissociations.
5. Rearrangements of the ions can occur in a similar fashion.
6. If the initial molecular ion has sufficient energy, the fragment ion will in turn have enough energy to undergo further decomposition.

2.2 Ionization

Besides some rare experimental setups the mass analyzer of any mass spectrometer can only handle charged species, i.e., ions that have been created from atoms or molecules, more seldom from radicals, zwitterions or clusters. It is the task of the ion source to perform this crucial step and there is a wide range of ionization methods in use to achieve this goal for a wide variety of analytes.

2.2.1 Electron Ionization

The classical procedure of ionization involves shooting energetic electrons on a neutral. This is called *electron ionization* (EI). Electron ionization has formerly been termed *electron impact ionization* or simply *electron impact* (EI). For EI, the neutral must previously have been transferred into the highly diluted gas phase, which is done by means of any sample inlet system suitable for the evaporation of the respective compound. In practice, the gas phase may be regarded highly diluted when the mean free path for the particles becomes long enough to make bimolecular interactions almost impossible within the lifetime of the particles concerned. This is easily achieved at pressures in the range of 10^{-4} Pa usually realized in electron ionization ion sources.

Here, the description of EI is restricted to what is essential for understanding the ionization process itself [18,19] and the consequences for the fate of the ions created. The practical aspects of EI and the interpretation of EI mass spectra are discussed later (Chaps. 5, 6).

2.2.1.1 Ions Generated by Electron Ionization

When a neutral is hit by an energetic electron carrying several tens of electron-volts (eV) of kinetic energy, some of the energy of the electron is transferred to the neutral. If the electron, in terms of energy transfer, collides very effectively with the neutral, the amount of energy transferred can effect ionization by ejection of one electron out of the neutral, thus making it a positive *radical ion*:

$$M + e^- \rightarrow M^{+\bullet} + 2e^- \qquad (2.1)$$

EI predominantly creates singly charged ions from the precursor neutral. If the neutral was a molecule as in most cases, it started as having an even number of electrons, i.e., it was an *even-electron* (*closed-shell*) *ion*. The molecular ion formed must then be a radical cation or an *odd-electron* (*open-shell*) *ion* as these species are termed, e.g., for methane we obtain:

$$CH_4 + e^- \rightarrow CH_4^{+\bullet} + 2e^- \qquad (2.2)$$

In the rare case the neutral was a radical, the ion created by electron ionization would be even-electron, e.g., for nitric oxide:

$$NO^\bullet + e^- \rightarrow NO^+ + 2e^- \qquad (2.3)$$

Depending on the analyte and on the energy of the primary electrons, doubly and even triply charged ions can also be observed. [20] In general, these are of low abundance.

$$M + e^- \rightarrow M^{2+} + 3e^- \qquad (2.4)$$

$$M + e^- \rightarrow M^{3+\bullet} + 4e^- \qquad (2.5)$$

While the doubly charged ion, M^{2+}, is an *even-electron ion*, the triply charged ion, $M^{3+\bullet}$, again is an *odd-electron ion*. In addition, there are several other events possible from the electron–neutral interaction, e.g., a less effective interaction will bring the neutral into an electronically excited state without ionizing it.

2.2.1.2 Ions Generated by Penning Ionization

Non-ionizing electron–neutral interactions create electronically excited neutrals. The ionization reactions occurring when electronically excited neutrals, e.g., noble gas atoms A*, collide with ground state species, e.g., some molecule M, can be divided into two classes. [21] The first process is *Penning ionization* (Eq. 2.6), [22] the second is *associative ionization* which is also known as the Hornbeck-Molnar process (Eq. 2.7). [23]

$$A^* + M \rightarrow A + M^{+\bullet} + e^- \qquad (2.6)$$

$$A^* + M \rightarrow AM^{+\bullet} + e^- \qquad (2.7)$$

Penning ionization occurs with the (trace) gas M having an ionization energy lower than the energy of the metastable state of the excited (noble gas) atoms A*. The above ionization processes have also been employed to construct mass spectrometer ion sources. [21,24] However, Penning ionization sources never escaped the realm of academic research to find widespread analytical application.

2.2.2 Ionization Energy

2.2.2.1 Definition of Ionization Energy

It is obvious that ionization of the neutral can only occur when the energy deposited by the electron–neutral collision is equal to or greater than the *ionization energy* (IE) of the corresponding neutral. Formerly, the ionization energy has been termed *ionization potential* (IP).

> **Definition:** The *ionization energy* (IE) is defined as the minimum amount of energy which has to be absorbed by an atom or molecule in its electronic and vibrational ground states form an ion that is also in its ground states by ejection of an electron.

2.2.2.2 Ionization Energy and Charge-Localization

Removal of an electron from a molecule can formally be considered to occur at a σ-bond, a π-bond or at a free electron pair with the σ-bond being the least favorable and the free electron pair being the most favorable position for *charge-localization* within the molecule. This is directly reflected in the IEs of molecules (Table 1.1). Nobel gases do exist as atoms having closed electron shells and therefore, they exhibit the highest IEs. They are followed by diatomic molecules with fluorine, nitrogen and hydrogen at the upper limit. The IE of methane is lower than that of molecular hydrogen but still higher than that of ethane and so on until the IE of long-chain alkanes approaches a lower limit. [25] The more atoms are contained within a molecule the easier it becomes to find a way for stabilization of the charge, e.g., by delocalization or hyperconjugation.

Molecules with π-bonds have lower IEs than those without, causing the IE of ethene to be lower than that of ethane. Again the IE is reduced further with increasing size of the alkene. Aromatic hydrocarbons can stabilize a single charge even better and expanding π-systems also help making ionization easier.

The electron charge is never really localized in a single orbital, but assuming so is often a good working hypothesis. In case of the *para*-tolyl ion, for example, it has been calculated that only about 36 % of the electron charge rest at the *para*-carbon atom (Fig. 2.1). [26] In addition, the ionic geometry looses the symmetry of the corresponding neutral molecule.

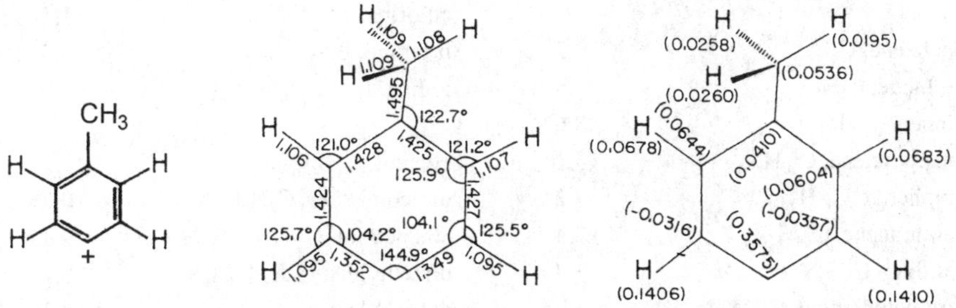

Fig. 2.1. Formula representation (left), calculated geometries (middle) and calculated distributions of formal charge (right) in the *para*-tolyl ion, $[C_7H_7]^+$. Reproduced from Ref. [26] with permission. © American Chemical Society, 1977.

Free electron pairs are a good source for an electron which is to be ejected and therefore, the IE of ethanol and dimethylether is lower than that of ethane. It has been shown that the IE of a poly-substituted alkane is almost the same as the IE of the structurally identical mono-substituted alkane which has the lowest value. [27] The other substituent, provided it is separated by at least two carbon atoms, exerts a very small effect upon the IE, e.g., the IE of dimethylsulfide, CH_3SCH_3, 8.7 eV, is almost the same as that of methionine, $CH_3SCH_2CH_2CH(NH_2)COOH$. Introduction of an oxygen decreases the IE less than nitrogen, sulfur or even selenium do, since these elements have lower electronegativities and thus, are even better sources of an electron.

The bottom line of IEs is reached when π-systems and heteroatoms are combined in the same molecule.

Note: Ionization energies of most molecules are in the range of 7–15 eV.

Table 2.1. Ionization energies of selected compounds[a]

Compound	IE[b] [eV]	Compound	IE[b] [eV]
hydrogen, H_2	15.4	helium, He	24.6
methane, CH_4	12.6	neon, Ne	21.6
ethane, C_2H_6	11.5	argon, Ar	15.8
propane, n-C_3H_8	10.9	krypton, Kr	14.0
butane, n-C_4H_{10}	10.5	xenon, Xe	12.1
pentane, n-C_5H_{12}	10.3		
hexane, n-C_6H_{14}	10.1	nitrogen, N_2	15.6
decane, n-$C_{10}H_{22}$	9.7	oxygen, O_2	12.1
		carbonmonoxide, CO	14.0
ethene, C_2H_4	10.5	carbondioxide, CO_2	13.8
propene, C_3H_6	9.7		
(E)-2-butene, C_4H_8	9.1	fluorine, F_2	15.7
		chlorine, Cl_2	11.5
benzene, C_6H_6	9.2	bromine, Br_2	10.5
toluene, C_6H_8	8.8	iodine, I_2	9.3
indene, C_9H_8	8.6		
naphthalene, $C_{10}H_8$	8.1	ethanol, C_2H_6O	10.5
biphenyl, $C_{12}H_{10}$	8.2	dimethylether, C_2H_6O	10.0
anthracene, $C_{14}H_{10}$	7.4	ethanethiol, C_2H_6S	9.3
aniline, C_6H_7N	7.7	dimethyldisulfide, C_2H_6S	8.7
triphenylamine, $C_{18}H_{15}N$	6.8	dimethylamine, C_2H_7N	8.2

[a] IE data extracted from Ref. [28] with permission. © NIST 2002.
[b] All values have been rounded to one digit.

2.3 Vertical Transitions

Electron ionization occurs extremely fast. The time needed for an electron of 70 eV to travel 1 nm distance, i.e., roughly half a dozen bonds, through a molecule is only about 2×10^{-16} s and even larger molecules can be traversed on the low femtosecond scale. The molecule being hit by the electron can be seen at rest because the thermal velocity of a few 100 m s^{-1} is negligible compared to the speed of the electron rushing through. Vibrational motions are slower by at least two orders of magnitude, e.g., even the fast C–H stretching vibration takes 1.1×10^{-14} s per cycle as can be calculated from its absorbance around 3000 cm^{-1} in infrared spectra. According to the *Born-Oppenheimer approximation*, the electronic motion and the nuclear motion can therefore be separated, i.e., the positions of the atoms and thus bond lengths do not change while ionization takes place. [29,30] In addition, the *Franck-Condon principle* states that the probability for an electronic transition is highest where the electronic wave functions of both ground state and ionized state have their maxima. [31,32] This gives rise to *vertical transitions*.

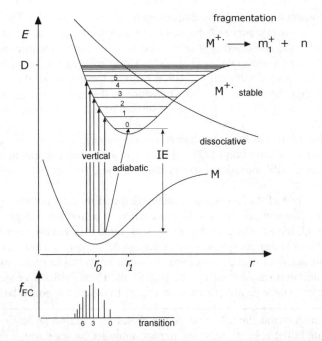

Fig. 2.2. Electron ionization can be represented by a vertical line in this diagram. Thus, ions are formed in a vibrationally excited state if the internuclear distance of the excited state is longer than in the ground state. Ions having internal energies below the dissociation energy D remain stable, whereas fragmentation will occur above. In few cases, ions are unstable, i.e., there is no minimum on their potential energy curve. The lower part schematically shows the distribution of Franck-Condon factors, f_{FC}, for various transitions.

The probability of a particular vertical transition from the neutral to a certain vibrational level of the ion is expressed by its *Franck-Condon factor*. The distribution of Franck-Condon factors, f_{FC}, describes the distribution of vibrational states for an excited ion. [33] The larger r_1 compared to r_0, the more probable will be the generation of ions excited even well above dissociation energy. Photoelectron spectroscopy allows for both the determination of adiabatic ionization energies and of Franck-Condon factors (Chap. 2.10.1).

The counterpart of the vertical ionization is a process where ionization of the neutral in its vibrational ground state would yield the radical ion also in its vibrational ground state, i.e., the $(0 \leftarrow 0)$ transition. This is termed *adiabatic ionization* and should be represented by a diagonal line in the diagram. The difference $IE_{vert} - IE_{ad}$ can lead to errors in ionization energies in the order of 0.1–0.7 eV. [7]

Independent of where the electron has formally been taken from, ionization tends to cause weakening of the bonding within the ion as compared to the precursor neutral. Weaker bonding means longer bond lengths on the average and this goes with a higher tendency toward dissociation of a bond. In terms of potential energy surfaces, the situation can be visualized by focusing on just one bond within the molecule or simply by discussing a diatomic molecule. The minimum of the potential energy curve of the neutral, which is assumed to be in its vibrational ground state, is located at shorter bond length, r_0, than the minimum of the radical ion in its ground state, r_1 (Fig. 2.2). Along with ionization of the neutral there has to be some vibrational excitation because the positions of the atoms are fixed during this short period, i.e., the transition is described by a vertical line in the diagram crossing the potential energy surface of the ion at some vibrationally excited level.

The characteristics of the ionization process as described above give the justification of the first assumption of QET. Further, it is obvious that electronic excitation goes along with vibrational excitation and thus, the second assumption of QET is also met.

The further fate of the ion depends on the shape of its potential energy surface. If there is a minimum and the level of excitation is below the energy barrier for dissociation, D, the ion can exist for a very long time. Ions having *internal energy* above the dissociation energy level will dissociate sooner or later causing fragment ions to occur in a mass spectrum. In some unfavorable cases, ions do not have any minimum on their energy surface at all. These will suffer spontaneous dissociation and consequently, there is no chance to observe a molecular ion.

Note: To understand the situation of the molecule imagine an apple through which a gun bullet is being shot: the impacting bullet passes through the apple, transfers an amount of energy, tears some of the fruit out and has left it by far when the perforated apple finally drops or breaks into pieces.

2.4 Ionization Efficiency and Ionization Cross Section

The ionization energy represents the absolute minimum required for ionization of the neutral concerned. This means in turn that in order to effect ionization, the impacting electrons need to carry at least this amount of energy. If this energy would then be quantitatively transferred during the collision, ionization could take place. Obviously, such an event is of rather low probability and therefore, the *ionization efficiency* is close to zero with electrons carrying just the IE of the neutral under study. However, a slight increase in electron energy brings about a steady increase in ionization efficiency.

Strictly speaking, every molecular species has an ionization efficiency curve of its own depending on the *ionization cross section* of the specific molecule. In case of methane, this issue has been studied repeatedly (Fig. 2.3). [18] The ionization cross section describes an area through which the electron must travel in order to effectively interact with the neutral and consequently, the ionization cross section is given in units of square-meters. Ionization cross section graphs are all of the same type exhibiting a maximum at electron energies around 70 eV (Chap. 5.1.3).

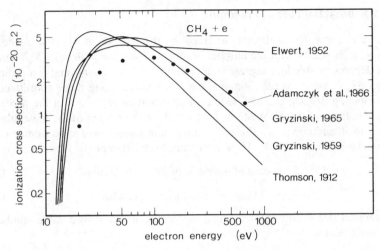

Fig. 2.3. Ionization cross sections for CH_4 upon electron ionization as obtained by several research groups. Reproduced from Ref. [18] with permission. © Elsevier Science, 1982.

2.5 Internal Energy and the Further Fate of Ions

When an analyte is transferred into the ion source by means of any sample introduction system it is in thermal equilibrium with this inlet device. As a result, the energy of the incoming molecules is represented by their thermal energy. Then, ionization changes the situation dramatically as comparatively large amounts of energy need to be "handled" by the freshly formed ion.

2.5.1 Degrees of Freedom

2.5.1.1 External Degrees of Freedom

Any atom or molecule in the gas phase has *external degrees of freedom*. Atoms and molecules can move along all three dimensions in space yielding three translational degrees of freedom. From the kinetic gas theory their average translational energy can easily be estimated as $^3/_2kT$ delivering 0.04 eV at 300 K and 0.13 eV at 1000 K. In case of diatomic and linear molecules we have to add two and for all other molecules three rotational degrees of freedom contributing another $^3/_2kT$ of energy making a total of 0.08 eV at room temperature and 0.26 eV at 1000 K. This does not change independently of the number of atoms of the molecule or of their atomic masses.

2.5.1.2 Internal Degrees of Freedom

Opposed to the external degrees of freedom, the number of *internal degrees of freedom*, *s*, increases with the number of atoms within the molecule, *N*. These internal degrees of freedom represent the number of vibrational modes a molecule can access. In other words, *N* atoms can each move along three coordinates in space yielding 3*N* degrees of freedom in total, but as explained in the preceding paragraph, three of them have to be subtracted for the motion of the molecule as a whole and an additional two (linear) or three (non-linear) have to be subtracted for rotational motion as a whole. Thus, we obtain for the number of vibrational modes

$$s = 3N - 5 \quad \text{in case of diatomic or linear molecules} \tag{2.8}$$

$$s = 3N - 6 \quad \text{in case of non-linear molecules.} \tag{2.9}$$

It is obvious that even relatively small molecules offer a considerable number of vibrational modes.

Example: The thermal energy distribution curves for 1,2-diphenylethane, $C_{14}H_{14}$, $s = 3 \times 28 - 6 = 78$, have been calculated at 75 and 200 °C. [34] Their maxima were obtained at about 0.3 and 0.6 eV, respectively, with almost no molecules reaching beyond twice that energy of maximum probability. At 200 °C, the most probable energy roughly corresponds to 0.008 eV per vibrational degree of freedom.

This tells us that excited vibrational states are almost fully unoccupied at room temperature and only the energetically much lower lying internal rotations are effective under these conditions. Upon electron ionization, the situation changes quite dramatically as can be concluded from the Franck-Condon principle and therefore, energy storage in highly excited vibrational modes becomes of key importance for the further fate of ions in a mass spectrometer. In case of an indene molecule having 45 vibrational modes, the storage of 10 eV would mean roughly 0.2 eV per vibration, i.e., roughly 20fold value of thermal energy, provided the energy is perfectly randomized among the bonds as postulated by the third assumption of QET.

2.5.2 Appearance Energy

When the neutral is hit by an electron, some of the energy transferred is converted into external degrees of freedom, i.e., translational motion and rotational motion as a whole. Most of the energy exchanged during this interaction creates electronic excitation along with ionization if at least the IE has been transferred. In that very special case the freshly generated molecular ion, $M^{+\bullet}$, would be in its vibrational ground state.

As explained by the Franck-Condon diagram, almost no molecular ions will be generated in their vibrational ground state. Instead, the majority of the ions created by EI is vibrationally excited and many of them are well above the dissociation energy level, the source of this energy being the 70 eV electrons. Dissociation of

$M^{+\bullet}$, or fragmentation as it is usually referred to in mass spectrometry, leads to the formation of a fragment ion, m_1^+, and a neutral, a process generally formulated as

$$M^{+\bullet} \rightarrow m_1^+ + n^\bullet \tag{2.10}$$

$$M^{+\bullet} \rightarrow m_1^{+\bullet} + n \tag{2.11}$$

Reaction 2.10 describes the loss of a radical, whereas reaction 2.11 corresponds to the loss of a molecule, thereby conserving the radical cation property of the molecular ion in the fragment ion. Bond breaking is a endothermal process and thus the potential energy of the fragment ion is usually located at a higher energy level (Fig. 2.4).

Definition: The amount of energy needed to be transferred to the neutral M to allow for the detection of the fragment ion m_1^+ is called *appearance energy* (AE) of that fragment ion. The old term *appearance potential* (AP) is still found in the literature.

In fact, the ions are not generated with one specific internal energy applying for all ions, but with a broad energy distribution $P_{(E)}$. One should not forget however, that such a distribution is only valid for a large number of ions as each individual one bears a defined internal energy, ranging from just above IE to well beyond 10 eV for certain ones. Fragmentation of those highly excited ions from the tail of the $P_{(E)}$ curve yields fragment ions having still enough energy for a second disso-ciation step, $m_1^+ \rightarrow m_2^+ + n'$ or even a third one. Each of the subsequent steps can in principle also be characterized by an appearance energy value.

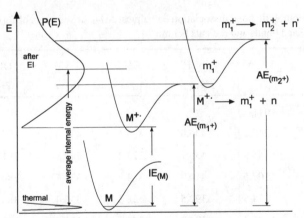

Fig. 2.4. Definition of appearance energy and visualization of changes in internal energy distributions, $P_{(E)}$, of relevant species upon electron ionization and subsequent fragmenta-tion. The energy scale is shown compressed for the ions.

2.5.3 Bond Dissociation Energies and Heats of Formation

Great efforts have been made to generate accurate and reliable ion thermochemistry data (Chap. 1.2.8). Once such data is available, it can be employed to elucidate fragmentation mechanisms and in addition, it is useful for obtaining some background on the energetic dimensions in mass spectrometry.

Heats of formation of neutral molecules, $\Delta H_{f(RH)}$, can be obtained from combustion data with high accuracy. Bond dissociation energies can either be derived for *homolytic bond dissociation*

$$R–H \rightarrow R^{\bullet} + H^{\bullet}, \ \Delta H_{Dhom} \tag{2.12}$$

or *heterolytic bond dissociation*

$$R–H \rightarrow R^{+} + H^{-}, \ \Delta H_{Dhet} \tag{2.13}$$

The homolytic bond dissociation enthalpies give the energy needed to cleave a bond of the neutral molecule which is in the gas phase in its vibrational and electronic ground states, to obtain a pair of radicals which also are not in excited modes. Homolytic bond dissociation enthalpies are found in the range of 3–5 eV (Table 2.2). The heterolytic bond dissociation energies apply for creation of a cation and an anion in their ground states from the neutral precursor which means that these values include the amount of energy needed for charge separation. They are in the order of 10–13 eV (Table 2.3). Due to the significantly altered bonding situation, breaking of a bond is clearly less demanding in molecular ions than in neutrals.

Table 2.2. Homolytic bond dissociation enthalpies, ΔH_{Dhom}, and heats of formation, $\Delta H_{f(X\bullet)}$, of some selected bonds and radicals [kJ mol^{-1}] [a]

		H$^{\bullet}$	CH$_3^{\bullet}$	Cl$^{\bullet}$	OH$^{\bullet}$	NH$_2^{\bullet}$
	$\Delta H_{f(X\bullet)}$	218.0	143.9	121.3	37.7	197.5
R$^{\bullet}$	$\Delta H_{f(R\bullet)}$					
H$^{\bullet}$	218.0	436.0	435.1	431.4	497.9	460.2
CH$_3^{\bullet}$	143.9	435.1	373.2	349.8	381.2	362.8
C$_2$H$_5^{\bullet}$	107.5	410.0	354.8	338.1	380.3	352.7
i-C$_3$H$_7^{\bullet}$	74.5	397.5	352.3	336.4	384.5	355.6
t-C$_4$H$_9^{\bullet}$	31.4	384.9	342.3	335.6	381.6	349.8
C$_6$H$_5^{\bullet}$	325.1	460.1	417.1	395.4	459.0	435.6
C$_6$H$_5$CH$_2^{\bullet}$	187.9	355.6	300.4	290.4	325.9	300.8
C$_6$H$_5$CO$^{\bullet}$	109.2	363.5	338.1	336.8	440.2	396.6

[a] Values from Ref. [35]

Table 2.3. Heterolytic bond dissociation enthalpies, ΔH_{Dhet}, and heats of formation of some molecules and ions [kJ mol^{-1}] [a]

Ion	ΔH_{Dhet}	$\Delta H_{f(R+)}$	$\Delta H_{f(RH)}$
proton, H$^+$	1674	1528	0.0
methyl, CH$_3^+$	1309	1093	−74.9
ethyl, C$_2$H$_5^+$	1129	903	−84.5
n-propyl, CH$_3$CH$_2$CH$_2^+$	1117	866	−103.8
i-propyl, CH$_3$CH$^+$CH$_3$	1050	802	−103.8
n-butyl, CH$_3$CH$_2$CH$_2$CH$_2^+$	1109	837	−127.2
sec-butyl, CH$_3$CH$_2$CH$^+$CH$_3$	1038	766	−127.2
i-butyl, (CH$_3$)$_2$CHCH$_2^+$	1109	828	−135.6
t-butyl, (CH$_3$)$_3$C$^+$	975	699	−135.6
phenyl, C$_6$H$_5^+$	1201	1138	82.8

[a] Values from Refs. [8,8,36,37,37,38]

Example: The minimum energy to form a CH$_3^+$ ion and a hydrogen radical from the methane molecular ion can be estimated from the heat of reaction, ΔH_r, of this process. According to Fig. 2.4, $\Delta H_r = AE_{(CH3+)} - IE_{(CH4)}$. In order to calculate the missing $AE_{(CH3+)}$ we use the tabulated values of $\Delta H_{f(H\cdot)} = 218.0$ kJ mol^{-1}, $\Delta H_{f(CH3+)} = 1093$ kJ mol^{-1}, $\Delta H_{f(CH4)} = -74.9$ kJ mol^{-1}, and $IE_{(CH4)} = 12.6$ eV $= 1216$ kJ mol^{-1}. First, the heat of formation of the methane molecular ion is determined:

$$\Delta H_{f(CH4+\cdot)} = \Delta H_{f(CH4)} + IE_{(CH4)} \tag{2.14}$$

$$\Delta H_{f(CH4+\cdot)} = -74.9 \text{ kJ mol}^{-1} + 1216 \text{ kJ mol}^{-1} = 1141.1 \text{ kJ mol}^{-1}$$

Then, the heat of formation of the products is calculated from

$$\Delta H_{f(prod)} = \Delta H_{f(CH3+)} + \Delta H_{f(H\cdot)} \tag{2.15}$$

$$\Delta H_{f(prod)} = 1093 \text{ kJ mol}^{-1} + 218 \text{ kJ mol}^{-1} = 1311 \text{ kJ mol}^{-1}$$

Now, the heat of reaction is obtained from the difference

$$\Delta H_r = \Delta H_{f(prod)} - \Delta H_{f(CH4+\cdot)} \tag{2.16}$$

$$\Delta H_r = 1311 \text{ kJ mol}^{-1} - 1141.1 \text{ kJ mol}^{-1} = 169.9 \text{ kJ mol}^{-1}$$

The value of 169.9 kJ mol^{-1} (1.75 eV) corresponds to $AE_{(CH3+)} = 14.35$ eV which is in good agreement with published values of about 14.3 eV. [28,39] In addition, this is only 40 % of the homolytic C–H bond dissociation enthalpy of the neutral methane molecule, thereby indicating the weaker bonding in the molecular ion.

The heat of formation of organic radicals and positive ions decreases with their size and even more important with their degree of branching at the radical or ionic site. A lower heat of formation is equivalent to a higher thermodynamic stability of the respective ion or radical. The corresponding trends are clearly expressed by the values given in Tables 2.2 and 2.3. This causes the fragmentation pathways of molecular ions proceeding by formation of secondary or tertiary radicals and/or ions to become dominant over those leading to smaller and/or primary radical and ionic fragments, respectively (Chap. 6.2.2).

Example: The effects of isomerization upon thermal stability of butyl ions, $C_4H_9^+$, are impressing. This carbenium ion can exist in four isomers with heats of formation that range from 837 kJ mol^{-1} in case of n-butyl over 828 kJ mol^{-1} for iso-butyl (also primary) to 766 kJ mol^{-1} for sec-butyl to 699 kJ mol^{-1} in case of t-butyl, meaning an overall increase in thermodynamic stability of 138 kJ mol^{-1} (Chap. 6.6.2) [36].

2.5.4 Randomization of Energy

The question remains as to how to validate the third and fourth assumptions of QET. The best evidence for randomization of internal energy over all vibrational modes of a molecular ion before any fragmentation occurs, is delivered by EI mass spectra themselves. If there was no randomization, fragmentation would occur directly at any bond that immediately suffers from the withdrawal of an electron. As a result, mass spectra would show an almost statistical bond breaking throughout the molecular ion. Instead, mass spectra reveal a great deal of selectivity of the molecular ion when choosing fragmentation pathways. This means the molecular ion explores many pathways up to their respective transition states and prefers the thermodynamically (and as we will see, also kinetically) more favorable ones. The same is true for fragment ions.

From the purely thermodynamic point of view the situation can be elucidated by considering a hypothetical molecular ion having some internal energy and being faced to the selection of a fragmentation pathway (Fig. 2.5).

a) The molecular ion $ABC^{+\bullet}$ has an internal energy E_{inta} being slightly above the activation energy, E_{02}, required to cross the transition state TS_2 leading to the formation of $A^\bullet$ and BC^+, but definitely more than needed to dissociate into AB^+ and $C^\bullet$. The difference between the energy content E_{inta} and E_{01} is termed *excess energy*, $E_{ex} = E_{int} - E_0$, of the transition state TS_1. In this case either ionic product would be observed, but the third pathway could not be accessed.

b) The molecular ion $ABC^{+\bullet}$ has an internal energy E_{intb} being clearly higher than any of the three activation energies. Here, formation of all possible products should occur.

A simple comparison of activation energies does not allow however, to predict the relative intensities of the ions AB^+ and BC^+ or of AB^+, BC^+ and A^+, respectively. In fact, from this model one would expect similar or even equal abundances of all fragment ions accessible. Actual mass spectra show greatly differing intensities of signals resulting from competing decomposition pathways. This is an important issue revealing some oversimplification of the preceding passage.

Note: Thermodynamic data such as heats of formation and activation energies alone are not sufficient to adequately describe the unimolecular fragmentations of excited ions.

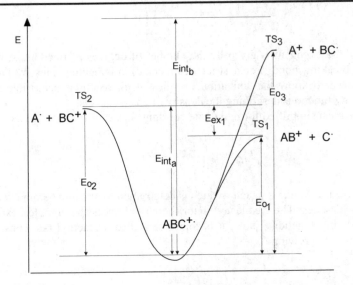

Fig. 2.5. Competition of fragmentation pathways strongly depends on the internal energy of the fragmenting ion and on the activation energies, E_0, of the transition states, i.e., the "energy barriers" of the respective reactions.

2.6 Rate Constants from QET

QET brings the dynamic aspects of ion fragmentation into focus. It describes the *rate constants* for the dissociation of isolated ions as a function of internal energy, E_{int}, and activation energy of the reaction, E_0. By doing so, it compensates for the shortcomings of the merely thermodynamic treatment above.

QET delivers the following expression for the unimolecular rate constant

$$k_{(E)} = \int_0^{E-E_0} \frac{1}{h} \cdot \frac{\rho^*_{(E_{int}, E_0, E_t)}}{\rho_{(E)}} \, dE_t \qquad (2.17)$$

In this equation, $\rho_{(E)}$ is the density of energy levels for the system with total energy E_{int}, and $\rho_{(E, E0, Et)}$ is the density of energy levels in the activated complex, i.e., transition state, with activation energy E_0 and translational energy E_t in the reaction co-ordinate. The reaction co-ordinate represents the bond which is actually being broken. The expression is slightly simplified by approximating the system by as many harmonic oscillators as there are vibrational degrees of freedom

$$k_{(E)} = \left(\frac{E_{int} - E_0}{E_{int}}\right)^{s-1} \frac{\prod\limits_{j=1}^{s} v_j}{\prod\limits_{i=1}^{s-1} v_i *} \qquad (2.18)$$

Then, the exponent is given by the number of degrees of freedom, s, minus 1 for the breaking bond. For a strict treatment of fragmenting ions by QET one would need to know the activation energies of all reactions accessible and the probability functions describing the density of energy levels.

In the most simplified form, the rate constant, $k_{(E)}$, can be expressed as

$$k_{(E)} = v \cdot \left(\frac{E_{int} - E_0}{E_{int}}\right)^{s-1} \qquad (2.19)$$

where v is a frequency factor which is determined by the number and density of vibrational states. The frequency factor thereby replaces the complex expression of probability functions. Now, it becomes clear that a reaction rate considerably increases with growing E_{ex}.

$$k_{(E)} = v \cdot \left(\frac{E_{ex}}{E_{int}}\right)^{s-1} \qquad (2.20)$$

Unfortunately, as with all oversimplified theories, there are limitations for the application of the latter equation to ions close to the dissociation threshold. In these cases, the number of degrees of freedom has to be replaced by an effective number of oscillators which is obtained by use of an arbitrary correction factor. [7] However, as long as we are dealing with ions having internal energies considerably above the dissociation threshold, i.e., where $(E - E_0)/E \approx 1$, the relationship is valid and can even be simplified to give the quasi-exponential expression

$$k_{(E)} = v \cdot e^{-(s-1)E_0/E} \qquad (2.21)$$

Example: For $v = 10^{15}$ s^{-1}, $s = 15$, $E_{int} = 2$ eV and $E_0 = 1.9$ eV the rate constant is calculated as 3.0×10^{-5} s^{-1}. For the same parameters but $E_{int} = 4$ eV we obtain $k = 6.3 \times 10^{10}$ s^{-1} being a 2.1×10^{15}-fold increase. This means that a reaction is extremely slow at small excess energies but becomes very fast as soon there is some substantial excess energy available.

2.6.1 Meaning of the Rate Constant

The rate constants of unimolecular reactions have the dimension *per second* (s^{-1}). This means the process can happen that often per second, e.g., $k = 6.3 \times 10^{10}$ s^{-1} being equivalent to 1.6×10^{-11} s *per fragmentation on the average*. It is important to notice an emphasis on "average", because rate constants are macroscopic and statistical in nature as they get their meaning only from looking at a very large number of reacting particles. A single ion will have a lifetime of 1.6×10^{-11} s *on the average* in this case; however, a specific one in consideration at might also decay much sooner or later, with the actual decay occurring at the speed of vibrational motions. The dimension s^{-1} also means there is no dependence on concentration as it is the case with second- or higher order reactions. This is because the ions are isolated in the gas phase, alone for their entire lifetime, and the only chance for change is by means of unimolecular reaction.

2.6.2 Typical $k_{(E)}$ Functions

Although the general shape of any $k_{(E)}$ function resembles the ionization efficiency curve to the left of the maximum, these must not be confused. At an excess energy close to zero, the rate constant is also close to zero but it rises sharply upon slight increase of the excess energy. However, there is an upper limit for the rate of a dissociation that is defined by the vibrational frequency of the bond to be cleaved. The fragments are not able to fly apart at a higher velocity than determined by their vibrational motion (Fig. 2.6).

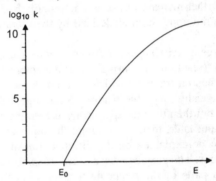

Fig. 2.6. General shape of a log k vs. E plot as determined by the simplified QET. At $E = E_0$ the reaction is extremely slow. A a slight increase in energy causes k to rise sharply.

2.6.3 Description of Reacting Ions Using $k_{(E)}$ Functions

Bearing the knowledge on rate constants in mind, Fig. 2.5 has to be interpreted in a different way. At an internal energy E_{inta} the molecular ion $ABC^{+\bullet}$ can easily cross the transition state TS_1 to form AB^+ and $C^\bullet$. The products of the second reaction can in principle be formed, even though the excess energy at TS_2 is so small that the product ion BC^+ will almost be negligible. The third pathway is still not accessible. At an internal energy E_{intb} the excess energy is assuredly high enough to allow for any of the three pathways with realistic rates to observe the products.

Nevertheless, due to the strong dependence of the rate constant on E_{ex}, the reaction over TS_3 will be by far the least important.

2.6.4 Direct Cleavages and Rearrangement Fragmentations

The reactions of excited ions are not always as straightforward as expected. Of course, the existence of multiple fragmentation pathways for an ion consisting of several tens of atoms brings about different types of reactions all of which certainly will not lead to the same $k_{(E)}$ function. [40]

The $k_{(E)}$ functions of two reactions of the same type, appear to be "parallel" in comparison, starting out at different activation energies (Fig. 2.7a), whereas different types of reactions will show a crossover of their $k_{(E)}$ functions at intermediate excess energy (Fig. 2.7b).

For example, case (a) is obtained when two homolytic bond cleavages are in competition. Homolytic bond cleavages are simple fragmentations of molecular ions creating an even-electron ion and a radical. One cleavage might require a somewhat higher activation energy than the other ($E_{02} > E_{01}$) because of the different bonds that are to be cleaved, but once having enough excess energy their rates will rise sharply. The more energy is pushed into the ion, the faster the bond rupture can occur. A further increase of excess energy only becomes ineffective when the rate approaches the upper limit as defined by the vibrational frequency of the bond to be cleaved.

Case (b) compares a *rearrangement fragmentation* (reaction 1) with a *homolytic bond cleavage* (reaction 2). During a rearrangement fragmentation the precursor ion expels a neutral fragment which is an intact molecule after having rearranged in an energetically favorable manner. Rearrangements have their onset at low excess energy, but then the rate approaches the limit relatively soon, whereas the cleavage starts out later, then overrunning the other at higher excess energy. The differences can be explained by the different transition states belonging to both types of reactions. The cleavage has a *loose transition state*, [34] i.e., there is no need for certain parts of the molecule to gain a specific position while the cleavage proceeds. The dissociation merely requires enough energy in the respective bond that the binding forces can be overcome. Once the bond is stretched too far, the fragments drift apart. The rearrangement demands less excess energy to proceed, because the energy for a bond rupture on one side is compensated for by the energy received when a new bond is formed on the accepting position. Compared to the simple cleavage, such a transition state is usually termed *tight transition state* (cf. Chap. 6.12.2). Then, the neutral is expelled in a second step. Such a reaction obviously depends on having the suitable conformation for rearrangement at the same time when sufficient energy is put into the bond to be cleaved. Furthermore, the second step has to follow in order to yield the products. In the end this means that there is no use in having more than enough energy until the ion reaches the conformation needed. Thus, after considering either basic type of fragmentation, the fifth assumption of QET is justified, because there is no reason why rearrangements should not be treated by means of QET.

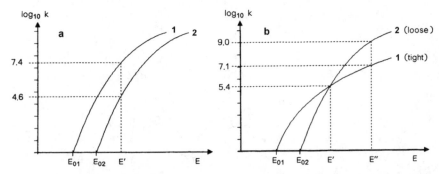

Fig. 2.7. Comparison of different types of reactions by their $k_{(E)}$ functions. Reactions of the same type show $k_{(E)}$ functions that are "parallel" (**a**), whereas $k_{(E)}$ functions of different types tend to cross over at intermediate excess energy (**b**). In **a** reaction **1** would proceed $10^{2.8}$ (630) times faster at an internal energy E' than would reaction **2**. In **b** both reactions have the same rate constant, $k = 10^{5.4}$ s^{-1}, at an internal energy E', whereas reaction **2** becomes $10^{1.9}$ (80) times faster at E''.

2.6.5 Practical Consequences of Internal Energy

Even comparatively small molecular ions can exhibit a substantial collection of first, second, and higher generation fragmentation pathways due their internal energy. [40] The fragmentation pathways of the same generation are competing with each other, their products being potential precursors for the next generation of fragmentation processes (Figs. 2.4 and 2.8).

The hypothetical molecular ion $ABCD^{+\bullet}$, for example, might show three possible first generation fragmentation pathways, two of them proceeding by rearrangement, one by simple cleavage of a bond. There are three first generation fragment ions which again have several choices each. The second generation fragment ions $Y^{+\bullet}$ and YZ^+ are even formed on two different pathways each, a widespread phenomenon. In an EI mass spectrum of ABCD, all ionic species shown would be detected (Chap. 5.1.4).

Fragmentation trees similar to that shown here can be constructed from any EI mass spectrum, providing ten thousands of examples for the sixth assumption of QET that fragment ions may again be subject to dissociation, provided their internal energy suffices.

> **Note:** The assumptions of QET have turned into basic statements governing the behavior of isolated ions in the gas phase and thus, in mass spectrometry in general.

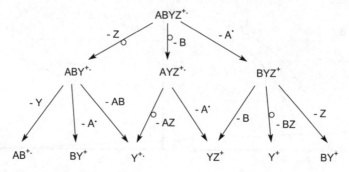

Fig. 2.8. Possible fragmentation pathways for hypothetical molecular ions ABCD$^{+\bullet}$ having internal energies typical for EI.

2.7 Time Scale of Events

A mass analyzer normally brings only those ions to detection that have been properly formed and accelerated by the ion source beforehand (Chap. 4). Therefore, a reaction needs to proceed within a certain period of time – the dwelltime of ions within the ion source – to make the products detectable in the mass spectrum, and for this purpose there is a need for some excess energy in the transition state.

The dwelltime of ions within the ion source is defined by the extraction voltages applied to accelerate and focus them into an ion beam and by the dimensions of that ion source. In standard EI ion sources the freshly formed ions dwell about 1 μs before they are forced to leave the ionization volume by action of the accelerating potential. [41] As the ions then travel at speeds of some 10^4 m s^{-1} they pass the mass analyzer in the order of 10–50 μs (Fig. 2.9). [9] Even though this illustration has been adapted for a double focusing magnetic sector mass spectrometer, an ion of m/z 100, and an acceleration voltage of 8 kV, the effective time scales for other types of instruments (quadrupole, time-of-flight) are very similar under their typical conditions of operation (Table 2.4).

Table 2.4. Typical ion flight times in different types of mass spectrometers

Mass Analyzer	Flight Path [m]	Acceleration Voltage [V]	Typical m/z	Flight Time [μs]
quadrupole	0.2	10	500	57
magnetic sector	2.0	5000	500	45
time-of-flight	2.0	20,000	2000	45

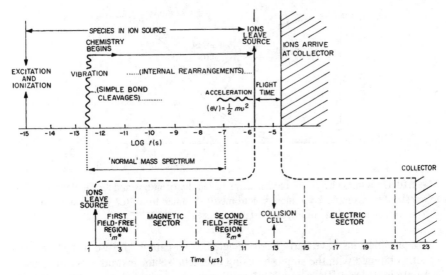

Fig. 2.9. The mass spectrometric time scale. It is important to note the logarithmic time scale for the ion source spanning over nine orders of magnitude. Reproduced from Ref. [9] with permission. © John Wiley & Sons Ltd., 1985.

2.7.1 Stable, Metastable, and Unstable Ions

The terminology for ions has been coined as a direct consequence of the mass spectrometric time scale. Non-decomposing molecular ions and molecular ions decomposing at rates below about 10^5 s^{-1} will reach the detector without fragmentation and are therefore termed *stable ions*. Consequently, ions dissociating at rates above 10^6 s^{-1} cannot reach the detector. Instead, their fragments will be detected and thus they are called *unstable ions*. A small percentage, however, decomposing at rates of 10^5–10^6 s^{-1} will just fragment on transit through the mass analyzer; those are termed *metastable ions* (Fig. 2.10) [4,5,9].

Definition:	*stable ions*, $k < 10^5$ s^{-1}
	metastable ions, 10^5 s^{-1} < k < 10^6 s^{-1}
	unstable ions, $k > 10^6$ s^{-1}

There is no justification for such a classification of ion stabilities outside the mass spectrometer because almost all ions created under the conditions inside a mass spectrometer would spontaneously react in the atmosphere or in solvents. Nevertheless, this classification is useful as far as ions isolated in the gas phase are concerned and is valid independently of the type of mass analyzer or ionization method employed.

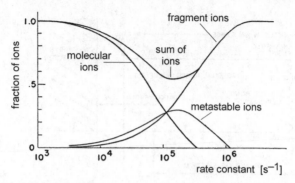

Fig. 2.10. Correlation between rate constants of ion dissociations and terminology referring to ion stability as employed in mass spectrometry. Adapted from Ref. [42] with permission. © The American Institute of Physics, 1959.

Metastable ions have early been detected as so-called "diffuse peaks" in mass spectra obtained with the single-focusing magnetic sector instruments of that time, but only in the mid 1940s had they been correctly interpreted as ions decomposing in transit. [43,44] It was also explained why the peaks from metastable ion decompositions are detected at non-integral m/z values. Instead, the peak corresponding to a fragment m_2^+ formed from m_1^+ upon decomposition in the field-free region in front of the magnetic sector is located at a magnet setting m* which is described by the relationship $m^* = m_2^2/m_1$. This equation and methods for detecting metastable ions are discussed in Chap. 4. Metastable ion spectra represent one of the indispensable tools for studying the mechanism and thermochemistry of ion dissociations. [4,5]

Example: The metastable decay of the o-nitrophenol molecular ion, m/z 139, by loss of NO to yield the $[M–NO]^+$ ion, m/z 109, has been studied on a single-focusing magnetic sector instrument (Fig. 2.11. [45] The mass spectrum shows a

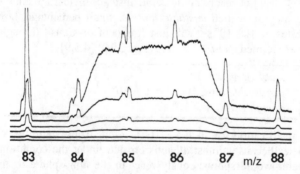

Fig. 2.11. Peak due to metastable NO loss of the o-nitrophenol molecular ion. The multiple traces correspond to different amplifier settings of a multi-channel recorder. Adapted from Ref. [45] with permission. © Verlag der Zeitschrift für Naturforschung, 1965.

flat-topped peak of low intensity expanding over three mass units).Some minor and narrow "regular" peaks corresponding to fragment ions formed within the ion source are observed beside and on top of it. The peak due to the metastable dissociation is centered at m/z 85.5 which can be explained by the simple calculation $m^* = m_2^2/m_1 = 109^2/139 = 85.5$.

2.7.2 Kinetic Shift

Fragment ion abundances as observed by means of any mass spectrometer strongly depend on ion lifetimes within the ion source and on ion internal energy distributions, i.e., kinetic aspects play an important role for the appearance of a mass spectrum. As explained by QET the rate constant of an ion dissociation is a function of excess energy in the transition state of the respective reaction, and there is a need of substantial excess energy to make the rate constant exceed the critical $10^6 \, s^{-1}$ needed to dissociate during the residence times within the ion source. Therefore, appearance and activation energies are overestimated from measurements by this amount of excess energy. This phenomenon is known as *kinetic shift* (Fig. 2.12). [34,42] Often, kinetic shifts are almost negligible (0.01–0.1 eV), but they can be as large as 2 eV, [34] e.g. an activation energy of 2.07 eV for the process $C_3H_6^{+\bullet} \rightarrow C_3H_5^+ + H^\bullet$ is accompanied by a kinetic shift of 0.19 eV. [46] As $k_{(E)}$ functions greatly differ between reactions and especially between homogeneous bond cleavages and rearrangement fragmentations, it is not an easy task to correct the experimental AEs by subtraction of E_{ex}. The influence of the kinetic shift can be minimized by increasing the ion source residence time and/or by increasing the detection sensitivity of the instrument.

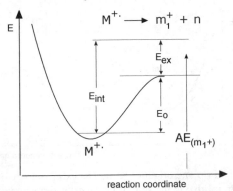

Fig. 2.12. Kinetic shift. The excess energy in the transition state affects AE measurements in the way that it always causes experimental values to be too high.

Note: The *kinetic shift* denotes the overestimation of AEs due to the contribution of excess energy in the transition state necessary to yield rate constants larger than $10^6 \, s^{-1}$. The determination of IEs does not suffer from kinetic shift as there are no kinetics involved in electron or photon ionization.

2.8 Activation Energy of the Reverse Reaction and Kinetic Energy Release

2.8.1 Activation Energy of the Reverse Reaction

By merely considering thermodynamic and kinetic models we may be able to understand how ions are formed and what parameters are effective to determine their further fate in the mass spectrometer. However, the potential energy surfaces considered did only reach up to the transition state (Fig. 2.5) Without explicitly mentioning it, we assumed the curves to stay on the same energetic level between transition state and the products of ion dissociation, i.e., the sum of heats of formation of the products would be equal to the energy of the transition state. This assumption is almost correct for homolytic bond cleavages in molecular ions because the recombination of an ion with a radical proceeds with almost negligible activation energy. [4,5] In other words, the *activation energy of the reverse reaction*, E_{0r} (or often simply called *reverse activation energy*) is then close to zero. This explains why the result of our simple estimation of the activation energy for hydrogen radical loss from the molecular ion of methane could be realistic (Chap. 2.5.3).

In case of rearrangement fragmentations the situation is quite different, because one of the products is an intact neutral molecule of comparatively high thermodynamic stability, i.e., having negative or at least low values of ΔH_f (Table 2.5). Once the transition state is crossed, the reaction proceeds by formation of products energetically much lower than the transition state. In case of the reverse reaction, this would require some or even substantial activation energy transfer to the fragments to allow their recombination, and thus $E_{0r} > 0$ (Fig. 2.13).

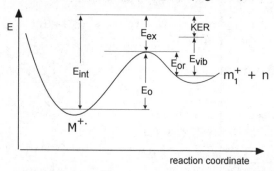

Fig. 2.13. Definition of E_{0r} and origin of KER. The excess energy of the decomposing ion in the transition state relative to the sum of the heats of formation of the ionic and neutral product is partitioned into vibrational excitation of the products plus KER.

Table 2.5. Gas phase heats of formation, ΔH_f, of some frequently eliminated molecules[a]

Molecule	ΔH_f [kJ mol^{-1}][b]	Molecule	ΔH_f [kJ mol^{-1}][b]
CO_2	–393.5	H_2CO	–115.9
HF	–272.5	HCl	–92.3
H_2O	–241.8	NH_3	–45.9
CH_3OH	–201.1	C_2H_4	52.5
CO	–110.5	HCN	135.1

[a] IE data extracted from Ref. [28] with permission. © NIST 2002.
[b] All values have been rounded to one digit.

2.8.2 Kinetic Energy Release

2.8.2.1 Source of Kinetic Energy Release

The total excess energy, E_{extot}, of the precursor ion relative to the heats of formation of the products in their ground state, comprises the excess energy in the transition state, E_{ex}, plus the activation energy of the reverse reaction, E_{0r}:

$$E_{extot} = E_{ex} + E_{0r} \qquad (2.22)$$

Although most ion fragmentations are endothermic, there is still a significant amount of energy to be redistributed among the reaction products. Much of E_{extot} is redistributed as vibrational energy, E_{vib}, among the internal modes, thereby supplying the energy for consecutive fragmentation of the fragment ion. Nonetheless, some of the energy is converted into translational motion of the fragments relative to their center of gravity. This portion of E_{extot} is released in the direction of the bond that is being cleaved, i.e., into diversion from each other. This is termed *kinetic energy release*, KER (Fig. 2.13). [4,5,9,47,48]

The larger the sum of $E_{ex} + E_{0r}$ the larger is the expected KER. Large reverse activation energy, especially when combined with repulsive electronic states in the transition state, will cause significant KER that can reach up to 1.64 eV [49-52], and significant KERs also have been observed in case of an exothermic fragmentation. [53] On the other hand, homolytic cleavages and ion–neutral complex-mediated reactions tend to proceed with very small KER in the range of 1–50 meV (Fig. 2.14).

Note: The importance of KER measurements results from the fact that the potential energy surface between transition state and products of a reaction can be reconstructed. [47] Thus, KER and AE data are complementary in determining the energy of the transition state.

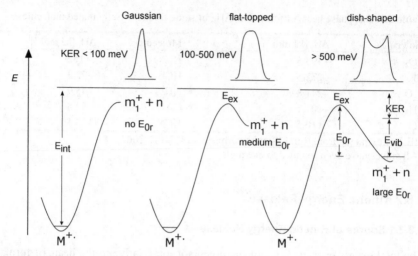

Fig. 2.14. Influence of the reverse activation energy on KER and thus, on peak shapes in metastable ion decompositions, suitable experimental setup as prerequisite. From left: no or small reverse barrier causes Gaussian peak shape, whereas medium E_{0r} yields flat-topped peaks and large E_{0r} causes dish-shaped peaks.

2.8.2.2 Energy Partitioning

The observed KER consists of two components, one from E_{ex} and one from E_{0r}. This splitting becomes obvious from the fact that even if there is no E_{0r}, a small KER is always observed, thus demonstrating partitioning of E_{ex} between E_{vib} and E_{trans*} (KER):

$$E_{ex} = E_{vib} + E_{trans*} \tag{2.23}$$

Diatomic molecular ion dissociations represent the only case where energy partitioning is clear as all excess energy of the decomposition has to be converted to translational energy of the products ($E_{ex} = E_{trans*}$). For polyatomic ions the partitioning of excess energy can be described by a simple empirical relationship between E_{ex} and the number of degrees of freedom, s: [6,54]

$$E_{trans*} = \frac{E_{ex}}{\alpha \cdot s} \tag{2.24}$$

with the empirical correction factor $\alpha = 0.44$. [54] According to Eq. 2.24 the ratio E_{trans*}/E_{ex} decreases as the size of the fragmenting ion increases. This influence has been termed *degrees of freedom effect* (DOF). [55-57]

Consequently, E_{trans*} becomes rather small for substantial values of s, e.g., 0.3 eV/(0.44·30) = 0.023 eV. Therefore, any observed KER in excess of E_{trans*} must originate from E_{0r} being the only alternative source. [53] The analogous partitioning of E_{0r} is described by:

$$E_{trans} = \beta E_{or} \tag{2.25}$$

with $\beta = 0.2–0.4$ and $^1/_3$ being a good approximation in many cases.

> **Note:** In practice, most of the observed KER with the exception of small KERs (< 50 meV), can be attributed to E_{trans} from E_{0r}, [53] and as a rule of thumb, $E_{trans} \approx 0.33\ E_{or}$

Example: The observed KER values and also peak shapes may change dramatically as E_{0r} decreases. Here, the chain length of a leaving alkene increases with an increase of the reacting alkyl substituent. Nevertheless, the mechanism of the reaction – McLafferty rearrangement of immonium ions (Chap. 6.7) – remains unaffected. [58]

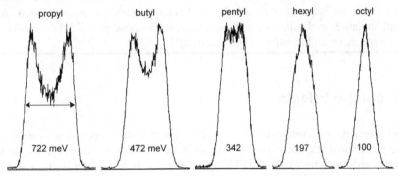

Fig. 2.15. Effect of decreasing E_{0r} on KER. The peaks from MIKE measurements belong to the m/z 58 ion product ion of alkene loss (ethene to hexene) from homologous iminium ions. The KER is determined from peak width at half height.

2.8.2.3 Determination of Kinetic Energy Release

What we just mentioned from a thermochemical point of view of a reaction has earlier been inferred from the observation that peaks due to metastable ion dissociations are much broader than "regular" peaks (Fig. 2.11). [45] This broadening is caused by the kinetic energy from KER grafted onto the kinetic energy of the ion beam as it passes through the mass analyzer. Due to the free rotation of the dissociating ion, there is no preferred orientation of this superimposed motion. Given a suitable experimental setup of the analyzer system that allows to obtain a kinetic energy spectrum, the x-component of KER, i.e., along the flight axis, can be calculated from the peak width of metastable ion decompositions. The best established one of these techniques is *mass-analyzed ion kinetic energy spectrometry* (MIKES) (Fig. 2.15 and Chap. 4.2). The peak width at half height is related to KER by: [4,5]

$$T = \frac{m_2 \, U_b \, e}{16n} \left(\frac{\Delta E}{E} \right)^2 \tag{2.26}$$

where T is the average kinetic energy released, m_2 is mass of the fragment ion, n the mass of the neutral, U_b the acceleration voltage, e the electron charge, E gives the position and ΔE the width of the peak on the kinetic energy scale. Correction of the width of the metastable peak, w_{50meta}, for the width of the main beam, w_{50main}, should be applied for smaller values of T: [59,60]

$$w_{50corr} = \sqrt{w_{50meta}^2 - w_{50main}^2} \tag{2.27}$$

Note: Before performing KER measurements, a "calibration" of the instrument(al parameters) against a well-established standard is recommended. Allylmethylether molecular ions, for example, decompose to yield three peaks of different shape and position in the spectrum. [61-63]

2.9 Isotope Effects

Most elements are composed of more than one naturally occurring *isotope*, i.e., having nuclei of the same *atomic number* but different *mass numbers* due to different numbers of neutrons. [64] The mass number of an isotope is given as a superscript preceding the element symbol, e.g., ^1H and ^2H (D) or ^{12}C and ^{13}C (Chap. 3).

Obviously, mass spectrometry is ideally suited for distinguishing between isotopic species, and isotopic labeling is used for mechanistic as well as analytical applications (Chap. 3.2.9). However, the effect of isotopic substitution is not only an effect on ionic mass, but "isotopic substitution can have several simultaneous effects, and this complication sometimes produces results which are at first sight curious." [65]

Any effect exerted by the introduction of isotopes are termed *isotope effects*. Isotope effects can be *intermolecular*, e.g., upon $D^{\bullet}$ loss from $CD_4^{+\bullet}$ versus $H^{\bullet}$ loss from $CH_4^{+\bullet}$, or *intramolecular*, e.g., upon $H^{\bullet}$ loss versus $D^{\bullet}$ loss from $CH_2D_2^{+\bullet}$.

2.9.1 Kinetic Isotope Effects

Kinetic isotope effects represent one particular type of an isotope effect. They are best rationalized when considering the potential energy diagram of both $H^{\bullet}$ and $D^{\bullet}$ loss from equivalent positions in the same isotopically labeled, e.g., deuterated, molecular ion (Fig. 2.16). [4,5,66] The diagram is basically symmetrical, the only differences arising from the zero-point energy (ZPE) terms. Since $H^{\bullet}$ and $D^{\bullet}$ are being lost from the same molecular ion, the single ZPE to start from is defined by this species. The transition states on the way to H⋅ and D⋅ loss possess ZPE terms associated with all the degrees of freedom of the dissociating ion except that involved in the reaction under consideration. The transition state corresponding to H⋅ loss must therefore have a lower ZPE because it contains a C—D bond that is not involved in the actual cleavage. This corresponds to a lower ZPE than is provided by the C—H bond-retaining transition state that leads to D⋅ loss. Therefore, an ion with an internal energy E_{int} has a larger excess energy E_{exH} when it is going to eliminate H⋅ than if it would cleave off a D⋅. The activation energy for H⋅ loss is thereby lowered making it proceed faster than D⋅ loss ($k_H/k_D > 1$).

The reason for the lower ZPE of the deuterium-retaining species is found in its lower vibrational frequencies due to the double mass of D as compared to H at almost identical binding forces. From classical mechanics the vibrational frequency v_D should therefore be lower by the inverse ratio of the square roots of their masses, i.e., $v_D/v_H \approx 1/1.41 \approx 0.71$. [67]

Example: Isotopic labeling does not only reveal the original position of a rearranging atom, but can also reveal the rate-determining step of multi-step reactions by its marked influence on reaction rates. Thus, the examination of H/D and $^{12}C/^{13}C$ isotope effects led to the conclusion that the McLafferty rearrangement of aliphatic ketones (Chap. 6.7) rather proceeds stepwise than concerted. [68]

> **Notes:** i) The isotope effects dealt with in mass spectrometry are usually *intramolecular kinetic isotope effects*, i.e., two competing fragmentations only differing in the isotopic composition of the products exhibit different rate constants k_H and k_D. [69] ii) The kinetic isotope effect is called *normal* if $k_H/k_D > 1$ and *inverse* if $k_H/k_D < 1$. iii) Isotope effects can also be observed on KER, [52,70] e.g. the KER accompanying H_2 loss from methylene immonium ion varies between 0.61 and 0.80 eV upon D labeling at various positions. [52]

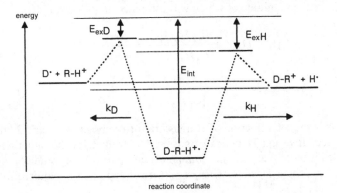

Fig. 2.16. Origin of kinetic isotope effects. [4,5,66] The change in vibrational frequencies, and thus in density of states causes somewhat higher activation energy and consequently smaller excess energy for the reaction of the deuterated bond, and thus reduces k_D.

2.9.1.1 Primary Kinetic Isotope Effects

Primary kinetic isotope effects are observed if the isotopic bond itself is being broken or formed during the reaction. It seems clear that there is no single well-defined value of the kinetic isotope effect of a reaction (Fig. 2.16) and it strongly depends on the internal energy of the decomposing ions. In the rare case where E_{int} is just above the activation energy for H˙ loss but still below that for D˙ loss, the isotope effect would be infinite. As ions in a mass spectrometer usually exhibit comparatively wide distributions of E_{int}, the probability for such a case is extremely low. However, kinetic isotope effects can be large in case of decomposing metastable ions ($k_H/k_D \approx 2$–4) because these ions possess only small excess energies. Such circumstances make them sensitive to the difference between E_{exD} and E_{exH}. On the other hand, ions decomposing in the ion source have usually high internal energies, and thus smaller isotope effects are observed ($k_H/k_D \approx 1$–1.5).

While the mass of H remarkably differs from that of D ($2\,u/1\,u = 2$), the relative increase in mass is much less for heavier elements such as carbon ($13\,u/12\,u \approx 1.08$) [68] or nitrogen ($15\,u/14\,u \approx 1.07$). As a result, kinetic isotope effects of those elements are particularly small and special attention has to be devoted in order for their proper determination.

2.9.1.2 Measurement of Isotope Effects

Mass spectrometry measures the abundance of ions versus their m/z ratio, and it is common practice to use the ratio $I_{mH}/I_{mD} = k_H/k_D$ as a direct measure of the isotope effect. The typical procedure for determining isotope effects from intensity ratios is to solve a set of simultaneous equations. [71-74]

Example: The transfer of a D compared to the transfer of an H during propene loss from partially labeled phenylproylethers is accompanied by an isotope effect, the approximate magnitude of which was estimated as follows: [74]

$$[1,1-D_2]: \quad \frac{[C_6H_6O]^+}{[C_6H_5OD]^+} = \frac{\alpha i}{1-\alpha} \tag{2.28}$$

$$[2,2,3,3,3-D_5]: \quad \frac{[C_6H_6O]^+}{[C_6H_5OD]^+} = \frac{(1-\alpha)i}{\alpha} \tag{2.29}$$

where α is the total fraction of H transferred from positions 2 and 3 and i is the isotope effect favoring H over D transfer. Unfortunately, the situation is more complicated if more than just one complementary pair is being studied, as an exact solution is no longer feasible. Then, a numerical approach is required to approximate the i value. [74,75]

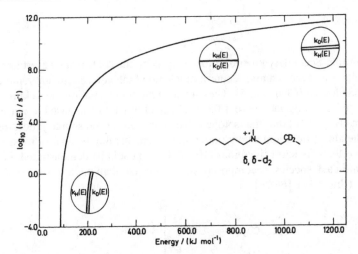

Fig. 2.17. Calculated $k_{H(E)}$ and $k_{D(E)}$ curves for the α-cleavage of deuterated amine molecular ions. The curves can be regarded as parallel over a small range of internal energies, but they are not in the strict sense. They may even cross to cause inverse isotope effects in the domain of highly excited ions. Reproduced from Ref. [76] with permission. © John Wiley & Sons, 1991.

Also, a rigorous treatment of isotope effects within the framework of QET reveals that the assumption $I_{mH}/I_{mD} = k_H/k_D$ represents a simplification. [69] It is only valid for when the species studied populate a small internal energy distribution, e.g., as metastable ions do, whereas wide internal energy distributions, e.g., those of ions fragmenting in the ion source after 70 eV electron ionization, may cause erroneous results. This is because the $k_{(E)}$ functions of isotopic reactions are not truly parallel, [76] but they fulfill this requirement over a small range of internal energies (Figs. 2.17 and 2.18)

2.9.1.3 Secondary Kinetic Isotope Effects

Secondary kinetic isotope effects are observed if an isotopic label is located adjacent to or remote from the bond that is being broken or formed during the reaction. Again, these depend on the internal energy of the decomposing ions. Secondary kinetic isotope effects, i_{sec}, are generally much smaller than their primary analogues.

Example: The ratio $[M–CH_3]^+/[M–CD_3]^+$ from isopropylbenzene molecular ions decomposing by benzylic cleavage (Chap. 6.4) varied from 1.02 for ion source fragmentations (70 eV EI) over 1.28 for metastable ions in the 1st FFR to 1.56 in the 2nd FFR, thus clearly demonstrating the dependence of the secondary kinetic isotope effect on internal energy. [77]

It is convenient to specify the value normalized *per heavier isotope present*, $i_{secnorm}$, because there can be a number of heavier isotopes, n_D, present: [66,76]

$$i_{sec\,norm} = \sqrt[n_D]{i_{sec}} \qquad (2.30)$$

Example: Secondary kinetic isotope effects on the α-cleavage of tertiary amine molecular ions occurred after deuterium labeling both adjacent to and remote from the bond cleaved (Chap. 6.2.5). They reduced the fragmentation rate relative to the nonlabeled chain by factors of 1.08–1.30 per D in case of metastable ion decompositions (Fig. 2.18), but the isotope effect vanished for ion source processes. [78] With the aid of field ionization kinetic measurements the reversal of these kinetic isotope effects for short-lived ions (10^{-11}–10^{-10} s) could be demonstrated, i.e., then the deuterated species decomposed slightly faster than their nonlabeled isotopomers (Fig. 2.17). [66,76]

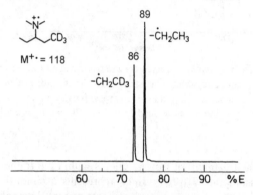

Fig. 2.18. Observation of secondary H/D isotope effects on the α-cleavage of tertiary amine molecular ions. For convenience, *m/z* labels have been added to the original energy scale of the MIKE spectrum. Adapted from Ref. [78] with permission. © American Chemical Society, 1988.

2.10 Determination of Ionization Energies and Appearance Energies

2.10.1 Conventional Determination of Ionization Energies

The experimental determination of ionization energies [46] is rather uncomplicated. One simply would have to read out the lower limit of the electron energy at a vanishing molecular ion signal. Unfortunately, doing so yields only coarse approximations of the real ionization energy of a molecule. The accuracy of IE data obtained by this simple procedure will be about ± 1 eV. One of the problems is located in the measurement of the electron energy itself. The electrons are thermally emitted from a hot metal filament (1600–2000 °C), and therefore, their total kinetic energy is not only defined by the potential applied to accelerate them, but also by their thermal energy distribution. [25] In addition, electron ionization preferably creates vibrationally excited ions (Chap. 2.3), because ionization is a threshold process, i.e., it will take place not just when the energy needed to accomplish the process is reached, but also for all higher energies. [79] This gives rise to a systematic error in that vertical IEs are obtained being higher than the adiabatic IEs one would like to measure. Further drawbacks are: i) Due to inhomogeneous electric fields within the ionization volume, the actual electron energy also depends on the location where ionization takes place. ii) The fact that the low electron acceleration voltages of 7–15 V have to be superimposed on the ion acceleration voltage of several kilovolts causes low precision of the voltage settings in commercial magnetic sector instruments. iii) There is additional thermal energy of the neutrals roughly defined by the temperature of the inlet system and the ion source.

2.10.1.1 Improved IE Accuracy from Data Post-Processing

Numerous approaches have been published to improve the accuracy of IE data. However, the uncertainty of electron energy remains, causing the ionization efficiency curves not to directly approach zero at IE. Instead of being linear, they bend close to the ionization threshold and exponentially approximate zero. Even though the electron energy scale of the instrument has been properly calibrated against IEs of established standards such as noble gases or solvents, IE data obtained from direct readout of the curve have accuracies of ± 0.3 eV (Fig. 2.19a).

To overcome the uncertainty of the actual onset of ionization, among several others, [80] the *critical slope method* has been developed. [25,81] It makes use of the fact that from theory realistic values of IE are expected at the position of the ionization efficiency curve where the slope of a semilog plot of the curve is

$$\frac{d}{dV}(\ln N_i) = \frac{n}{n+1}\frac{1}{kT} \tag{2.31}$$

with N_i being the total number of ions produced at an electron acceleration voltage V, and with an empirical value of $n = 2$ (Fig. 2.19b).

2.10.2 Experimental Improvements of IE Accuracy

If reliable thermochemical data [28,82] is required, the above influences have to be substantially reduced. [46] One way is to use an *electron monochromator* (accuracy up to ± 0.1 eV). [83,84] An electron monochromator is a device for selecting nearly monoenergetic electrons from an electron beam. [85] Alternatively, one can employ *photoionization* (PI) instead of EI:

$$M + h\nu \rightarrow M^{+\bullet} + e^- \tag{2.32}$$

Photoionization yields even more accurate results (± 0.05 eV) than the electron monochromator. [86] In any case, the half width of the electron or photon energy distribution becomes small enough to detect detailed structural features of the ionization efficiency curves such as electronic transitions. Both techniques have been widely employed to obtain ionization energy data (Table 1.1).

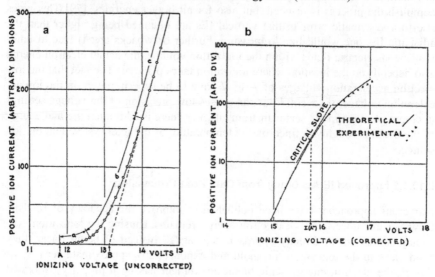

Fig. 2.19. Ionization efficiency curve of argon plotted on a linear scale (**a**) and as semilog plot (**b**). Extrapolation of the linear portion of **a** gives erroneous IEs, whereas the x-position of the tangent of an empirical critical slope to the semilog plot yields accuracies of ± 0.05 eV. Reproduced from Ref. [25] by permission. © American Chemical Society, 1948.

2.10.3 Photoelectron Spectroscopy and Derived Modern Methods

Photoelectron spectroscopy (PES, a non-mass spectral technique) [87] has proven to be very useful in providing information not only about ionization potentials, but also about the electronic and vibrational structure of atoms and molecules. Energy resolutions reported from PES are in the order of 10–15 meV. The resolution of PES still prevents the observation of rotational transitions, [79] and to overcome these limitations, PES has been further improved. In brief, the principle of *zero kinetic energy photoelectron spectroscopy* (ZEKE-PES or just ZEKE, also a non-mass spectral technique) [89-91] is based on distinguishing excited ions from ground state ions.

First, imagine a neutral interfering with a photon carrying some meV more than its IE. The neutral is going to expel a kinetic electron travelling away due to this slight excess energy, E_{kinel}, defined by:

$$E_{\mathrm{kinel}} = h\nu - \mathrm{IE} \tag{2.33}$$

Next, consider the situation if $h\nu = \mathrm{IE}$. Here, the electron can just be released, but it cannot travel away from the freshly formed ion. Waiting for some short delay (1 µs) now allows to separate *zero kinetic energy electrons* from others in space. Applying an extraction voltage along a given drift direction will then add a different amount of kinetic energy to the kinetic electrons, as these have traveled further, than to those ejected at threshold, i.e., they differ in speed. A time-of-flight measurement of the electrons therefore produces one signal for each group of electrons. (For another application of this concept cf. Chap. 4.2.5).

Example: The electronic and vibrational states of the oxygen molecular ion could be perfectly resolved by PES (Fig. 2.20), thus allowing to directly read out the Franck-Condon factors and to identify the (0 ← 0) transitions corresponding to adiabatic ionization. [88]

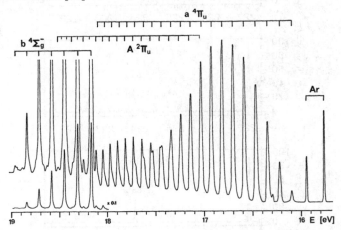

Fig. 2.20. High resolution photoelectron spectrum of O_2, showing overlapping vibrational progressions from transitions to different electronic states of the ion (range of IE not shown). Reproduced from Ref. [88] with permission. © Royal Swedish Academy of Sciences, 1970.

The main disadvantage of PES and ZEKE experiments results from the detection of electrons making the measurements sensitive to impurities, because the electrons could arise from these instead of the intended sample. This can be circumvented by detecting the ions produced instead, and the corresponding technique is known as *mass-analyzed threshold ionization* (MATI). [92] In MATI experiments, the neutrals are excited in a field-free environment by means of a tunable light source (usually a multi-photon laser process) very close to ionization threshold. Eventually occurring prompt ions are removed after about 0.1 μs by a weak positive electric field. Then, the near-threshold Rydberg species are ionized by applying a negative electric field pulse also effecting acceleration of those ions towards a time-of-flight mass analyzer. [92-94] Thus, we are finally back to a real mass spectral technique. Different from ZEKE, the mass selectivity of MATI allows not only for the study of molecules, [92,94] but also to deal with dissociating complexes and clusters. [79,95,96]

Example: The ionization spectra at the first ionization threshold of pyrazine as obtained by PI, MATI and ZEKE-PES are clearly different (Fig. 2.21). [92] The

PI spectrum is a plot of ion current *versus* wavelength of the probe laser. While the PI spectrum shows a simple rise of the curve at IE, MATI and ZEKE yield a sharp peak at ionization threshold plus additional signals from lower vibrational ionization thresholds.

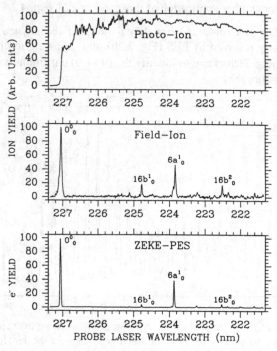

Fig. 2.21. Comparison of PI, MATI and ZEKE spectra of pyrazine. Reproduced from Ref. [92] with permission. © American Institute of Physics, 1991.

2.10.4 Determination of Appearance Energies

The techniques used for the determination of appearance energies are essentially identical to those described above for IEs. However, even when using the most accurately defined electron or photon energies, great care has to be taken when AEs are to be determined because of the risk of overestimation due to kinetic shift. Provided that there is no reverse activation energy for the reaction under study, the AE value also delivers the sum of heats of formation of the dissociation products. If substantial KER is observed, the AE may still be used to determine the activation energy of the process.

2.10.5 Breakdown Graphs

Employing the above techniques, one can examine the fragmentations of a molecular ion as a function of internal energy by constructing a so-called *breakdown graph*. This is essentially done by plotting the ion intensities of interest, i.e., those of a certain *m/z*, versus electron energy or versus ion internal energy if the IE has been subtracted before. [97] Typically, a molecular ion can access a considerable number of fragmentation pathways as soon as there are some 1–3 eV of internal energy available. Breakdown graphs allow to compare the energetic demands of those different fragmentation pathways. In addition, breakdown graphs help to correlate ion internal energy distributions derived from other methods such as photoelectron spectroscopy [98] with mass spectral data.

Example: For the case of 4-methyl-1-pentene, the breakdown graph, the internal energy distribution from the photoelectron spectrum, and the 70 eV EI mass spectrum are compared (Fig. 2.22). [99] From the fragmentation threshold to about

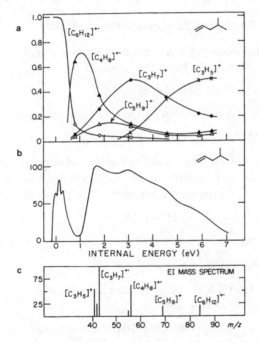

Fig. 2.22. Relationship of breakdown graph (**a**), internal energy distribution from PES (**b**), and mass spectrum of 4-methyl-1-pentene (**c**). Reproduced from Ref. [99] by permission. © John Wiley & Sons, 1982.

2 eV of internal energy the breakdown graph is dominated by the $[C_4H_8]^{+\bullet}$ ion, *m/z* 56. In the range 2–4.5 eV the $[C_3H_7]^+$ ion, *m/z* 43, becomes most prominent. However, in the spectrum, $[C_4H_8]^{+\bullet}$ is of only 60 % of the intensity of $[C_3H_7]^+$. It is obvious from the graphs that the 0.5–1 eV energy region, where $[C_4H_8]^{+\bullet}$ is the predominating fragment ion, corresponds to a region of the internal energy

distribution which has a low ion population. This explains why $[C_3H_7]^+$ constitutes the base peak of the spectrum. Beyond 5 eV internal energy, the $[C_3H_5]^+$ ion, *m/z* 41, becomes the most prominent fragment ion.

2.11 Gas Phase Basicity and Proton Affinity

Not all ionization methods rely on such strictly unimolecular conditions as EI does. Chemical ionization (CI, Chap. 7), for example, makes use of reactive collisions between ions generated from a reactant gas and the neutral analyte to achieve its ionization by some bimolecular process such as proton transfer. The question which reactant ion can protonate a given analyte can be answered from *gas phase basicity* (GB) or *proton affinity* (PA) data. Furthermore, proton transfer, and thus the relative proton affinities of the reactants, play an important role in many ion-neutral complex-mediated reactions (Chap. 6.12).

2.11.1 Definition of Gas Phase Basicity and Proton Affinity

Proton affinity and gas *phase basicity* are thermodynamic quantities. Consider the following gas phase reaction of a (basic) molecule, B:

$$B_g + H_g^+ \rightarrow [BH]_g^+ \tag{2.34}$$

The tendency of B to accept a proton is then quantitatively described by

$$-\Delta G_r^0 = GB_{(B)} \quad \text{and} \quad -\Delta H_r^0 = PA_{(B)}, \tag{2.35}$$

i.e. the gas phase basicity $GB_{(B)}$ is defined as the negative free energy change for the proton transfer, $-\Delta G_r^0$, whereas the proton affinity $PA_{(B)}$ is the negative enthalpy change, $-\Delta H_r^0$, for the same reaction. [100,101] From the relation

$$\Delta G^0 = \Delta H^0 - T \Delta S^0 \tag{2.36}$$

we obtain the expression

$$PA_{(B)} = GB_{(B)} - T \Delta S^0 \tag{2.37}$$

with the entropy term $T\Delta S^0$ usually being relatively small (25–40 kJ mol^{-1}). Furthermore, in case of an equilibrium

$$[AH]^+ + B \quad A + [BH]^+ \tag{2.38}$$

with the equilibrium constant K_{eq} for which we have

$$K_{eq} = [BH^+]/[AH^+] \cdot [A]/[B] \tag{2.39}$$

the gas phase basicity is related to K_{eq} by [102,103]

$$GB_{(B)} = -\Delta G^0 = RT \ln K_{eq} \tag{2.40}$$

2.11.2 Determination of Gas Phase Basicities and Proton Affinities

The methods for the determination of GBs and PAs make use of their relation to K_{eq} (Eq. 2.40) and the shift of K_{eq} upon change of $[AH]^+$ or B, respectively. [101,103] Basically, the value of GB or PA is bracketed by measuring K_{eq} with a series of several reference bases ranging from lower to higher GB than the unknown. There are two methods we should address in brief, a detailed treatment of the topic being beyond the scope of the present book, however.

2.11.2.1 Kinetic Method

The *kinetic method* [102,104-106] compares the relative rates of the competitive dissociations of a proton-bound adduct $[A-H-B]^+$ formed by admitting a mixture of A and B to a CI ion source. [104,105] There, the proton-bound adduct $[A-H-B]^+$ is generated amongst other products such as $[AH]^+$ and $[BH]^+$. Using standard tandem MS techniques, e.g., MIKES, the cluster ion $[A-H-B_{ref}]^+$ is selected and allowed to undergo metastable decomposition:

$$[AH]^+ + B_{ref} \leftarrow [A-H-B_{ref}]^+ \rightarrow A + [B_{ref}H]^+ \qquad (2.41)$$

The relative intensities of the products $[AH]^+$ and $[B_{ref}H]^+$ are then used as a measure of relative rate constants of the competing reactions. In case the PA of the unknown was equal to that of the reference, both peaks would be of equal intensity. As this will almost never be the case, a series of reference bases is employed instead, and PA is determined by interpolation. The value of $PA_{(A)}$ is obtained from a plot of $\ln[AH]^+/[B_{ref}H]^+$ versus $PA_{(B)}$ at $\ln[AH]^+/[B_{ref}H]^+ = 0$.

2.11.2.2 Thermokinetic Method

The *thermokinetic method* [107,108] uses the measurement of the forward rate constant of the equilibrium

$$[AH]^+ + B_{ref} \quad A + [B_{ref}H]^+ \qquad (2.42)$$

The thermokinetic method takes advantage of the correlation observed between k_{exp} and ΔG_2^0 through the relationship

$$\frac{k_{exp}}{k_{coll}} = \frac{1}{1 + e(\Delta G_2^0 + \Delta G_a^0)/RT} \qquad (2.43)$$

where k_{coll} is the collision rate constant and ΔG_a^0 a term close to RT. The GB of the unknown is then obtained from $\Delta G_2^0 = GB - GB_{ref}$. The task to establish a proper value of the reaction efficiency $RE = k_{exp}/k_{coll}$, is solved by plotting the experimental values of RE versus $GB_{(B)}$ and interpolating these points with a parametric function. Although this can be done with high accuracy, it is still a matter of debate which value of RE yields the most realistic GB, suggestions being $RE = 0.1–0.5$. [101]

Example: GB and PA of cyclohexanecarboxamide were determined by either experimental method. The kinetic method based on both metastable dissociation (MI) and collision-induced dissociation (CID, Chap. 2.12.1) yielded $GB = 862 \pm 7$ kJ mol^{-1} and $PA = 896 \pm 5$ kJ mol^{-1}, while the thermokinetic method gave $GB = 860 \pm 5$ kJ mol^{-1} and $PA = 891 \pm 5$ kJ mol^{-1}, i.e., both methods yield comparable results (Fig. 2.23).

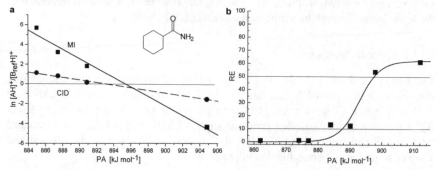

Fig. 2.23. Determination of the proton affinity of cyclohexanecarboxamide (**a**) by the kinetic method, and (**b**) by the thermokinetic method. The horizontal lines show the indicative values $\ln[AH]^+/[B_{ref}H]^+ = 0$ in (**a**) and *RE*s according to different authors in (**b**). [101] Adapted from Ref. [109] by permission. © IM Publications, 2003.

Table 2.6. Selected proton affinities and gas phase basicities [28,100,110]

Molecule	PA$_{(B)}$ [kJ mol^{-1}]	GB$_{(B)}$ [kJ mol^{-1}]
H_2	424	396
CH_4	552	527
C_2H_6	601	558
H_2O	697	665
$H_2C=O$	718	685
$CH_3CH=CH_2$	751	718
C_6H_6 (benzene)	758	731
$(CH_3)_2C=CH_2$	820	784
$(CH_3)_2C=O$	823	790
$C_{14}H_{10}$ (phenanthrene)	831	802
C_4H_8O (tetrahydrofurane)	831	801
$C_2H_5OC_2H_5$	838	805
NH_3	854	818
CH_3NH_2	896	861
C_5H_5N (pyridine)	924	892
$(CH_3)_3N$	942	909

2.12 Tandem Mass Spectrometry

Tandem mass spectrometry summarizes the numerous techniques where mass-selected ions (MS1) are subjected to a second mass spectrometric analysis (MS2). [111,112] Dissociations in transit through the mass analyzer may either occur spontaneously (*metastable*, Chaps. 2.7.1, 2.8.2) or can result from intentionally supplied additional activation, i.e., typically from collisions with neutrals. Below, we will chiefly discuss the collision process as such and its consequences for the further fate of the ions. Instrumental aspects of tandem MS are included in Chap. 4 and applications are presented in Chaps. 7-12.

> **Note:** *Tandem mass spectrometry* is also known as MS/MS or MS^2. The latter terminology elegantly allows for the expansion to multi-stage experimental set-ups, e.g., MS^3, MS^4 or generally MS^n.

2.12.1 Collision-Induced Dissociation

Even though collisions of ions with neutral gas atoms or molecules appear to be perfectly contradictory to the conditions of high vacuum, most mass spectrometers are equipped or can be upgraded to allow for their study. Consequently, fundamental and analytical studies make use of activating or reactive collisions within the mass spectrometer. The most prominent collision technique is *collision-induced dissociation* (CID); [113,114] the terms *collisionally activated dissociation* (CAD) or *collisional activation* (CA) also have been in use. CID allows for the fragmentation of gaseous ions that were perfectly stable before the activating event. Thus, CID is especially useful for the structure elucidation of ions of low internal energy, e.g., for those created by soft ionization methods (Chaps. 7-11).

2.12.1.1 Effecting Collisions in a Mass Spectrometer

CID is generally realized by passing an ion beam through a *collision cell* where the collision gas (He, N_2, Ar) is regulated at a pressure considerably above that of the surrounding high vacuum. This can be achieved by introducing the gas via a needle valve into a comparatively tight volume only having a narrow entrance and exit for the ion beam (Fig. 2.24). A vacuum pump is placed near by to remove effusing gas, thereby creating a *differentially pumped* region, because there is no more laminar flow at some 10^{-4} Pa. Instead, expansion of the gas is diffusion-controlled. The reading of a pressure gauge can be used to reproduce the pressure adjustment, but it never shows the real pressure inside the cell. [9]

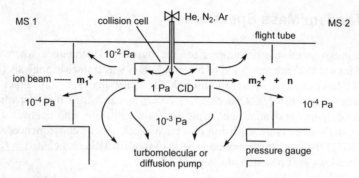

Fig. 2.24. Schematic of a collision cell for CID experiments in a beam instrument.

2.12.1.2 Energy Transfer During Collisions

The collision of an ion AB^+ carrying some kiloelectronvolts of kinetic energy with a neutral N takes about 10^{-15} s. This allows to apply the assumptions of QET analogously to electron ionization [5,115-118] (Chap. 2.1.1). The collision-induced dissociation of AB^+ can therefore be regarded as a two-step process. [119] First, the activated species AB^{+*} is formed. Second, after randomization of the internal energy AB^{+*} dissociates along any fragmentation pathway available at this specific level of internal energy:

$$AB^+ + N \rightarrow AB^{+*} + N \rightarrow A^+ + B + N \tag{2.44}$$

The internal energy E_{AB+*} is composed of the internal energy prior to the collision, E_{AB+}, and of the amount of energy Q transferred during the collision:

$$E_{AB+*} = E_{AB+} + Q \tag{2.45}$$

Thus, the collision marks a new start of the time scale for the activated ion. As $Q > E_{AB+}$ generally holds valid, the internal energy prior to the collision is of minor relevance – though not generally negligible – for the behavior of the activated ion. As may be expected, the CID spectra of stable molecular ions exhibit marked similarity to the 70 eV EI spectra of the respective compounds. [113,114]

Example: Except for the intensities relative to the precursor ion, the B/E linked scan CID spectrum of toluene molecular ion, m/z 92, closely resembles the 70 eV EI mass spectrum of toluene (Fig. 2.25; for EI cf. Chap. 6.4.3). Here, all fragments are due to CID because the molecular ion was generated by field ionization and did not show any metastable decomposition, i.e., $E_{AB+} = 0$ and $E_{AB+*} = Q$.

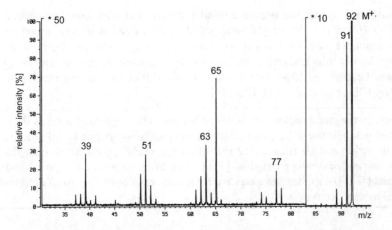

Fig. 2.25. CID spectrum of toluene molecular ion, m/z 92. $E_{lab} = 10$ kV, B/E linked scan, collision gas He at about 50 % transmission.

The absolute upper limit for the value of Q is defined by the *center-of-mass collision energy*, E_{CM}, [115,116]

$$E_{CM} = E_{LAB} \frac{m_N}{m_N + m_{AB}} \tag{2.46}$$

where m_N is the mass of the neutral, m_{AB} the mass of the ion, and E_{LAB} the ion kinetic energy in the laboratory frame of reference. By regarding a polyatomic ion as separated into the atom B actually involved in the collision process and the remainder A, the maximum of Q is calculated to have a lower value than E_{CM}. Assuming central collisions we obtain: [116]

$$E_{\text{int max}} = 4E_{LAB}\, m_A\, m_B \left(\frac{m_N}{m_{AB}(m_B + m_N)} \right)^2 \tag{2.47}$$

However, most collisions are not "head-on", but occur at some angle θ. Increasing m_{AB} makes E_{intmax} decrease, whereas larger m_N is beneficial for energy transfer. In CID-MIKES the center of the peak is shifted to the low mass side, i.e., to the low ion translational energy side, because the uptake Q originates from a loss in E_{LAB}. [120-125] The relationship between ΔE_{LAB} and Q can be expressed as: [122]

$$Q = \frac{\Delta E_{LAB}(m_{AB} + m_N)}{m_N} - \left[\left(\frac{2m_{AB}E_{LAB}}{m_N} \right) \left(1 - \sqrt{\frac{E_{LAB} - \Delta E_{LAB}}{E_{LAB}}} \cos\theta \right) \right] \tag{2.48}$$

Several conclusions can be drawn from Eq. 2.48: i) Q has a broad distribution (0–15 eV) due to variations in θ and typically is in the order of some electronvolts [126]; ii) up to a certain ionic mass the neutral penetrates the incident ion, i.e., the activated ion leaves the neutral behind (forward-scattered ion), but beyond the limit ($> 10^2$ atoms) the neutral is expelled in the direction of ion motion (backward-scattered ion) [122]; iii) at about m/z 1500 $\Delta E_{LAB} = Q$, i.e., the translational energy loss is fully converted into vibrational excitation of the ion; and iv) Q decreases beyond m/z 1500, thus explaining the difficulties in fragmenting heavy ions by CID (Fig. 2.26). [118,122,127]

> **Note:** In magnetic sector and TOF instruments, He is typically used as the collision gas because E_{intmax} from *high-energy collisions* (keV) is still comparatively large, and He reduces the risk of charge exchange due to its high IE. In the *low collision energy regime* [116] of quadrupole and ion trapping instruments (1–200 eV), heavier gases are frequently employed (N_2, Ar, Xe) to make CID more effective.

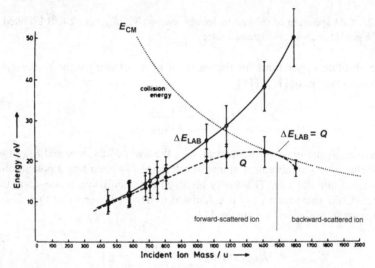

Fig. 2.26. Illustration of the relationship between E_{CM}, ΔE_{LAB}, and Q. Adapted from Ref. [122] with permission. © Verlag der Zeitschrift für Naturforschung, 1984.

2.12.1.3 Single and Multiple Collisions in CID

Generally, the collision gas pressure is adjusted in order to achieve a certain attenuation of the beam of mass-selected ions that are going to be collided. As the so-called *main beam* becomes increasingly attenuated, the probability for multiple collisions rises and so does the yield of fragments resulting from high activation energy processes. In a typical high collision energy experiment a transmission of 90 % for the main beam means that 95 % of the colliding ions undergo single collisions, about 25 % of the ions encounter double and some percent even triple and quadruple collisions at 50 % transmission (Fig. 2.27). [9] To achieve sufficient activation of the ions in the low collision energy regime, elongated collision cells are employed where Q is accumulated from numerous collision events.

Note: It has turned out that medium transmission is optimal for structure elucidation. Too strong reduction of the main beam favors ion losses due to scattering, charge exchange ($M^{+\bullet} + N \rightarrow M + N^{+\bullet}$) or charge stripping processes ($M^{+\bullet} + N \rightarrow M^{2+} + N^{+\bullet}$) instead of delivering additional structural information.

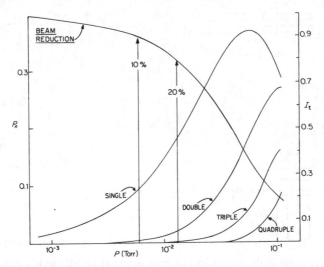

Fig. 2.27. Total collision probability P_n and fractions of single, double, triple and quadruple collisions versus collision gas pressure. The transmission of the main beam I_t is given on the right ordinate. Values are for an ion of collision cross section 5×10^{-16} cm^2 and 1 cm collision path; 10^{-2} Torr = 1.33 Pa. Reproduced from Ref. [9] with permission. © John Wiley & Sons, 1985.

2.12.2 Other Methods of Ion Activation

2.12.2.1 Surface-Induced Dissociation

Instead of using gaseous atoms as the collision partners, collisions with solid surfaces can be employed to induce dissociation of the incident ions. This technique has become known as *surface-induced dissociation* (SID). [128] Ions of some ten electronvolts kinetic energy are collided with a solid surface typically at an angle of 45° to effect SID. The evolving fragments are mass-analyzed by means of a linear quadrupole mass analyzer at right angles to the incident ion beam (Fig. 2.28). [128-130] Such an experimental setup allows for the control of the energy of the incident ions, and therefore for the adjustment of the degree of fragmentation.

It has been shown that SID spectra are very similar to high as well as to low-energy CID spectra. [128] The absence of collision gas presents an advantage of

SID over CID because losses of resolution due to high background pressure are avoided. SID has been successfully employed for structure elucidation of protonated peptides, [131] and a SID mode of operation has even been implemented with a quadrupole ion trap. [132] However, apart from the quadrupole ion trap SID requires substantial modifications of the instrumental hardware; circumstances that made SID lag behind the countless applications of CID.

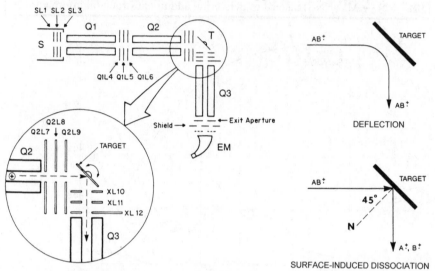

Fig. 2.28. Apparatus (left) and modes of operation (right) for SID with a modified triple-quadrupole mass spectrometer. Reproduced from Ref. [129] with permission. © Elsevier Science, 1987.

2.12.2.2 Infrared Multiphoton Dissociation

The energy received from multiple photon absorption may also be used to activate and dissociate otherwise stable gaseous ions: [133]

$$m_1^+ + x \times h\nu \rightarrow m_2^+ + n \qquad (2.49)$$

Infrared multiphoton dissociation (IRMPD) is a technique that can be conveniently applied to trapped ions, i.e., in *Fourier transform ion cyclotron resonance* (FT-ICR) analyzers. IRMPD normally employs a continuous wave carbon dioxide laser of 10.6 μm wavelength having a power of 25-40 W which is passed into the ICR cell through a ZnSe or BaF_2 window. [134,135] It is a great advantage of combining FT-ICR with IRMPD that the amount of energy put onto the ions can be varied via the duration of laser irradiation, typically in the range of 5–300 ms. [136] This allows for the application of IRMPD to small ions [133] as well as to medium-sized [136] or high-mass ions. [135]

2.12.2.3 Electron Capture Dissociation

For an ion, the cross section for *electron capture* (EC) roughly increases with the square of the ionic charge. [137] This makes multiply charged ions as produced by *electrospray ionization* (ESI, Chap. 11) the ideal targets for this process, e.g.:

$$[M+11H]^{11+} + e^- \rightarrow [M+11H]^{10+\bullet} \qquad (2.50)$$

The energy from partial neutralization becomes the ion's internal energy, thus causing its dissociation, so-called *electron capture dissociation* (ECD). Again, this is a technique for trapped ions only. To effectively achieve ECD the electrons must have energies < 0.2 eV. Therefore, they are supplied analogously to EI from a carefully regulated heated filament [138] (Chap. 5.2.3) externally mounted to the ICR cell. Although a very recent discovery, [137,139] ECD is now widely applied in biomolecule sequencing by means of ESI-FT-ICR-MS. [138,140] Presumably the main reason for the numerous ECD applications arises from the fact that ECD yields information complementary to CID [141] and IRMPD. [134]

> **Note:** As one electron charge is neutralized upon EC, the precursor ion for ECD must at least be a doubly charged positive even-electron ion to yield a singly charged radical ion for subsequent dissociation.

2.12.3 Reactive Collisions

2.12.3.1 Neutralization-Reionization Mass Spectometry

In CID, charge exchange between the ions and the collision gas is an unwanted side reaction. However, it may become useful when employed in combination with a subsequent reionization step as realized in *neutralization-reionization mass spectrometry* (NR-MS). [142-147] In NR-MS the precursor ion of some kiloelectronvolts kinetic energy is mass-selected in MS1 and passed through a first collision cell containing a reducing collision gas. A certain fraction of these ions will be reduced by charge exchange to become neutrals. As these neutrals basically retain their initial kinetic energy and direction, they leave the first collision cell along with the precursor ion beam. Remaining ions can then be removed from the beam by electrostatic deflection. Having traveled for some microseconds along a short path through the field-free region, the neutrals are subjected to ionizing collisions in a second collision cell. Finally, the mass spectrum of the reionized species is detected by means of MS2.

For the purpose of neutralization it is possible to employ i) noble gases of low IE such as Xe (12.1 eV), [144] ii) metal vapors effusing from an oven, e.g., Hg (10.4 eV), Zn (9.3 eV) and Cd (9.0 eV), [142,148] or iii) volatile organic molecules, e.g., benzene (9.2 eV) or triphenylamine (6.8 eV). [149] Reionization of the neutrals in the second collision cell can be achieved i) using O_2 (12.1 eV) or He (24.6 eV), [148,149] or ii) some standard ionization method such as electron ionization or field ionization. [146]

The applications of NR-MS are numerous. [147,150-152] In particular, NR-MS plays to its strength where the existence of short-lived and otherwise unstable species has to be proved. [153-155]

Example: The NR mass spectrum of acetone closely resembles its 70 eV EI mass spectrum (Chap. 6.2.1), thereby demonstrating that the molecular ion basically retains the structure of the neutral (Fig. 2.29). [144] However, the isomeric $C_3H_6O^{+\bullet}$ ions formed by McLafferty rearrangement of 2-hexanone molecular ion are expected to have enol structure (Chap. 6.7.1), and thus the corresponding NR mass spectrum is easily distinguished from that of acetone.

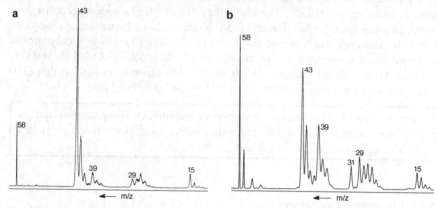

Fig. 2.29. NR-MS (Xe–He) of acetone (**a**) and its enol (**b**) from neutralization of the corresponding ions. Adapted from Ref. [144] with permission. © Elsevier Science, 1985.

2.12.3.2 Ion-Molecule Reactions

When gaseous ions collide with neutrals at *thermal energy* rather than at multi- or even kiloelectronvolt energy other bimolecular reactions than the mere charge exchange may take place.

Proton transfer is one of the prominent representatives of an *ion-molecule reaction* in the gas phase. It is employed for the determination of GBs and PAs (Chap. 2.11.2) by either method: the kinetic method makes use of the dissociation of proton-bound heterodimers, and the thermokinetic method determines the equilibrium constant of the acid-base reaction of gaseous ions. In general, proton transfer plays a crucial role in the formation of protonated molecules, e.g., in positive-ion chemical ionization mass spectrometry (Chap. 7).

More recently, the catalytic activities of a large pool of transition-metal carbene complexes have been screened by means of ion-molecule reactions in tandem-MS experiments. [156-158] Different from the concepts and methods discussed so far, the latter experiments are not designed to study the fundamentals of mass spectrometry. Instead, sophisticated methods of modern mass spectrometry are now employed to reveal the secrets of other complex chemical systems.

Reference List

1. Porter, C.J.; Beynon, J.H.; Ast, T. The Modern Mass Spectrometer. A Complete Chemical Laboratory. *Org. Mass Spectrom.* **1981**, *16*, 101-114.
2. Schwarz, H. The Chemistry of Naked Molecules or the Mass Spectrometer As a Laboratory. *Chem. Unserer Zeit* **1991**, *25*, 268-278.
3. Kazakevich, Y. Citation Used by Permission. *http://hplc. chem. shu. edu/* **1996**, Seton Hall University, South Orange, NJ.
4. Cooks, R.G.; Beynon, J.H.; Caprioli, R.M. *Metastable Ions;* Elsevier: Amsterdam, 1973.
5. Levsen, K. *Fundamental Aspects of Organic Mass Spectrometry;* 1st ed.; Verlag Chemie: Weinheim, 1978.
6. Franklin, J.L. Energy Distributions in the Unimolecular Decomposition of Ions, in *Gas Phase Ion Chemistry*, 1st ed.; Bowers, M.T., editor; Academic Press: New York, 1979; Vol. 1, Chapter 7, pp. 272-303.
7. Beynon, J.H.; Gilbert, J.R. Energetics and Mechanisms of Unimolecular Reactions of Positive Ions: Mass Spectrometric Methods, in *Gas Phase Ion Chemistry*; Bowers, M.T., editor; Academic Press: New York, 1979; Vol. 2, Chap. 13, 153-179.
8. Vogel, P. The Study of Carbocations in the Gas Phase, in *Carbocation Chemistry*, Elsevier: Amsterdam, 1985; Chap. 2, 61-84.
9. Holmes, J.L. Assigning Structures to Ions in the Gas Phase. *Org. Mass Spectrom.* **1985**, *20*, 169-183.
10. Lorquet, J.C. Basic Questions in Mass Spectrometry. *Org. Mass Spectrom.* **1981**, *16*, 469-481.
11. Lorquet, J.C. Landmarks in the Theory of Mass Spectra. *Int. J. Mass Spectrom.* **2000**, *200*, 43-56.
12. Wahrhaftig, A.L. Ion Dissociations in the Mass Spectrometer, in *Advances in Mass Spectrometry*, Waldron, J.D., editor; Pergamon Press: Oxford, 1959; pp. 274-286.
13. Wahrhaftig, A.L. Unimolecular Dissociations of Gaseous Ions, in *Gaseous ion Chemistry and Mass Spectrometry*, Futrell, J.H., editor; John Wiley and Sons: New York, 1986; 7-24.
14. Rosenstock, H.M.; Krauss, M. Quasi-Equilibrium Theory of Mass Spectra, in *Mass Spectrometry of Organic Ions*, 1st ed.; McLafferty, F.W., editor; Academic Press: London, 1963; 1-64.
15. Bohme, D.K.; Mackay, G.I. Bridging the Gap Between the Gas Phase and Solution: Transition in the Kinetics of Nucleophilic Displacement Reactions. *J. Am. Chem. Soc.* **1981**, *103*, 978-979.
16. Speranza, M. Gas Phase Ion Chemistry Versus Solution Chemistry. *Int. J. Mass Spectrom. Ion Proc.* **1992**, *118/119*, 395-447.
17. Rosenstock, H.M.; Wallenstein, M.B.; Wahrhaftig, A.L.; Eyring, H. Absolute Rate Theory for Isolated Systems and the Mass Spectra of Polyatomic Molecules. *Proc. Natl. Acad. Sci. U. S. A.* **1952**, *38*, 667-678.
18. Märk, T.D. Fundamental Aspects of Electron Impact Ionization. *Int. J. Mass Spectrom. Ion Phys.* **1982**, *45*, 125-145.
19. Märk, T.D. Electron Impact Ionization, in *Gaseous ion Chemistry and Mass Spectrometry*, Futrell, J.H., editor; John Wiley and Sons: New York, 1986; 61-93.
20. Meyerson, S.; Van der Haar, R.W. Multiply Charged Organic Ions in Mass Spectra. *J. Chem. Phys.* **1962**, *37*, 2458-2462.
21. Jones, E.G.; Harrison, A.G. Study of Penning Ionization Reactions Using a Single-Source Mass Spectrometer. *Int. J. Mass Spectrom. Ion Phys.* **1970**, *5*, 137-156.
22. Penning, F.M. Ionization by Metastable Atoms. *Naturwissenschaft.* **1927**, *15*, 818.
23. Hornbeck, J.A.; Molnar, J.P. Mass-Spectrometric Studies of Molecular Ions in the Noble Gases. *Phys. Rev.* **1951**, *84*, 621-625.
24. Faubert, D.; Paul, G.J.C.; Giroux, J.; Betrand, M.J. Selective Fragmentation and Ionization of Organic Compounds Using an Energy-Tunable Rare-Gas Metastable Beam Source. *Int. J. Mass Spectrom. Ion Proc.* **1993**, *124*, 69-77.
25. Honig, R.E. Ionization Potentials of Some Hydrocarbon Series. *J. Chem. Phys.* **1948**, *16*, 105-112.
26. Cone, C.; Dewar, M.J.S.; Landman, D. Gaseous Ions. MINDO/3 Study of the Rearrangement of Benzyl Cation to Tropylium. *J. Am. Chem. Soc.* **1977**, *99*, 372-376.
27. Svec, H.J.; Junk, G.A. Electron-Impact Studies of Substituted Alkanes. *J. Am. Chem. Soc.* **1967**, *89*, 790-796.
28. NIST NIST Chemistry Webbook. *http://webbook. nist. gov/* **2002**.

29. Born, M.; Oppenheimer, J.R. Zur Quantentheorie der Molekeln. *Annalen der Physik* **1927**, *84*, 457-484.
30. Seiler, R. Born-Oppenheimer Approximation. *Int. J. Quant. Chem.* **1969**, *3*, 25-32.
31. Franck, J. Elementary Processes of Photochemical Reactions. *Trans. Faraday Soc.* **1925**, *21*, 536-542.
32. Condon, E.U. Theory of Intensity Distribution in Band Systems. *Phys. Rev.* **1926**, *28*, 1182-1201.
33. Dunn, G.H. Franck-Condon Factors for the Ionization of H_2 and D_2. *J. Chem. Phys.* **1966**, *44*, 2592-2594.
34. McLafferty, F.W.; Wachs, T.; Lifshitz, C.; Innorta, G.; Irving, P. Substituent Effects in Unimolecular Ion Decompositions. XV. Mechanistic Interpretations and the QET. *J. Am. Chem. Soc.* **1970**, *92*, 6867-6880.
35. Egger, K.W.; Cocks, A.T. Homopolar- and Heteropolar BDEs and Heats of Formation of Radicals and Ions in the Gas Phase. I. Data on Organic Molecules. *Helv. Chim. Acta* **1973**, *56*, 1516-1536.
36. Lossing, F.P.; Semeluk, G.P. Free Radicals by Mass Spectrometry. XLII. Ionization Potentials and Ionic Heats of Formation for C_1-C_4 Alkyl Radicals. *Can. J. Chem.* **1970**, *48*, 955-965.
37. Lossing, F.P.; Holmes, J.L. Stabilization Energy and Ion Size in Carbocations in the Gas Phase. *J. Am. Chem. Soc.* **1984**, *106*, 6917-6920.
38. Cox, J.D.; Pilcher, G. *Thermochemistry of Organic and Organometallic Compounds*; Academic Press: London, 1970.
39. Chatham, H.; Hils, D.; Robertson, R.; Gallagher, A. Total and Partial Electron Collisional Ionization Cross Sections for Methane, Ethane, Silane, and Disilane. *J. Chem. Phys.* **1984**, *81*, 1770-1777.
40. McAdoo, D.J.; Bente, P.F.I.; Gross, M.L.; McLafferty, F.W. Metastable Ion Characteristics. XXIII. Internal Energy of Product Ions Formed in Masspectral Reactions. *Org. Mass Spectrom.* **1974**, *9*, 525-535.
41. Meier, K.; Seibl, J. Measurement of Ion Residence Times in a Commercial Electron Impact Ion Source. *Int. J. Mass Spectrom. Ion Phys.* **1974**, *14*, 99-106.
42. Chupka, W.A. Effect of Unimolecular Decay Kinetics on the Interpretation of Appearance Potentials. *J. Chem. Phys.* **1959**, *30*, 191-211.
43. Hipple, J.A. Detection of Metastable Ions With the Mass Spectrometer. *Phys. Rev.* **1945**, *68*, 54-55.
44. Hipple, J.A.; Fox, R.E.; Condon, E.U. Metastable Ions Formed by Electron Impact in Hydrocarbon Gases. *Phys. Rev.* **1946**, *69*, 347-356.
45. Beynon, J.H.; Saunders, R.A.; Williams, A.E. Formation of Metastable Ions in Mass Spectrometers With Release of Internal Energy. *Z. Naturforsch.* **1965**, *20A*, 180-183.
46. Rosenstock, H.M. The Measurement of IPs and APs. *Int. J. Mass Spectrom. Ion Phys.* **1976**, *20*, 139-190.
47. Holmes, J.L.; Terlouw, J.K. The Scope of Metastable Peak Shape Observations. *Org. Mass Spectrom.* **1980**, *15*, 383-396.
48. Williams, D.H. A Transition State Probe. *Acc. Chem. Res.* **1977**, *10*, 280-286.
49. Williams, D.H.; Hvistendahl, G. Kinetic Energy Release in Relation to Symmetry-Forbidden Reactions. *J. Am. Chem. Soc.* **1974**, *96*, 6753-6755.
50. Williams, D.H.; Hvistendahl, G. Kinetic Energy Release As a Mechanistic Probe. The Role of Orbital Symmetry. *J. Am. Chem. Soc.* **1974**, *96*, 6755-6757.
51. Hvistendahl, G.; Williams, D.H. Partitioning of Reverse Activation Energy Between Kinetic and Internal Energy in Reactions of Organic Ions. *J. Chem. Soc. , Perkin Trans. 2* **1975**, 881-885.
52. Hvistendahl, G.; Uggerud, E. Secondary Isotope Effect on Kinetic Energy Release and Reaction Symmetry. *Org. Mass Spectrom.* **1985**, *20*, 541-542.
53. Kim, K.C.; Beynon, J.H.; Cooks, R.G. Energy Partitioning by Mass Spectrometry. Chloroalkanes and Chloroalkenes. *J. Chem. Phys.* **1974**, *61*, 1305-1314.
54. Haney, M.A.; Franklin, J.L. Correlation of Excess Energies of Electron Impact Dissociations With the Translational Energies of the Products. *J. Chem. Phys.* **1968**, *48*, 4093-4097.
55. Cooks, R.G.; Williams, D.H. The Relative Rates of Fragmentation of Benzoyl Ions Generated Upon Electron Impact from Different Precursors. *Chem. Commun.* **1968**, 627-629.
56. Lin, Y.N.; Rabinovitch, B.S. Degrees of Freedom Effect and Internal Energy Partitioning Upon Ion Decomposition. *J. Phys. Chem.* **1970**, *74*, 1769-1775.
57. Bente III., P.F.; McLafferty, F.W.; McAdoo, D.J.; Lifshitz, C. Internal Energy of Product Ions Formed in Mass Spectral Reactions. The Degrees of Freedom Effect. *J. Phys. Chem.* **1975**, *79*, 713-721.

58. Gross, J.H.; Veith, H.J. Unimolecular Fragmentations of Long-Chain Aliphatic Iminium Ions. *Org. Mass Spectrom.* **1993,** *28,* 867-872.

59. Ottinger, C. Fragmentation Energies of Metastable Organic Ions. *Phys. Lett.* **1965,** *17,* 269-271.

60. Baldwin, M.A.; Derrick, P.J.; Morgan, R.P. Correction of Metastable Peak Shapes to Allow for Instrumental Broadening and the Translational Energy Spread of the Parent Ion. *Org. Mass Spectrom.* **1976,** *11,* 440-442.

61. Bowen, R.D.; Wright, A.D.; Derrick, P.J. Unimolecular Reactions of Ionized Methyl Allyl Ether. *Org. Mass Spectrom.* **1992,** *27,* 905-915.

62. Cao, J.R.; George, M.; Holmes, J.L. Fragmentation of 1- and 3-Methoxypropene Ions. Another Part of the $[C_4H_8O]^{+\cdot}$ Cation Radical Potential Energy Surface. *J. Am. Chem. Soc. Mass Spectrom.* **1992,** *3,* 99-107.

63. Gross, J.H.; Veith, H.J. Unimolecular Fragmentations of Long-Chain Aliphatic Iminium Ions. *Org. Mass Spectrom.* **1993,** *28,* 867-872.

64. Todd, J.F.J. Recommendations for Nomenclature and Symbolism for Mass Spectroscopy Including an Appendix of Terms Used in Vacuum Technology. *Int. J. Mass Spectrom. Ion Proc.* **1995,** *142,* 211-240.

65. Robinson, P.J.; Holbrook, K.A. Unimolecular Reactions, in *Unimolecular Reactions,* John Wiley & Sons: London, 1972; Chapter 9.

66. Ingemann, S.; Hammerum, S.; Derrick, P.J.; Fokkens, R.H.; Nibbering, N.M.M. Energy-Dependent Reversal of Secondary Isotope Effects on Simple Cleavage Reactions: Tertiary Amine Radical Cations With Deuterium at Remote Positions. *Org. Mass Spectrom.* **1989,** *24,* 885-889.

67. Lowry, T.H.; Schueller Richardson, K. Isotope Effects, in *Mechanism and Theory in Organic Chemistry,* 1st ed.; Harper and Row: New York, 1976; Chapter 1.7.

68. Stringer, M.B.; Underwood, D.J.; Bowie, J.H.; Allison, C.E.; Donchi, K.F.; Derrick, P.J. Is the McLafferty Rearrangement of Ketones Concerted or Stepwise? The Application of Kinetic Isotope Effects. *Org. Mass Spectrom.* **1992,** *27,* 270-276.

69. Derrick, P.J. Isotope Effects in Fragmentation. *Mass Spectrom. Rev.* **1983,** *2,* 285-298.

70. Hvistendahl, G.; Uggerud, E. Deuterium Isotope Effects and Mechanism of the Gas-Phase Reaction $[C_3H_7]^+ \rightarrow [C_3H_5]^+ + H_2$. *Org. Mass Spectrom.* **1986,** *21,* 347-350.

71. Howe, I.; McLafferty, F.W. Unimolecular Decomposition of Toluene and Cycloheptatriene Molecular Ions. Variation of the Degree of Scrambling and Isotope Effect with Internal Energy. *J. Am. Chem. Soc.* **1971,** *93,* 99-105.

72. Bertrand, M.; Beynon, J.H.; Cooks, R.G. Isotope Effects Upon Hydrogen Atom Loss from Molecular Ions. *Org. Mass Spectrom.* **1973,** *7,* 193-201.

73. Lau, A.Y.K.; Solka, B.H.; Harrison, A.G. Isotope Effects and H/D Scrambling in the Fragmentation of Labeled Propenes. *Org. Mass Spectrom.* **1974,** *9,* 555-557.

74. Benoit, F.M.; Harrison, A.G. Hydrogen Migrations in Mass Spectrometry. I. The Loss of Olefin From Phenyl-*n*-Propyl Ether Following EI and CI. *Org. Mass Spectrom.* **1976,** *11,* 599-608.

75. Veith, H.J.; Gross, J.H. Alkene Loss From Metastable Methyleneimmonium Ions: Unusual Inverse Secondary Isotope Effect in Ion-Neutral Complex Intermediate Fragmentations. *Org. Mass Spectrom.* **1991,** *26,* 1097-1105.

76. Ingemann, S.; Kluft, E.; Nibbering, N.M.M.; Allison, C.E.; Derrick, P.J.; Hammerum, S. Time-Dependence of the Isotope Effects in the Unimolecular Dissociation of Tertiary Amine Molecular Ions. *Org. Mass Spectrom.* **1991,** *26,* 875-881.

77. Nacson, S.; Harrison, A.G. Dependence of Secondary Hydrogen/Deuterium Isotope Effects on Internal Energy. *Org. Mass Spectrom.* **1985,** *20,* 429-430.

78. Ingemann, S.; Hammerum, S.; Derrick, P.J. Secondary Hydrogen Isotope Effects on Simple Cleavage Reactions in the Gas Phase: The α-Cleavage of Tertiary Amine Cation Radicals. *J. Am. Chem. Soc.* **1988,** *110,* 3869-3873.

79. Urban, B.; Bondybey, V.E. Multiphoton Photoelectron Spectroscopy: Watching Molecules Dissociate. *Phys. Chem. Chem. Phys.* **2001,** *3,* 1942-1944.

80. Nicholson, A.J.C. Measurement of Ionization Potentials by Electron Impact. *J. Chem. Phys.* **1958,** *29,* 1312-1318.

81. Barfield, A.F.; Wahrhaftig, A.L. Determination of Appearance Potentials by the Critical Slope Method. *J. Chem. Phys.* **1964,** *41,* 2947-2948.

82. Levin, R.D.; Lias, S.G. Ionization Potential and Appearance Potential Measurements, 1971-1981. *National Standard Reference Data Series* **1982**, *71*, 634 pp.

83. Harris, F.M.; Beynon, J.H. Photodissociation in Beams: Organic Ions, in *Gas Phase Ion Chemistry - Ions and Light*; Bowers, M.T., editor; Academic Press: New York, 1985; Vol. 3, Chapter 19, 99-128.

84. Dunbar, R.C. Ion Photodissociation, in *Gas Phase Ion Chemistry*, 1st ed.; Bowers, M.T., editor; Academic Press: New York, 1979; Vol. 2, Chapter 14, 181-220.

85. Maeda, K.; Semeluk, G.P.; Lossing, F.P. A Two-Stage Double-Hemispherical Electron Energy Selector. *Int. J. Mass Spectrom. Ion Phys.* **1968**, *1*, 395-407.

86. Traeger, J.C.; McLoughlin, R.G. A PI Study of the Energetics of the $C_7H_7^+$ Ion Formed from C_7H_8 Precursors. *Int. J. Mass Spectrom. Ion Phys.* **1978**, *27*, 319-333.

87. Turner, D.W.; Al Jobory, M.I. Determination of IPs by Photoelectron Energy Measurement. *J. Chem. Phys.* **1962**, *37*, 3007-3008.

88. Edqvist, O.; Lindholm, E.; Selin, L.E.; Åsbrink, L. Photoelectron Spectrum of Molecular Oxygen. *Phys. Scr.* **1970**, *1*, 25-30.

89. Müller-Dethlefs, K.; Sander, M.; Schlag, E.W. Two-Color PI Resonance Spectroscopy of Nitric Oxide: Complete Separation of Rotational Levels of Nitrosyl Ion at the Ionization Threshold. *Chem. Phys. Lett.* **1984**, *112*, 291-294.

90. Müller-Dethlefs, K.; Sander, M.; Schlag, E.W. A Novel Method Capable of Resolving Rotational Ionic States by the Detection of Threshold Photoelectrons with a Resolution of 1.2 cm^{-1}. *Z. Naturforsch.* **1984**, *39a*, 1089-1091.

91. Schlag, E.W. *ZEKE Spectroscopy;* Cambridge University Press: Cambridge, 1998.

92. Zhu, L.; Johnson, P. Mass Analyzed Threshold Ionization Spectroscopy. *J. Chem. Phys.* **1991**, *94*, 5769-5771.

93. Weickhardt, C.; Moritz, F.; Grotemeyer, J. Time-of-Flight Mass Spectrometry: State-of-the-Art in Chemical Analysis and Molecular Science. *Mass Spectrom. Rev.* **1997**, *15*, 139-162.

94. Gunzer, F.; Grotemeyer, J. New Features in the Mass Analyzed Threshold Ionization (MATI) Spectra of Alkyl Benzenes. *Phys. Chem. Chem. Phys.* **2002**, *4*, 5966-5972.

95. Peng, X.; Kong, W. ZEKE and MATI of $Na(NH_3)_n$ Complexes. *J. Chem. Phys.* **2002**, *117*, 9306-9315.

96. Haines, S.R.; Dessent, C.E.H.; Müller-Dethlefs, K. MATI of Phenol×CO: Intermolecular Binding Energies of a Hydrogen-Bonded Complex. *J. Chem. Phys.* **1999**, *111*, 1947-1954.

97. Lavanchy, A.; Houriet, R.; Gäumann, T. The Mass Spectrometric Fragmentation of *n*-Heptane. *Org. Mass Spectrom.* **1978**, *13*, 410-416.

98. Meisels, G.G.; Chen, C.T.; Giessner, B.G.; Emmel, R.H. Energy-Deposition Functions in Mass Spectrometry. *J. Chem. Phys.* **1972**, *56*, 793-800.

99. Herman, J.A.; Li, Y.-H.; Harrison, A.G. Energy Dependence of the Fragmentation of Some Isomeric $C_6H_{12}^{+\cdot}$ Ions. *Org. Mass Spectrom.* **1982**, *17*, 143-150.

100. Lias, S.G.; Liebman, J.F.; Levin, R.D. Evaluated Gas Phase Basicities and Proton Affinities of Molecules; Heats of Formation of Protonated Molecules. *J. Phys. Chem. Ref. Data* **1984**, *13*, 695-808.

101. Harrison, A.G. The GBs and PAs of Amino Acids and Peptides. *Mass Spectrom. Rev.* **1997**, *16*, 201-217.

102. Kukol, A.; Strehle, F.; Thielking, G.; Grützmacher, H.-F. Methyl Group Effect on the Proton Affinity of Methylated Acetophenones Studied by Two Mass Spectrometric Techniques. *Org. Mass Spectrom.* **1993**, *28*, 1107-1110.

103. McMahon, T.B. Thermochemical Ladders: Scaling the Ramparts of Gaseous Ion Energetics. *Int. J. Mass Spectrom.* **2000**, *200*, 187-199.

104. Cooks, R.G.; Kruger, T.L. Intrinsic Basicity Determination Using Metastable Ions. *J. Am. Chem. Soc.* **1977**, *99*, 1279-1281.

105. Cooks, R.G.; Wong, P.S.H. Kinetic Method of Making Thermochemical Determinations: Advances and Applications. *Acc. Chem. Res.* **1998**, *31*, 379-386.

106. Cooks, R.G.; Patrick, J.S.; Kotiaho, T.; McLuckey, S.A. Thermochemical Determinations by the Kinetic Method. *Mass Spectrom. Rev.* **1994**, *13*, 287-339.

107. Bouchoux, G.; Salpin, J.Y. Gas-Phase Basicity and Heat of Formation of Sulfine $CH_2=S=O$. *J. Am. Chem. Soc.* **1996**, *118*, 6516-6517.

108. Bouchoux, G.; Salpin, J.Y.; Leblanc, D. A Relationship Between the Kinetics and Thermochemistry of Proton Transfer Re-

actions in the Gas Phase. *Int. J. Mass Spectrom. Ion Proc.* **1996,** *153,* 37-48.

109. Witt, M.; Kreft, D.; Grützmacher, H.-F. Effects of Internal Hydrogen Bonds Between Amide Groups: Protonation of Alicyclic Diamides. *Eur. J. Mass Spectrom.* **2003,** *9,* 81-95.

110. Lias, S.G.; Bartmess, J.E.; Liebman, J.F.; Holmes, J.L.; Levin, R.D.; Mallard, W.G. Gas-Phase Ion and Neutral Thermochemistry. *J. Phys. Chem. Ref. Data* **1988,** *17, Supplement 1,* 861 pp.

111. *Tandem Mass Spectrometry;* McLafferty, F.W., editor; John Wiley & Sons: New York, 1983.

112. Busch, K.L.; Glish, G.L.; McLuckey, S.A. *Mass Spectrometry/Mass Spectrometry;* Wiley VCH: New York, 1988.

113. McLafferty, F.W.; Bente, P.F.I.; Kornfeld, R.; Tsai, S.-C.; Howe, I. Collisional Activation Spectra of Organic Ions. *J. Am. Chem. Soc.* **1973,** *95,* 2120-2129.

114. Levsen, K.; Schwarz, H. Collisional Activation MS - a New Probe for Structure Determination of Ions in the Gaseous Phase. *Angew. Chem.* **1976,** *88,* 589-601.

115. Levsen, K.; Schwarz, H. Gas-Phase Chemistry of Collisionally Activated Ions. *Mass Spectrom. Rev.* **1983,** *2,* 77-148.

116. Bordas-Nagy, J.; Jennings, K.R. Collision-Induced Decomposition of Ions. *Int. J. Mass Spectrom. Ion Proc.* **1990,** *100,* 105-131.

117. McLuckey, S.A. Principles of Collisional Activation in Analytical MS. *J. Am. Chem. Soc. Mass Spectrom.* **1992,** *3,* 599-614.

118. Shukla, A.K.; Futrell, J.H. Tandem Mass Spectrometry: Dissociation of Ions by Collisional Activation. *J. Mass Spectrom.* **2000,** *35,* 1069-1090.

119. Guevremont, R.; Boyd, R.K. Are Derrick Shifts Real? An Investigation by Tandem MS. *Rapid. Commun. Mass Spectrom.* **1988,** *2,* 1-5.

120. Bradley, C.D.; Derrick, P.J. CID of Peptides. An Investigation into the Effect of Collision Gas Pressure on Translational Energy Losses. *Org. Mass Spectrom.* **1991,** *26,* 395-401.

121. Bradley, C.D.; Derrick, P.J. CID of Large Organic Ions and Inorganic Cluster Ions. Effects of Pressure on Energy Losses. *Org. Mass Spectrom.* **1993,** *28,* 390-394.

122. Neumann, G.M.; Sheil, M.M.; Derrick, P.J. CID of Multiatomic Ions. *Z. Naturforsch.* **1984,** *39A,* 584-592.

123. Alexander, A.J.; Thibault, P.; Boyd, R.K. Target Gas Excitation in CID: a Reinvestigation of Energy Loss in CA of Molecular Ions of Chlorophyll-a. *J. Am. Chem. Soc.* **1990,** *112,* 2484-2491.

124. Kim, B.J.; Kim, M.S. Peak Shape Analysis for CID-MIKES. *Int. J. Mass Spectrom. Ion Proc.* **1990,** *98,* 193-207.

125. Vékey, K.; Czira, G. Translational Energy Loss and Scattering in CID Processes. *Org. Mass Spectrom.* **1993,** *28,* 546-551.

126. Wysocki, V.H.; Kenttämaa, H.; Cooks, R.G. Internal Energy Distributions of Isolated Ions After Activation by Various Methods. *Int. J. Mass Spectrom. Ion Proc.* **1987,** *75,* 181-208.

127. Bradley, C.D.; Curtis, J.M.; Derrick, P.J.; Wright, B. Tandem MS of Peptides Using a Magnetic Sector/Quadrupole Hybrid-the Case for Higher Collision Energy and Higher Radio-Frequency Power. *Anal. Chem.* **1992,** *64,* 2628-2635.

128. Mabud, Md.A.; Dekrey, M.J.; Cooks, R.G. SID of Molecular Ions. *Int. J. Mass Spectrom. Ion Proc.* **1985,** *67,* 285-294.

129. Bier, M.E.; Amy, J.W.; Cooks, R.G.; Syka, J.E.P.; Ceja, P.; Stafford, G. A Tandem Quadrupole Mass Spectrometer for the Study of SID. *Int. J. Mass Spectrom. Ion Proc.* **1987,** *77,* 31-47.

130. Wysocki, V.H.; Ding, J.-M.; Jones, J.L.; Callahan, J.H.; King, F.L. SID in Tandem Quadrupole Mass Spectrometers: a Comparison of Three Designs. *J. Am. Chem. Soc. Mass Spectrom.* **1992,** *3,* 27-32.

131. Dongré, A.R.; Somogyi, A.; Wysocki, V.H. SID: an Effective Tool to Probe Structure, Energetics and Fragmentation Mechanisms of Protonated Peptides. *J. Mass Spectrom.* **1996,** *31,* 339-350.

132. Lammert, S.A.; Cooks, R.G. SID of Molecular Ions in a Quadrupole Ion Trap Mass Spectrometer. *J. Am. Chem. Soc. Mass Spectrom.* **1991,** *2,* 487-491.

133. Woodin, R.L.; Bomse, D.S.; Beauchamp, J.L. Multiphoton Dissociation of Molecules With Low Power Continuous Wave Infrared Laser Radiation. *J. Am. Chem. Soc.* **1978,** *100,* 3248-3250.

134. Håkansson, K.; Cooper, H.J.; Emmet, M.R.; Costello, C.E.; Marshall, A.G.; Nilsson, C.L. ECD and IRMP MS/MS of an *N*-Glycosylated Tryptic Peptic to Yield Complementary Sequence Information. *Anal. Chem.* **2001,** *73,* 4530-4536.

135. Little, D.P.; Senko, M.W.; O'Conner, P.B.; McLafferty, F.W. IRMPD of Multiply-

Charged Ions for Biomolecule Sequencing. *Anal. Chem.* **1994**, *66*, 2809-2815.

136. Watson, C.H.; Baykut, G.; Eyler, J.R. Laser Photodissociation of Gaseous Ions Formed by Laser Desorption. *Anal. Chem.* **1987**, *59*, 1133-1138.

137. Zubarev, R.A.; Kelleher, N.L.; McLafferty, F.W. ECD of Multiply Charged Protein Cations. A Nonergodic Process. *J. Am. Chem. Soc.* **1998**, *120*, 3265-3266.

138. Axelsson, J.; Palmblad, M.; Håkansson, K.; Håkansson, P. ECD of Substance P Using a Commercially Available FT-ICR Mass Spectrometer. *Rapid. Commun. Mass Spectrom.* **1999**, *13*, 474-477.

139. Cerda, B.A.; Horn, D.M.; Breuker, K.; Carpenter, B.K.; McLafferty, F.W. Electron Capture Dissociation of Multiply-Charged Oxygenated Cations. A Nonergodic Process. *Eur. Mass Spectrom.* **1999**, *5*, 335-338.

140. Håkansson, K.; Emmet, M.R.; Hendrickson, C.L.; Marshall, A.G. High-Sensitivity ECD Tandem FT-ICR-MS of Microelectrosprayed Peptides. *Anal. Chem.* **2001**, *73*, 3605-3610.

141. Leymarie, N.; Berg, E.A.; McComb, M.E.; O'Conner, P.B.; Grogan, J.; Oppenheim, F.G.; Costello, C.E. Tandem MS for Structural Characterization of Proline-Rich Proteins: Application to Salivary PRP-3. *Anal. Chem.* **2002**, *74*, 4124-4132.

142. Danis, P.O.; Wesdemiotis, C.; McLafferty, F.W. Neutralization-Reionization Mass Spectrometry (NRMS). *J. Am. Chem. Soc.* **1983**, *105*, 7454-7456.

143. Gellene, G.I.; Porter, R.F. Neutralized Ion-Beam Spectroscopy. *Acc. Chem. Res.* **1983**, *16*, 200-207.

144. Terlouw, J.K.; Kieskamp, W.M.; Holmes, J.L.; Mommers, A.A.; Burgers, P.C. The Neutralization and Reionization of Mass-Selected Positive Ions by Inert Gas Atoms. *Int. J. Mass Spectrom. Ion Proc.* **1985**, *64*, 245-250.

145. Holmes, J.L.; Mommers, A.A.; Terlouw, J.K.; Hop, C.E.C.A. The Mass Spectrometry of Neutral Species Produced From Mass-Selected Ions by Collision and by Charge Exchange. Experiments With Tandem Collision Gas Cells. *Int. J. Mass Spectrom. Ion Proc.* **1986**, *68*, 249-264.

146. Blanchette, M.C.; Holmes, J.L.; Hop, C.E.C.A.; Mommers, A.A. The Ionization of Fast Neutrals by EI in the 2.FFR of a Mass Spectrometer of Reversed Geometry. *Org. Mass Spectrom.* **1988**, *23*, 495-498.

147. McLafferty, F.W. Neutralization-Reionization MS. *Int. J. Mass Spectrom. Ion Proc.* **1992**, *118/119*, 221-235.

148. Blanchette, M.C.; Bordas-Nagy, J.; Holmes, J.L.; Hop, C.E.C.A.; Mommers, A.A.; Terlouw, J.K. NR-MS; a Simple Metal Vapor Cell for VG ZAB-2F Mass Spectrometers. *Org. Mass Spectrom.* **1988**, *23*, 804-807.

149. Zang, M.Y.; McLafferty, F.W. Organic Neutralization Agents for NR-MS. *J. Am. Chem. Soc. Mass Spectrom.* **1992**, *3*, 108-112.

150. Zhang, M.-Y.; McLafferty, F.W. Quantitative Analysis of Isomeric Ion Mixtures. *Org. Mass Spectrom.* **1992**, *27*, 991-994.

151. Zagorevskii, D.V.; Holmes, J.L. NR-MS Applied to Organometallic and Coordination Chemistry. *Mass Spectrom. Rev.* **1994**, *13*, 133-154.

152. Zagorevskii, D.V.; Holmes, J.L. NR-MS Applied to Organometallic and Coordination Chemistry (Update: 1994-1998). *Mass Spectrom. Rev.* **1999**, *18*, 87-118.

153. Weiske, T.; Wong, T.; Krätschmer, W.; Terlouw, J.K.; Schwarz, H. Proof of the Existence of an He@C_{60} Structure Via Gas-Phase Neutralization of [HeC_{60}]⁺. *Angew. Chem., Int. Ed.* **1992**, *31*, 183-185.

154. Keck, H.; Kuchen, W.; Tommes, P.; Terlouw, J.K.; Wong, T. The Phosphonium Ylide CH_2PH_3 Is Stable in Gas Phase. *Angew. Chem. , Int. Ed.* **1992**, *31*, 86-87.

155. Schröder, D.; Schwarz, H.; Wulf, M.; Sievers, H.; Jutzi, P.; Reiher, M. Evidence for the Existence of Neutral P_6: A New Allotrope of Phosphorus. *Angew. Chem. , Int. Ed.* **1999**, *38*, 313-315.

156. Adlhart, C.; Hinderling, C.; Baumann, H.; Chen, P. Mechanistic Studies of Olefin Metathesis by Ruthenium Carbene Complexes Using ESI-MS/MS. *J. Am. Chem. Soc.* **2000**, *122*, 8204-8214.

157. Hinderling, C.; Adlhart, C.; Chen, P. Mechanism-Based High-Throughput Screening of Catalysts. *Chimia* **2000**, *54*, 232-235.

158. Volland, M.A.O.; Adlhart, C.; Kiener, C.A.; Chen, P.; Hofmann, P. Catalyst Screening by ESI-MS/MS: Hofmann Carbenes for Olefin Metathesis. *Chem. Eur. J.* **2001**, *7*, 4621-4632.

3 Isotopes

Mass spectrometry is based upon the separation of charged ionic species by their mass-to-charge ratio, m/z. Within the general chemical context however, we are not used to taking into concern the isotopes of the elemental species involved in a reaction. The molecular mass of tribromomethane, $CHBr_3$, would therefore be calculated to 252.73 g mol^{-1} using the relative atomic masses of the elements as listed in most periodic tables. In mass spectrometry we have to leave this custom behind. Because the mass spectrometer does not separate by elements but by isotopic mass, there is no signal at m/z 252.73 in the mass spectrum of tribromomethane. Instead, major peaks are present at m/z 250, 252, 254 and 256 accompanied by some minor others.

In order to successfully interpret a mass spectrum, we have to know about the isotopic masses and their relation to the atomic weights of the elements, about isotopic abundances and the isotopic patterns resulting therefrom and finally, about high-resolution and accurate mass measurements. These issues are closely related to each other, offer a wealth of analytical information, and are valid for any type of mass spectrometer and any ionization method employed. (The kinetic aspect of isotopic substitution are discussed in Chap. 2.9.)

3.1 Isotopic Classification of the Elements

An element is specified by the number of protons in its nucleus. This equals the *atomic number* of the respective element, and thus determines its place within the periodic table of the elements. The atomic number is given as a subscript preceding the elemental symbol, e.g., $_6C$ in case of carbon. Atoms with nuclei of the same atomic number differing in the number of neutrons are termed *isotopes*. One isotope differs from another isotope of the same element in that it possesses a different number of neutrons, i.e., by the *mass number* or *nucleon number*. The mass number m is the sum of the total number of protons and neutrons in an atom, molecule, or ion. [1] The mass number of an isotope is given as superscript preceding the elemental symbol, e.g., ^{12}C.

> **Note:** The *mass number* must not be confused with the *atomic number* of an element. For the heavier atoms there can be isotopes of the same mass number belonging to different elements, e.g., the most abundant isotopes of both $_{18}Ar$ and $_{20}Ca$ have mass number 40.

3.1.1. Monoisotopic Elements

Some elements do exist in only one naturally occurring stable isotope, and there-fore, they are termed *monoisotopic elements*. Among these, fluorine (^{19}F), sodium (^{23}Na), phosphorus (^{31}P) and iodine (^{127}I) belong to the more prominent examples in organic mass spectrometry, but there are several more such as beryllium (^{9}Be), aluminum (^{27}Al), scandium (^{45}Sc), manganese (^{55}Mn), cobalt (^{59}Co), arsenic (^{75}As), yttrium (^{89}Y), niobium (^{93}Nb), rhodium (^{103}Rh), cesium (^{133}Cs), gold (^{197}Au) and some heavier transition metals, too. The monoisotopic elements are also referred to as A or X elements (see below). [2,3]

3.1.2 Di-isotopic Elements

Several elements naturally exist in two isotopes. Within the context of mass spec-trometry it is useful to deal with them as a class of their own. Nevertheless, the term *di-isotopic element* is not an official one. These elements can even be sub-classified into those having one isotope that is 1 u heavier than the most abundant isotope and those having one isotope that is 2 u heavier than the most abundant isotope. (For the unit u cf. Chap. 3.1.4.2). The first group has been termed A+1 or X+1 elements, the latter ones have been termed A+2 or X+2 elements, respec-tively. [2,3] If we do not restrict our view to the elements typically encountered in organic mass spectrometry, the class of X–1 elements with one minor isotope of 1 u lower mass than the most abundant one should be added.

3.1.2.1 X+1 Elements

Prominent examples of X+1 elements are hydrogen (^{1}H, ^{2}H $\equiv$ D), carbon (^{12}C, ^{13}C) and nitrogen (^{14}N, ^{15}N). Deuterium (D) is of low abundance (0.0115 %) and therefore, hydrogen is usually treated as monoisotopic or X element, which is a valid approximation, even if a hundred hydrogens are contained in a molecule.

3.1.2.2 X+2 Elements

Among the X+2 elements, chlorine (^{35}Cl, ^{37}Cl) and bromine (^{79}Br, ^{81}Br) are com-monly known, but copper (^{63}Cu, ^{65}Cu), gallium (^{69}Ga, ^{71}Ga), silver (^{107}Ag, ^{109}Ag), indium (^{113}In, ^{115}In) and antimony (^{121}Sb, ^{123}Sb) also belong to this group. Even though occurring in more than two isotopes, some additional elements such as oxygen, sulfur and silicon, can be regarded as X+2 elements for practical reasons. As long as there are only a few oxygen atoms present in an empirical formula, oxygen might alternatively be treated as an X element, because of the low abun-dances of ^{17}O and ^{18}O.

3.1.2.3 X–1 Elements

The elements lithium (^{6}Li, ^{7}Li), boron (^{10}B, ^{11}B) and vanadium (^{50}V, ^{51}V) come together with a lighter isotope of lower abundance than the heavier one and thus, they can be grouped together as X–1 elements (cf. Chap. 3.1.4).

3.1.3 Polyisotopic Elements

The majority of elements belongs to the *polyisotopic elements* because they consist of three or more isotopes showing a wide variety of isotopic distributions.

3.1.3.1 Representation of Isotopic Abundances

Isotopic abundances are listed either as their sum being 100 % or with the abundance of the most abundant isotope normalized to 100 %. The latter is used throughout this book because this is consistent with the custom of reporting mass spectra normalized to the base peak (Chap. 1). The isotopic classifications and *isotopic compositions* of some common elements are listed below (Table 3.1). A full table of the elements is included in the Appendix.

> **Note:** Care has to be taken when comparing isotopic abundances from different sources as they might be compiled using the one *or* the other procedure of normalization.

Table 3.1. Isotopic classifications and isotopic compositions of some common elements © IUPAC 2001 [4,5]

Classification	Atomic Symbol	Atomic No.	Mass No.	Isotopic Composition	Isotopic Mass [u]	Relative Atomic Mass [u]
(X)[a]	H	1	1	100	1.007825	1.00795
			2	0.0115	2.014101	
X	F	9	19	100	18.998403	18.998403
X	Na	11	23	100	22.989769	22.989769
X	P	15	31	100	30.973762	30.973762
X	I	53	127	100	126.904468	126.904468
X+1	C	6	12	100	12.000000[b]	12.0108
			13	1.08	13.003355	
X+1	N	7	14	100	14.003070	14.00675
			15	0.369	15.000109	
(X+2)[a]	O	8	16	100	15.994915	15.9994
			17	0.038	16.999132	
			18	0.205	17.999116	

(X+2)[a]	Si	14	28	100	27.976927	28.0855
			29	5.0778	28.976495	
			30	3.3473	29.973770	
(X+2)[a]	S	16	32	100	31.972071	32.067
			33	0.80	32.971459	
			34	4.52	33.967867	
			36	0.02	35.967081	
X+2	Cl	17	35	100	34.968853	35.4528
			37	31.96	36.965903	
X+2	Cu	29	63	100	62.929601	63.546
			65	44.57	64.927794	
X+2	Br	35	79	100	78.918338	79.904
			81	97.28	80.916291	
X+2	Ag	47	107	100	106.905094	107.8682
			109	92.90	108.904756	
X–1	Li	3	6	8.21	6.015122	6.941
			7	100	7.016004	
X–1	B	5	10	24.8	10.012937	10.812
			11	100	11.009306	
poly	Sn	50	112	2.98	111.904822	118.711
			114	2.03	113.902782	
			115	1.04	114.903346	
			116	44.63	115.901744	
			117	23.57	116.902954	
			118	74.34	117.901606	
			119	26.37	118.903309	
			120	100	119.902197	
			122	14.21	121.903440	
			124	17.77	123.905275	
poly	Xe	54	124	0.33	123.905896	131.29
			126	0.33	125.904270	
			128	7.14	127.903530	
			129	98.33	128.904779	
			130	15.17	129.903508	
			131	78.77	130.905082	
			132	100	131.904154	
			134	38.82	133.905395	
			136	32.99	135.907221	

[a] Classification in parentheses means "not in the strict sense".
[b] Standard of atomic mass scale.

Bar graph representations are much better suited for visualization of isotopic compositions than tables, and in fact they exactly show how such a distribution would appear in a mass spectrum (Fig. 3.1). This appearance gives rise to the term *isotopic pattern*.

Note: Some authors use the term *isotopic cluster*, which is incorrect, as *cluster* refers to an associate of more atoms, molecules or ions of the same species, sometimes associated to one other species, e.g., $[Ar_n]^{+\bullet}$, $[(H_2O)_nH]^+$, and $[I(CsI)_n]^-$ are cluster ions.

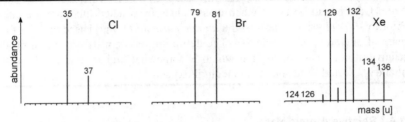

Fig. 3.1. Isotopic patterns of chlorine, bromine and xenon. The bar graph representations of the isotopic distributions have the same optical appearance as mass spectra.

3.1.4 Calculation of Atomic, Molecular and Ionic Mass

3.1.4.1 Nominal Mass

In order to calculate the approximate mass of a molecule we are used to summing up integer masses of the elements encountered, e.g., for CO_2 we calculate the mass by 12 u + 2 × 16 u = 44 u. The result of this simple procedure is not particularly precise but provides acceptable values for simple molecules. This is called *nominal mass*. [3]

The nominal mass is defined as the *integer mass* of the most abundant naturally occurring stable isotope of an element. [3] The nominal mass of an element is often equal to the integer mass of the lowest mass isotope of that element, e.g., for H, C, N, O, S, Si, P, F, Cl, Br, I (Table 3.1). The nominal mass of an ion is the sum of the nominal masses of the elements in its empirical formula.

Example: To calculate the nominal mass of $SnCl_2$, the masses of ^{120}Sn and ^{35}Cl have to be used, i.e., 120 u + 2 × 35 u = 190 u. While, the ^{35}Cl isotope represents the most abundant as well as the lowest mass isotope of chlorine, ^{120}Sn is the most abundant, but not the lowest mass isotope of tin which is ^{112}Sn.

Note: When dealing with nominal mass, mass number and nominal mass both have the same numerical value. However, the *mass number* is dimensionless and must not be confused with *nominal mass* in units of u (below).

3.1.4.2 Isotopic Mass

The *isotopic mass* is the exact mass of an isotope. It is very close to but not equal to the nominal mass of the isotope (Table 3.1). The only exception is the carbon isotope ^{12}C which has an isotopic mass of 12.000000 u. The *unified atomic mass*

[u] is defined as $^1/_{12}$ of the mass of one atom of nuclide ^{12}C which has been assigned to 12 u exactly by convention; 1 u = 1.66055×10^{-27} kg. [1,3,6,7] This convention dates back to 1961. The issue of isotopic mass will be addressed in detail later in this chapter.

> **Note:** Care has to be taken when mass values from dated literature are cited. Prior to 1961 physicists defined the *atomic mass unit* [amu] based on $^1/_{16}$ of the mass of one atom of nuclide ^{16}O. The definition of chemists was based on the relative atomic mass of oxygen which is somewhat higher resulting from the nuclides ^{17}O and ^{18}O contained in natural oxygen.

3.1.4.3 Relative Atomic Mass

The *relative atomic mass* or the *atomic weight* as it is also often imprecisely termed is calculated as the weighted average of the naturally occurring isotopes of an element. [3]The weighted average M_r is calculated from

$$M_r = \frac{\sum\limits_{i=1}^{i} A_i \cdot m_i}{\sum\limits_{i=1}^{i} A_i} \tag{3.1}$$

with A_i being the abundances of the isotopes and m_i their respective isotopic masses. [8] For this purpose, the abundances can be used in any numerical form or normalization as long as they are used consistently.

Example: The relative atomic mass of chlorine is 35.4528 u. However, there is no atom of this mass. Instead, chlorine is composed of ^{35}Cl (34.968853 u) and ^{37}Cl (36.965903 u) making up 75.78 % and 24.22 % of the total or having relative abundances of 100 % and 31.96 %, respectively (cf. Table 3.1 and Fig 3.1). According to Eq. 3.1, we can now calculate the relative atomic mass of chlorine $M_r = (100 \times 34.968853 \text{ u} + 31.96 \times 36.965903 \text{ u})/(100 + 31.96) = 35.4528$ u.

> **Note:** The assumption of a ^{13}C content of 1.1 % which is equal to a ^{13}C/^{12}C ratio of 0.0111 for carbon has proven to be useful for most mass-spectrometric applications.

3.1.4.4 Monoisotopic Mass

The exact mass of the most abundant isotope of an element is termed *monoisotopic mass*. [3] The monoisotopic mass of a molecule is the sum of the monoisotopic masses of the elements in its empirical formula. As mentioned before, the monoisotopic mass is not necessarily the naturally occurring isotope of lowest mass. However, for the common elements in organic mass spectrometry the monoisotopic mass is obtained using the mass of the lowest mass isotope of that element, because this is also the most abundant isotope of the respective element (Chap. 3.1.4.1).

3.1.4.5 Relative Molecular Mass

The *relative molecular mass*, M_r, or *molecular weight* is calculated from the relative atomic masses of the elements contributing to the empirical formula. [3]

3.1.4.6 Exact Ionic Mass

The *exact mass of a positive ion* formed by the removal of one or more electrons from a molecule is equal to its monoisotopic mass minus the mass of the electron(s), m_e. [1] For negative ions, the electron mass (0.000548 u) has to be added accordingly.

Example: The exact mass of the carbon dioxide molecular ion, $CO_2^{+\bullet}$, is calculated as 12.000000 u + 2 × 15.994915 u − 0.000548 u = 43.989282 u.

3.1.5 Natural Variations in Relative Atomic Mass

The masses of isotopes can be measured with accuracies better than *parts per billion* (ppb), e.g., m_{40Ar} = 39.9623831235 ± 0.000000005 u. Unfortunately, determinations of abundance ratios are less accurate, causing errors of several *parts per million* (ppm) in relative atomic mass. The real limiting factor, however, comes from the variation of isotopic abundances from natural samples, e.g., in case of lead which is the final product of radioactive decay of uranium, the atomic weight varies by 500 ppm depending on the Pb/U ratios in the lead ore. [8]

For organic mass spectrometry the case of carbon is of much greater importance. Carbon is ubiquitous in metabolic processes and the most prominent example of variations in the $^{13}C/^{12}C$ isotopic ratio is presented by the different pathways of CO_2 fixation during photosynthesis, causing $^{13}C/^{12}C$ ratios of 0.01085–0.01115. Petroleum, coal and natural gas yield very low $^{13}C/^{12}C$ ratios of 0.01068–0.01099 and carbonate minerals on the other side define the upper limit at about 0.01125. [8] Even proteins of different origin (plants, fish, mammals) can be distinguished by their $^{13}C/^{12}C$ ratios. [9]

Isotope ratio mass spectrometry (IR-MS) makes use of these facts to determine the origin or the age of a sample. For convenience, the minor changes in isotopic ratios are expressed using the *delta notation* stating the deviation of the isotopic ratio from a defined standard in parts per thousand (‰). [8,10] The delta value of carbon, for example, is calculated from

$$\delta^{13}C \; (‰) = [(^{13}C/^{12}C_{sample})/(^{13}C/^{12}C_{standard}) - 1] \times 1000 \qquad (3.2)$$

The internationally accepted value for $^{13}C/^{12}C_{standard}$ is 0.0112372 from a belemnite fossil from the Pee Dee formation in South Carolina (PDB standard). Naturally occurring $^{13}C/^{12}C$ ratios range from about $\delta^{13}C$ = −50 to +5 ‰.

3.2 Calculation of Isotopic Distributions

As long as we are dealing with molecular masses in a range up to some 10^3 u, it is possible to separate ions which differ by 1 u in mass. The upper mass limit for their separation depends on the resolution of the instrument employed. Consequently, the isotopic composition of the analyte is directly reflected in the mass spectrum.

Even if the analyte is chemically perfectly pure it represents a mixture of different isotopic compositions, provided it is not composed of monoisotopic elements only. Therefore, a mass spectrum is normally composed of superimpositions of the mass spectra of all isotopic species involved. [11] The *isotopic distribution* or *isotopic pattern* of molecules containing one chlorine or bromine atom is listed in Table 3.1. But what about molecules containing two or more diisotopic or even polyisotopic elements? While it may seem, at the first glance, to complicate the interpretation of mass spectra, isotopic patterns are in fact an ideal source of analytical information.

3.2.1 X+1 Element Carbon

In the mass spectrum of methane (Fig. 1.2), there is a tiny peak at m/z 17 that has not been mentioned in the introduction. As one can infer from Table 3.1 this should result from the ^{13}C content of natural carbon which belongs to the X+1 elements in our classification.

Example: Imagine a total of 1000 CH_4 molecules. Due to a content of 1.1 % ^{13}C, there will be 11 molecules containing ^{13}C instead of ^{12}C; the remaining 989 molecules have the composition $^{12}CH_4$. Therefore, the ratio of relative intensities of the peaks at m/z 16 and m/z 17 is defined by the ratio 989/11 or after the usual normalization by 100/1.1.

In a more general way, carbon consists of ^{13}C and ^{12}C in a ratio r that can be formulated as $r = c/(100 - c)$ where c is the abundance of ^{13}C. Then, the probability to have only ^{12}C in a molecular ion M consisting of w carbons, i.e., the probability of monoisotopic ions P_M is given by [12]

$$P_M = \left(\frac{100-c}{100}\right)^w \tag{3.3}$$

The probability of having exactly one ^{13}C atom in an ion with w carbon atoms is therefore

$$P_{M+1} = w\left(\frac{c}{100-c}\right)\left(\frac{100-c}{100}\right)^w \tag{3.4}$$

and the ratio P_{M+1}/P_M is given as

$$\frac{P_{M+1}}{P_M} = w\left(\frac{c}{100-c}\right) \tag{3.5}$$

In case of a carbon-only molecule such as the buckminster fullerene C_{60}, the ratio P_{M+1}/P_M becomes $60 \times 1.1/98.9 = 0.667$. If the monoisotopic peak at m/z 720 due to $^{12}C_{60}$ is assigned 100 %, the M+1 peak due to $^{12}C_{59}{}^{13}C$ will have 66.7 % relative intensity.

Note: As mentioned above, the exact abundances depend on the actual $^{13}C/^{12}C$ ratio. Without causing significant error, we can estimate the M+1 peak intensity in percent in a simplified manner by multiplying the number of carbon atoms by 1.1 %, e.g., $60 \times 1.1 \% = 66 \%$.

There is again a certain probability for one of the remaining 59 carbon atoms to be ^{13}C instead of ^{12}C. After simplification of Beynon's approach [12] to be used for one atomic species only, the probability that there will be an ion containing two ^{13}C atoms is expressed by

$$\frac{P_{M+2}}{P_M} = \frac{w}{2}\left(\frac{c}{100-c}\right)(w-1)\left(\frac{c}{100-c}\right) = \frac{w(w-1)c^2}{2(100-c)^2} \tag{3.6}$$

For C_{60} the ratio P_{M+2}/P_M now becomes $(60 \times 59 \times 1.1^2)/(2 \times 98.9^2) = 0.219$, i.e., the M+2 peak at m/z 722 due to $^{12}C_{58}{}^{13}C_2$ ions will have 21.9 % relative to the M peak, which definitely can not be neglected. By extension of the equation, the P_{M+3}/P_M ratio for the third isotopic peak becomes

$$\frac{P_{M+3}}{P_M} = \frac{w}{6}\left(\frac{c}{100-c}\right)(w-1)\left(\frac{c}{100-c}\right)(w-2)\left(\frac{c}{100-c)}\right) \tag{3.7}$$

$$= \frac{w(w-1)(w-2)c^3}{6(100-c)^3}$$

The relative abundances obtained from Eq. 3.7 by using 1.1 % for the ^{13}C content are in good agreement with those tabulated, e.g., for the $^{12}C_{57}{}^{13}C_3$ ion at m/z 723 we obtain $P_{M+3}/P_M = (60 \times 59 \times 58 \times 1.1^3)/(6 \times 98.9^3) = 0.047$ (Table 3.2 and Fig. 3.2). This tells us that the isotopic peaks due to ^{13}C do contribute significantly to the appearance of a mass spectrum, if the number of carbon atoms in a molecule is not very small.

The calculation of isotopic patterns as just shown for the carbon-only molecule C_{60} can be done analogously for any X+1 element. Furthermore, the application of this scheme is not restricted to molecular ions, but can also be used for fragment ions. Nevertheless, care has be taken to assure that the presumed isotopic peak is not partially or even completely due to a different fragment ion, e.g., an ion containing one hydrogen more than the presumed X+1 composition.

Note: It is very helpful to read out the P_{M+1}/P_M ratio from a mass spectrum to calculate the approximate number of carbon atoms. Provided no other element contributing to M+1 is present, an M+1 intensity of 15 %, for example, indicates the presence of 14 carbons. (For the risk of overestimation due to auto-protonation cf. Chap. 7.2.1)

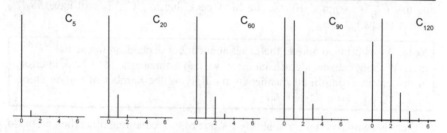

Fig. 3.2. Calculated isotopic patterns for carbon. Note the steadily expanding width of the pattern as X+2, X+3, X+4,... become visible. At about C_{90} the X+1 peak reaches the same intensity as the X peak. At higher carbon number it becomes the base peak of the pattern.

Table 3.2. Calculated isotopic distributions for carbon based on the ^{13}C content cccording to IUPAC. © IUPAC 1998. [4]

Number of carbons	X+1	X+2	X+3	X+4	X+5
1	1.1	0.00			
2	2.2	0.01			
3	3.3	0.04			
4	4.3	0.06			
5	5.4	0.10			
6	6.5	0.16			
7	7.6	0.23			
8	8.7	0.33			
9	9.7	0.42			
10	10.8	0.5			
12	13.0	0.8			
15	16.1	1.1			
20	21.6	2.2	0.1		
25	27.0	3.5	0.2		
30	32.3	5.0	0.5		
40	43.2	9.0	1.3	0.1	
50	54.1	14.5	2.5	0.2	0.1
60	65.0	20.6	4.2	0.6	0.2
90	97.2	46.8	14.9	3.5	0.6
120[a]	100.0	64.4	27.3	8.6	2.2

[a] The X peak has an abundance of 77.0 % in that case.

It is interesting how the width of the isotopic pattern increases as X+2, X+3, X+4 and so forth become detectable. In principle, the isotopic pattern of C_w expands up to X+w, because even the composition $^{13}C_w$ is possible. As a result, the isotopic pattern of w atoms of a di-isotopic element consists at least theoretically of $w+1$ peaks. However, the probability for the extreme combinations is negligible and even somewhat more probable combinations are of no importance as long they are below about 0.1 %. In practice, the interpretation of the carbon isotopic pattern is limited by experimental errors in relative intensities rather than by detection limits for peaks of low intensity. Such experimental errors can be due to poor signal-to-noise ratios (Chap. 5.2.2.3), autoprotonation (Chap.7) or interference with other peaks.

At about C_{90} the X+1 peak reaches the same intensity as the X peak and at higher carbon number w, it becomes the base peak of the pattern, because the probability that an ion contains at least one ^{13}C becomes larger than that for the monoisotopic ion. A further increase in w makes the X+2 signal stronger than the X and X+1 peak and so on.

3.2.1.1 Isotopic Molecular Ion

The *isotopic molecular ion* is a molecular ion containing one or more of the less abundant naturally occurring isotopes of the atoms that make up the molecular structure. [1] This term can be generalized for any non-monoisotopic ion. Thus, *isotopic ions* are those ions containing one or more of the less abundant naturally occurring isotopes of the atoms that make up the ion.

3.2.1.2 Most Abundant Mass

The position of the most intensive peak of an isotopic pattern is termed *most abundant mass*. [3] For example, the most abundant mass in case of C_{120} is 1441 u corresponding to M+1 (Fig. 3.2). The most abundant mass is of high relevance if large ions are concerned (Chap. 3.4.3).

3.2.2 Binomial Approach

The above stepwise treatment of X+1, X+2 and X+3 peaks has the advantage that it can be followed easier, but it bears the disadvantage that an equation needs to be solved for each individual peak. Alternatively, one can calculate the relative abundances of the isotopic species for a di-isotopic element from a binomial expression. [2,13,14] In the term $(a + b)^n$ the isotopic abundances of both isotopes are given as a and b, respectively, and n is the number of this species in the molecule.

$$(a+b)^n = a^n + na^{n-1}b + n\,(n-1)a^{n-2}b^2\,/\,(2!) \qquad (3.8)$$
$$+ n(n-1)(n-2)a^{n-3}b^3\,/\,(3!)+...$$

For $n = 1$ the isotopic distribution can of course be directly obtained from the isotopic abundance table (Table 3.1 and Fig. 3.1) and in case of $n = 2$, 3 or 4 the expression can easily be solved by simple multiplication, e.g.,

$$(a+b)^2 = a^2 + 2ab + b^2 \tag{3.9}$$

$$(a+b)^3 = a^3 + 3a^2b + 3ab^2 + b^3$$

$$(a+b)^4 = a^4 + 4a^3b + 6a^2b^2 + 4ab^3 + b^4$$

Again, we obtain $w+1$ terms for the isotopic pattern of w atoms. The binomial approach works for any di-isotopic element, regardless of whether it belongs to X+1, X+2 or X–1 type. However, as the number of atoms increases above 4 it is also no longer suitable for manual calculations.

3.2.3 Halogens

The halogens Cl and Br occur in two isotopic forms, each of them being of significant abundance, whereas F and I are monoisotopic (Table 3.1). In most cases there are only a few Cl and/or Br atoms contained in a molecule and this predestinates the binomial approach for this purpose.

Example: The isotopic pattern of Cl_2 is calculated from Eq. 3.9 with the abundances $a = 100$ and $b = 31.96$ as $(100 + 31.96)^2 = 10000 + 6392 + 1019$. After normalization we obtain $100 : 63.9 : 10.2$ as the relative intensities of the three peaks. Any other normalization for the isotopic abundances would give the same result, e.g., $a = 0.7578$, $b = 0.2422$. The calculated isotopic pattern of Cl_2 can be understood from the following practical consideration: The two isotopes ^{35}Cl and ^{37}Cl can be combined in three different ways: i) $^{35}Cl_2$ giving rise to the monoisotopic composition, ii) $^{35}Cl^{37}Cl$ yielding the first isotopic peak which is here X+2, and finally iii) $^{37}Cl_2$ giving the second isotopic peak X+4. The combinations with a higher number of chlorine atoms can be explained accordingly.

It proves helpful to have the more frequently found isotopic distributions at hand. For some Cl_x, Br_y and Cl_xBr_y combinations these are tabulated in the Appendix. Tables are useful for the construction of isotopic patterns from "building blocks". Nevertheless, as visual information is easier to compare with a plotted spectrum these patterns are also shown below (Fig. 3.3). In case of Cl and Br the peaks are always separated from each other by 2 u, i.e., the isotopic peaks are located at X+2, 4, 6 and so on.

If there are two bromine or four chlorine atoms contained in the empirical formula, the isotopic peaks become more intensive than the monoisotopic peak, because the second isotope is of much higher abundance than in case of the ^{13}C isotope.

> **Note:** For a rapid estimation of the isotopic patterns of chlorine and bromine the approximate isotope ratios $^{35}Cl/^{37}Cl = 3 : 1$ and $^{79}Br/^{81}Br = 1 : 1$ yield good results. Visual comparison to calculated patterns is also well suited (Fig. 3.3).

For halogens the
isotopic peaks
are separated by
2 u.

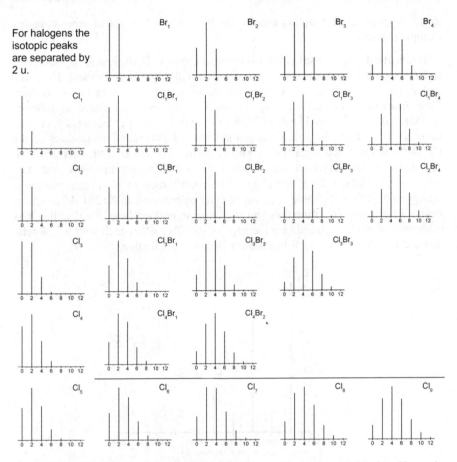

Fig. 3.3. Calculated isotopic patterns for combinations of bromine and chlorine. The peak shown at zero position corresponds to the monoisotopic ion at m/z X. The isotopic peaks are then located at m/z = X+2, 4, 6, ... The numerical value of X is given by the mass number of the monoisotopic combination, e.g., 70 u for Cl_2.

3.2.4 Combinations of Carbon and Halogens

So far we have treated the X+1 and the X+2 elements separately, which is not how they are encountered in most analytes. The combination of C, H, N and O with the halogens F, Cl, Br and I covers a large fraction of the molecules one usually has to deal with. When regarding H, O and N as X elements, which is a valid approximation for not too large molecules, the construction of isotopic patterns can be conveniently accomplished. By use of the isotopic abundance tables of the elements or of tables of frequent combinations of these as provided in this chapter or in the Appendix, the building blocks can be combined to obtain more complex isotopic patterns.

Example: Let us construct the isotopic pattern of $C_9N_3Cl_3$ restricting ourselves to the isotopic contributions of C and Cl, i.e., with N as an X element. Here, the isotopic pattern of chlorine can be expected to be dominant over that of carbon. First, from the cubic form of Eq. 3.9 the Cl_3 pattern is calculated as follows: $(0.7578 + 0.2422)^3 = 0.435 + 0.417 + 0.133 + 0.014$ and after normalization this becomes $100 : 95.9 : 30.7 : 3.3$. Of course, using the tabulated abundances of the Cl_3 distribution in the Appendix would be faster. The result is then plotted with 2 u distance, beginning at the nominal mass of the monoisotopic ion, i.e., $9 \times 12 \ u + 3 \times 14 \ u + 3 \times 35 \ u = 255 \ u$. The contribution of C_9 to the pattern is mainly at X+1 (9.7 %, Table 3.2), whereas its contribution to X+2 (0.4 %) is negligible in this simple estimation and therefore is omitted here. Finally, the X+1 contribution of C_9 is placed into the gaps of the Cl_3 pattern each with 9.7 % relative to the preceding peak of the chlorine isotopic distribution (Fig. 3.4).

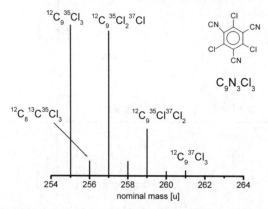

Fig. 3.4. The isotopic pattern of $C_9N_3Cl_3$ as constructed in the example above. The first ^{13}C isotopic peaks are located between the X+2, 4 and 6 peaks of the dominant Cl_3 pattern. Nitrogen is treated as X element and has been omitted from the peak labels for clarity.

3.2.5 Polynomial Approach

The polynomial approach is the logical expansion of the binomial approach. It is useful for the calculation of isotopic distributions of polyisotopic elements or for formulas composed of several non-monoisotopic elements. [2,14] In general, the isotopic distribution of a molecule can be described by a product of polynominals

$$\left(a_1 + a_2 + a_3 + ...\right)^m \left(b_1 + b_2 + b_3 + ...\right)^n \left(c_1 + c_2 + c_3 + ...\right)^o + ... \qquad (3.10)$$

where a_1, a_2, a_3 etc. represent the individual isotopes of one element, b_1, b_2, b_3 etc. represent those of another and so on until all elements are included. The exponents m, n, o etc. give the number of atoms of these elements as contained in the empirical formula.

Example: According to Eq. 3.10 the complete isotopic distribution of stearic acid trichloromethylester, $C_{19}H_{35}O_2Cl_3$, is obtained from the polynomial expression

$$\left(A_{12C} + A_{13C}\right)^{19} \left(A_{1H} + A_{2H}\right)^{35} \left(A_{16O} + A_{17O} + A_{18O}\right)^2 \left(A_{35Cl} + A_{37Cl}\right)^3$$

with A_x representing the relative abundances of the isotopes involved for each element. The problem with the calculation of isotopic patterns resides in the enormous number of terms obtained for larger molecules. Even for this simple example, the number of terms would be $(2)^{19} \times (2)^{35} \times (3)^2 \times (2)^3 = 1.297 \times 10^{18}$. The number is dramatically reduced if like terms are collected which describe the same isotopic composition regardless where the isotopes are located in the molecule. However, manual calculations are prone to become tedious if not impractical; computer programs now simplify the process. [15,16]

> **Note:** Mass spectrometers usually are delivered with the software for calculating isotopic distributions. Such programs are also offered as internet-based or shareware solutions. While such software is freely accessible, it is still necessary to obtain a thorough understanding of isotopic patterns as a prerequisite for the interpreting mass spectra.

3.2.6 Oxygen, Silicon and Sulfur

In a strict sense, oxygen, silicon, and sulfur are polyisotopic elements. Oxygen exists as isotopes ^{16}O, ^{17}O and ^{18}O, sulfur as ^{32}S, ^{33}S, ^{34}S and ^{36}S and silicon as ^{28}Si, ^{29}Si and ^{30}Si.

For oxygen the abundances of ^{17}O (0.038 %) and ^{18}O (0.205 %) are so low that the occurrence of oxygen usually cannot be detected from the isotopic pattern in routine spectra because the experimental error in relative intensities tends to be larger than the contribution of ^{18}O. Therefore, oxygen is frequently treated as an X type element although X+2 would be a more appropriate but practically unuseful classification. On the other hand, a substantial number of oxygen atoms as in oligosaccharides, for example, contributes to the X+2 signal.

Especially in the case of sulfur, the simplification to deal with it as an X+2 element can be accepted as long as only a few sulfur atoms are present in a molecule. However, the contribution of 0.8 % from ^{33}S to the X+1 is similar to that of ^{13}C (1.1 % per atom). If the X+1 peak is used for the estimation of the number of carbons present, ^{33}S will cause an overestimation of the number of carbon atoms by roughly one carbon per sulfur.

The problem is more serious in the case of silicon: the ^{30}Si isotope contributes "only" 3.4 % to the X+2 signal, whereas ^{29}Si gives 5.1 % to X+1. Neglecting ^{29}Si would cause an overestimation of the carbon number by 5 per Si present, which is not acceptable.

The isotopic patterns of sulfur and silicon are by far not as prominent as those of chlorine and bromine, but as pointed out, their contributions are sufficiently important (Fig. 3.5).

The relevance of oxygen and sulfur isotopic patterns is nicely demonstrated by the cluster ion series in fast atom bombardment (FAB) spectra of concentrated sulfuric acid, where the comparatively large number of sulfur and oxygen atoms gives rise to distinct isotopic patterns in the mass spectrum (Chap. 9).

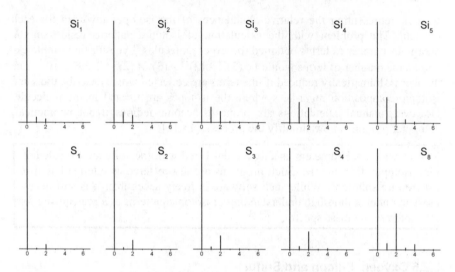

Fig. 3.5. Calculated isotopic patterns for combinations of elemental silicon and sulfur. The peak shown at zero position corresponds to the monoisotopic ion at m/z X. The isotopic peaks are then located at m/z = X+1, 2, 3, ...

Note: The preferred procedure to reveal the presence of S and Si in a mass spectrum is to examine the X+2 intensity carefully: this signal's intensity will be too high to be caused by the contribution of $^{13}C_2$ alone, even if the number of carbons has been obtained from X+1 without prior subtraction of the S or Si contribution.

Example: The isotopic pattern related to the elemental composition of ethyl propyl thioether, $C_5H_{12}S$, (Chap. 6.12) is shown below. The contributions of ^{33}S and ^{13}C to the M+1 and of ^{34}S and $^{13}C_2$ to the M+2 signal are indicated (Fig. 3.6). If the M+1 peak resulted from ^{13}C alone, it would indicate rather the presence of 6 carbon atoms, which in turn would require an M+2 intensity of only 0.1 % instead of the observed 4.6 %. The introduction of Si to explain the isotopic pattern would still fit the M+2 intensity with comparatively low accuracy. For M+1, the situation would be quite different. As ^{29}Si alone demands 5.1 % at M+1, there would be no or 1 carbon maximum allowed to explain the observed M+1 intensity.

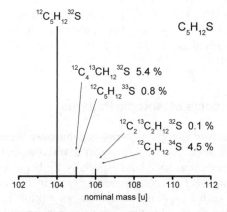

Fig. 3.6. Calculated isotopic pattern of the molecular ion of ethyl propyl thioether, $C_5H_{12}S$ with the respective contributions of ^{33}S and ^{13}C to the M+1 and of ^{34}S and $^{13}C_2$ to the M+2 signal indicated.

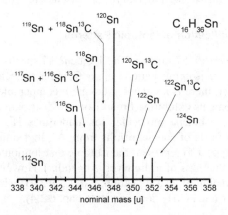

Fig. 3.7. Calculated isotopic pattern of tetrabutyltin, $C_{16}H_{36}Sn$, with labels to indicate major isotopic contributions.

3.2.7 Polyisotopic Elements

The treatment of polyisotopic elements does not require other techniques as far as calculation or construction of isotopic patterns are concerned. However, the appearance of isotopic patterns can differ largely from what has been considered so far and it is worth mentioning their peculiarities.

Example: The presence or absence of the polyisotopic element tin (Table 3.1) can readily be detected from its characteristic isotopic pattern. In case of tetra-butyltin, $C_{16}H_{36}Sn$, the lowest mass isotopic composition is $^{12}C_{16}H_{36}{}^{112}Sn$, 340 u. Due to the 16 carbon atoms, the ^{13}C isotopic abundance is about 17.5 %. This is superimposed on the isotopic pattern of elemental Sn, which becomes especially obvious at 345 and 347 u (Fig. 3.7). The bars labeled with a tin isotope alone are almost solely due to $^xSn^{12}C$ species. Tin neither has an isotope ^{121}Sn nor ^{123}Sn. Therefore, the contributions at 349 and 351 u must be due to $^{120}Sn^{13}C$ and $^{122}Sn^{13}C$, respectively.

3.2.8 Practical Aspects of Isotopic Patterns

Some practical aspects of isotopic patterns as commonly observed in mass spectra should now be considered. Although it seems trivial, the first step is the recognition of an isotopic pattern as such, and especially for beginners this is not always easy. Especially if signals from compounds differing by two or four hydrogens are superimposed or if such a superimposition can not a priori be excluded, a careful stepwise check of the observed pattern has to be performed to avoid misinterpretation of mass spectral data. Similar care has to be taken when isotopically labeled compounds are involved. Next, potential pitfalls for the novice, the correct treatment, and the benefits of isotopic patterns are discussed.

3.2.8.1 False Identification of Isotopic Patterns

The peaks in the m/z 50–57 range of the 1-butene EI spectrum could be misinterpreted as a complex isotopic pattern if no formula were available on the plot (Fig. 3.8). However, there is no element having a comparable isotopic pattern and in addition, all elements exhibiting broad isotopic distributions have much higher mass. Instead, the 1-butene molecular ion undergoes $H^•$, H_2 and multiple H_2 losses. The m/z 57 peak, of course, results from ^{13}C. In a similar fashion the peaks at m/z 39 and 41 appear to represent the isotopic distribution of iridium, but this is impossible due to the mass of iridium (cf. Appendix). However, these peaks originate from the formation of an allyl cation, $C_3H_5^+$, m/z 41, which fragments further by loss of H_2 to form the $C_3H_3^+$ ion, m/z 39 (Chap. 6.2.4).

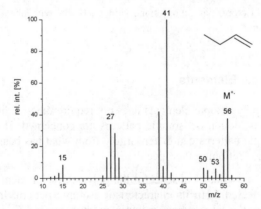

Fig. 3.8. EI mass spectrum of 1-butene. Adapted with permission. © NIST, 2002.

3.2.8.2 Handling of Isotopic Patterns in Mass Spectra

Bearing in mind the knowledge on hydrogen losses, one now could try to fit this scheme to the EI mass spectrum of tribromomethane (Fig. 3.9). Here, the situation is quite different and almost all we can see is the result of the bromine isotopic distribution. By referring to Fig. 3.3 one can easily identify the patterns of Br_3, Br_2, and Br in this spectrum. As a matter of fact, the molecular ion must contain the full number of bromine atoms (m/z 250, 252, 254, 256). The primary fragment ion due to $Br^{\bullet}$ loss will then show a Br_2 pattern (m/z 171, 173, 175). Subsequent elimination of HBr leads to CBr^+ at m/z 91, 93. Alternatively, the molecular ion can eliminate Br_2 to form $CHBr^{+\bullet}$, m/z 92, 94, overlapping with the m/z 91, 93 dublet, or it may loose CHBr to yield $Br_2^{+\bullet}$, causing the peaks at m/z 158, 160, 162. The peaks at m/z 79, 81 are due to Br^+ and those at m/z 80, 82 result from $HBr^{+\bullet}$ formed by CBr_2 loss from the molecular ion.

At this point, there is no need to worry about *why* bromoform behaves like this upon electron ionization. Instead it is sufficient to accept the occurrence of such fragments and to focus on the consequences in the appearance of the mass spectrum (Chap. 6.1.4).

Note: Proof of the identity of isotopic patterns requires careful comparison with those theoretically expected. Furthermore, the mass differences must be consistent with the mass of the presumed neutral loss. In no case can the pattern be observed in signals corresponding to lower mass than given by the sum of all isotopes included.

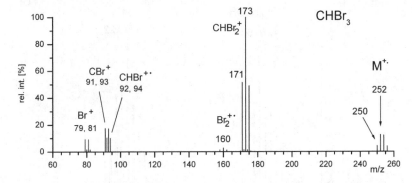

Fig. 3.9. EI mass spectrum of tribromomethane. Adapted with permission. © NIST, 2002.

Note: In order to calculate the mass difference between peaks belonging to different isotopic patterns it is strongly recommended to proceed from the monoisotopic peak of one group to the monoisotopic peak of the next. Accordingly, the mass difference obtained also belongs to the loss of a monoisotopic fragment. Otherwise, there is a risk of erroneously omitting or adding hydrogens in a formula.

Example: From the tribromomethane spectrum (Fig. 3.9) a mass difference of 79 u is calculated between m/z 250 and m/z 171, which belongs to ^{79}Br, thus identifying the process as a loss of a bromine radical. Starting from the $CH^{79}Br_2^{81}Br$ isotopic ion at m/z 252 would yield the same information if ^{81}Br was used for the calculation. Use of ^{79}Br would be misleading and suggest the loss of H_2Br.

3.2.8.3 Information from Complex Isotopic Patterns

If the isotopic distribution is very broad and/or there are elements encountered that have a lowest mass isotope of very low abundance, recognition of the mono-isotopic peak would become rather uncertain. However, there are ways to cope with that situation.

Example: In mass spectra of tin compounds, the ^{112}Sn isotopic peak could easily be overseen or simply be superimposed by background signals (Fig. 3.7). Here, one should identify the ^{120}Sn peak from its unique position within the characteristic pattern before stepping through the spectrum from peak to peak. For all other elements contained in the respective ions still the lowest mass isotope would be used in calculations.

Example: Ruthenium exhibits a wide isotopic distribution where the ^{102}Ru isotope can be used as a marker during assignment of mass differences. Moreover, the strong isotopic fingerprint of Ru makes it easily detectable from mass spectra and even compensates for a lack of information resulting from moderate mass accuracy (Fig. 3.10).

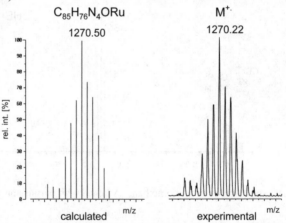

Fig. 3.10. Calculated and experimental (FD-MS, cf. Chap. 8.5.4) isotopic pattern of a ruthenium carbonyl porphyrin complex. The isotopic pattern supports the presumed molecular composition. The label is attached to the peak corresponding to the ^{102}Ru-containing ion. Adapted from Ref. [17] with permission. © IM Publications, 1997.

> **Note:** If the isotopic distribution is broad and/or there are elements encountered that have a lowest mass isotope of very low abundance, it is recommended to base calculations on the most abundant isotope of the respective element.

For those preferring monoisotopic spectra, a paper dealing with a least squares analysis to simplify multi-isotopic mass spectra is recommended. [18]

3.2.9 Isotopic Enrichment and Isotopic Labeling

3.2.9.1 Isotopic Enrichment

If the abundance of a particular nuclide is increased above the natural level in an ion, the term *isotopically enriched ion* is used to describe any ion enriched in the isotope. [1] The degree of *isotopic enrichment* is best determined by mass spectrometry.

Example: Isotopic enrichment is a standard means to enhance the response of an analyte in nuclear magnetic resonance (NMR). Such measures gain importance if extremely low solubility is combined with a large number of carbons, as is often the case with [60]fullerene compounds. [19] The molecular ion signals, $M^{+\bullet}$, of C_{60} with natural isotopic abundance and of ^{13}C-enriched C_{60} are shown below (Fig. 3.11; for EI-MS of [60]fullerenes cf. Refs. [20-22]). From these mass spectra, the ^{13}C enrichment can be determined by use of Eq. 3.1. For C_{60} of natural isotopic abundance we obtain $M_{rC60} = 60 \times 12.0108$ u $= 720.65$ u. Applying Eq. 3.1 to the isotopically enriched compound yields $M_{r13C-C60} = (35 \times 720 + 65 \times 721 + 98 \times 722 + 100 \times 723 + 99 \times 724 + 93 \times 725 + ...)u/(35 + 65 + 98 + 100 + 99 + 93 + ...) = 724.10$ u. (Integer mass and intensity values are used here for clarity.) This result is equivalent to an average content of 4.1 ^{13}C atoms per [60]fullerene molecule which on the average means 3.45 ^{13}C atoms more than the natural content of 0.65 ^{13}C atoms per molecule.

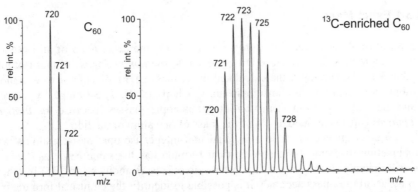

Fig. 3.11. Comparison of the molecular ion signals, $M^{+\bullet}$, of [60]fullerene with natural isotopic abundance and of ^{13}C-enriched C_{60}. By courtesy of W. Krätschmer, Max Planck Institute for Nuclear Physics, Heidelberg.

3.2.9.2 Isotopic Labeling

If the abundance of a particular nuclide is increased above the natural level at one or more (specific) positions within an ion, the term *isotopically labeled ion* is used to describe such an ion. Among other applications, *isotopic labeling* is used in order to track metabolic pathways, to serve as internal standard for quantitative analysis, or to elucidate fragmentation mechanisms of ions in the gas phase. In mass spectrometry, the non-radiating isotopes ^{2}H (deuterium, D) and ^{13}C are preferably employed and thus, a rich methodology to incorporate isotopic labels has been developed. [23] Isotopic labeling is rather a mass spectrometric research tool than mass spectrometry is a tool to control the quality of isotopic labeling. As a result, isotopic labeling is used in many applications and thus many such examples will be found throughout the book.

3.3 High-Resolution and Accurate Mass

High-resolution and accurate mass measurements are closely related to each other because the obtainable mass accuracy also depends on sufficiently resolved peaks. Nevertheless, they should not be confused, as performing a measurement at *high resolution* alone does not equally imply measuring the *accurate mass*.

Measurements at high resolution are more frequently being used to determine accurate mass and to identify empirical formulas. While accurate mass measurements were limited to electron ionization up to the 1970s and early 1980s, new developments such as high-resolving time-of-flight (TOF) and Fourier transform ion cyclotron resonance (FT-ICR) analyzers have made it possible since the late 1990s to obtain such information from any ionization method. [24] Nowadays, the more widespread application of such methodologies demand for a thorough understanding of their potential and limitations.

3.3.1 Exact Mass

As we have already seen, the *isotopic mass* also is the *exact mass* of an isotope. The isotopic mass is very close but not equal to the *nominal mass* of that isotope (Table 3.1). Accordingly, the calculated exact mass of a molecule or of a monoisotopic ion equals its monoisotopic mass (Chap. 3.1.4). The isotope ^{12}C represents the only exception from non-integer isotopic masses, because the *unified atomic mass* [u] is defined as $^1/_{12}$ of the mass of one atom of nuclide ^{12}C.

As a consequence of these individual non-integer isotopic masses, almost no combination of elements in an empirical formula has the same *calculated exact mass*, or simply *exact mass* as it is often referred to, as another one. [25] In other words, at infinite mass accuracy it is possible to identify the empirical formula by mass spectrometry alone.

Example: the molecular ions of nitrogen, $N_2^{+\bullet}$, carbon monoxide, $CO^{+\bullet}$, and ethene, $C_2H_4^{+\bullet}$, have the same nominal mass of 28 u, i.e., they are so-called *isobaric ions*. The isotopic masses of the most abundant isotopes of hydrogen, carbon, nitrogen and oxygen are 1.007825 u, 12.000000 u, 14.003070 u and 15.994915 u, respectively. Using these values, the calculated ionic masses are 28.00559 u for $N_2^{+\bullet}$, 27.99437 u for $CO^{+\bullet}$, and 28.03075 u for $C_2H_4^{+\bullet}$. This means they differ by some "*millimass units*" (mmu) from each other, and none of these isobaric ions has precisely 28.00000 u (Chap. 3.3.4 and Chap. 6.9.6).

Terms related to exact mass:

- *Isobaric ions* are ions having the same nominal mass. However, their exact mass is different. Isomers are (almost) perfect isobars.
- For historical reasons, 10^{-3} u is referred to as 1 *millimass unit* (mmu). The use of mmu is widespread because of its convenience in dealing with small differences in mass, although the mmu is not an SI unit.

3.3.2 Deviations from Nominal Mass

As just demonstrated, exact mass values show some deviation from nominal mass that can be on either side, higher or lower, depending on the isotopes encountered. While the matter itself can be easily understood, existing terminology in this specific field is comparatively vague.

Note: Commonly, the term *mass defect*, defined as the difference between the exact mass and the integer mass, is used to describe this deviation. [3] Application of this concept leads to *positive* and *negative mass* defects, respectively. In addition, the association of something being "defective" with certain isotopic masses can be misleading.

3.3.2.1 Mass Deficiency

The term *mass deficiency* better describes the fact that the exact mass of an isotope or a complete molecule is lower than the corresponding nominal mass. In case of ^{16}O, for example, the isotopic mass is 15.994915 u, being 5.085 mmu deficient as compared to the nominal value.

Most isotopes are more or less mass deficient with a tendency towards larger mass deficiencies for the heavier isotopes, e.g., M_{35Cl} = 34.96885 u (–31.15 mmu), M_{79Br} = 78.91834 u (–81.66 mmu) and M_{127I} = 129.90447 u (–95.53 mmu). The increasing mass deficiency of the heavier isotopes is caused by the increasing nuclear binding energy as explained by relativistic theory. This is consistent with the fact that the radioactive isotopes of thorium and uranium have isotopic masses above the nominal value, thus reflecting their comparatively labile nuclei (see Appendix). The numerical values of mass deficiency, however, are merely defined by the "arbitrary" setting of ^{12}C as the standard in the atomic mass scale.

3.3.2.2 Mass Sufficiency

Among the elements frequently encountered in mass spectrometry, only the isotopes of the five elements preceding carbon in the Periodic Table plus nitrogen, i.e., H, He, Li, Be, B and N, exhibit isotopic masses larger than their nominal value. Such a negative mass deficiency is sometimes called *mass sufficiency*. Among the mass-sufficient isotopes, ^1H is the most important one, because each hydrogen adds 7.825 mmu. Thereby, it significantly contributes to the mass of larger hydrocarbon molecules. In general, the ubiquitous occurrence of hydrogen in empirical formulas of organic molecules causes most of them to exhibit considerable mass sufficiency which again decreases with the number of mass-deficient isotopes, e.g., from halogens, oxygen or metals.

Example: If the deviations from nominal mass of different oligomers are plotted as a function of nominal mass, only pure carbon molecules as represented by fullerenes are always located on the *x*-axis. Hydrocarbons have the highest mass sufficiency due to the large number of hydrogens causing roughly 1 u added per 1000 u in molecular mass. Halogenated oligomers, on the other hand, are more or less mass deficient and those oligomers containing some oxygen are located in between (Fig. 3.12).

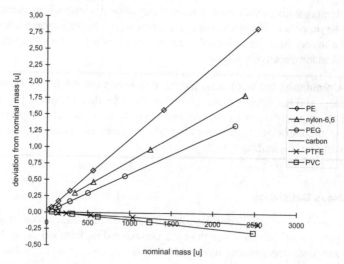

Fig. 3.12. Deviation from nominal mass for some oligomers as a function of nominal mass. PE: polyethylene, PEG: polyethyleneglycol, PTFE: polytetrafluoroethylene, PVC: polyvinylchloride.

Note: The use of nominal mass is limited to the low mass range. Above about 500 u the first decimal of isotopic mass can be larger than .5 causing it to be rounded up to 501 u instead of the expected value of 500 u. This will in turn lead to severe misinterpretation of a mass spectrum (Chap. 6).

3.3.2.3 Deltamass

Recently, the term *deltamass* has been coined and defined as the mass value following the decimal point. [26] This elegantly circumvents the problems related to the mass deficiency/sufficiency terminology. Unfortunately, it is only valid in the context for which it has been developed, i.e., to describe mass deviations of peptides from average values. Beyond this context, ambiguities arise from its rigorous application because i) real mass deficiency would then be expressed the same way as larger values of mass sufficiency, e.g., in case of ^{127}I as 0.9060 u and in case of $C_{54}H_{110}$ as 0.8608 u, respectively, and ii) deviations of more than 1 u would be expressed by the same numerical value as those of less than 1 u.

Example: Provided a correctly calibrated mass spectrometer is used, the magnitude of mass deficiency can be exploited to obtain a first idea of what class of compound is being analyzed. At a sufficient level of sophistication, the deviation from nominal mass can even give more detailed information. [27] Peptides consist of amino acids and therefore, their elemental compositions are rather similar independent of their size or sequence. This results in a characteristic relationship of formulas and deltamass values. Phosphorylation and more pronounced glycosylation give rise to lower deltamass values, because they introduce mass-deficient atoms (P, O) into the molecule. The large number of hydrogens associated with lipidation, on the other side, contributes to a deltamass above normal level. On the average, an unmodified peptide of 1968 u, for example, shows a deltamass of 0.99 u, whereas a glycosylated peptide of the same nominal mass will have a value of 0.76 u. Therefore, the deltamass can be employed to obtain information on the type of covalent protein modification. [26] (cf. Chap. 3.4.3.)

3.3.2.4 Kendrick Mass Scale

The *Kendrick mass scale* is based on the definition $M_{(CH_2)} = 14.0000$ u. [28] Conversion from the IUPAC mass scale to the Kendrick mass scale is achieved by multiplying each mass by $14.00000/14.01565 = 0.99888$. The intention of the Kendrick mass scale is to provide an effective means of data reduction, an issue gaining importance again as a result of the steadily increasing *resolution* and *mass accuracy* (next paragraphs) of modern mass spectrometers.

Example: The composition of complex hydrocarbon systems consisting of numerous compound classes with some 30 homologues each can be displayed by breaking an ultrahigh-resolution broadband mass spectrum into segments of 1 u which are aligned on the abscissa. Then, the Kendrick mass defect of the isobaric ions contained in the respective segment is plotted onto the ordinate. The resulting graph preserves not only the "coarse" spacing, e.g., about 1 u between odd and even mass values, but also the "fine structure", i.e., different mass defects for different elemental compositions across each segment. [29] Several thousand elemental compositions in the spectrum of petroleum crude oil [29] or diesel fuel [30] may be resolved visually in a single graphical display, allowing for the visual identification of various compound classes and/or alkylation series (Fig. 3.13).

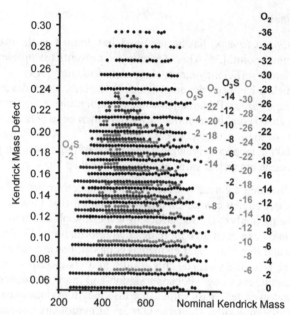

Fig. 3.13. Kendrick mass defect versus nominal Kendrick mass for odd-mass $^{12}C_x$ ions ([M–H]$^-$ ions). The compound classes (O, O_2, O_3S and O_4S) and the different numbers of rings plus double bonds (Chap. 6.4.4) are separated vertically. Horizontally, the points are spaced by CH_2 groups along a homologous series. [29] By courtesy of A. G. Marshall, NHFL, Tallahassee.

3.3.3 Mass Accuracy

The *mass accuracy* is defined as the difference between *measured accurate mass* and calculated exact mass. The mass accuracy can be stated as absolute units of u (or mmu) or as relative mass accuracy in ppm, i.e., absolute mass accuracy divided by the mass it is determined for. As mass spectrometers tend to have similar absolute mass accuracies over a comparatively wide range, absolute mass accuracy represents a more meaningful way of stating mass accuracies than the more trendy use of ppm.

> **Note:** *Part per million* (ppm) is simply a relative measure as are percent (%), permill (parts per thousand, ‰), or *parts per billion* (ppb).

Example: A magnetic sector mass spectrometer allows for an absolute mass accuracy of 2–5 mmu in scanning mode over a range of about m/z 50–1500. At m/z 1200 an error of 3.5 mmu corresponds to inconspicuous 2.9 ppm, whereas the same error yields 70 ppm at m/z 50 which seems to be unacceptably large.

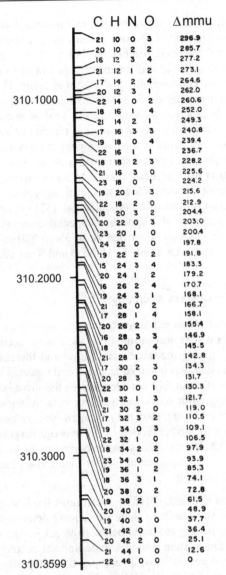

	C	H	N	O	Δmmu
	21	10	0	3	296.9
	20	10	2	2	285.7
	16	12	3	4	277.2
	21	12	1	2	273.1
	17	14	2	4	264.6
	20	12	3	1	262.0
310.1000	22	14	0	2	260.6
	18	16	1	4	252.0
	21	14	2	1	249.3
	17	16	3	3	240.8
	19	18	0	4	239.4
	22	16	1	1	236.7
	18	18	2	3	228.2
	21	16	3	0	225.6
	23	18	0	1	224.2
	19	20	1	3	215.6
	22	18	2	0	212.9
	18	20	3	2	204.4
	20	22	0	3	203.0
	23	20	1	0	200.4
	24	22	0	0	197.8
	19	22	2	2	191.8
	15	24	3	4	183.3
310.2000	20	24	1	2	179.2
	16	26	2	4	170.7
	19	24	3	1	168.1
	21	26	0	2	166.7
	17	28	1	4	158.1
	20	26	2	1	155.4
	16	28	3	3	146.9
	18	30	0	4	145.5
	21	28	1	1	142.8
	17	30	2	3	134.3
	20	28	3	0	131.7
	22	30	0	1	130.3
	18	32	1	3	121.7
	21	30	2	0	119.0
	17	32	3	2	110.5
	19	34	0	3	109.1
	22	32	1	0	106.5
310.3000	18	34	2	2	97.9
	23	34	0	0	93.9
	19	36	1	2	85.3
	18	36	3	1	74.1
	20	38	0	2	72.8
	19	38	2	1	61.5
	20	40	1	1	48.9
	19	40	3	0	37.7
	21	42	0	1	36.4
	20	42	2	0	25.1
	21	44	1	0	12.6
310.3599	22	46	0	0	0

Fig. 3.14. Exact masses of isobaric ions of nominal mass 310 u. The Δm scale is plotted in mmu relative to the saturated hydrocarbon, $C_{22}H_{46}$, at 310.3599 u. Adapted from Ref. [13] with permission. © University Science Books, 1993.

Given infinite mass accuracy we should be able to identify the empirical formula of any ion from its exact mass. The emphasis is on *infinite mass accuracy*, however, and the meaning of "infinite" is perfectly illustrated by an example from McLafferty's book (Fig. 3.14). [13] Even though the compositions are restricted to $C_xH_yN_{0-3}O_{0-4}$, there is a considerable number of isobaric ions of nominal mass

310 u with the calculated exact mass of the saturated hydrocarbon $C_{22}H_{46}$ of 310.3599 u at the upper limit. The exact masses of all other compositions are increasingly lower until the scale reaches the bottom line at 296.9 mmu less with $C_{21}H_{10}O_3$ at 310.0630 u. While it is comparatively simple to distinguish these extremes, there are other combinations just 1.5 mmu apart. The situation becomes drastically worse as more elements and fewer limitations of their number must be taken into account. In practice, there is a strong need of restricting oneself to certain elements and a maximum and/or minimum number of certain isotopes to assure a high degree of confidence in the assignment of formulas.

Isotopic patterns provide a prime source of such additional information. Combining the information from accurate mass data and experimental peak intensities with calculated isotopic patterns allows to significantly reduce the number of potential elemental compositions of a particular ion. [31] Otherwise, even at an extremely high mass accuracy of 1 ppm the elemental composition of peptides, for example, can only be uniquely identified up to about 800 u, i.e., an error of less than 0.8 mmu is required even if only C, H, N, O and S are allowed. [27,32,33]

3.3.3.1 Specification of Mass Accuracy

It has been stated that measured accurate masses when used to assign molecular formulae should always be accompanied by their mass accuracies. [34] Ideally, this can be done by giving the mean mass value and the corresponding error in terms of standard deviation as obtained from several repeated measurements of the same ion. [35] This is definitely not identical to the error which is usually provided with the listing from mass spectrometer data systems, where an error is given as the difference of calculated and measured mass value.

The theoretically expected reduction of the average mass error with the square root of the number of determinations has experimentally been verified: [36] single-scan mass errors were significantly larger than those obtained from the average of multiple measurements.

Example: The $[M\text{-}Cl]^+$ ion, $[CHCl_2]^+$, represents the base peak in the EI spectrum of chloroform. The results of three subsequent determinations for the major peaks of the isotopic pattern are listed below (Fig. 3.15). The typical printout of a mass spectrometer data system provides experimental accurate mass and relative intensity of the signal and an error as compared to the calculated exact mass of possible compositions. For the $[^{12}CH^{35}Cl_2]^+$ ion, the experimental accurate mass values yield an average of 82.9442 ± 0.0006 u. The comparatively small absolute error of 0.6 mmu corresponds to a relative error of 7.5 ppm.

```
[ Elemental Composition ]
Data   : JMS18120_009            Date : 14-May-2002 08:20
Sample: CHCl3
Elements : C 10/1(12C 10/0, 13C 1/0), H 10/0, Cl 4/0(35Cl 4/0, 37Cl 4/0)
Mass Tolerance      : 10mmu
Unsaturation (U.S.) : -200.0 - 200.0

Observed m/z Int%  Err[ppm / mmu]   U.S.  Composition
  82.9433  100.0   +27.5 /  +2.3    1.0   13C 35Cl 2
                   -26.4 /  -2.2    0.5   12C H 35Cl 2

  84.9406   62.5   +29.5 /  +2.5    1.0   13C 35Cl 37Cl
                   -23.2 /  -2.0    0.5   12C H 35Cl 37Cl

  86.9379   10.1   +31.6 /  +2.7    1.0   13C 37Cl 2
                   -19.9 /  -1.7    0.5   12C H 37Cl 2

Observed m/z Int%  Err[ppm / mmu]   U.S.  Composition
  82.9449  100.0   +45.8 /  +3.8    1.0   13C 35Cl 2
                    -8.1 /  -0.7    0.5   12C H 35Cl 2

  84.9422   66.0   +48.0 /  +4.1    1.0   13C 35Cl 37Cl
                    -4.7 /  -0.4    0.5   12C H 35Cl 37Cl

  86.9404   11.2   +60.3 /  +5.2    1.0   13C 37Cl 2
                    +8.9 /  +0.8    0.5   12C H 37Cl 2

Observed m/z Int%  Err[ppm / mmu]   U.S.  Composition
  82.9447  100.0   +43.7 /  +3.6    1.0   13C 35Cl 2
                   -10.2 /  -0.8    0.5   12C H 35Cl 2

  84.9419   65.1   +44.6 /  +3.8    1.0   13C 35Cl 37Cl
                    -8.1 /  -0.7    0.5   12C H 35Cl 37Cl

  86.9391   10.8   +45.1 /  +3.9    1.0   13C 37Cl 2
                    -6.3 /  -0.6    0.5   12C H 37Cl 2
```

Fig. 3.15. Printout of elemental compositions of $[M–Cl]^+$ ions of chloroform as obtained from three subsequent measurements (70 eV EI, $R = 8000$). The error for each proposal is listed in units of ppm and mmu. The column U.S. lists "unsaturation", i.e., the number of rings and/or double bonds (Chap. 6.4.4).

3.3.3.2 Technical Aspects of Exact Mass Calculations

On one side, the isotopic masses provided in this book are listed with six decimal places corresponding to an accuracy of 10^{-3} mmu (Tab. 3.1 and Appendix). On the other side, the realistic magnitude of usually encountered mass errors in organic mass spectrometry is in the order of 1 mmu.

The number of decimal places one should employ in mass calculations depends on the purpose they are used for. In the m/z range up to about 500 u, the use of isotopic mass with four decimal places will provide sufficient accuracy. Above, at least five decimal places are required, because the increasing number of atoms results in an unacceptable multiplication of many small mass errors. Beyond 1000 u, even six decimal places should be employed. However, the final results of these calculations may be reported with only four decimal places, because this is sufficient for most applications. If mass accuracies of significantly less than 1 mmu are to be expected, the use of six decimal places becomes necessary in any case.

3.3.4 Resolution

Good mass accuracy can only be obtained from sufficiently sharp and evenly shaped signals that are well separated from each other. The ability of an instrument to perform such a separation of neighboring peaks is called *resolving power*. It is obtained from the peak width expressed as a function of mass.

The term *resolution* essentially describes the same; the small difference being that its definition is based on the resulting signals. The resolution R is defined as the ratio of the mass of interest, m, to the difference in mass, Δm, as defined by the width of a peak at a specific fraction of the peak height. [1]

$$R = \frac{m}{\Delta m} \qquad (3.11)$$

Two neighboring peaks are assumed to be sufficiently separated when the valley separating their maxima has decreased to 10 % of their intensity. Hence, this is known as *10 % valley definition of resolution, $R_{10\%}$*. The 10 % valley conditions are fulfilled if the peak width at 5 % relative height equals the mass difference of the corresponding ions, because then the 5 % contribution of each peak to the same point of the m/z axis adds up to 10 % (Fig. 3.16).

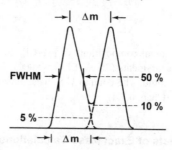

Fig. 3.16. Illustration of the 10 % valley and of the full width at half maximum (FWHM) definitions of resolution. Peak heights are not to scale.

With the advent of linear quadrupole analyzers the *full width at half maximum (FWHM) definition* of resolution became widespread especially among instruments manufacturers. It is also commonly used for time-of-flight and quadrupole ion trap mass analyzers. With Gaussian peak shapes, the ratio of R_{FWHM} to $R_{10\%}$ is 1.8. The practical consequences of resolution for a pair of peaks at different m/z are illustrated below (Fig. 3.17).

Note: The attributive *low-resolution* (LR) is generally used to describe spectra obtained at $R = 500$–2000. *High-resolution* (HR) is appropriate for $R > 5000$. However, there is no exact definition of these terms.

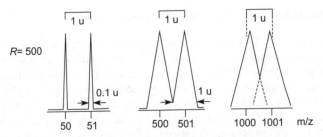

Fig. 3.17. Signals at m/z 50, 500 and 1000 as obtained at $R = 500$. At m/z 1000 the peak maxima are shifted towards each other due to superimposing of the peaks; this also approximates the result of $R_{FWHM} = 900$.

Example: The changes in the electron ionization spectra of residual air are well suited to demonstrate the effect of increasing resolution (Fig. 3.18). Setting $R = 1000$ yields a peak width of 28 mmu for the m/z 28 signal. An increase to $R = 7000$ perfectly separates the minor contribution of CO^+, m/z 27.995, from the predominating $N_2^{+\cdot}$ at m/z 28.006 (The CO^+ ion rather results from fragmenting $CO_2^{+\cdot}$ ions than from carbon monoxide in laboratory air.)

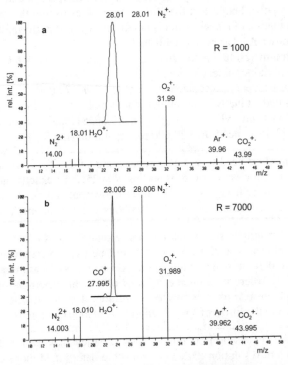

Fig. 3.18. EI mass spectra of residual air (**a**) at $R = 1000$ and (**b**) at $R = 7000$. The relative intensities are not affected by the different settings of resolution. The decimal places of the mass labels give an estimate of achievable mass accuracies under the respective conditions.

3.3.4.1 Influence of Resolution on Relative Peak Intensity

Increasing resolution does not affect the relative intensities of the peaks, i.e., the intensity ratios for m/z 28 : 32 : 40 : 44 in the spectrum of air basically remain constant (Fig. 3.18). However, an increase of resolution is usually obtained at the cost of transmission of the analyzer, thereby reducing the absolute signal intensity (Chap. 4.3.4). Accordingly, isotopic patterns are not affected by increasing resolution up to $R \approx 10,000$; beyond, there can be changes in isotopic patterns due to the separation of different isotopic species of the same nominal mass (Chap. 3.4).

> **Note:** If a peak is not present in the low-resolution mass spectrum, there is no reason why it should be detectable from the high-resolution spectrum as long as the same ionization method is employed. Instead, use of a more suitable ionization method is recommended. (And sometimes one has to accept the truth that the sample is not what it was expected to be.)

3.3.4.2 Experimental Determination of Resolution

In principle, resolution is always determined from the peak width of some signal at a certain relative height and therefore, any peak can serve this purpose. As the exact determination of a peak width is not always easy to perform, the use of some dublets of known Δm has been established.

The minimum resolution to separate CO^+ from $N_2^{+\bullet}$ is $28/0.011 \approx 2500$. The dublet from pyridine molecular ion, $C_5H_5N^{+\bullet}$, m/z 79.0422, and from the first isotopic peak of benzene molecular ion, $^{13}CC_5H_6^{+\bullet}$, m/z 79.0503, demands $R = 9750$ to be separated. Finally, the dublet composed of the first isotopic ion of $[M–CH_3]^{+\bullet}$ ion from xylene, $^{13}CC_6H_7^+$, m/z 92.0581, and toluene molecular ion, $C_7H_8^{+\bullet}$, m/z 92.0626, needs $R = 20,600$ for separation (Fig 3.19).

> **Note:** There is no need to use a more accurate value of m than nominal mass and likewise, there is no use of reporting $R = 2522.52$ exactly as obtained for the $CO^+/N_2^{+\bullet}$ pair. It is fully sufficient to know that setting $R = 3000$ is sufficient for one specific task or that $R = 10,000$ is suitable for another.

With good magnetic sector instruments resolutions up to $R = 10,000$ can routinely be employed, and $R = 15,000$ may still be used. Nevertheless, resolutions larger than 10,000 are rarely employed in accurate mass measurements, and these are of use only where interferences of ions of the same nominal m/z need to be avoided. With an instrument in perfect condition, it is possible to achieve higher resolutions up to the limits of specification ($R \approx 60,000$) on intensive peaks.

> **Note:** As the signals become noisy, mass accuracy may even suffer from too high settings of resolution. Often centroids are determined more accurate from smooth and symmetrically shaped peaks at moderate HR conditions. One should also be aware of the fact that the position of a peak of 0.1 u width needs to be determined to $^1/_{50}$ of its width to obtain 2 mmu accuracy.

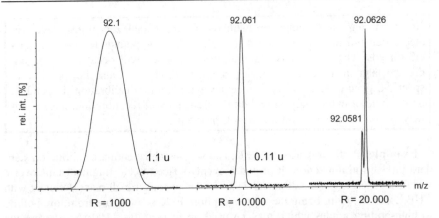

Fig. 3.19. The *m/z* 92 peak from a mixture of xylene and toluene at different settings of resolution. At $R = 10,000$ some separation of the lower mass ion can already be presumed from a slight asymmetry of the peak. $R = 20,600$ is needed to fully separate $^{13}CC_6H_7^{+\bullet}$, *m/z* 92.0581, from $C_7H_8^{+\bullet}$, *m/z* 92.0626. The *m/z* scale is the same for all of the signals.

3.3.5 Mass Calibration

So far, the concepts of exact mass, mass accuracy and resolution have been introduced without considering the means by which *accurate mass measurements* can be realized. The key to this problem is *mass calibration*. Resolution alone can separate ions of different *m/z* value, but it does not automatically include the information where on the *m/z* axis the respective signals precisely are located.

3.3.5.1 External Mass Calibration

Any mass spectrometer requires mass calibration before use. However, the procedures to perform it properly and the number of calibration points needed may largely differ between different types of mass analyzers. Typically, several peaks of well-known *m/z* values evenly distributed over the mass range of interest are necessary. These are supplied from a well-known *mass calibration compound* or *mass reference compound*. Calibration is then performed by recording a mass spectrum of the *calibration compound* and subsequent correlation of experimental *m/z* values to the *mass reference list*. Usually, this conversion of the mass reference list to a calibration is accomplished by the mass spectrometer's data system. Thereby, the mass spectrum is recalibrated by interpolation of the *m/z* scale between the assigned calibration peaks to obtain the best match. The mass calibration obtained may then be stored in a *calibration file* and used for future measurements without the presence of a calibration compound. This procedure is termed *external mass calibration*.

Note: The numerous ionization methods and mass analyzers in use have created a demand for a large number of calibration compounds to suit their specific needs. Therefore, mass calibration compounds will occasionally be addressed later in the chapters on ionization methods. It is also not possible to specify a general level of mass accuracy with external calibration. Depending on the type of mass analyzer and on the frequency of recalibration, mass accuracy can be as high as 1 mmu or as low as 0.5 u.

Example: Perfluorokerosene, PFK, is a well-established mass calibration standard in electron ionization. It provides evenly spaced $C_xF_y^+$ fragment ions over a wide mass range (Figs. 3.20 and 3.21). The major ions are all mass deficient with CHF_2^+, m/z 51.0046, being the only exception. PFK is available from low-boiling to high-boiling grades which may be used up to m/z 700–1100. Apart from the highest boiling grades, PFK is suitable for introduction by means of the reference inlet system (Chap. 5.3.3), a property making it very attractive for internal calibration as well.

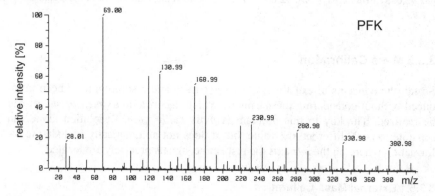

Fig. 3.20. Partial 70 eV EI mass spectrum of perfluorokerosene, PFK. The peaks are evenly spaced over a wide m/z range. In the low m/z range peaks from residual air do occur.

3.3.5.2 Internal Mass Calibration

If high-resolution measurements are performed in order to assign elemental compositions, *internal mass calibration* is almost always required. The calibration compound can be introduced from a second inlet system or be mixed with the analyte before the analysis. Mixing calibration compounds with the analyte requires some operational skills in order not to suppress the analyte by the reference or vice versa. Therefore, a separate inlet to introduce the calibration compound is advantageous. This can be achieved by introducing volatile standards such as PFK from a reference inlet system in electron ionization, by use of a dual-target probe in fast atom bombardment, or by use of a second sprayer in electrospray ionization.

Internal mass calibration typically yields mass accuracies as high as 0.1–0.5 mmu with Fourier transform ion cyclotron resonance, 0.5–5 mmu with magnetic sector, and 2–10 mmu with time-of-flight analyzers.

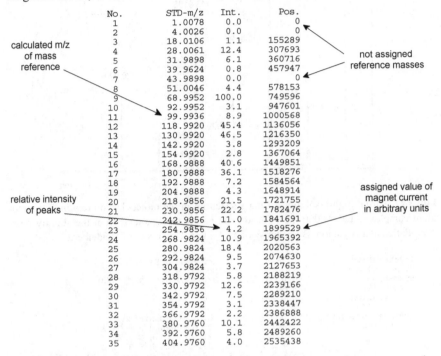

No.	STD-m/z	Int.	Pos.
1	1.0078	0.0	0
2	4.0026	0.0	0
3	18.0106	1.1	155289
4	28.0061	12.4	307693
5	31.9898	6.1	360716
6	39.9624	0.8	457947
7	43.9898	0.0	0
8	51.0046	4.4	578153
9	68.9952	100.0	749596
10	92.9952	3.1	947601
11	99.9936	8.9	1000568
12	118.9920	45.4	1136056
13	130.9920	46.5	1216350
14	142.9920	3.8	1293209
15	154.9920	2.8	1367064
16	168.9888	40.6	1449851
17	180.9888	36.1	1518276
18	192.9888	7.2	1584564
19	204.9888	4.3	1648914
20	218.9856	21.5	1721755
21	230.9856	22.2	1782476
22	242.9856	11.0	1841691
23	254.9856	4.2	1899529
24	268.9824	10.9	1965392
25	280.9824	18.4	2020563
26	292.9824	9.5	2074630
27	304.9824	3.7	2127653
28	318.9792	5.8	2188219
29	330.9792	12.6	2239166
30	342.9792	7.5	2289210
31	354.9792	3.1	2338447
32	366.9792	2.2	2386888
33	380.9760	10.1	2442422
34	392.9760	5.8	2489260
35	404.9760	4.0	2535438

calculated m/z of mass reference

relative intensity of peaks

not assigned reference masses

assigned value of magnet current in arbitrary units

Fig. 3.21. Reproduction of a PFK calibration table of a magnetic sector instrument covering the m/z 1–405 range. In order to expand the PFK reference peak list to the low m/z range, ^1H, ^4He and peaks from residual air are included, but for intensity reasons ^1H, ^4He and CO_2 have not been assigned in this particular case.

Note: Mass accuracy is highly dependent on many parameters such as resolving power, scan rate, scanning method, signal-to-noise ratio of the peaks, peak shapes, overlap of isotopic peaks at same nominal mass, mass difference between adjacent reference peaks etc. An error of ± 5 mmu for routine applications is a conservative estimate and thus the experimental accurate mass should lie within this error range independent of the ionization method and the instrument used. [37] There is no reason that the correct (expected) composition has to be the composition with the smallest error.

Example: The high-resolution spectrum in the molecular ion range of a zirconium complex is typified by the isotopic pattern of zirconium and chlorine (Fig. 3.22). ^{90}Zr represents the most abundant isotope of zirconium which is accompanied by ^{91}Zr, ^{92}Zr, ^{94}Zr and ^{96}Zr, all of them having considerable abun-

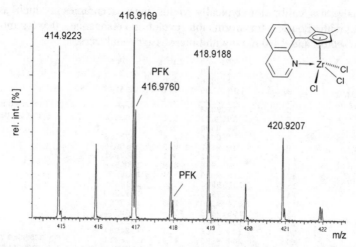

Fig. 3.22. Partial high-resolution EI mass spectrum in the molecular ion region of a zirco-nium complex. At $R = 8000$ the PFK ion can barely be separated from the slightly more mass deficient analyte ion. By courtesy of M. Enders, University of Heidelberg.

```
Inlet : Direct                      Ion Mode : EI+
RT : 1.94 min                       Scan#: 5
Elements : C 40/0, H 40/0, N 2/0, Cl 3/0(35Cl 3/0, 37Cl 3/0), Zr 1/0
Mass Tolerance      : 8mmu
Unsaturation (U.S.) : 0.0 - 100.0

Observed m/z  Int%    Err[ppm / mmu]    U.S.  Composition
   418.9188    7.5      -8.2 /  -3.4    24.5  C 26 H 2 35Cl 3
                        +0.2 /  +0.1    27.5  C 25 H N 2 Zr
                        -4.2 /  -1.7    22.5  C 24 H 6 35Cl Zr
                       -12.5 /  -5.3     9.5  C 15 H 14 N 2 35Cl 2 37Cl Zr
                        +1.8 /  +0.8    10.0  C 16 H 14 N 35Cl 37Cl 2 Zr
                       +16.2 /  +6.8    10.5  C 17 H 14 37Cl 3 Zr

   416.9169    8.9      -6.2 /  -2.6    22.5  C 23 H 2 N 2 37Cl 3
                        +3.1 /  +1.3    28.0  C 26 H N Zr
                        -9.7 /  -4.0    10.0  C 16 H 14 N 35Cl 2 37Cl Zr
                        +9.1 /  +3.8    15.5  C 18 H 9 N 2 37Cl 2 Zr
                        +4.8 /  +2.0    10.5  C 17 H 14 35Cl 37Cl 2 Zr

   414.9223    7.5      -0.4 /  -0.2    22.5  C 23 H 2 N 2 35Cl 37Cl 2
                       -16.2 /  -6.7    22.5  C 25 H 4 37Cl 3
                       +14.1 /  +5.8    23.0  C 24 H 2 N 37Cl 3
                        -3.9 /  -1.6    10.0  C 16 H 14 N 35Cl 3 Zr
                       -15.3 /  -6.3    15.0  C 19 H 11 N 35Cl 37Cl Zr
                       +15.0 /  +6.2    15.5  C 18 H 9 N 2 35Cl 37Cl Zr
                       +10.6 /  +4.4    10.5  C 17 H 14 35Cl 2 37Cl Zr
                        -0.8 /  -0.3    15.5  C 20 H 11 37Cl 2 Zr
```

Fig. 3.23. Listing of possible elemental compositions of the zirconium complex shown in the preceding figure. The error for each proposal is listed in units of ppm and mmu. The column U.S. lists "unsaturation", i.e., the number of rings and/or double bonds (Chap. 6.4.4). By courtesy of M. Enders, University of Heidelberg.

dances. If the peak at m/z 414.9223 represents the monoisotopic ion, only the ele-mental composition containing ^{90}Zr and ^{35}Cl can be correct. Therefore, the for-mula $C_{16}H_{14}NCl_3Zr$ can be undoubtedly identified from the composition list (Fig. 3.23). The X+2 and X+4 compositions should mainly be due to $^{35}Cl_2{}^{37}Cl$ and

$^{35}Cl_{37}Cl_2$, respectively, and thus these can be identified in the next step. The numbers of all other elements must remain the same, i.e., here $C_{16}H_{14}N$ must be part of any formula. In this example, $R = 8000$ is the minimum to separate the PFK reference peak at m/z 417 from that of the analyte. If separation could not have been achieved, the mass assignment would have been wrong because the m/z 417 peak would then be centered at a weighted mass average of its two contributors. Alternatively, such a peak may be omitted from both reference list and composition list.

> **Note:** The assignment of empirical formulae from accurate mass measurements always must be in accordance with the experimentally observed and the calculated isotopic pattern. Contradictions strongly point towards erroneous interpretation of the mass spectrum.

3.3.5.3 Compiling Mass Reference Lists

Once the mass spectrum of a calibration standard is known and the elemental composition of the ions that are to be included in the mass reference list are established by an independent measurement, a reference list may be compiled. For this purpose, the listed reference masses should be calculated using six decimal places. Otherwise, there is a risk of erroneous reference values, especially when masses of ion series are calculated by multiplication of a subunit. Such a tasks are well suited for spreadsheet applications on a personal computer.

Example: Cesium iodide is frequently used for mass calibration in fast atom bombardment (FAB) mass spectrometry (Chap. 9) because it yields cluster ions of the general formula $[Cs(CsI)_n]^+$ in positive-ion and $[I(CsI)_n]^-$ in negative-ion mode. For the $[Cs(CsI)_{10}]^+$ cluster ion, m/z 2730.9 is calculated instead of the correct value m/z 2731.00405 by using only one decimal place instead of the exact values $M_{133Cs} = 132.905447$ and $M_{127I} = 126.904468$. The error of 0.104 u is acceptable for LR work, but definitely not acceptable if accurate mass measurements have to be performed.

3.3.5.4 Role of the Electron Mass

The question remains whether the mass of the electron m_e (0.548 mmu) has really to been taken into account in accurate mass work as demanded by the IUPAC convention (Chap. 3.1.4). This issue was almost only of academic interest as long as mass spectrometry never yielded mass accuracies better than several mmu. Nowadays, extremely accurate FT-ICR instruments become more widespread in use, and thus, the answer depends on the intended application: The electron mass has to be included in calculations if the result is expected to report the accurate mass of the ion with highest accuracy. Here, neglecting the electron mass would cause an systematic error of the size of m_e. This cannot be tolerated when mass measurement accuracies in the order of 1 mmu or better are to be achieved.

3.4 Interaction of Resolution and Isotopic Patterns

3.4.1 Multiple Isotopic Compositions at Very High Resolution

When starting our consideration of isotopic patterns, it was done in terms of nominal mass, which makes it much easier to understand how isotopic patterns are formed. The procedure of summing up all isotopic abundances contributing to the same nominal mass is correct as long as very high resolution is not employed. In that case, following such a simplified protocol is acceptable, because generally isobaric ions resulting from different isotopic compositions of the same elemental composition are very similar in mass. [16]

During the introduction of resolution it should have become clear that very high resolution is capable of separating different isotopic compositions of the same nominal mass, thereby giving rise to multiple peaks on the same nominal m/z (Figs. 3.18, 3.19, 3.22). In molecules with masses of around 10,000 u, a single unresolved isotopic peak may consist of as many as 20 different isotopic compositions. [38]

m/z	composition	rel.int. [%]	
256.12836	C16.H20.O.Si	100.00000	monoisotopic
257.12790	C16.H20.O.29Si	5.10957	1st isotopic
257.13171	C15.13C.H20.O.Si	17.92663	
257.13257	C16.H20.17O.Si	0.03709	
258.12518	C16.H20.O.30Si	3.38468	2nd isotopic
258.13126	C15.13C.H20.O.29Si	0.91597	
258.13214	C16.H20.17O.29Si	0.00190	
258.13260	C16.H20.18O.Si	0.20449	
258.13507	C14.13C2.H20.O.Si	1.50639	
258.13593	C15.13C.H20.17O.Si	0.00665	
259.12854	C15.13C.H20.O.30Si	0.60676	3rd isotopic
259.12939	C16.H20.17O.30Si	0.00126	
259.13214	C16.H20.18O.29Si	0.01045	
259.13461	C14.13C2.H20.O.29Si	0.07697	
259.13550	C15.13C.H20.17O.29Si	0.00034	
259.13596	C15.13C.H20.18O.Si	0.03666	
259.13843	C13.13C3.H20.O.Si	0.07876	
259.13928	C14.13C2.H20.17O.Si	0.00056	
260.12943	C16.H20.18O.30Si	0.00692	4th isotopic
260.13190	C14.13C2.H20.O.30Si	0.05099	
260.13275	C15.13C.H20.17O.30Si	0.00023	
260.13550	C15.13C.H20.18O.29Si	0.00187	
260.13797	C13.13C3.H20.O.29Si	0.00402	
260.13885	C14.13C2.H20.17O.29Si	0.00003	
260.13931	C14.13C2.H20.18O.Si	0.00308	
260.14175	C12.13C4.H20.O.Si	0.00287	
260.14264	C13.13C3.H20.17O.Si	0.00003	

Fig. 3.24. Listing of the theoretical isotopic distribution of $C_{16}H_{20}OSi$ at infinite resolution. The contribution of ^2H is not considered, and isotopic peaks above m/z 260 are omitted due to their minor intensities.

Example: The isotopic pattern of $C_{16}H_{20}OSi$ is made up of a series of compositions among which the monoisotopic ion at m/z 256.1283 is the most abundant (Fig. 3.24). For the first isotopic peak at m/z 257 the major contribution derives from ^{13}C, but ^{29}Si and ^{17}O also play a role. Actually, ^{2}H should also be considered, but has been omitted due to its extremely low isotopic abundance. As none of the isobars has exactly the same m/z as any other, their intensities will not add up in one common peak. Instead, they are detected side by side provided sufficient resolving power is available. The pair $^{13}C^{12}C_{15}H_{20}{}^{16}OSi$, m/z 257.13171, and $^{12}C_{16}H_{20}{}^{17}OSi$, m/z 257.13257, roughly necessitates 3×10^{5} resolution ($R = 257/0.00086 = 299{,}000$). The second isotopic peak at m/z 258 consists of six and the third isotopic peak at m/z 259 even of eight different compositions. Again, it is the merit of FT-ICR-MS that resolutions in the order of several 10^{5} are now available. However, such *ultrahigh-resolution*, as this is termed, is still not a routine application of those instruments (Chap. 4).

Example: Peptides often contain sulfur from cysteine. Provided there are at least two cysteines in the peptide molecule, the sulfur can be incorporated as thiol group (SH, reduced) or sulfur bridge (S–S, oxidized). Often, both forms are contained in the same sample. At ultrahigh-resolution, the contributions of these compositions to the same nominal m/z can be distinguished. The ultrahigh-resolution matrix-assisted laser desorption/ionization (MALDI) FT-ICR mass spectrum of native and reduced [D-Pen^{2,5}]enkephalin gives an example of such a separation (Fig. 3.25). [39] The left expanded view shows fully resolved peaks due to ^{34}S and $^{13}C_{2}$ isotopomers of the native and the all-^{12}C peak of the reduced compound at m/z 648. The right expansion reveals the $^{13}C_{1}{}^{34}S$ peak of the native plus the $^{13}C_{1}$ signal of the reduced form at m/z 649. Here, R_{FWHM} is more than $9 \cdot 10^{5}$.

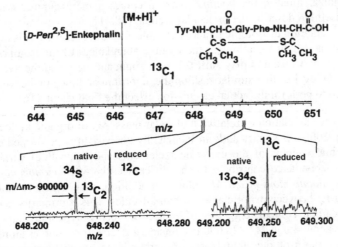

Fig. 3.25. Ultrahigh-resolution MALDI-FT-ICR mass spectrum of native (S–S) and reduced (2·SH) [D-Pen^{2,5}]enkephalin. The expanded m/z views of the second and third isotopic peak show fully mass-resolved signals. Reproduced from Ref. [39] with permission. © American Chemical Society, 1997.

3.4.2 Multiple Isotopic Compositions and Accurate Mass

It has been pointed out that routine accurate mass measurements are conducted at resolutions which are too low to separate isobaric isotopic compositions in most cases. Unfortunately, coverage of multiple isotopic compositions under the same signal distort the peak shape. This effect causes a loss of mass accuracy when elemental compositions have to be determined from such multicomponent peaks, e.g., if the monoisotopic peak is too weak as the case with many transition metals. The observed decrease in mass accuracy is not dramatic and the loss of mass accuracy is counterbalanced by the information derived from the isotopic pattern. However, it can be observed that mass accuracy decreases, e.g., from 2–3 mmu on monoisotopic peaks to about 4–7 mmu on multicomponent signals.

3.4.3 Isotopic Patterns of Large Molecules

3.4.3.1 Isotopic Patterns of Large Molecules at Sufficient Resolution

Terms such as *large molecules* or *high mass* are subject to steady change in mass spectrometry as new techniques for analyzing high-mass ions are being developed or improved. [40] Here, the focus is on masses in the range of 10^3–10^4 u.

With increasing m/z the center of the isotopic pattern, i.e., the *average molecular mass*, is shifted to values higher than the *monoisotopic mass*. The center, i.e., the average mass, may not be represented by a real peak, but it tends to be close to the peak of *most abundant* mass (Fig. 3.26). The monoisotopic mass is of course still related to a real signal, but it may be of such a low intensity that it is difficult to recognize. Finally, the *nominal mass* becomes a mere number which is no longer useful to describe the molecular weight. [38,41]

The calculation of isotopic patterns of molecules of several 10^3 u is not a trivial task, because slight variations in the relative abundances of the isotopes encountered gain relevance and may shift the most abundant mass and the average mass up or down by 1 u. In a similar fashion the algorithm and the number of iterations employed to perform the actual calculation affect the final result. [16]

Note: The calculation of relative molecular mass, M_r, of organic molecules exceeding 2000 u is significantly influenced by the basis it is performed on. Both the atomic weights of the constituent elements and the natural variations in isotopic abundance contribute to the differences between *monoisotopic-* and *relative atomic mass-based* M_r values. In addition, they tend to characteristically differ between major classes of biomolecules. This is primarily because of molar carbon content, e.g., the difference between polypeptides and nucleic acids is about 4 u at $M_r = 25,000$ u. Considering terrestrial sources alone, variations in the isotopic abundance of carbon lead to differences of about 10–25 ppm in M_r which is significant with respect to mass measurement accuracy in the region up to several 10^3 u. [41]

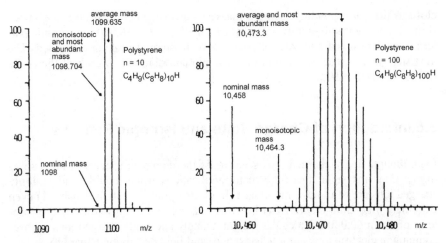

Fig. 3.26. Calculated isotopic patterns of large polystyrene ions. Adapted from Ref. [38] with permission. © American Chemical Society, 1983.

3.4.3.2 Isotopic Patterns of Large Molecules at Low Resolution

It is certainly desirable to have at least sufficient resolution to resolve isotopic patterns to their nominal mass contributions. However, not every mass analyzer is capable of doing so with any ion it can pass through. Such conditions often occur when ions of several thousand u are being analyzed by quadrupole, time-of-flight or quadrupole ion trap analyzers, and hence it is useful to know about the changes in spectral appearance and their effect on peak width and detected mass. [42]

Example: The isotopic pattern of the $[M+H]^+$ ion, $[C_{254}H_{378}N_{65}O_{75}S_6]^+$, of bovine insulin has been calculated for $R_{10\%} = 1000$, 4000 and 10,000. At $R = 1000$ the isotopic peaks are not resolved (Fig. 3.27). An envelope smoothly covering the isotopic peaks is observed instead. It is even slightly wider than the real isotopic pattern. The maximum of this envelope is in good agreement with the calculated average mass, i.e., the molecular weight (Eq. 3.1). At $R = 4000$ the isotopic peaks become sufficiently resolved to be recognized as such. The m/z values are very

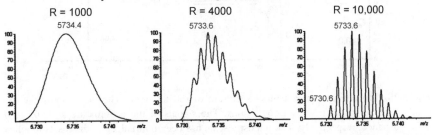

Fig. 3.27. The isotopic pattern calculated for $[M+H]^+$ of bovine insulin at different resolutions ($R_{10\%}$). Note that the envelope at $R = 1000$ is wider than the real isotopic pattern.

close to the corresponding isotopic masses; however, there can be some minor shifts due to their still significant overlap. Finally, at $R = 10,000$ the isotopic pattern is well resolved and interferences between isotopic peaks are avoided. The next step, i.e., to resolve the multiple isotopic contributions to each of the peaks would require $R > 10^6$.

3.5 Interaction of Charge State and Isotopic Patterns

Even though singly charged ions seem to be the dominant species in mass spectrometry at first sight, there are many applications where doubly and multiply charged ions are of utmost importance. In electrospray ionization (Chap. 11) even extremely high *charge states* can be observed, e.g., up to 60-fold in case of proteins of about 60,000 u molecular weight. Doubly and triply charged ions are also common in electron ionization (Chaps. 5, 6) and field desorption (Chap. 8).

The effect of higher charges state are worth a closer consideration. With z increasing from 1 to 2, 3 etc., the numerical value of m/z is reduced by a factor of 2, 3 etc., i.e., the ion will be detected at lower m/z than the corresponding singly charged ion of the same mass. In general, the entire m/z scale is compressed by a factor of z if $z > 1$. Consequently, the isotopic peaks are then located at $^1/_z$ u distance. Vice versa, the charge state is obtained from the reciprocal value of the distance between adjacent peaks, e.g., peaks at $^1/_3$ u correspond to triply charged ions. The reasons for the compression of the m/z scale are discussed later (Chap. 4).

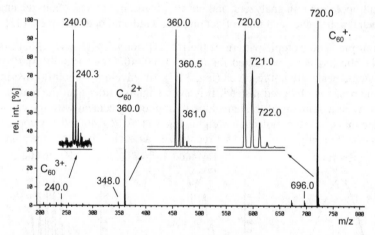

Fig. 3.28. The electron impact mass spectrum of [60]fullerene. The insets show the expanded signals of $M^{+\bullet}$, M^{2+} and $M^{3+\bullet}$ ions. The signals of the patterns are at 1, 0.5 and 0.33 u distance, respectively. The intensity scale has been normalized in the insets to allow for easier comparison of the isotopic patterns. By courtesy of W. Krätschmer, Max Planck Institute for Nuclear Physics, Heidelberg.

Example: The EI mass spectrum of C_{60} also shows an abundant doubly charged molecular ion, C_{60}^{2+}, at m/z 360 with its isotopic peaks located at 0.5 u distance and a $C_{60}^{3+\bullet}$ signal at m/z 240 of very low intensity (Fig. 3.28). [22] The isotopic pattern remains the same for all of them. As a consequence of the compressed m/z scale, the doubly charged C_{58}^{2+} fragment ion is detected at m/z 348. More examples of multiply charged ions can be found throughout the book.

> **Note:** Isotopic distributions remain unaffected by the charge state of an ion as far as the relative intensities are concerned. However, the distance between the peaks is reduced to $^1/_z$ u, thus allowing multiply charged ions to be easily distinguished from singly charged ions and their charge state to be determined.

Reference List

1. Todd, J.F.J. Recommendations for Nomenclature and Symbolism for Mass Spectroscopy Including an Appendix of Terms Used in Vacuum Technology. *Int. J. Mass Spectrom. Ion Proc.* **1995**, *142*, 211-240.

2. Yergey, J.A. A General Approach to Calculating Isotopic Distributions for Mass Spectrometry. *Int. J. Mass Spectrom. Ion Phys.* **1983**, *52*, 337-349.

3. Sparkman, O.D. *Mass Spec Desk Reference;* Global View Publishing: Pittsburgh, 2000.

4. IUPAC Isotopic Composition of the Elements 1997. *Pure Appl. Chem.* **1998**, *70*, 217-235.

5. IUPAC; Coplen, T.P. Atomic Weights of the Elements 1999. *Pure Appl. Chem.* **2001**, *73*, 667-683.

6. Price, P. Standard Definitions of Terms Relating to Mass Spectrometry. A Report From the Committee on Measurements and Standards of the Amercian Society for Mass Spectrometry. *J. Am. Soc. Mass Spectrom.* **1991**, *2*, 336-348.

7. Busch, K.L. Units in Mass Spectrometry. *Spectroscopy* **2001**, *16*, 28-31.

8. Platzner, I.T. Applications of Isotope Ratio Mass Spectrometry, in *Modern Isotope Ratio Mass Spectrometry*, John Wiley & Sons: Chichester, 1997; pp. 403-447.

9. Beavis, R.C. Chemical Mass of Carbon in Proteins. *Anal. Chem.* **1993**, *65*, 496-497.

10. Carle, R. Isotopen-Massenspektrometrie: Grundlagen und Anwendungsmöglichkeiten. *Chem. Unserer Zeit* **1991**, *20*, 75-82.

11. Busch, K.L. Isotopes and Mass Spectrometry. *Spectroscopy* **1997**, *12*, 22-26.

12. Beynon, J.H. The Compilation of a Table of Mass and Abundance Values, in *Mass spectrometry and its applications to organic chemistry*, 1st ed.; Elsevier: Amsterdam, 1960; Chapter 8.3, pp. 294-302.

13. McLafferty, F.W.; Turecek, F. *Interpretation of Mass Spectra;* 4th ed.; University Science Books: Mill Valley, 1993.

14. Margrave, J.L.; Polansky, R.B. Relative Abundance Calculations for Isotopic Molecular Species. *J. Chem. Educ.* **1962**, 335-337.

15. Hsu, C.S. Diophantine Approach to Isotopic Abundance Calculations. *Anal. Chem.* **1984**, *56*, 1356-1361.

16. Kubinyi, H. Calculation of Isotope Distributions in Mass Spectrometry. A Trivial Solution for a Non-Trivial Problem. *Anal. Chim. Acta* **1991**, *247*, 107-119.

17. Frauenkron, M.; Berkessel, A.; Gross, J.H. Analysis of Ruthenium Carbonyl-Porphyrin Complexes: a Comparison of Matrix-Assisted Laser Desorption/Ionization Time-of-Flight, Fast-Atom Bombardment and Field Desorption Mass Spectrometry. *Eur. Mass Spectrom.* **1997**, *3*, 427-438.

18. Brauman, J.I. Least Squares Analysis and Simplification of Multi-Isotope Mass Spectra. *Anal. Chem.* **1966**, *38*, 607-610.

19. Giesa, S.; Gross, J.H.; Hull, W.E.; Lebedkin, S.; Gromov, A.; Krätschmer, W.; Gleiter, R. $C_{120}OS$: the First Sulfur-Containing Dimeric [60]Fullerene Derivative. *Chem. Commun.* **1999**, 465-466.

20. Luffer, D.R.; Schram, K.H. Electron Ionization Mass Spectrometry of Synthetic C_{60}. *Rapid Commun. Mass Spectrom.* **1990**, *4*, 552-556.

21. Srivastava, S.K.; Saunders, W. Ionization of C_{60} (Buckminsterfullerene) by Electron Impact. *Rapid Commun. Mass Spectrom.* **1993**, *7*, 610-613.

22. Scheier, P.; Dünser, B.; Märk, T.D. Production and Stability of Multiply-Charged C_{60}. *Electrochem. Soc. Proc.* **1995**, *95-10*, 1378-1394.

23. Thomas, A.F. *Deuterium Labeling in Organic Chemistry;* Appleton-Century-Crofts: New York, 1971.

24. Busch, K.L. The Resurgence of Exact Mass Measurement with FTMS. *Spectroscopy* **2000**, *15*, 22-27.

25. Beynon, J.H. Qualitative Analysis of Organic Compounds by Mass Spectrometry. *Nature* **1954**, *174*, 735-737.

26. Lehmann, W.D.; Bohne, A.; von der Lieth, C.W. The Information Encrypted in Accurate Peptide Masses-Improved Protein Identification and Assistance in Glycopeptide Identification and Characterization. *J. Mass Spectrom.* **2000**, *35*, 1335-1341.

27. Zubarev, R.A.; Håkansson, P.; Sundqvist, B. Accuracy Requirements for Peptide Characterization by Monoisotopic Molecular Mass Measurements. *Anal. Chem.* **1996**, *68*, 4060-4063.

28. Kendrick, E. Mass Scale Based on CH_2 = 14.0000 for High-Resolution Mass Spectrometry of Organic Compounds. *Anal. Chem.* **1963**, *35*, 2146-2154.

29. Hughey, C.A.; Hendrickson, C.L.; Rodgers, R.P.; Marshall, A.G. Kendrick Mass Defect Spectrum: A Compact Visual Analysis for Ultrahigh-Resolution Broadband Mass Spectra. *Anal. Chem.* **2001**, *73*, 4676-4681.

30. Hughey, C.A.; Hendrickson, C.L.; Rodgers, R.P.; Marshall, A.G. Elemental Composition Analysis of Processed and Unprocessed Diesel Fuel by Electrospray Ionization Fourier Transform Ion Cyclotron Resonance Mass Spectrometry. *Energy Fuels* **2001**, *15*, 1186-1193.

31. Roussis, S.G.; Proulx, R. Reduction of Chemical Formulas from the Isotopic Peak Distributions of High-Resolution Mass Spectra. *Anal. Chem.* **2003**, *75*, 1470-1482.

32. Zubarev, R.A.; Demirev, P.A.; Håkansson, P.; Sundqvist, B.U.R. Approaches and Limits for Accurate Mass Characterization of Large Biomolecules. *Anal. Chem.* **1995**, *67*, 3793-3798.

33. Clauser, K.R.; Baker, P.; Burlingame, A. Role of Accurate Mass Measurement (±10 Ppm) in Protein Identification Strategies Employing MS or MS/MS and Database Searching. *Anal. Chem.* **1999**, *71*, 2871-2882.

34. Gross, M.L. Accurate Masses for Structure Confirmation. *J. Am. Soc. Mass Spectrom.* **1994**, *5*, Editorial.

35. Sack, T.M.; Lapp, R.L.; Gross, M.L.; Kimble, B.J. A Method for the Statistical Evaluation of Accurate Mass Measurement Quality. *Int. J. Mass Spectrom. Ion Proc.* **1984**, *61*, 191-213.

36. Kilburn, K.D.; Lewis, P.H.; Underwood, J.G.; Evans, S.; Holmes, J.; Dean, M. Quality of Mass and Intensity Measurements From a High Performance Mass Spectrometer. *Anal. Chem.* **1979**, *51*, 1420-1425.

37. Bristow, A.W.T.; Webb, K.S. Intercomparison Study on Accurate Mass Measurement of Small Molecules in Mass Spectrometry. *J. Am. Soc. Mass Spectrom.* **2003**, *14*, 1086-1098.

38. Yergey, J.; Heller, D.; Hansen, G.; Cotter, R.J.; Fenselau, C. Isotopic Distributions in Mass Spectra of Large Molecules. *Anal. Chem.* **1983**, *55*, 353-356.

39. Solouki, T.; Emmet, M.R.; Guan, S.; Marshall, A.G. Detection, Number, and Sequence Location of Sulfur-Containing Amino Acids and Disulfide Bridges in Peptides by Ultrahigh-Resolution MALDI-FTICR Mass Spectrometry. *Anal. Chem.* **1997**, *69*, 1163-1168.

40. Matsuo, T.; Sakurai, T.; Ito, H.; Wada, Y. "High Masses". *Int. J. Mass Spectrom. Ion Proc.* **1991**, *118/119*, 635-659.

41. Pomerantz, S.C.; McCloskey, J.A. Fractional Mass Values of Large Molecules. *Org. Mass Spectrom.* **1987**, *22*, 251-253.

42. Werlen, R.C. Effect of Resolution on the Shape of Mass Spectra of Proteins: Some Theoretical Considerations. *Rapid Commun. Mass Spectrom.* **1994**, *8*, 976-980.

4 Instrumentation

"A modern mass spectrometer is constructed from elements which approach the state-of-the-art in solid-state electronics, vacuum systems, magnet design, precision machining, and computerized data acquisition and processing." [1] This is and has ever been a fully valid statement about mass spectrometers.

Under the headline of instrumentation we shall mainly discuss the different types of mass analyzers in order to understand their basic principles of operation, their specific properties and their performance characteristics. Of course, this is only one aspect of instrumentation; hence topics such as ion detection and vacuum generation will be addressed in brief. As a matter of fact, sample introduction is more closely related to particular ionization methods than to the type of mass analyzer employed, and therefore, this issue is treated in the corresponding chapters on ionization methods. The order of appearance of the mass analyzers in this chapter neither reflects the ever-changing percentage they are employed in mass spectrometry nor does it strictly follow a time line of their invention. Instead, it is attempted to follow a trail of easiest understanding.

From its very beginnings to the present almost any physical principle ranging from time-of-flight to cyclotron motion has been employed to construct mass-analyzing devices (Fig 4.1). Some of them became extremely successful at the time they were invented, for others it took decades until their potential had fully been recognized. The basic types of mass analyzers employed for analytical mass spectrometry are summarized below (Tab. 4.1).

Table 4.1. Common mass analyzers

Type	Acronym	Principle
Time-of-flight	TOF	Time dispersion of a pulsed ion beam; separation by time-of-flight
Magnetic sector	B	Deflection of a continuous ion beam; separation by momentum in magnetic field due to Lorentz force
Linear quadrupole	Q	Continuous ion beam in linear radio frequency quadrupole field; separation due to stability of trajectories
Linear quadrupole ion trap	LIT	Continuous ion beam and trapped ions; storage and eventually separation in linear radio frequency quadrupole field due to stability of trajectories
Quadrupole ion trap	QIT	Trapped ions; separation in three-dimensional radio frequency quadrupole field due to stability of trajectories
Ion cyclotron resonance	ICR	Trapped ions; separation by cyclotron frequency (Lorentz force) in magnetic field

Fig. 4.1. Mass spectrometer islands. A cartoon by C. Brunnée. Reproduced from Ref. [2] with permission. © Elsevier Science, 1987.

The properties of an ideal mass analyzer are well described, [2] but despite the tremendous improvements made, still no mass analyzer is perfect. To reach a deeper insight into the evolution of mass spectrometers the articles by Beynon, [3] Habfast and Aulinger, [4,5] Brunnée [6,7], Chapman et al. [8] and McLuckey [9] are recommended for further reading. In recent years, miniature mass analyzers have gained interest for *in situ* analysis, [10] e.g., in environmental [11] or biochemical applications, [12] for process monitoring, for detection of chemical warfare agents, for extraterrestrial applications, [13] and to improve Space Shuttle safety prior to launch. [14]

4.1 Creating a Beam of Ions

Consider an ion that is brought into or generated within an electric field as realized between two oppositely charged plates of a capacitor. Such an ion will be accelerated towards the plate of opposite charge sign. If the attracting plate has a hole or a slit, a ion beam is produced by this simple ion source. Assuming the spread in ion kinetic energy to be small as compared to the total ion kinetic energy, i.e., $\Delta E_{kin} \ll E_{kin}$, the beam can be regarded monoenergetic. The actual charge of an ion may be either positive or negative depending on the ionization method employed. Changing the polarity of the plates switches from the extraction of positive ions to negative ions or vice versa. For practical reasons, the attracting electrode is usually grounded and the pushing plate is set to high voltage. Doing so allows to keep the whole mass analyzer grounded, thereby contributing substantially to safety of operation (Fig. 4.2). The extraction of ions and the shape of the ion beam can largely be improved if the acceleration voltage is applied in two or more successive steps instead of a single step.

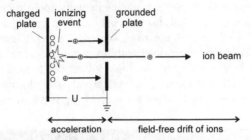

Fig. 4.2. Schematic of a simple ion source. The ionizing event is preferably located close to the charged plate. After a neutral has been ionized (positive in this illustration), it is attracted by the opposite plate. Those ions passing through a hole of the grounded electrode create an ion beam emerging into the field-free region behind. The ion beam produced by such a primitive ion source is not parallel, but has some angular spread.

4.2 Time-of-Flight Instruments

4.2.1 Introduction to Time-of-Flight

The construction of a *time-of-flight* (TOF) analyzer has been published in 1946 by Stephens. [15] The principle of TOF is quite simple: ions of different *m/z* are dispersed in time during their flight along a field-free drift path of known length. Provided all the ions start their journey at the same time or at least within a sufficiently short time interval, the lighter ones will arrive earlier at the detector than the heavier ones. This demands that they emerge from a pulsed ion source which can be realized either by pulsing ion packages out of a continuous beam or more conveniently by employing a true pulsed ionization method.

Soon, other groups shared Stephens' concept [16] and increasingly useful TOF instruments were constructed [17-19] leading to their first commercialization by Bendix in the mid 1950s. These first generation TOF instruments were designed for gas chromatography-mass spectrometry (GC-MS) coupling. [20,21] Their performance was poor as compared to modern TOF analyzers, but the specific advantage of TOF over the competing magnetic sector instruments was the rate of spectra per second they were able to deliver (Fig. 4.3). In GC-MS the TOF analyzer soon became superseded by linear quadrupole analyzers and it took until the late 1980s for their revival [22,23] – the success of pulsed ionization methods, especially of matrix-assisted laser desorption/ionization (MALDI), made it possible (Chap. 10).

MALDI generated a great demand for a mass analyzer ideally suited to be used in conjunction with a pulsed ion source and capable of transmitting ions of extremely high mass up to several 10^5 u. [24] Since then, the performance of TOF instruments has tremendously increased. [25,26] TOF analyzers were adapted for use with other ionization methods and are now even strong competitors to the well-established magnetic sector instruments in many applications. [25,27]

The main advantages of TOF instruments are: i) in principle, the m/z range is unlimited; [28,29] ii) from each ionizing event, e.g., a single laser shot in MALDI, a complete mass spectrum is obtained within several tens of microseconds; iii) the transmission of a TOF analyzer is very high, giving rise to high sensitivity; iv) the construction of a TOF instrument is comparatively straightforward and inexpensive; and v) recent instruments allow for accurate mass measurements and tandem MS experiments.

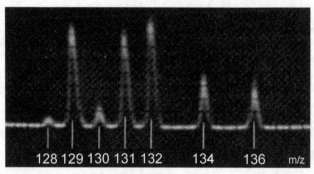

Fig. 4.3. Photograph of the oscillographic output of the electron ionization TOF spectrum of xenon on a Bendix TOF-MS. The dark lines are a grid on the oscillographic screen. (For the isotopic pattern of Xe cf. Fig. 3.1.) Adapted from Ref. [20] with permission. © Pergamon Press, 1959.

4.2.2 Basic Principle of TOF Instruments

4.2.2.1 Velocity of Ions

Independent of the ionization method, the electric charge q of an ion of mass m_i is equal to an integer number z of electron charges e, and thus $q = ez$. The energy uptake E_{el} by moving through a voltage U is given by

$$E_{el} = qU = ezU \qquad (4.1)$$

Thereby, the former potential energy of a charged particle in an electric field is converted into kinetic energy E_{kin}, i.e., into translational motion

$$E_{el} = ezU = \frac{1}{2}m_i v^2 = E_{kin} \qquad (4.2)$$

Assuming the ion was at rest before, which is correct in a first approximation, the velocity attained is obtained by rearranging Eq. 4.2 into

$$v = \sqrt{\frac{2ezU}{m_i}} \qquad (4.3)$$

i.e., v is inversely proportional to the square root of mass.

Example: From Eq. 4.3 we obtain the velocity of the [60]fullerene molecular ion, $C_{60}^{+\bullet}$, after acceleration by 19.5 kV as

$$v = \sqrt{\frac{2 \cdot 1.6022 \times 10^{-19}\,C \times 19,500V}{1.1956 \times 10^{-24}\,kg}} = 72,294\,m\,s^{-1}$$

An ion velocity of 72,294 m s^{-1} appears comparatively high, although it is only 0.0024 % of the speed of light. The acceleration voltages and thus, the ion velocities are the highest in TOF-MS, although magnetic sector instruments are operated with kiloelectron volt ion beams, too (Chap. 4.3). The other types of mass analyzers require ions to be entering at much lower kinetic energies.

Note: Equation 4.3 describes the velocity of any ion after acceleration in an electric field, and therefore it is valid not only for TOF-MS, but for any part of a mass spectrometer handling beams of ions.

4.2.2.2 Time-of-Flight

It does not take a great leap to the idea of measuring the time for an ion of unknown m/z to travel a distance s after having been accelerated by a voltage U. The relationship between velocity and time t needed to travel the distance s is

$$t = \frac{s}{v} \qquad (4.4)$$

which upon substitution of v with Eq. 4.3 becomes

$$t = \frac{s}{\sqrt{\frac{2ezU}{m_i}}} \qquad (4.5)$$

Equation 4.5 delivers the time needed for the ion to travel the distance s at constant velocity, i.e., in a *field-free* environment after the process of acceleration has been completed. Rearrangement of Eq. 4.5 reveals the relationship between the instrumental parameters s and U, the experimental value of t and the ratio m_i/z

$$\frac{m_i}{z} = \frac{2\,e\,U\,t^2}{s^2} \qquad (4.6)$$

It is also obvious from Eq. 4.5 that the time to drift through a fixed length of field-free space is proportional to the square root of m_i/z

$$t = \frac{s}{\sqrt{2eU}} \sqrt{\frac{m_i}{z}} \qquad (4.7)$$

and thus, the time interval Δt between the arrival times of ions of different m/z is proportional to $s \cdot (m_i/z_1^{1/2} - m_i/z_2^{1/2})$.

> **Note:** Here, the ratio m_i/z denotes ion mass [kg] per number of electron charges. The index i at the mass symbol is used to avoid confusion with the *mass-to-charge ratio*, m/z, as used to specify the position of a peak on the abscissa of a mass spectrum. There, m/z is mass number devided by the number of electron charges due to the current convention (Chap. 1.2.3). [30,31]

Example: According to Eq. 4.7 the $C_{60}^{+\bullet}$ ion, m/z 720, of the preceding example will travel through a field-free flight path of 2.0 m in 27.665 µs. Its isotopomer $^{13}C^{12}C_{59}^{+\bullet}$, m/z 721, needs slightly longer. The proportionality to the square root of m/z gives (use of m_i/z values yields identical results as the dimension is cancelled)

$$\frac{t_{721}}{t_{720}} = \frac{\sqrt{721}}{\sqrt{720}} = 1.000694$$

Thus, $t_{721} = 27.684$ µs. This corresponds to a difference in time-of-flight of 19 ns under these conditions (cf. Fig. 4.7).

The proportionality of time-of-flight to the square root of m/z causes Δt for a given $\Delta m/z$ to decrease with increasing m/z: under otherwise the same conditions Δt per 1 u is calculated as 114 ns at m/z 20, 36 ns at m/z 200, and just 11 ns at m/z 2000. Therefore, the realization of a time-of-flight mass analyzer depends on the ability to measure short time intervals with sufficient accuracy. [32-34] At this point it becomes clear that the performance of the early TOF analyzers – among other reasons – suffered from the too slow electronics of their time. It took until the mid-1990s to overcome this barrier.

4.2.2.3 Time-of-Flight of Multiply Charged Ions

The effect of a charge state $z > 1$ on time-of-flight allows to explain the consequences of higher charge states for isotopic patterns (Chap. 3.5). As z increases to 2, 3 etc., the numerical value of m/z is reduced by a factor of 2, 3 etc., i.e., the ion will be detected at lower m/z than the corresponding singly-charged ion of the same mass. According to Eq. 4.7, the time-of-flight is reduced by a factor of 1.414 (the square root of 2) for doubly-charged ions which is the same time-of-flight as for a singly-charged ion of half of its mass. Accordingly, the time-of-flight is reduced by a factor of 1.732 (the square root of 3), for triply-charged ions corresponding to a singly-charged ion of one third of its mass.

> **Note:** In case of multiply-charged ions, the m/z scale is compressed by a factor of z equal to the charge state of the ion. Isotopic distributions remain unaffected as far as the relative intensities are concerned. As the distance between isotopic peaks is reduced to $^1/_z$ u, the charge state can readily be assigned.

4.2.3 Linear Time-of-Flight Analyzer

4.2.3.1 Setup of the Linear TOF

Restricting its use to laser desorption/ionization (LDI) and matrix-assisted laser desorption/ionization (MALDI, Chap. 10), a simple TOF instrument can be set up as follows (Fig. 4.4). The analyte is supplied as a thin layer on a *sample holder* or *target* upon which a pulsed laser is focused. The acceleration voltage U is applied between this target and a grounded counter electrode. Ions formed and desorbed during the laser pulse are continuously extracted and accelerated as they emerge from the target into the gas phase. When leaving the acceleration region s_0 the ions should possess equal kinetic energies. They drift down a field-free flight path s in the order of 1–2 m and finally hit the detector. Typically, *microchannel plate* (MCP) detectors are employed to compensate for the angular spread of the ion beam (Chap. 4.8.3). A fast *time-to-analog converter* (8 bit ADC) transforms the analog output of the detector for computer-based data storage and data processing. Such an instrumental setup where the ions are travelling on a straight line from the point of their creation to the detector is called *linear TOF*.

In principle, any other ionization method can be combined with a TOF analyzer even if it is not an intrinsically pulsed technique, provided there are means to extract ions in a pulsed manner from such an ion source (Chap. 4.2.6).

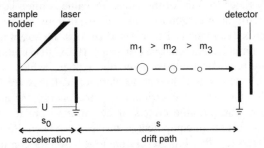

Fig. 4.4. Scheme of a linear TOF instrument. After being created during a laser pulse onto the sample layer, ions are continuously accelerated by a voltage U. While drifting along the field-free drift path s, they are dispersed in time. Lighter ions reach the detector first.

The drift time t_d as calculated by means of Eq. 4.7 is not fully identical to the total time-of-flight. Obviously, the time needed for acceleration of the ions t_a has to be added. Furthermore, a short period of time t_0 may be attributed to the laser pulse width and the process of desorption/ionization, which is typically in the order of a few nanoseconds. Thus, the total time-of-flight t_{total} is given by

$$t_{total} = t_0 + t_a + t_d \qquad (4.8)$$

An exact mathematical treatment of the TOF analyzers as needed for the construction of such instruments of course has to include all contributions to the total time-of-flight. [32,33]

4.2.3.2 Properties of the Linear TOF

The transmittance of a linear TOF analyzer approaches 90 % because ion losses are solely caused by collisional scattering due to residual gas or by poor spatial focusing of the ion source. With a sufficiently large detector surface located in not a too large distance from the ion source exit, a very high fraction of the ions will be collected by the detector.

Even metastable decompositions during the flight do not reduce the intensity of a molecular ion signal, because the fragments formed conserve the velocity of the fragmenting ion and are therefore detected at the same time-of-flight as their intact precursors. In such a case, the ionic *and* the neutral fragment cause a response at the detector. These properties make the linear TOF analyzer the ideal device for analysis of easily fragmenting and/or high mass analytes.

> **Note:** The detection of neutrals formed by metastable fragmentation in linear TOF analyzers is one of the few exceptions where neutrals are handled and give rise to a signal in mass spectrometry.

The example in Chap. 4.2.2 demonstrates that the differences in time-of-flight of ions differing by 1 u in mass are in the order of about 10 ns at about m/z 1000. The time needed for desorption/ionization is roughly 100 ns in case of standard UV lasers, but depends on the duration of the laser pulse. Consequently, the spread in starting times for ions of the same m/z value is often larger than the difference in time-of-flight of neighboring m/z values, thereby limiting the resolution.

Another effect is even worse: Laser-desorbed ions, for example, possess comparatively large initial kinetic energies of some 10 eV that are superimposed on the kinetic energy provided by the acceleration voltage, which is typically in the order of 10–30 kV. Obviously, higher acceleration voltages are advantageous in that they diminish the relative contribution of the energy spread caused by the ionization process. The possible effects of deviations from the assumption that ions initially are at rest on an equipotential surface have been illustrated (Fig. 4.5). [34] All these effects together limit resolution to $R \approx 500$ for continuous extraction linear TOF analyzers.

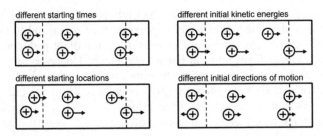

Fig. 4.5. Effects of initial time, space, and kinetic energy distributions on mass resolution in TOF-MS. Adapted from Ref. [34] with permission. © American Chemical Society, 1992.

4.2.4 Reflector Time-of-Flight Analyzer

4.2.4.1 Principle of the Reflector

The *reflector* or *reflectron* has been developed by Mamyrin. [35] In the reflector TOF analyzer – often abbreviated *ReTOF* – the reflector acts as an ion mirror that focuses ions of different kinetic energies in time. Its performance is improved by using two-stage or even multistage reflector designs.

Commonly, reflector instruments are also equipped with a detector behind the reflector allowing linear mode operation simply by switching off the reflector voltage. A complete mathematical treatment of single-stage and two-stage reflectors is found in the literature. [32,33,35] Here, we will restrict to a qualitative explanation of the reflector.

A simple reflector consist of a retarding electric field located behind the field-free drift region opposed to the ion source. In practice, a reflector is comprised of a series of rings or less preferably grids at increasing potential. The reflection voltage U_r is set to about 1.05–1.10 times the acceleration voltage U in order to insure that all ions are reflected within the homogeneous portion of the electric field of the device (Fig. 4.5). The ions penetrate the reflectron until they reach zero kinetic energy and are then expelled from the reflector in opposite direction. The kinetic energy of the leaving ions remains unaffected, however their flight paths vary according to their differences in kinetic energy. Ions carrying more kinetic energy will fly deeper into the decelerating field, and thus spend more time within the reflector than less energetic ions. Thereby, the reflector achieves a correction in time-of-flight that substantially improves the resolving power of the TOF analyzer. [32-34] In addition, the reflector provides (imperfect) focusing with respect to angular spread of the ions leaving the source and it corrects for their spatial distribution. [32,35] Adjusting the reflector at a small angle with respect to the ions exiting from the source allows the reflector detector to be placed adjacent to the ion source (Mamyrin design). Alternatively, a detector with a central hole to transmit the ions leaving the ion source has to be used in *coaxial reflectrons* (Fig. 4.8).

Although the reflector almost doubles the flight path and hence the dispersion in time-of-flight, this effect is of lower importance than its capability to compensate for initial energy spread. Simple elongation of the flight path could also be achieved with longer flight tubes in linear instruments. However, too long flight paths may even decrease the overall performance of TOF analyzers due to loss of ions by angular spread of the ion beam and scattering of ions after collisions with residual gas. To compensate for the latter effects, *ion guide wires* have been incorporated in some long TOF instruments. [36]

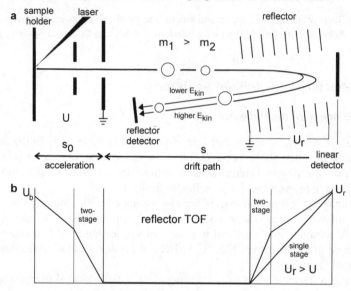

Fig. 4.6. Schematic of a ReTOF (**a**) and potentials along the instrument (**b**).

4.2.4.2 Comparison of Linear and Reflector TOF

The ability of the ReTOF to compensate for the initial energy spread of ions largely increases the resolving power of TOF instruments. While a typical continuous extraction TOF instrument in linear mode cannot resolve isotopic patterns of analytes above about m/z 500, it will do when operated in reflector mode (Fig. 4.7). At substantially higher m/z, the ReTOF still fails to resolve isotopic patterns, even though its esolution is still better than that of a linear TOF analyzer.

In case of metastable fragmentations, the ReTOF behaves different from the linear TOF. If fragmentation occurs between ion source and reflector, the ions will be lost by the reflector due to their change in kinetic energy. Only fragments still having kinetic energies close to that of the precursor, e.g., $[M+H-NH_3]^+$ in case of a peptide of about m/z 2000, are transmitted due to the energy tolerance of the reflector. However, such ions are not detected at correct m/z thereby giving rise to a "tailing" of the signal. Ions fragmenting in transit from reflector to detector are treated the same way as ions in the linear TOF. Thus, the elongated flight path of the ReTOF, which is equivalent to more time for fragmentation, and the above property complicate the detection of very labile analytes in the ReTOF.

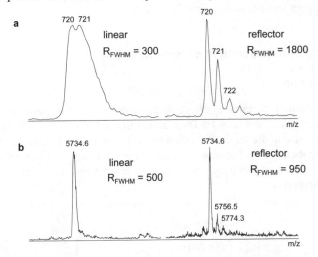

Fig. 4.7. The molecular ion signal of C_{60} at m/z 720 (**a**) and the quasimolecular ion signal of bovine insulin, m/z 5734.6, (**b**) as obtained on a TOF instrument in linear mode (left) and reflector mode (right). All other experimental parameters remained unchanged.

4.2.4.3 Curved-Field Reflectron

The so-called *curved-field reflectron* offers advantages in speed and simplicity of operation when metastable fragmentations are to be studied [37-39] as in the case of MALDI-TOF peptide sequencing (Chaps. 4.2.7 and 10). It provides a retarding field that increases with its depth by steadily increasing the voltage difference between a comparatively large number of lenses (Fig. 4.8). The curved-field reflectron acts as a divergent ion mirror causing some more ion losses than two-stage reflectors. Instruments with curved-field reflectrons are commercially available (Shimadzu AXIMA series).

Example: The curved-field reflectron of a coaxial ReTOF instrument occupies a large portion (D) of the total flight path ($s = l_1 + D$) and uses 86 lens elements

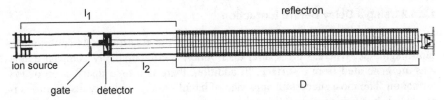

Fig. 4.8. Coaxial curved-field reflectron TOF spectrometer. The total length is about 1 m. Adapted from Ref. [37] with permission. © John Wiley & Sons, 1995.

whose voltages are set by 85 precision potentiometers located between them. [37] The distance l_1 between ion source and reflector entrance provides sufficient field-free flight path for metastable dissociations.

4.2.5 Further Improvement of Resolution

4.2.5.1 Vacuum

Improved vacuum conditions effect an elongated mean free path for the ions and thus a lower risk of collision on their transit through the TOF analyzer. The background pressure in the analyzer is directly reflected by the resolution. [40] Despite improvements of resolving power in the order of a factor of two can be realized (Fig. 4.9), enhanced pumping systems alone are not able to effect a breakthrough in resolving power.

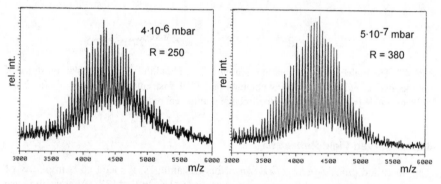

Fig. 4.9. Dependence of resolution on analyzer pressure: linear MALDI-TOF spectra of polyethylene glycol 400 recorded at a pressure of 4×10^{-4} Pa (left) and 5×10^{-5} Pa (right). Adapted from Ref. [40] with permission. © John Wiley & Sons, 1994.

> **Note:** With the exception of the quadrupole ion trap, which uses buffer gas to damp ion trajectories, a reduced background pressure, i.e., better vacuum, is also beneficial for all other types of mass analyzers.

4.2.5.2 Using a Delay Before Extraction

In case of MALDI-TOF-MS the energy spread of the emerging ions is of significant magnitude, whereas the spatial distribution is comparatively compact as the ions are generated from a surface. In addition, there is a time distribution of ion formation. The most successful approach of handling wide velocity distributions is presented by ion sources allowing a delay between generation and extraction/acceleration of the ions. *Time-lag focusing* dates back to 1955 [18] and since then has been adapted to the needs of MALDI-TOF-MS by several groups. [36,41-

43] (The concept of using a time delay to separate charged species in space is also employed in ZEKE-PES and MATI, Chap. 2.10.3.)

> **Note:** Unfortunately, establishing patents and trademarks has caused redundant names for almost the same thing: *Time Lag Focusing* (TLF, Micromass); *Delayed Extraction* (DE, Applied Biosystems); *Pulsed Ion Extraction* (PIE, Bruker Daltonik).

The principle of the method is to generate ions in a field-free environment, thereby allowing them to separate in space according to their different initial velocities. Then, after a variable delay of some hundreds of nanoseconds the acceleration voltage is switched on with a fast pulse. This procedure also ensures that laser-induced reactions have terminated before ion acceleration begins. Using a two-stage acceleration ion source allows the electric field between target (repeller) and extraction plate P_1 to be varied. The remaining fraction of the acceleration voltage is applied in the second stage. At the onset of extraction, ions with high initial velocities have traveled farther than slower ones, and therefore experience only a fraction of the extraction voltage dV between target and extraction plate P_1. The voltage of the second stage is the same for all ions. As a result, the fastest ions receive less energy from the accelerating field than the slowest. Thereby, such ion sources compensate for the initial energy distribution. [36]

The optimum settings of the delay time and of the ratio of pulsed to fixed voltage (V_1/V_2) depend on m/z with a tendency towards longer delays and/or a larger V_1/V_2 ratio for heavier ions. In practice, the variation of potentials in pulsed ion extraction with time can be handled as follows (Fig. 4.10): target and extraction plate P_1 are held at the same potential during the delay time, thereby creating a field-free region d_1. After the delay has passed, P_1 switches from V_1 by a variable value dV to V_2 causing extraction of the ions.

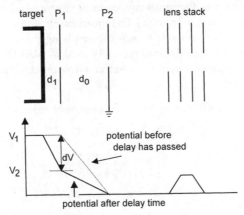

Fig. 4.10. Variation of potentials in pulsed ion extraction with time. The lens stack acts as angular focusing device for the ion beam. By courtesy of Bruker Daltonik, Bremen.

Example: Regardless of the manufacturer of the hardware, the effect of a time lag on resolution is quite dramatic. The resolving power of linear instruments is improved by a factor of 3–4 and reflector instruments become better by a factor of about 2–3. [36] The advantages are obvious by comparison of the molecular ion signal of C_{60} as obtained from a ReTOF instrument with continuous extraction (Fig. 4.7) and from the same instrument after upgrading with PIE (Fig. 4.12), or by examination of MALDI-TOF spectra of substance P, a low mass peptide, as obtained in continuous extraction mode and after PIE upgrade of the same instrument (Fig. 4.11).

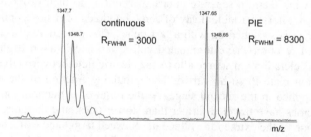

Fig. 4.11. MALDI-TOF spectra of substance P, a low mass peptide, as obtained from a Bruker Biflex ReTOF in continuous extraction mode (left) and after PIE upgrade (right).

4.2.5.3 Analog-to-Digital Conversion

In a TOF analyzer, the ion packages hit the detector in very short time intervals making the signal change extremely fast in time. This causes the *analog-to-digital conversion* to become technically demanding. It can easily be shown that the final peak shape directly depends on the speed of the *analog-to-digital converter* (ADC), [40] or vice versa, the resolving power of advanced TOF analyzers creates a need for high-speed ADCs. In actual TOF instruments, 8-bit ADCs with a digitization rate of 4 GHz are available to suit the resolving powers of their analyzers (Fig. 4.12). In case of 8-bit ADCs as typically used in MALDI-TOF instruments, the intensity of the detector output is converted into a numerical value of 0–255.

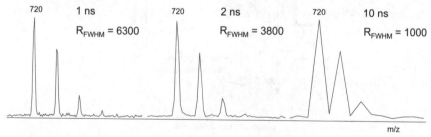

Fig. 4.12. The molecular ion peak of [60]fullerene at different settings of the dwell time per data point of the analog-to-digital converter. At 1 ns per datapoint (1 GHz) the peaks are well resolved and resolution is limited by the analyzer, at 2 ns (500 MHz) some broadening occurs and at 10 ns (100 MHz) peak shapes are reduced to triangles.

Note: The *digitization rate* or *sampling rate* may be reported as dwell time per data point, e.g., 1 ns, or as sampling frequency, e.g., 1 GHz.

4.2.5.4 Dynamic Range

The *dynamic range* is the ratio obtained by deviding the intensity of the most intense signal by that of the weakest while both are correctly detected in the same spectrum. To improve the small dynamic range delivered by 8-bit ADCs (0–255) several tens to a few hundred single spectra are usually summed up. Slow scanning instruments are typically provided with 16–20-bit ADCs corresponding to intensity values of 0–65535 (2^{16}–1) and 0–1048575 (2^{20}–1), respectively.

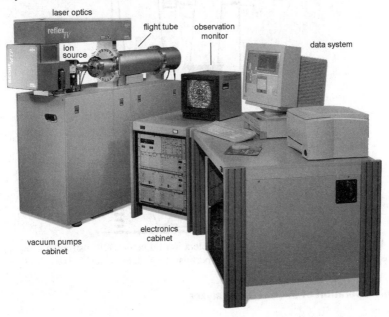

Fig. 4.13. Bruker Reflex IV™ MALDI-ReTOF instrument. With kind permission of Bruker Daltonik, Bremen.

4.2.6 Orthogonal Acceleration TOF

Up to here, the discussion has focused on MALDI-TOF equipment, because MALDI is a pulsed ionization method. MALDI initiated tremendous improvements of TOF analyzers which makes it attractive to combine such compact but powerful analyzers with any other non-pulsed ionization method desirable. The major breakthrough came from the design of the *orthogonal acceleration* TOF analyzer (oaTOF), i.e., a TOF analyzer into which pulses of ions are extracted orthogonally from a continuous ion beam (Fig. 4.14). [44,45] The oaTOF analyzer can be of ReTOF type as shown, or of linear type. Although different problems have to be overcome for each ionization method, the oaTOF analyzer is in principle suited for any of them. [23,46,47] Recently, an oaTOF analyzer has even been adapted to a MALDI ion source. [48]

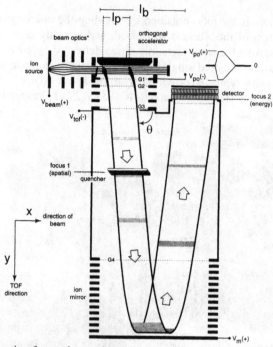

Fig. 4.14. Schematic of an orthogonal acceleration reflector TOF instrument. See text for discussion. Reproduced from Ref. [47] by permission. © John Wiley & Sons, 2000.

4.2.6.1 Operation of the oaTOF Analyzer

Ions leaving the ion source are focused to form an almost parallel ion beam, i.e., to minimize ion motion orthogonal to the ion beam axis. Then, the ion beam enters the orthogonal accelerator (x-axis). The ion kinetic energy in this beam is only in the range of 10–100 eV. A package of ions of length l_p is pushed out orthogonally from its initial direction by a sharp pulse, and is then accelerated into the TOF analyzer (y-axis) by a voltage of 5–10 kV. The design of the TOF analyzer itself is analogous to those discussed before, but its diameter is much wider because the velocity of the ions in x-direction, v_{beam}, is not affected by the orthogonal push-out process. Therefore, the detector needs to be comparatively wide in x-direction to enable ions extracted anywhere from the whole path l_p to hit its surface. In practice, the detector needs to be several centimeters in length. The angle θ between the x-direction and the flight axis into the TOF analyzer is given by

$$\theta = \tan^{-1} \sqrt{\frac{-V_{tof}}{V_{beam}}} = \tan^{-1} \left(\frac{v_{tof}}{v_{beam}} \right) \qquad (4.9)$$

where V_{tof} and V_{beam} are the acceleration voltages for ion beam and orthogonal accelerator, respectively, and v_{tof} and v_{beam} are the corresponding ion velocities.

While one ion package travels through the TOF analyzer and is being dispersed in time, the accelerator is refilled with new ions from the ion source (the time-of-flight to cover an m/z range up to about 5000 by the oaTOF is about 100 μs). During the drift time of an ion through the oaTOF, it travels a distance l_b. The ratio of the length of the orthogonal accelerator l_p to l_b determines the efficiency of the mass analyzer in terms of the ratio of ions used to ions created. This efficiency is known as the *duty cycle* of an instrument. The duty cycle of an oaTOF instrument is at an optimum if the time-of-flight to pass the TOF analyzer is slightly longer than the time needed for the continuous ion beam to refill the orthogonal accelerator. As soon as the heaviest ions have reached the detector, the next package is pulsed into the analyzer, giving rise to 10,000 complete mass spectra per second. In order to reduce the amount of data and to improve the signal-to-noise ratio, summation of several single spectra is done by an acquisition processing unit before the spectra are passed to a computer for data storage and post-processing.

4.2.6.2 Time-to-Digital Conversion

Time-to-digital converters (TDC) are often employed in oaTOF instruments instead of analog-to-digital converters, because of the much higher speed of TDCs. Speed in terms of sampling rate is needed to make use of the high resolution of such instruments (up to $R_{FWHM} = 20,000$), and speed in terms of data flow is a prerequisite to handle the enormous number of spectra per second. However, a TDC offers only 1 bit dynamic range (value 0 or 1). Thus, summation of many single spectra is also necessary to obtain a higher dynamic range via improved signal statistics. Nevertheless, the result is still unsatisfactory for quantitative analysis and for correct isotopic patterns, because a TDC cannot distinguish whether one, two or many ions have hit the detector simultaneously to produce the actual signal. Some recently developed oaTOF instruments are therefore equipped with 8-bit ADCs to provide dynamic range of up to 10^4 after summation of several single spectra (JEOL AccuTOF, Bruker BioTOF series).

4.2.6.3 Applications of oaTOF Instruments

The advantages of oaTOF analyzers are: i) high sensitivity due to very good duty cycle and high transmission of TOF analyzers, ii) high rate of spectra per second even after pre-averaging, iii) high mass-resolving power, iv) mass accuracies up to 5 ppm allowing for accurate mass measurements, and v) compact design. Therefore, it is not astonishing that oaTOF instruments are currently becoming widespread for a wide range of applications. Electrospray ionization (ESI) oaTOF instruments represent the majority of these systems (for applications cf. Chaps. 10–12), [49] but gas chromatography-mass spectrometry (GC-MS) instruments are also available. In particular for *fast GC* applications [50,51] and for HR-GC-MS [52] oaTOF systems are advantageous.

4.2.7 Tandem MS on TOF Instruments

Tandem mass spectrometry or *MS/MS* summarizes the numerous techniques where mass-selected ions (MS1) are subjected to a second mass spectrometric analysis (MS2). [53,54] Ion dissociations in transit through the mass analyzer may either occur spontaneously (*metastable*, Chaps. 2.7.1, 2.8.2) or can result from intentionally supplied additional activation, i.e., typically from collisions with neutrals. The methods to activate or react otherwise stable ions in the *field-free region* (FFR) between MS1 and MS2 are discussed in Chap. 2.12 and applications of MS/MS on TOF instruments are shown in Chaps. 10–12.

In order to perform two consecutive mass-analyzing steps, two mass analyzers may be mounted in tandem. This technique is applied with beam transmitting devices, i.e., TOF, sector and quadrupole analyzers can be combined that way (*tandem-in-space*, Fig. 4.15). Alternatively, a suitable mass analyzer may be operated by combining selection, activation, and analysis in the very same place. Quadrupole ion trap (QIT) and ion cyclotron resonance (ICR) instruments can perform such *tandem-in-time* experiments.

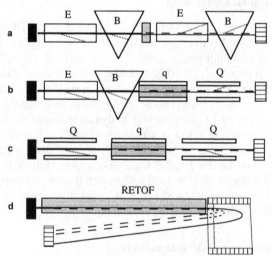

Fig. 4.15. Tandem-in-space setups for different beam instruments: magnetic four-sector instrument (**a**), magnetic sector–quadrupole hybrid (**b**), triple quadrupole (**c**), and ReTOF (**d**). The linestyles indicate —— stable precursor ions, - - - transmitted ions, and ······· non-transmitted fragment ions. Shaded areas show the region of analytically useful ion dissociations. Reproduced from Ref. [55] with permission. © Elsevier Science, 1994.

4.2.7.1 Tandem MS in a ReTOF Analyzer

Methods for the detection of metastable ion dissociations in ReTOF-MS in combination with *secondary ion mass spectrometry* (SIMS) and ^{252}Cf *plasma desorption* (^{252}Cf-PD, Chap. 9.5.2) mass spectrometry were known before the advent of MALDI. [56-59]

Consider an ion m_1^+ decomposing in transit through a field-free region. Its kinetic energy is then distributed among the product ion m_2^+ and the neutral fragment n according to their relative contribution to the mass of the precursor ion

$$E_{kin(m2+)} = E_{kin(m1+)} \frac{m_{i2}}{m_{i1}} \quad \text{and} \quad E_{kin(n)} = E_{kin(m1+)} \left(1 - \frac{m_{i2}}{m_{i1}}\right) \qquad (4.10)$$

While the kinetic energy changes upon dissociation, the ion velocity remains constant, as already noted in the discussion of the linear TOF. This causes fragment ions generated on the flight to the reflector to have kinetic energies lower than intact ions. Fortunately, a reflector is capable to deal with ions of down to 70–90 % of the energy it has been adjusted (0.7–$0.9 \times E_{kin(m1+)}$). By stepwise reduction of the reflector potential, partial spectra each covering several percent of the precursor ion mass can be acquired. [55,60] Pasting these pieces together yields a spectrum of the product ions formed by metastable dissociation of m_1^+. To cover the range from m_1 to $0.1 \times m_{i1}$ the reflector must be stepped down from its potential V_0 to $0.1 \times V_0$ in some 10–20 steps.

The ReTOF analyzer alone supplies only the field-free region and MS2, i.e., the metastable dissociations of all precursor ions leaving the ion source would be detected simultaneously. The precursor ion selection (MS1) – although technically demanding – follows a simple principle: first, ions below the selected precursor m/z value are electrostatically deflected, then the deflector is off to pass the precursor ion through, and finally the high voltage is switched on again to deflect heavier ions. As the deflector is located comparatively close to the ion source it acts as a coarse TOF gate providing poor precursor ion resolution. A collision cell may be placed near by to allow for CID.

Despite the comparatively poor precursor ion resolution and being a time- and sample-consuming procedure, MS/MS on the ReTOF works so well that this method has become established as one of the major tools of biomolecule sequencing for ions in the m/z 500–3000 range (examples are shown in Chap. 10).

Note: In particular the MALDI-TOF community has coined some sort of an own terminology, e.g., *in-source decay* (ISD) for all fragmentations occurring within the ion source, *post-source decay* (PSD) instead of metastable ion dissociation and *fragment analysis and structural TOF* (FAST) for the specific mode of operation of a ReTOF to detect metastable ions.

4.2.7.2 Tandem MS in True Tandem TOF Instruments

Several true tandem TOF instruments have been designed in order to provide better precursor ion selection. [61-65] Two instruments of that type have become commercially available (Applied Biosystems 4700 Proteomics Analyzer [64] and Bruker Daltonik Ultraflex [63]) Although differing in detail, the basic idea of these TOF/TOF instruments is to employ a short linear TOF as MS1 and a high resolving ReTOF as MS2 which is separated from TOF1 by a timed ion selector for precursor ion selection. TOF1 is operated at comparatively low acceleration voltage, whereas TOF2 is desigend to analyze ions of about 20 keV kinetic energy. The difference in kinetic energy is provided by a second acceleration stage located behind the collision cell (Fig. 4.16). By lifting all ions by a certain amount of kinetic energy, their relative spread in kinetic energy is reduced, e.g., a precursor ion of 5 keV yields fragments having 0.5–5 keV. Addition of 15 keV to all fragment ions lifts them to 15.5–20 keV. Given a reflector of sufficient energy acceptance they can be analyzed without tedious stepping of the reflector voltage.

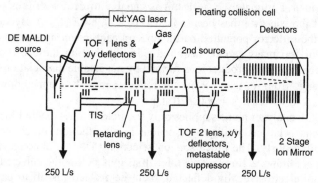

Fig. 4.16. Tandem TOF analyzer with linear TOF1 and ReTOF2. The ions from TOF1 are passed into ReTOF2 by a timed ion selector. Reproduced from Ref. [64] with permission. © Elsevier Science, 2002.

4.3 Magnetic Sector Instruments

4.3.1 Introduction to Magnetic Sector Instruments

Magnetic sector instruments paved the road to organic mass spectrometry. "Whereas mass spectrometry was still very an art rather than a science in 1940, the picture changed drastically during the war years. Vacuum and electronic techniques had matured." [66] The evolution of the early instruments took until the 1950s to make magnetic sector instruments commercially available. [7,8] These pioneering instruments were by no means small or easy to use. [66-68] Nevertheless, they provided a kind of analytical information chemists had been seeking for long. They became faster, more accurate, and higher resolving from year to year.

[3,5,9,69] The first instruments used a single magnetic sector (symbol B) to effect separation of the ions. Later, the introduction of *double-focusing instruments* having an *electrostatic sector* or *electrostatic analyzer* (ESA, symbol E) in addition defined a standard which is still valid.

With few exceptions, magnetic sector instruments are comparatively large devices capable of high resolution and accurate mass determination, and suited for a wide variety of ionization methods. Double-focusing sector instruments are the choice of MS laboratories with a large chemical diversity of samples. In recent years, there is a tendency to substitute these machines by TOF or by Fourier transform ion cyclotron resonance (FT-ICR) instruments.

4.3.2 Principle of the Magnetic Sector

4.3.2.1 Magnetic Sector as a Momentum Separator

The Lorentz Force Law can be used to describe the effects exerted onto a charged particle entering a constant magnetic field. The Lorentz Force F_L depends on the velocity v, the magnetic field B, and the charge q of an ion. In the simplest form the force is given by the scalar equation [3,4,70,71]

$$F_L = qvB \qquad (4.11)$$

Equation 4.11 is valid if v and B (both are vectors) are perpendicular to each other. Otherwise, the relationship becomes

$$F_L = qvB(\sin\alpha) \qquad (4.11a)$$

where α is the angle between v and B. Figure 4.17 demonstrates the relationship between the direction of the magnetic field, the direction of the ionic motion, and the direction of the resulting Lorentz force. Each of them is at right angles to the others. An ion of mass m and charge q travelling at a velocity v in a direction perpendicular to a homogeneous magnetic field will follow a circular path of radius r_m that fulfills the condition of equilibrium of F_L and centripetal force F_c

$$F_L = qvB = \frac{m_i v^2}{r_m} = F_c \qquad (4.12)$$

Upon rearrangement we obtain the radius r_m of this circular motion

$$r_m = \frac{m_i v}{qB} \qquad (4.13)$$

This shows the working principle of a magnetic sector the radius r_m depends on the momentum mv of an ion, and therefore the momentum depends on m/z.

Note: The magnetic sector is a momentum analyzer rather than a direct mass analyzer as commonly assumed.

Finally, dispersion in momentum causes a dependence of r_m on the square root of mass becoming obvious by substitution of v from Eq. 4.3 and $q = ez$

$$r_m = \frac{m_i}{ezB}\sqrt{\frac{2eU}{m_i}} = \frac{1}{zB}\sqrt{\frac{2m_iU}{e}} \qquad (4.13a)$$

Alternatively, the ratio m_i/q can be expressed by rearranging Eq. 4.13

$$\frac{m_i}{q} = \frac{r_m B}{v} \qquad (4.14)$$

which upon substitution of v as performed above becomes

$$\frac{m_i}{q} = \frac{r_m B}{\sqrt{\dfrac{2qU}{m_i}}} \quad \Rightarrow \quad \frac{m_i}{q} = \frac{r_m^2 B^2}{2U} \qquad (4.14a)$$

For singly charged ions ($z = 1$, $q = e$) we obtain the more widespread form

$$\frac{m_i}{e} = \frac{r_m^2 B^2}{2U} \qquad (4.14b)$$

> **Note:** Equation 4.14b has been known as the *basic equation of mass spectrometry*. Nowadays, there is no more justification for a single basic equation of mass spectrometry because of the various mass analyzers employed.

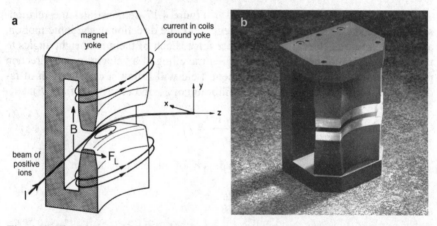

Fig. 4.17. The Right Hand Rule (*I* thumb, *B* index finger, F_L middle finger) to determine the direction of the Lorentz Force (**a**); the current corresponds to the direction where positive charges move, i.e., the figure directly applies for positive ions. (**b**) A real magnet yoke without coils and flight tube. With kind permission of Thermo Electron (Bremen) GmbH, (left) and Waters Corporation, MS Technologies, Manchester, UK (right).

4.3.2.2 Direction Focusing by the Magnetic Sector

The focusing action of a homogeneous magnetic field on a beam of ions having the same m/z and the same kinetic energy can best be illustrated by a 180° sector (Fig. 4.18). If the beam is divergent by a half-angle α, the collector slit must be $\alpha^2 r_m$ wide to pass all ions after suffering 180° deflection. This is because the ions come to a first order, i.e., imperfect, focus as they all traverse the magnetic field at the same radius but not all of them entered the field at right angles. Ions of different m/z fly at a different radius, e.g., the lighter ions of m/z_1 hit the wall while ions of m/z_2 reach the collector slit. To allow for detection of various masses, such an analyzer could either be equipped with a photographic plate in the focal plane to become a so-called *mass spectrograph*, or it could be designed with variable magnetic field to detect different masses at the same point by bringing them subsequently at the collector slit. Indeed, such a 180° geometry with *scanning* magnetic field has been used by Dempster. [72] Later, the term *mass spectrometer* has been coined for this type of instruments. [68]

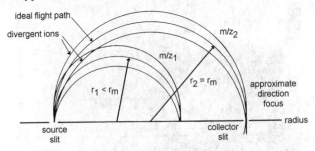

Fig. 4.18. Direction focusing properties of a 180° magnetic sector on a diverging beam of ions of the same m/z and the same kinetic energy and effect on ions of different m/z. In this illustration, B has to come out of the plane towards the reader for positive ions.

Among other complications, the 180° design demands for large and heavy magnetic sectors. It is by far more elegant to employ magnetic sectors of smaller angles (Figs. 4.19, 4.20). An optimized magnetic sector alone can provide resolutions of $R = 2000-7000$ depending on its radius. The limitation arises from the fact that ions emerging from the ion source are not really monoenergetic. This way, ions of different m/z can obtain the same momentum and thus cause overlap of adjacent ion beams at the detector.

Example: Equation 4.13a describes the radius r_m in the magnetic field. Obviously, the value r_m remains constant as long as $m_i U = const$. If the instrument is set to pass an ion of say m/z 500 and 3000 eV kinetic energy, it will simultaneously allow the passage for ions of m/z 501 having 2994 eV or of m/z 499 having 3006 eV of kinetic energy. This is why obtaining higher resolution requires small kinetic energy distributions.

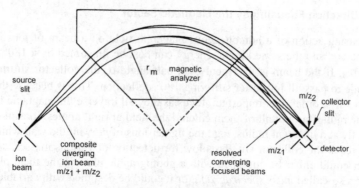

Fig. 4.19. A 90° magnetic sector illustrating m/z separation and direction focusing in a plane (angles are shown exaggerated). Reprinted from Ref. [1] with permission. © American Association for the Advancement of Science, 1979.

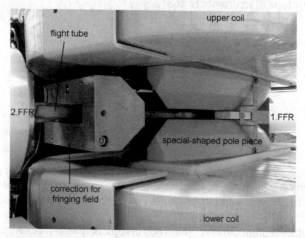

Fig. 4.20. Flight tube passing through the gap of the magnetic sector of a JEOL JMS-700 instrument seen from the ESA side. The shapes of the pole pieces of the yoke and the additional blocks around the tube are designed to minimize fringing fields. In addition the pole faces are rotated to increase the mass range.

4.3.3 Double-Focusing Sector Instruments

4.3.3.1 Principle of the Electrostatic Sector

The *electrostatic sector* or *electrostatic analyzer* (ESA) produces a radial electric field between two oppositely charged plates extending over the ESA angle ϕ (Fig. 4.21). An ion passes the ESA midway on a circular path if

$$F_e = qE = \frac{m_i v^2}{r_e} = F_c \qquad (4.15)$$

where F_e represents the electric force, E the electric field strength, and r_e the radius of the ESA. Rearrangement of Eq. 4.15 demonstrates that the ESA acts as an energy dispersive device

$$r_e = \frac{m_i v^2}{qE} = \frac{m_i v^2}{ezE} \qquad (4.16)$$

> **Note:** The ESA effects energy dipersion. Thus, the kinetic energy distribution of an ion beam can be reduced. The ESA does not allow for mass separation among monoenergetic ions.

Upon substitution of v with Eq. 4.3 one obtains the simple relationship

$$r_e = \frac{2U}{E} \qquad (4.17)$$

to describe the radius of the ESA. As with the magnetic sector before, the ESA has direction focusing properties in one plane (Fig. 4.21). Ions entering the ESA in the middle and at right angles to the field boundaries pass through on a path of equipotential, whereas ions with a velocity component towards one of the capacitor plates are brought into focus at the focal length l_e. To understand this, imagine an ion drifting towards the outer plate having the same charge sign as the ion. As it approaches the plate it is decelerated by the opposed electric field and finally reflected towards the center of the beam. With its radial component of v inverted it crosses the ideal path at l_e. In an analogous fashion an ion approaching the inner plate becomes accelerated by the attractive force. The resulting higher velocity causes an increase in centripetal force, and thereby effects a correction of the flight path in the appropriate sense.

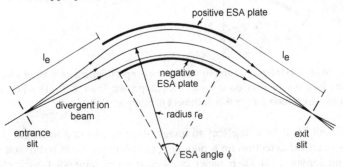

Fig. 4.21. Direction focusing of a radial electric field. Ions of appropriate kinetic energy are focused at the exit slit. Divergent ions pass the ESA close to either plate. Here, the electric potentials are set to transmit positive ions. The image distance l_e depends on the ESA angle.

4.3.3.2 Double-Focusing

Energy focusing is achieved by combining a magnetic sector and an electric sector in a way that the energy dispersion of the magnetic field is just compensated by the energy dispersion of the electric field. Additional direction focusing is obtained, if the radii and angles of these fields and their mutual alignment does not diminish the focusing properties of each of them. Then, an ion optical system is obtained that is able of focusing ions on a single image point, although these were emerging from the ion source in (slightly) different directions and with (slightly) different kinetic energies. This is called *double-focusing*. Double-focusing can improve the resolving power of a magnetic sector instrument more than ten times. The following passages present examples of double-focusing geometries that either have been milestones in instrument design [3,5] or still are incorporated in modern mass spectrometers.

4.3.3.3 Forward Geometry

The EB design by Mattauch and Herzog combines a 31° 50' ESA with a 90° magnetic sector producing an image plane that allows for simultaneous photographic detection of a comparatively large *m/z* range. [73] This mass spectrograph attains double-focusing over the total image plane resulting in resolving powers of $R > 10,000$. Thus, it became the basis for numerous commercial instruments (Fig. 4.22).

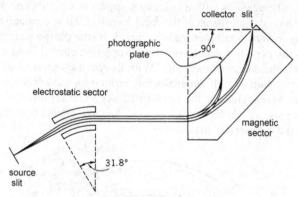

Fig. 4.22. Mattauch-Herzog double-focusing mass spectrograph providing direction and momentum focusing of ions on a plane. [73] Reprinted from Ref. [1] with permission. © American Association for the Advancement of Science, 1979.

A different EB design to effect an image plane of 140 mm in length with a linear mass scale for detection on a photographic plate has been published by Bainbridge and Jordan. [74] Their paper is especially recommended, as it also nicely illustrates the use of photoplates in these days.

Another famous type of EB arrangement has become known as *Nier-Johnson geometry* (Fig. 4.23). [75] Here, the ion beam first passes through the electrostatic analyzer producing energy-resolved beams without mass dispersion in the plane of an intermediate slit located at the focal point of the ESA. It then passes through a magnetic analyzer to achieve mass dispersion of the ions. Different from the Mattauch-Herzog *mass spectrograph*, the Nier-Johnson *mass spectrometer* was constructed to be used in conjunction with a *scanning magnet* to focus one specific *m/z* after the other onto a point detector. This presented a major advantage for the use of UV recorders or electronic data acquisition as well as for the achievement of better ion optics.

> **Note:** Typically, on magnetic sector mass spectrometers a spectrum is produced by varying the strength of the magnetic field to successively pass through ions of different *m/z*. This is termed *magnetic field scan*. [31]
> For scans *linear in time*, the *scan rate* is given in units of u s^{-1}, e.g., 500 u s^{-1}.
> For scans *exponential in time*, the *scan rate* is reported in units of s per decade, e.g., 10 s/decade means 10 s from *m/z* 30 to 300 or from *m/z* 100 to 1000.

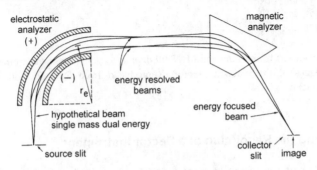

Fig. 4.23. Nier-Johnson geometry (EB). [75] Direction and velocity focusing are shown exaggerated. Reprinted from Ref. [1] with permission. © American Association for the Advancement of Science, 1979.

Nowadays, forward geometry instruments are often constructed to be used in combination with an (optional) array detector, e.g., the JEOL HX-110 (EB), the Thermo Finnigan MAT 900 (EB) and the Micromass Autospec (EBE) instruments can be equipped in that way. The array detector is then located at the focus plane of the magnet, but different from the Mattauch-Herzog design it only covers a comparatively small *m/z* range simultaneously (Chap. 4.8).

4.3.3.4 Reversed Geometry

Successful constructions of *reversed geometry*, i.e., BE instead of *forward* EB design, have been presented by the MAT 311 in the mid 1970s and shortly after by the VG Analytical ZAB-2F instrument [76] based on a proposal by Hintenberger and König. [77] The first generation ZABs were based on a magnetic sector of 30 cm radius followed by an ESA of 38 cm radius. The specified resolving power (of a new instrument) was R > 70,000.

Modern BE geometry instruments are presented by the Thermo Finnigan MAT 90 and 95 series (Fig. 4.24), the JEOL JMS-700, the AMD Intectra AMD 403S, and basically, but not strictly as it has EBE geometry, [78] the Micromass Auto-spec.

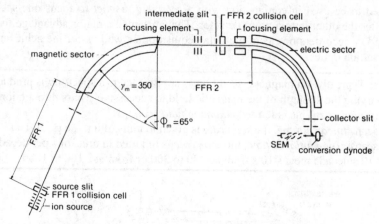

Fig. 4.24. Schematic of the Finnigan MAT 90 double-focusing mass spectrometer with rotated pole faces of the magnetic sector. Reproduced from Ref. [2] with permission. © Elsevier Science, 1987.

4.3.4 Setting the Resolution of a Sector Instrument

The ion optics of a sector instrument are the analogue to a cylindrical lens in light optics. Accordingly, the reduction of an aperture can be used to obtain a sharper image, i.e., to increase resolution (Chap 3.3.4). Slits are used instead of circular apertures to comply with the cylindrical properties of the ion optical system. The settings of the *object slit* (*source slit, entrance slit*) and the *image slit* (*collector slit, exit slit, detector slit*) are most important. Intermediate slits may be used in addition. Unfortunately, closing slits also means cutting off ions from the beam, and thus reduction of the transmission of the mass analyzer. In the ideal case, the improvement of resolution by a factor of 10 goes along with a reduction in transmission to 10 %, in practice the effect is often even less.

Example: The influence of relative slit width on peak shape and resolution is demonstrated on the second isotopic peak of toluene molecular ion, $^{13}C_2^{12}C_5H_8^{+\cdot}$, m/z 94 (Fig. 4.25). With the entrance slit at 50 μm and the exit slit at 500 μm the peak is flat-topped (left), because a narrow beam from the entrance sweeps over the wide open detector slit keeping the intensity constant as the scan proceeds until the beam passes over the other edge of the slit. Closing the exit slit to 100 μm increases resolution to 2000 without affecting the peak height (middle), but reduces the peak area by a factor of 4 in accordance with an increase in resolution by the same factor. Further reduction of the exit slit width to 30 μm improves resolution at the cost of peak height (right). (Any sector instrument must behave alike, otherwise, cleaning or other maintenance are required.)

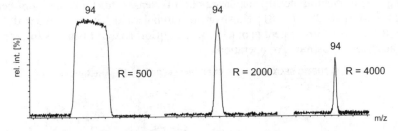

Fig. 4.25. The influence of relative slit width settings on peak shape and resolution on a magnetic sector instrument. The peak shape first changes from flat-topped (left) to Gaussian (middle) and finally resolution improves at cost of peak height (right).

> **Note:** The ultimate resolution of a magnetic sector mass spectrometer is reached when the slits are closed to a width of a few micrometers. Often, the slit height is also reduced, e.g., from 5 to 1 mm. In daily work, the resolution will be set to fit the actual task, e.g., $R = 1000–2000$ for low resolution work, $R = 3000–5000$ if accurate mass determination at high scan rates is needed (GC-MS, Chap. 12) or isotopic patterns of high mass analytes have to be resolved, or $R = 7000–15,000$ in slow-scanning accurate mass measurements.

4.3.5 Further Improvement of Sector Instruments

To further improve the performance of a sector instrument in terms of scan speed, resolution, transmission, and mass range, the construction, in particular that of the magnet, needs some additional refining.

The rapid change of a magnetic field suffers from hysteresis, i.e., the magnetic flux does not exactly follow the change of the electric current through the coils in time, but lags behind due to the induction of eddy currents. On one side, this causes problems in creating a scan perfectly linear in time, on the other this prevents high scan rates as required for GC-MS. Lamination of the yoke substantially reduces these problems [79] and is part of all modern sector instruments.

Another problem is due to fringing fields. To reduce these defocusing effects at the entrance and exit of the field, the pole pieces of the yoke of the electromagnet generally have specially shaped edges (Fig. 4.20).

In order to extend the mass range one either has to increase the field strength or the radius of the magnet. However, there are limitations of the field strength with non-superconducting magnets at about 2.4 T. Instead of simply enlarging the radius, the pole faces of the magnet can be rotated to preserve a compact design by reduction of its focal length (Figs. 4.20, 4.24, 4.27, 4.28). Alternatively, inhomogeneous fields can be applied. For details refer to the article by Brunnée. [2]

Furthermore, pure EB and BE designs are somewhat limited concerning the transmittance through a narrow magnet gap as compared to instrument designs using some quadrupole (q) or hexapole (h) lenses to form the ion beam (Figs. 4.24, 4.26–28). [80,81] However, quadrupole and hexapole lenses do not alter the effective instrument geometry, i.e., a qqBqE analyzer behaves like a pure BE analyzer in any mode of operation.

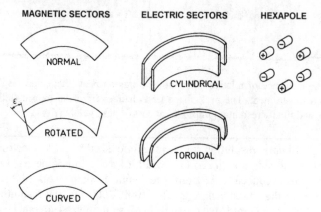

Fig. 4.26. Types and shapes of ion optical elements used in magnetic sector instruments. By courtesy of Thermo Electron (Bremen) GmbH.

4.3.6 Tandem MS with Magnetic Sector Instruments

There are multiple ways of detecting metastable and collision-induced dissociations with magnetic sector instruments. [82] In fact, the phenomenon was discovered with this particular type of mass analyzer (Chap. 2.7.1). [83,84]

4.3.6.1 Dissociations in the FFR Preceding the Magnetic Sector

Dissociation in a *field-free region* (FFR) not only causes partitioning of ion kinetic energy (Eq. 4.10), but also goes with partitioning of momentum p

$$P_{(m2+)} = P_{(m1+)} \frac{m_{i2}}{m_{i1}} \qquad (4.18)$$

As we have conservation of velocity, i.e., $v_1 = v_2 \equiv v$, the momentum of a fragment ion m_2^+ formed in a FFR preceding the magnetic sector is different from that of such a fragment ion arising from the ion source. The ion formed by metastable ion dissociation thus passes the magnet as if it had the virtual mass m*

$$m^* = \frac{m_2^{\,2}}{m_1} \qquad (4.19)$$

Fig. 4.27. Photograph of a JEOL JMS-700 double focusing magnetic sector instrument.

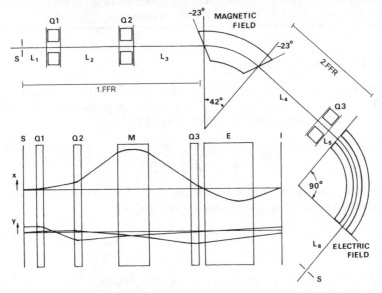

Fig. 4.28. Schematic design of the above instrument. Horizontal (x) and vertical (y) trajectories are also shown. Quadrupole lenses are used to improve the transmission of the magnetic sector, thus resulting in a qqBqE geometry. Adapted from Ref. [80] with permission. © Elsevier Science, 1985.

Note: This explains the occurence of "diffuse" peaks due to metastable ion dissociations at fractional m/z values in the B scan spectra of B and EB instruments (Chap. 2.7.1). [83,84] In turn, the mass spectra obtained from BE instruments do not show any metastable ion peaks in normal operation.

4.3.6.2 Mass-Analyzed Ion Kinetic Energy Spectra

Mass-analyzed ion kinetic energy spectra (MIKES) [85,86] can be measured on BE geometry instruments only. The precursor ion is selected by the magnet and the fragments from dissociations of m_1^+ in the 2.FFR are analyzed by the ESA due to their kinetic energy. This is possible, because the kinetic energy of the precursor is distributed among the product ion and the neutral. Derived from Eq. 4.10 we have

$$\frac{E_2}{E_1} = \frac{m_{i2}}{m_{i1}} = \frac{m_{i2}v^2}{m_{i1}v^2} \tag{4.20}$$

Thus, scanning of the electric field (*E* scan) yields an energy spectrum which allows for the determination of the *kinetic energy release* (KER) from the peak width (Chap. 2.8.2). The MIKE technique provides good precursor ion resolution, but poor product ion resolution according to the influence of KER on peak shapes.

> **Note:** In MIKES E_1 represents the full electric field necessary to transmit the precursor ion m_1^+ through the ESA, and E_1 is often denoted as starting value E_0. Then, the abscissa of MIKE spectra is divided in units of $E/E_0 = m_2/m_1$.

4.3.6.3 B/E = Constant Linked Scan

Ions decomposing in the 1.FFR of BE and EB instruments can be detected using the *B/E = constant* linked scan. [87] Due to the proportionality of *B* and *p* (Eq. 4.13) ions are transmitted through the magnet if

$$\frac{B_2}{B_1} = \frac{m_{i2}v}{m_{i1}v} = \frac{p_2}{p_1} = \frac{m_{i2}}{m_{i1}} \tag{4.21}$$

For their subsequent passage through the ESA, Eq. 4.20 has to be satisfied. Therefore, we have for the passage through both fields

$$\frac{B_2}{B_1} = \frac{E_2}{E_1} = \frac{m_{i2}}{m_{i1}} \tag{4.22}$$

which defines the conditions for a scan as *B/E = constant*. Thus, *B* and *E* have to be scanned together, i.e., in a *linked* fashion. The *B/E = constant* linked scan provides good fragment ion resolution (R ≈ 1000) but poor precursor ion resolution (R ≈ 200). As with all linked scans, there is a risk of artefact peaks [88-91] because linked scan techniques represent no true tandem MS where MS1 and MS2 are clearly separate. The use of two complementary scan modes is therefore suggested to avoid ambiguities. [89]

4.3.6.4 Additional Linked Scan Functions

Whereas TOF instruments solely allow for the detection of product ions of a selected precursor, sector instruments offer additional modes of operation: i) to exclusively identify product ions of a particular precursor ion, so-called *precursor ion scans*, [92,93] or ii) to detect only ions formed by loss of a specific neutral mass, so-called *constant neutral loss* (CNL) scan. [94] This can be achieved by some technically more demanding linked scans (Table 4.2). [95-98]

Table 4.2. Common scan laws for the detection of metastable ion dissociations on magnetic sector mass spectrometers[a]

Selected Mass	Scan Law	KER from Peak Width?	Analyzer and FFR	Properties
m_1	$B = B_{m1}$, and $E/E_0 = m_2/m_1$	yes	BE 2.FFR	mass-analyzed ion kinetic energy spectrum (MIKES), E is scanned
m_1	$B/E = B_{0(1)}/E_0$ i.e., $B/E = const.$	no	BE or EB 1.FFR	linked scan, poor precursor but good product ion resolution
m_2	$B^2/E = B_2^2/E_0$ i.e., $B^2/E = const.$	yes	BE or EB 1.FFR	linked scan, poor product but good precursor ion resolution, demands precise control of B
m_2	$B^2 \cdot E = B_{0(2)} \cdot E_0$ i.e., $B^2 \cdot E = const.$	no	BE 2.FFR	linked scan, resolution controlled by that of B, needs precise control of B
m_n	$(B/E) \cdot [1-(E/E_0)]^{1/2}$ i.e., $E/E_0 = m_2/m_1$ $= 1-(m_n/m_1)$	no	BE or EB 1.FFR	constant neutral loss (CNL), linked scan

a) All scans listed use constant acceleration voltage U. Scanning of U offers additional scan modes. However, scanning of U over a wide range causes detuning of the ion source.

Example: The $B^2E = const.$ linked scan [93] has been employed to quantify the caffeine content of coffee, black tea, and caffeinated cola softdrinks. [99] Caffeine, $M^{+\bullet} = 194$, was determined by spiking the sample with a known concentration of [D_3]caffeine, $M^{+\bullet} = 197$, as internal standard. Both molecular ions dissociate to form a fragment ion at m/z 109 which was selected as m_2^+. Then, the precursor ion scan showed either molecular ion, m_1^+ and [D_3]m_1^+, as precursor of the ion at m/z 109. The ratio of peak intensities was taken as a measure for the relative concentration of analyte and labeled standard (Fig. 4.29).

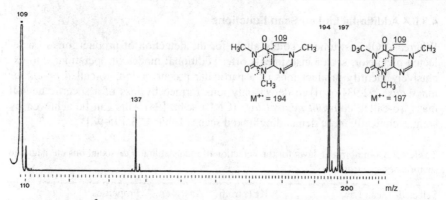

Fig. 4.29. CID B^2E spectrum of the ion m/z 109 of a caffeinated cola softdrink with [D₃]caffeine as internal standard. Different Cola brands yielded 73–158 mg l^{-1}. Adapted from Ref. [99] with permission. © John Wiley & Sons, 1983.

> **Note:** *Precursor ion scanning* suggests some sort of measuring into the past. One should be aware that also in precursor ion scans the product ions are detected, but this is accomplished in a way that only fragments of a selected precursor ion mass can reach the detector, hence the term.

4.3.6.5 Multi-Sector Instruments

Multi-sector instruments, typically four-sector machines, have been developed in the 1980s to combine high precursor ion resolution with high fragment ion resolution. [100,101] Their major field of application was sequencing of biomolecules by FAB-CID-MS/MS (Chap. 9.4.6).

Commercial representatives are the JEOL HX110/HX110A (EBEB, Fig. 4.30), the JEOL MStation-T (BEBE), or the Micromass AutospecT (EBEBE). As industrial laboratories are governed by high-throughput and other economic aspects, modern TOF/TOF, QqTOF and FT-ICR instruments have almost replaced these impressing "dinosaurs" weighing some 4–5 tons and having about 3×5 m footprint. Numerous custom-built four-sector instruments are still operated in laboratories devoted to gas phase ion chemistry. [102]

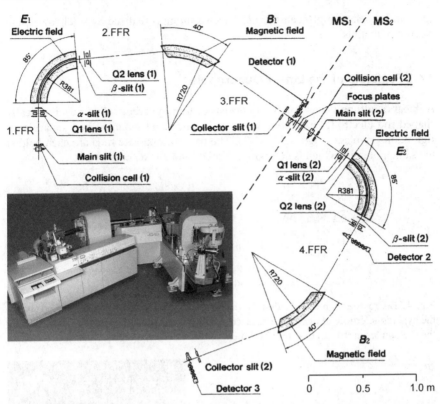

Fig. 4.30. Schematic and photograph (inset) of the JEOL HX110/HX110A tandem sector instrument with EBEB ion optics. By courtesy of JEOL, Tokyo.

4.4 Linear Quadrupole Instruments

4.4.1 Introduction to the Linear Quadrupole

Since the Nobel Prize-awarded discovery of the mass-analyzing and ion-trapping properties of two- and three-dimensional electric quadrupole fields [103,104] and the concomitant construction of a *quadrupole* (Q) *mass spectrometer*, [105,106] this type of instrument has steadily gained importance. Chiefly starting from GC-MS applications, where rapid scanning devices were urgently required, quadrupole analyzers made their way into the MS laboratories, [107-109] although the early systems offered poor resolving power and low mass range, e.g., *m/z* 1–200. Modern quadrupole instruments cover the *m/z* 2000–4000 range with good resolving power and represent some kind of standard device in LC-MS. The advantages of quadrupoles are that they i) have high transmission, ii) are light-weighted, very compact and comparatively low-priced, iii) have low ion acceleration voltages, and iv) allow high scan speeds, since scanning is realized by solely sweeping electric potentials.

4.4.2 Principle of the Linear Quadrupole

A linear quadrupole mass analyzer consists of four hyperbolically or cyclindrically shaped rod electrodes extending in the z-direction and mounted in a square configuration (xy-plane, Figs. 4.31, 4.32). The pairs of opposite rods are each held at the same potential which is composed of a DC and an AC component.

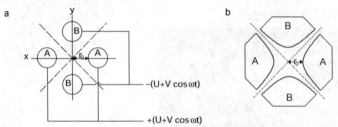

Fig. 4.31. Cross section of a quadrupole (**a**) for the cyclindrical approximation and (**b**) for the hyperbolic profile of the rods. The electric field is zero along the dotted lines, i.e., along the asymptotes in (**b**). (**a**) Courtesy of Waters Corp., MS Technologies, Manchester, UK.

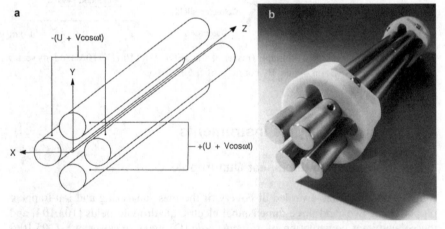

Fig. 4.32. Schematic (**a**) and photograph (**b**) of a linear quadrupole mass analyzer. By courtesy of JEOL, Tokyo (**a**) and Waters Corp., MS Technologies, Manchester, UK. (**b**).

As an ion enters the quadrupole assembly in z-direction, an attractive force is exerted on it by one of the rods with its charge actually opposite to the ionic charge. If the voltage applied to the rods is periodic, attraction and repulsion in both the x- and y-directions are alternating in time, because the sign of the electric force also changes periodically in time. If the applied voltage is composed of a DC voltage U and a radiofrequency (RF) voltage V with the frequency ω the total potential Φ_0 is given by

$$\Phi_0 = U + V \cos \omega t \tag{4.23}$$

Thus, the equations of motion are

$$\frac{d^2x}{dt^2} + \frac{e}{m_i r_0^2}(U + V \cos \omega t)x = 0 \tag{4.24}$$

$$\frac{d^2y}{dt^2} - \frac{e}{m_i r_0^2}(U + V \cos \omega t)y = 0$$

In case of an inhomogenous periodic field such as the above quadrupole field, there is a small average force which is always in the direction of the lower field. The electric field is zero along the dotted lines in Fig. 4.31, i.e., along the asymptotes in case of the hyperbolic electrodes. It is therefore possible that an ion may traverse the quadrupole without hitting the rods, provided its motion around the z-axis is stable with limited amplitudes in the xy-plane. Such conditions can be derived from the theory of the Mathieu equations, as this type of differential equations is called. Writing Eq. 4.24 dimensionless yields

$$\frac{d^2x}{d\tau^2} + (a_x + 2q_x \cos 2\tau)x = 0 \tag{4.25}$$

$$\frac{d^2y}{d\tau^2} + (a_y + 2q_y \cos 2\tau)y = 0$$

The parameters a and q can now be obtained by comparison with Eq. 4.24

$$a_x = -a_y = \frac{4eU}{m_i r_0^2 \omega^2}, \quad q_x = -q_y = \frac{2eV}{m_i r_0^2 \omega^2}, \quad \tau = \frac{\omega t}{2} \tag{4.26}$$

For a given set of U, V and ω the overall ion motion can result in a stable trajectory causing ions of a certain m/z value or m/z range to pass the quadrupole. Ions oscillating within the distance $2r_0$ between the electrodes will have stable trajectories. These are transmitted through the quadrupole and detected thereafter. The path stability of a particular ion is defined by the magnitude of the RF voltage V and by the ratio U/V.

By plotting the parameter a (ordinate, time invariant field) versus q (abscissa, time variant field) one obtains the stability diagram of the two-dimensional quadrupole field. This reveals the existence of regions where i) both x- and y-trajectories are stable, ii) either x- or y-trajectories are stable, and iii) no stable ion motion occurs (Fig. 4.33). [109] Among the four stability regions of the first category, region I is of special interest for the normal mass-separating operation of the linear quadrupole (Fig. 4.34). [103,104]

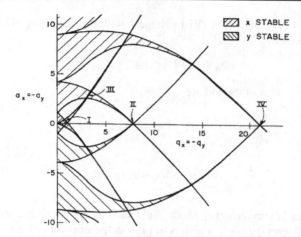

Fig. 4.33. Stability diagram for a linear quadrupole analyzer showing four stability regions (I–IV) for *x*- and *y*-motion. Reproduced from Ref. [109] with permission. © John Wiley & Sons Inc., 1986.

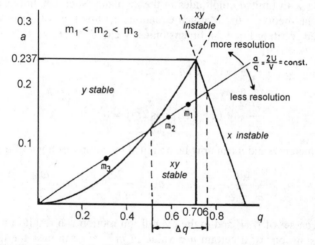

Fig. 4.34. Detail of the upper half of region I of the stability diagram for a linear quadrupole analyzer. Reproduced from Ref. [104] with permission. © World Scientific Publishing, 1993.

If the ratio *a/q* is chosen so that $2U/V = 0.237/0.706 = 0.336$, the *xy*-stability region shrinks to one point, the apex, of the diagram (cf. Eq. 4.26, Fig. 4.34). By reducing *a* at constant *q*, i.e., reducing *U* relative to *V*, an increasingly wider *m/z* range can be transmitted simultaneously. Sufficient resolution is achieved as long as only a small *m/z* range remains stable, e.g., one specific $m/z \pm 0.5$ for *unit resolution* (Chap. 4.4.3). Thus, the width (Δq) of the stable region determines the resolution (Fig. 4.35). By varying the magnitude of *U* and *V* at constant *U/V* ratio an $U/V = constant$ linked scan is obtained allowing ions of increasingly higher m/z values to travel through the quadrupole.

Overall, the quadrupole analyzer rather acts as a mass filter than as a momentum (B sector) or energy (ESA) spectrometer; hence the widespread use of the term *quadrupole mass filter*.

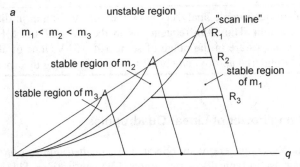

Fig. 4.35. Scanning of linear quadrupoles means performing a $U/V = constant$ linked scan. Resolution is adjusted by variation of the a/q ratio: higher a/q ratio means higher resolution, and is represented by a steeper "scan line"; $R_1 > R_2 > R_3$.

Note: Scanning of any quadrupole means shifting the whole stability diagram along a "scan line", because each m/z value has a stability diagram of its own (Figs. 4.35, 4.43). The representation of a scan by a "scan line" would only be correct in case of infinite resolution, i.e., if the apices were connected. Any real resolution is represented by a horizontal line across the stability region, where only ions falling in the region above that line are transmitted.

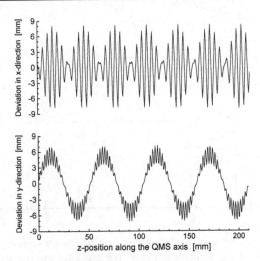

Fig. 4.36. Projection of a 3D trajectory simulation of a stable ion onto the *x*- and *y*-coordinate. Reproduced from Ref. [110] with permission. © Elsevier Science, 1998.

Ion trajectory simulations allow for the visualization of the ion motions while travelling through a quadrupole mass analyzer (Fig. 4.36). Furthermore, the optimum number of oscillations to achieve a certain level of performance can be determined. It turns out that best performance is obtained when ions of about 10 eV kinetic energy undergo a hundred oscillations. [110]

> **Note:** Standard quadrupole analyzers have rods of 10–20 mm in diameter and 15–25 cm in length. The radiofrequency is in the order of 1–4 MHz, and the DC and RF voltages are in the range of some 10^2–10^3 V. Ions of about 10 eV kinetic energy undergo approximately 100 oscillation during their passage.

4.4.3 Resolving Power of Linear Quadrupoles

Quadrupole analyzers generally are operated at so-called *unit resolution* normally restricting their use to typical low resolution (LR) applications. [111,112] At unit resolution adjacent peaks are just separated from each other over the entire *m/z* range, i.e., $R = 20$ at *m/z* 20, $R = 200$ at *m/z* 200, and $R = 2000$ at *m/z* 2000 (Fig. 4.37).

The resolution as adjusted by the U/V ratio cannot arbitrarily be increased, but is ultimately limited by the mechanical accuracy with which the rods are constructed and supported ($\pm$ 10µm). [111] Above an *m/z* value characteristic of each quadrupole assembly, any further improvement of resolution can only be achieved at the cost of significantly reduced transmission. High-performance quadrupoles allowing for about 10-fold unit resolution have only recently been developed. [112]

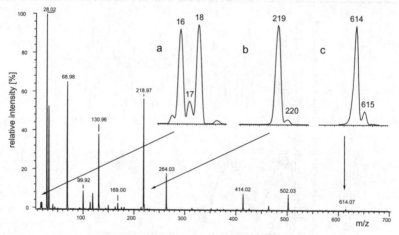

Fig. 4.37. EI mass spectrum of perfluorotributylamine (mass calibrant FC43) to demonstrate unit resolution of a quadrupole analyzer. The expanded views **a–c** show peaks separated to almost identical degree.

4.4.3.1 Performance of Cylindrical versus Hyperbolic Rods

Theoretically, each electrode should have a hyperbolic cross section for optimized geometry of the resulting quadrupole field, and thus for optimized performance. [103,104] However, cyclindrical rods are often employed instead, for ease of manufacture. By adjusting the radius of the rods carefully ($r = 1.1468r_0$), a hyperbolic field may be approximated. [113] However, even slight distortions of the ideal quadrupole field either from interference with external fields or due to low mechanical precision or inadequate shape of the device cause severe losses of transmission and resolution. [114] The expected advantages of hyperbolic rods [115] have been demonstrated by ion trajectory calculations: [110,116] circular rods cause a reduction in macromotion frequency because of an increased residence time of the ions in close vicinity to the rods; this in turn means reduced resolution.

4.4.3.2 High-Resolution with Quadrupole Mass Analyzers

Besides optimization of mechanical accuracy, the resolving power of quadrupoles can be improved by innovative modes of operation. The operation as a multiple pass system with ion reflection at either end extends the flight path and thus the number of RF cycles. The same effect can be obtained from increased radiofrequency and for ions travelling slower through the device. [111] Alternatively, the quadrupole may be operated in an other than the first stability region, e.g., in the second or fourth; [111,117] doing so requires higher ion kinetic energies of about 750 eV.

4.4.4 RF-Only Quadrupoles

Upon setting the DC voltage U to zero, the quadrupole becomes a wide band pass for ions. In the stability diagram this mode of operation is represented by an operation line equivalent to the q-axis (Fig. 4.34). Such devices are commonly known as RF-only quadrupoles (q). Hexapoles and octopoles are used analogously, because these possess better wide band pass characteristics. This property led to the widespread application of electric quadrupoles, hexapoles, and octopoles as *ion guides* and *collision cells*.

Ion guides are needed to transfer ions of low kinetic energy from one region to another without substantial losses, [118] e.g., from an atmospheric pressure ion source (ESI, APCI) to the entrance of a mass analyzer (Chap. 11).

Collisional cooling of ions travelling slowly through RF-only multipoles can be effected during transit if some collision gas is present. [119] Collisional cooling also substantially reduces the axial motion of the ions causing a confinement of the ion beam toward the central axis of the quadrupole which in turn increases the transmission efficiency through an exit aperture. This phenomenon is known as *collisional focusing*. [120]

Quadrupole or multipole collision cells [121,122] are employed in *multi quadrupole mass spectrometers*. Different from magnetic sector or TOF instruments having collimated beams of energetic ions, the ions are exiting from quadrupoles in almost any direction on the *xy*-plane. Thus, real field-free regions as employed in sector or TOF instruments have to be replaced by ion-guiding collision cells to allow CID of ions of 5–100 eV kinetic energy. RF-only collision cells need not to be straight: starting from the Finnigan TSQ700 bent geometries were introduced, and 90° cells (Finnigan TSQ Quantum) or 180° cells (Varian 1200L) are nowadays common to reduce the footprint of the instruments.

> **Note:** The initial kinetic energy of slow ions can be lost upon several collisions, thereby stopping their motion along the cell. [119,120] Under such conditions, the continuous ion current into the cell is the only impetus to push the ions through as a result of space-charge effects. The resulting dwell time of about 10 ms allows up to about 5000 (reactive) collisions to take place at some 5 Pa collision gas (or reagent gas) pressure in an octopole collision cell. [123]

4.4.5 Tandem MS with Quadrupole Analyzers

4.4.5.1 Triple Quadrupole Mass Spectrometers

Triple quadrupole mass spectrometers, QqQ, have almost become a standard analytical tool for LC-MS/MS applications, in particular when accurate quantitation is desired (Chap. 12). Ever since their introduction [124-126] they have continuously been improved in terms of mass range, resolution, and sensitivity (Fig. 4.38). [127-129]

To operate triple quadrupole mass spectrometers in the MS/MS mode, Q_1 serves as MS1, the intermediate RF-only device, q_2, acts as "field-free region" for metastable dissociations or more often as collision cell for CID experiments, and Q_3 is used to analyze the fragment ions exiting from q_2. Typically, the mass-selected ions emerging from Q_1 are accelerated by an offset of some ten electronvolts into q_2 where the collision gas (N_2, Ar) is provided at a pressure of 0.1–0.3 Pa. Careful optimization of all parameters allows major improvements of CID efficiency and resolution to be made. [130] If MS/MS is not intended, either Q_1 or Q_3 may be set to RF-only mode, thereby reducing its function to that of a simple flight tube with ion-guiding capabilities. The instrument then behaves as though it was a single quadrupole mass spectrometer.

> **Note:** At first sight, there is no difference whether Q_1 or Q_3 is switched to RF-only for MS mode. However, for EI it seems better to operate Q_3 in RF-only mode. Otherwise, the ion source would effectively extend up to the entrance of Q_3 making fragment ions more abundant due to elongated time for dissociations. Soft ionization methods do not show such differences.

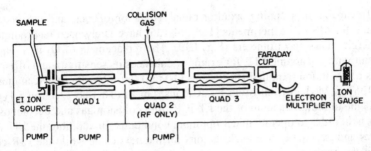

Fig. 4.38. Schematic of a triple quadrupole mass spectrometer. Reproduced from Ref. [125] with permission. © Elsevier Science, 1979.

4.4.5.2 Scan Modes for Tandem MS with Triple Quadrupole Instruments

In triple quadrupole instruments Q_1 and Q_3 are operated independently as MS1 and MS2, respectively, making MS/MS a straightforward matter. The experimental setups for product ion, precursor ion, and neutral loss scanning are summarized in Table 4.3.

Table 4.3. Scan modes of triple quadrupole instruments.

Scan Mode[a]	Operation of Q_1	Operation of q_2	Operation of Q_3
product ion, define m_1	no scan, select m_1	metastable or CID	scan up to m_1 to collect its fragments
precursor ion, define m_2	scan from m_2 upwards to cover potential precursors	metastable or CID	no scan, select m_2
constant neutral loss, define Δm	scan desired range	metastable or CID	scan range shifted by Δm to low mass

a) Masses for reaction $m_1^+ \rightarrow m_2^+ + n$.

4.4.6 Linear Quadrupole Ion Traps

Collisional cooling may bring translational ion motion along the axis of an RF-only multipole to a halt, thereby enabling ion storage within that multipole. [119] Placing electrodes of higher potential near the front and back ends of RF-only multipoles creates a trapping potential within the multipole. [131] Such devices are known as *linear (quadrupole) ion traps* (LIT).

Since a few years, LITs belong to a rapidly expanding area of instrument development. Recently, LITs have been established to collect ions externally before injecting them in bunches into an FT-ICR [132] or TOF analyzer. [131,133] Even extensive H/D exchange in the gas phase prior to mass analysis can be accomplished this way.

LITs capable of scanning, axial or radial excitation of ions, and precursor ion selection for MS/MS experiments [118,134-136] have lately been incorporated in commercial mass spectrometers (Fig. 4.39). The replacement of Q_3 in a QqQ instrument with a scanning LIT, for example, enhances its sensitivity and offers new modes of operation (Applied Biosystems Q-Trap). Introduction of a scanning LIT [118,135] as MS1 in front of an FT-ICR instrument (Thermo Electron LTQ-FT) shields the ultrahigh vacuum of the FT-ICR from collision gas and decomposition products in order to operate under optimum conditions. In addition, the LIT accumulates and eventually mass-selects ions for the next cycle while the ICR cell is still busy with the previous ion package.

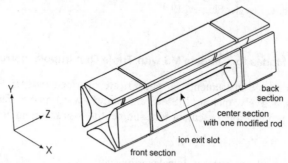

Fig. 4.39. Schematic of a linear quadrupole ion trap with scanning capability. Applying higher potential to the front and back sections creates a trapping potential for ions in the center section. Ions are exiting through the slot in one of the four rods. Adapted from Ref. [135] by permission. © Elsevier Science, 2002.

> **Note:** Here, we are going beyond the domain of the classical mass spectrometric time scale (Chap. 2.7). In ion trapping devices, ions are stored for milliseconds to seconds, i.e., 10^3–10^6 times longer than their lifetimes in beam instruments.

4.5 Three-Dimensional Quadrupole Ion Trap

4.5.1 Introduction to the Quadrupole Ion Trap

The *quadrupole ion trap* (QIT) creates a three-dimensional RF quadrupole field to store ions within defined boundaries. Its invention goes back to 1953, [103-105] however, it took until the mid-1980s to access the full analytical potential of quadrupole ion traps. [137-140] The first commercial quadrupole ion traps were incorporated in GC-MS benchtop instruments (Finnigan MAT ITD and ITMS). Electron ionization was effected inside the trap by admitting the GC effluent and a beam of electrons directly into the storage volume of the trap. Later, external ion sources became available, and soon a large number of ionization methods could be fitted to the QIT analyzer. [141-143] Modern QITs cover ranges up to about *m/z* 3000 with fast scanning at unit resolution, and in addition, offer *"zoom scans"* over smaller *m/z* ranges at higher resolution. Accurate mass measurements with QITs are to be expected in the near future. [140] Moreover, the tandem-in-time capabilities of QITs can be employed to conveniently perform MS^n experiments, [138,139] and their compact size is ideal for field applications. [11]

> **Note:** Paul himself preferred to call the device *"Ionenkäfig"* (ion cage) rather than the nowadays accepted term *quadrupole ion trap* because it does not actively act to catch ions from outside. The acronym QUISTOR derived from *quadrupole ion store* was also widespread in use.

4.5.2 Principle of the Quadrupole Ion Trap

The QIT consists of two hyperbolic electrodes serving as end caps and a ring electrode that replaces two of the linear quadrupole rods, i.e., it could theoretically be obtained by rotating a linear quadrupole with hyperbolic rods through 360° (Fig. 4.40, 4.41). Thus, a section through the *rz*-plane of the QIT closely resembles that of the entrance of linear quadrupole with hyperbolic rods (cf. Fig. 4.31b). [107,141] However, the angle between the asymptotes enclosing the ring electrode is 70.5° (2·arctan(1/√2)) instead of 90°. The end caps are electrically connected and the DC and RF potentials are applied between them and the ring electrode. The working principle of the QIT is based on creating stable trajectories for ions of a certain *m/z* or *m/z* range while removing unwanted ions by colliding them with the walls or by axial ejection from the trap due to their unstable trajectories. [137]

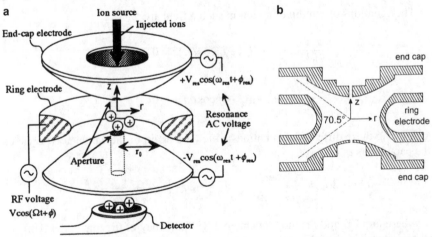

Fig. 4.40. Schematic of a quadrupole ion trap. (**a**) QIT with external ion source (illustration stretched in *z*-direction) and (**b**) section in the *rz*-plane (in scale). (**a**) Reproduced from Ref. [144] by permission. © John Wiley & Sons, 2000.

Fig. 4.41. Electrodes of the Finnigan MAT ITS40 quadrupole ion trap. By courtesy of Thermo Electron (Bremen) GmbH.

For the QIT, the electric field has to be considered in three dimensions. Let the potential Φ_0 be applied to the ring electrode (xy-plane) while $-\Phi_0$ is applied to the hyperbolic end caps. Then, the field can be described in cylindrical coordinates by the expression [137,141]

$$\Phi_{x,y,z} = \frac{\Phi_0}{r_0^2}(r^2 \cos^2\theta + r^2 \sin^2\theta - 2z^2) \tag{4.27}$$

Because of $\cos^2 + \sin^2 = 1$ this reduces to

$$\Phi_{x,y,z} = \frac{\Phi_0}{r_0^2}(r^2 - 2z^2) \tag{4.28}$$

The equations of motion of an ion in such a field are

$$\frac{d^2z}{dt^2} - \frac{4e}{m_i r_0^2}(U - V \cos\omega t)z = 0 \tag{4.29}$$

$$\frac{d^2r}{dt^2} + \frac{2e}{m_i r_0^2}(U - V \cos\omega t)r = 0$$

Solving these differential equations which are again of the Mathieu type yields the parameters a_z and q_z

$$a_z = -2a_r = -\frac{16eU}{m_i r_0^2 \omega^2}, \quad q_z = -2q_r = -\frac{8eV}{m_i r_0^2 \omega^2} \tag{4.30}$$

where $\omega = 2\pi f$ and f is the fundamental RF frequency of the trap (≈ 1 MHz). To remain stored in the QIT, an ion has to be simultaneously stable in the r and z directions. The occurrence of stable ion trajectories is determined by the stability parameters β_r and β_z which depend on the parameters a and q. The borders of the first stability region are defined by $0 < \beta_r, \beta_z < 1$. [145]

A stability diagram can be drawn where the stability region closest its origin is of greatest importance for the operation of the QIT (Fig. 4.42). At a given U/V ratio, ions of different m/z are found along a straight line crossing the stability region. Ions of higher m/z are located nearer to the origin than lighter ones. The regions of stability as plotted in the a/q plane are represented as envelopes of characteristic shape. Ions with their m/z value inside the boundaries are stored in the QIT. The lower limit of the trapped m/z range is defined by $q_z = 0.908$.

4.5.2.1 Visualization of Ion Motion in the Ion Trap

The way the three-dimensional quadrupole field acts to keep ions within a certain volume, i.e., within a potential well some electron volts in depth, can be illustrated by a mechanical analogue: A ball has to be prevented from rolling from a saddle by rotating the saddle just right to bring the ball back to the middle before it can leave the surface via one of the steeply falling sides (Fig. 4.43). Paul demonstrated the dynamic stabilization of up to three steel balls by such a device in his Noble lecture. [103,104]

The trajectories of low-mass ions in a QIT were shown to be similar to those observed for charged aluminum dust particles. [146-149] Wuerker recorded Lissajous trajectories, superimposed by the RF drive frequency, as a photomicrograph (Fig. 4.44). [146] The complex motion of the ions is the result of the two superimposed secular oscillations in r and z direction.

The use of a light buffer gas (0.1 Pa He) to dampen the ion motion towards the center of the trap significantly enhances resolution and sensitivity of the QIT. [150]

4.5.3 Operation of the Quadrupole Ion Trap

4.5.3.1 Mass-Selective Stability Mode

The whole range of ions is generated within or admitted to the QIT, but solely ions of one particular m/z are trapped at a time by setting appropriate parameters of the QIT. Then, the stored ions are pulsed out of the storage volume by applying a negative pulse to one of the endcaps. [151,152] Thus, they hit the detector located behind a hole in the center of one of the endcaps. A full-scan mass spectrum is obtained by addition of several hundred single steps, one for each nominal m/z value. This is the so-called *mass-selective stability* mode of the QIT. [153,154] The mass-selective stability mode is no longer in use, because it is too slow and provides poor sensitivity as most ions are wasted.

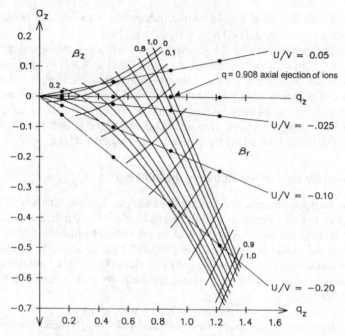

Fig. 4.42. Stability diagram for the quadrupole ion trap. The points collected on a common line mark the a/q values of a set of ions. Each line results from different settings of the U/V ratio. Reproduced from Ref. [137] by permission. © John Wiley and Sons, 1989.

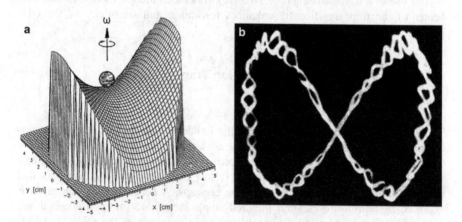

Fig. 4.43. Visualization of ion motion in the ion trap. (**a**) Mechanical analogue of the QIT. (**b**) Photograph of ion trajectories of charged aluminum particles in a quadrupole ion trap. (**a**) Reproduced from Ref. [104] with permission. © World Scientific Publishing, 1993. (**b**) Reproduced from Ref. [146] with permission. © American Institute of Physics, 1959.

4.5.3.2 Mass-Selective Instability Mode

First, the full m/z range of interest is trapped within the QIT. The trapped ions may either be created inside the QIT or externally. Then, with the end caps grounded an RF-voltage scan (V) is applied to the ring electrode causing consecutive ejection of ions in the order of their m/z values. This is known as *mass-selective instability (ejection)* mode. [150,155] It can be represented in the stability diagram by a horizontal line from the origin to the point of axial ejection at $q_z = 0.908$. The timing sequence is shown in Fig. 4.44.

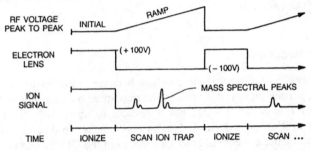

Fig. 4.44. Timing sequence used for mass-selective instability mode (about 1.5 cycles shown). With an external ion source the ionization time is replaced by the ion injection pulse. Reproduced from Ref. [150] by permission. © Elsevier Science, 1984.

4.5.3.3 Resonant Ejection

Modern QITs employ the effects of *resonant ejection* to remove ions of successively increasing m/z value from the storage volume, i.e., to achive a scan. In an ideal QIT, the motions of the ions in radial and axial directions are mutually independent. Their oscillations can be described by a radial and an axial secular frequency, each of them being a function of the Mathieu trapping parameters a and q. If a supplementary RF voltage which matches the axial secular frequency is applied to the end caps, resonant ejection of ions occurs at $q < 0.908$ (Fig. 4.42). [156] Excitation occurs when the frequency of a supplementary RF signal matches the secular frequency of a trapped ion in z direction. The secular frequency components in axial direction (ω_z) are given by $\omega_z = (n + \beta_z/2)\Omega$, where Ω represents the angular frequency, n is an integer, and β_z is determined by the working point of an ion within the stability diagram. [157] In the special case when $\beta_z = 1$ and $n = 0$, the fundamental secular frequency is exactly half of the RF drive frequency applied between ring electrode and end caps.

Example: To effect resonant ejection for the set of $\beta_z = 0.5$ and $n = 0$ we have $\omega_z = (0 + 0.5/2) = 0.25\Omega$, i.e., $^1/_4$ of the RF drive frequency has to be applied to eject ions at the $\beta_z = 0.5$ borderline. By scanning the voltage of the RF drive frequency upwards, ions of inreasing m/z ratio are successively ejected.

Scans based on resonant ejection may either be carried out in a forward, i.e., from low to high mass, or a reverse manner. This allows for the selective storage of ions of a certain m/z value by elimination of ions below and above that m/z value from the trap. Thus, it can serve for precursor ion selection in tandem MS experiments. [156,158] Axial excitation can also be used to cause collision-induced dissociation (CID) of the ions as a result of numerous low-energy collisions with the helium buffer gas that is present in the trap in order to dampen the ion motion. [150,156] A substantial increase of the mass range is realized by reduction of both the RF frequency of the modulation voltage and the physical size of the QIT. [154,159,160]

4.5.3.4 Axial Modulation

Ion trapping devices are sensitive to overload because of the detrimental effects of coulombic repulsion on ion trajectories. The maximum number of ions that can be stored in a QIT is about 10^6–10^7, but it reduces to about 10^3–10^4 if unit mass resolution in an RF scan is desired. *Axial modulation*, a sub-type of resonant ejection, allows to increase the number of ions stored in the QIT by one order of magnitude while maintaining unit mass resolution. [160,161] During the RF scan, the modulation voltage with a fixed amplitude and frequency is applied between the end caps. Its frequency is chosen slightly below $1/2$ of the fundamental RF frequency, because for $\beta_z \leq 1$, e.g., $\beta_z = 0.98$, we have $\omega_z = (0 + 0.98/2) = 0.49 \times \Omega$. At the stability boundary, ion motion is in resonance with this modulation voltage, and thus ion ejection is facilitated. Axial modulation basically improves the mass-selective instability mode of operation.

If resolution is not important, scanning of QITs can be very fast, a property that can be employed to make a pre-scan. Then, the result of the pre-scan is used to adjust the number of ions inside the QIT close to the optimum for the subsequent analytical scan. Such an *automatic gain control* (AGC) [138,162] gives increased sensitivity at low sample flow and avoids overload of the QIT at high sample flow.

Provided sufficiently high scan rates are also available whilst resolution is preserved, the pre-scan can be omitted. Instead, a trend analysis based on a set of two or three preceding analytical scans can be performed. This procedure avoids wasting of ions and results in further optimization of the filling level of the QIT. The exploitation of the phenomenon of nonlinear resonances turned out to be of key importance for the realization of this method.

Example: Tandem mass spectrometric experiments in quadrupole ion traps are performed by combining the techniques of resonant ejection, and forward and reverse scanning to achieve an optimum in precursor ion selection, ion activation, and fragment ion scanning (Fig. 4.45). [156]

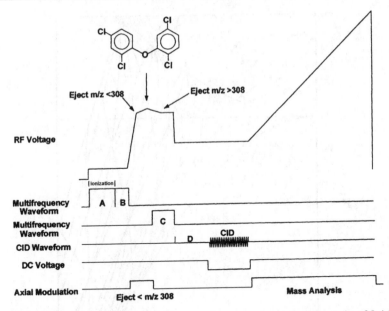

Fig. 4.45. Complex scan function used for a tandem mass spectrometric study of 2,4,3',6'-tetrachlorodiphenylether. Reproduced from Ref. [156] by permission. © John Wiley and Sons, 1997.

4.5.3.5 Nonlinear Resonances

In any real ion trap higher multipole fields, in particular octopole fields, are induced by deviations from the ideal electrode structure. The trapping potential may then be represented as a sum of an ideal quadrupole field and weak higher order field contributions. [145,163,164] Application of an excitation voltage across the end caps induces dipole and hexapole fields in addition. Those higher order fields in the QIT may have beneficial effects such as increase in mass resolution in the resonant ejection mode, but may also result in losses of ions due to nonlinear resonances. [165] Nonlinear resonances have been known for long, [166,167] but useful theoretical descriptions were only recently developed. [145,164,168,169] The condition for the appearance of instabilities is related to certain frequencies through the stability parameters β_r and β_z and the integer multiples n_r and n_z. The locations of instability spread like a net over the stability diagram and have been experimentally verified with astonishing accuracy (Fig. 4.46). [145] Excitation of ions with a suitable frequency can cause their fast ejection from the trap due to the sudden shift to nonlinear stability. Nonlinear resonances can thus be exploited to realize very fast scans of the QIT (26,000 u s^{-1}) while maintaining good resolution, [169] e.g., irradiation of $0.33 \times \Omega$ amplifies hexapole resonances and causes sudden ejection at the $\beta_z = 0.66$ borderline.

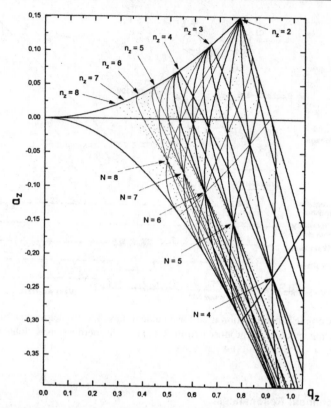

Fig. 4.46. Plot of theoretical lines of instability corresponding to the relation $n_r/2\beta_r + n_z/2\beta_z = 1$ for different orders $N = n_r + n_z$. Strong resonances are represented by thick lines, weak ones by dashed lines. Reproduced from Ref. [145] by permission. © Elsevier Science, 1996.

4.5.4 External Ion Sources for the Quadrupole Ion Trap

Chemical ionization (CI) mass spectra were first obtained by using the mass-selective instability mode of the QIT. [154,155,170] The reagent gas was admitted into the QIT, ionized and then allowed to react with the analyte.

With external ion sources it became feasible to interface any ionization method to the QIT mass analyzer. [171] However, commercial QITs are chiefly offered for two fields of applications: i) GC-MS systems with EI and CI, because they are either inexpensive or capable of MS/MS to improve selectivity of the analysis (Chap. 12); and ii) instruments equipped with atmospheric pressure ionization (API) methods (Chap. 11) offering higher mass range, and some 5-fold unit resolution to resolve isotopic patterns of multiply charged ions (Fig. 4.47). [149,149,162,172,173]

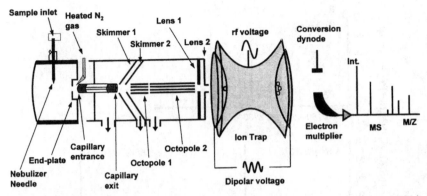

Fig. 4.47. Schematic of a quadrupole ion trap instrument equipped with an external ESI ion source. Reproduced from Ref. [173] by permission. © John Wiley & Sons, 2000.

4.5.6 Tandem MS with the Quadrupole Ion Trap

Ion traps are *tandem-in-time* instruments, i.e., they perform the steps of precursor ion selection, ion activation and acquisition of fragment ion spectra in the very same place. This advantageous property allows the multiple use of a single QIT to perform not only MS^2 but also MS^3 and higher order MS^n experiments – indeed a very economic concept. Depending on the abundance of the initial precursor ion, its fragmentation behavior – and of course, on the performance of the QIT – MS^6 experiments are possible. [138] However, in contrast to tandem-in-space instruments, tandem-in-time instruments do not support constant neutral loss and precursor ion scans.

In the QIT, MS^n is accomplished by using appropriate scan functions for the fundamental RF and the auxiliary modulation voltage. [138,154,162] At a sufficient level of sophistication, e.g., by combining slow and fast forward and reverse RF voltage scans with suitable settings of the auxiliary voltage, monoisotopic precursor ions can even be isolated in case of triply charged ions. [174] The resonance excitation provided by moderate auxiliary voltages can be employed to effect low-energy CID of the trapped ions due to activating collisions with the buffer gas. A full description of the numerous approaches to continuously improving scans can be found in the literature. [159,174-177]

Example: MS^n on a QIT was used for the identification of beauverolides, cyclic peptides from the fermentation broth of *Beauveria bassiana*, a fungus of pathogenic activity on insects. [178] All MS^n (ESI-CID-QIT) experiments started from singly charged $[M+H]^+$ precursor ions (Fig. 4.48).

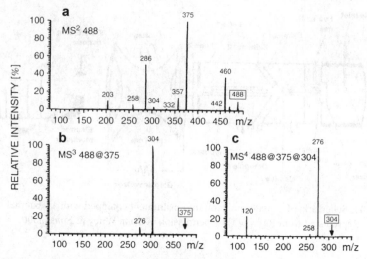

Fig. 4.48. Sequence of ESI-CID-QIT mass spectra of a beauverolide: (a) MS^2 of $[M+H]^+$, m/z 488, selected out of a full scan spectrum, (b) MS^3 of m/z 375 selected from (a), and (c) MS^4 of m/z 304 selected from (b). Adapted from Ref. [178] by permission. © John Wiley & Sons, 2001.

4.6 Fourier Transform Ion Cyclotron Resonance

4.6.1 Introduction to Ion Cyclotron Resonance

The development that led to modern *Fourier transform ion cyclotron resonance* (FT-ICR) mass spectrometers began in 1932. [179,180] It was demonstrated that the angular frequency of the circular motion of ions moving perpendicular to a homogeneous magnetic field is independent of the radius the ions are travelling on. Applying a transverse alternating electric field of the same frequency allows to achieve tremendous acceleration of the circulating ions while the radius of their orbit increases. Later, the working principle of ICR accelerators was applied to construct an ICR mass spectrometer. [181,182] In the first ICR instruments, magnetic field strength and frequency were fixed and the m/z value was obtained from the energy absorption, i.e., the number of half cycles, until the ions struck an electrometer plate. ICR mass spectrometers measuring the power absorption from the exciting oscillator followed and were commercialized in the mid 1960s by Varian. Starting from their application to gas phase ion chemistry, [183] ICR instruments made their way in analytical mass spectrometry. [184] However, it was the introduction of FT-ICR in 1974 that initiated the major breakthrough. [185,186] Ever since, the performance of FT-ICR instruments has steadily improved [187,188] to reach unprecedented levels of resolution and mass accuracy when superconducting magnets are employed. [189-192]

Note: Modern FT-ICR mass spectrometers offer ultrahigh resolving power ($R = 10^5$–10^6) [193,194] and highest mass accuracy ($\Delta m = 10^{-4}$–10^{-3} u, cf. examples in Chaps. 3.3.2 and 3.4.1), attomol detection limits (with nanoESI or MALDI sources), high mass range and MS^n capabilities. [195]

4.6.2 Principle of Ion Cyclotron Resonance

As we know from the discussion of magnetic sectors, an ion of velocity v entering a uniform magnetic field B perpendicular to its direction will move on a circular path by action of the Lorentz force (Chap. 4.3.2), the radius r_m of which is determined by Eq. 4.13:

$$r_m = \frac{m_i v}{qB} \tag{4.13}$$

Upon substitution with $v = r_m \omega$ the angular frequency ω_c becomes:

$$\omega_c = \frac{qB}{m_i} \tag{4.31}$$

Hence, the cyclotron angular frequency ω_c is independent of the ion's initial velocity, but a function of its mass, charge, and the magnetic field. By applying a transverse electric field alternating at the cyclotron frequency f_c ($\omega_c = 2\pi f_c$) the ions are accelerated. Such a field can be applied by a pair of RF electrodes placed on opposite sides of the orbit. As the ions accelerate, the radius of their orbit increases, and the resulting overall motion is a spiral (Fig. 4.49). [180,188] For lighter ions, the spiral reaches the same radius with fewer cycles than in case of heavier ones, i.e., the spiral is steeper, because low-mass ions need less energy than high-mass ions to accelerate to a certain velocity. The working principle of the first-generation ICR instruments essentially was to perform an energy scan: the m/z value was obtained from the number of half cycles, until the ions struck an electrometer plate at $r = r_{cell}$. [181,182]

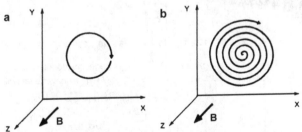

Fig. 4.49. Motion of positive ions in a uniform magnetic field B. (**a**) The radius is a function of ion velocity, but the frequency f_c of circulation is not. (**b**) Excitation of the ions by an RF electric field oscillating at their cyclotron resonance frequency. Adapted from Ref. [196] by permission. © John Wiley & Sons, 1986.

The disadvantages of this concept are clear: i) mass accuracy and resolution are limited to $1/N_c$ (N_c = number of half cycles); ii) the electric signal for ion detection is solely due to neutralization of the ions, and there is no amplification as obtained with multiplier-type detectors used with all other analyzers; and iii) the ions are removed from the cell upon detection making MS/MS impossible.

4.6.3 Fourier Transform Ion Cyclotron Resonance

FT-ICR circumvents the above disadvantages of scanning ICR instruments. The FT-ICR experiment requires full temporal separation of excitation and subsequent detection of the trapped ions. Detection is based on the measurement of image currents in the detector plates. An image current is induced by each ion package when repeatedly passing the detector plates at its individual cyclotron frequency, i.e., detection in FT-ICR means "listening to the circulating ions". The transient *free induction decay* (FID) is recorded, and afterwards, the FID is converted from the *time domain* to the *frequency domain* by means of Fourier transformation. This means that the complex FID caused by superimposition of many single frequencies is deconvoluted to reveal the single contributing frequencies and their respective amplitudes. Using Eq. 4.31, the frequencies are converted to *m/z* values; their amplitudes now representing the abundances of the corresponding ions (Fig. 4.50). [180,185,188,197,198] Some advantages of FT-ICR-MS are obvious: i) the $1/N_c$ limit vanishes because every ion makes some 10^4–10^8 cycles during the detection interval; ii) sensitivity improves because the ions give rise to a detectable image charge during each passage of a detector plate; and iii) ion detection is non-destructive, i.e., ions are not lost upon detection giving the opportunity to perform MS/MS experiments. In addition, elongated recording of the FID allows for an extremely precise determination of all cyclotron frequencies, thereby yielding the highest values of resolution and mass accuracy available. [193,194,199]

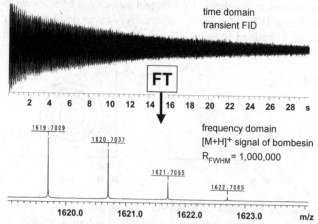

Fig. 4.50. Illustration of the effect of Fourier transformation. The longer the detection interval the more accurate the result will be. By courtesy of Bruker Daltonik, Bremen.

> **Note:** Occasionally, the acronym FTMS is used instead of FT-ICR-MS. Of course, ICR without Fourier transformation would not have the tremendous success it has, but Fourier transformation alone cannot separate ions according to m/z, and hence there is no FTMS.

4.6.4 Experimental Setup of FT-ICR-MS

For FT-ICR-MS, the ICR cell itself must not necessarily differ from one used in scanning ICR-MS (Fig. 4.51). Two of the four side walls (x-axis) of the ICR cell are connected to the RF power supply during the period of excitation. Then, the image current induced in the detector plates (y-axis) is recorded as transient signal for some period of time (0.5–30 s). The excitation of the ions within the ICR cell has to stop at a level low enough to avoid wall collisions of the lightest ions to be measured. [191,200,201]

4.6.4.1 Cyclotron Motion in FT-ICR-MS

Regardless of whether the ions are created inside the cell or whether they come from an external ion source, they will not be at rest because of their thermal energy. This gives rise to a circular micromotion of the ions in the xy-plane as they are entering the magnetic field. Upon excitation, the circular micromotion is superimposed by the cyclotron motion, i.e., the RF excitation field forms coherent ion packages composing of ions of the same m/z value. As the initial kinetic energy of the ions is small as compared to the energy uptake from the RF field, it is of minor importance for the experiment. [201] Nonetheless, the complexity of the overall motion affects frequency-to-mass calibration if accurate results are required. [202]

Example: Consider a singly charged ion of 100 u at thermal energy. Assuming a temperature of 300 K, its average velocity (Boltzmann distribution) is about 230 m s^{-1}. In a 3 T magnetic field it will circulate at $r_m \approx 0.08$ mm. To increase the radius to 1 cm, Eq. 4.13 demands the velocity to rise by a factor of 125, i.e., to 28,750 m s^{-1}. Rearranging Eq. 4.3 delivers $eU = v2m_i/2$, and thus we calculate a kinetic energy of about 430 eV. Thus, the translational energy of ions in an ICR cell is definitely high enough to effect activating collisions for CID experiments.

> **Note:** In ICR cells, the ions circulate like separate swarms of birds rather than like matter in the rings of Saturn. If ions of the same m/z non-coherently circulated at the same frequency and radius, but occupied the total orbit rather than a small sector of it, there would be no image current induced upon their passage at the detector plates.

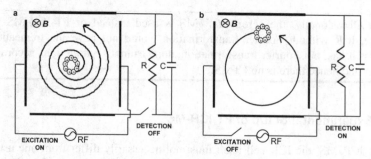

Fig. 4.51. The sequence of excitation (**a**) and detection (**b**) in FT-ICR-MS. The ionic micro-motion is indicated by the small circle of ions. The cell is shown along the direction of the magnetic field.

4.6.4.2 Axial Trapping in FT-ICR Instruments

The z-dimension of the cell seems to be of no importance for the function of an (FT-)ICR mass spectrometer. However, the z-component of thermal energy and the kinetic energy of ion injection into the ICR cell in case of an external ion source both would lead to rapid loss of the trapped ions along that axis. It is there-fore important to establish a trapping potential in z-direction. The simplest way is to place DC trapping electrodes at the "open" sides of a cubic or cylindrical cell. [182] However, due to field imperfections other designs have been developed. [200,201] These include segmented end caps on cyclindrical cells (*infinity cell* [203]), cyclindrical cells devided into three segments along the z-axis [204] analogous to the LIT mass spectrometer (Fig. 4.39), and many others. [205]

4.6.4.3 Magnetic Field Strengths in FT-ICR-MS

FT-ICR instruments from any vendor (Bruker Daltonik, IonSpec, Thermo Electron) are exclusively equipped with superconducting magnets, with 7 T and 9.4 T being the favorite field strenghts. The advantages of increased field strength are numerous: resolving power and scan speed, for example, increase linearly with B. Moreover, upper mass limit, ion trapping time, ion energy, and number of trapped ions even increase with B^2.

4.6.5 Excitation Modes in FT-ICR-MS

In the ICR cell, there is a stringent correlation of cyclotron frequency f_c and m/z value. For simplicity, the very first FT-ICR experiment was therefore performed with an excitation pulse of a fixed f_c tailored to fit the model analyte, methane molecular ion. [185] However, useful measurements require the simultaneous ex-citation of all ions in the cell, and this in turn demands for a large RF bandwidth.

Example: Rewriting Eq. 4.31 to obtain the cyclotron frequency f_c delivers:

$$f_c = \frac{qB}{2\pi m_i}$$

(4.31a)

We may now calculate f_c for a singly charged ion of m/z 1500 (2.49×10^{-24} kg) in a 7 T magnet as $f_c = (1.66 \cdot 10^{-19} \text{ C} \times 7 \text{ T})/(2\pi \times 2.49 \times 10^{-24} \text{ kg}) = 71,600$ Hz. For an ion of m/z 15 (2.49×10^{-26} kg) we get a 100 times higher f_c, i.e., to excite ions in the range m/z 15–1500, the ICR frequency band has to span from 72 kHz to 7.2 MHz.

Several excitation methods have been developed to cope with this technically non-trivial task. The differences between these methods can easily be judged from the shape of the frequency-domain excitation waveforms (Fig. 4.52). [188,201] The simplest approach is to irradiate a single frequency (narrow band) for some time, i.e., as a *rectangular pulse*, but this selectively excites ions of a single mass. [185] Applying such a narrow band excitation as a short pulse yields some expansion of the range. The effect of a *frequency sweep* or *"chirp" excitation* is much better, [186] however, it gives distortions close to the borders of the range. The best results are obtained from *stored waveform inverse Fourier transform* (SWIFT) excitations. [206] The ideal excitation waveform is tailored to the needs of the intended experiment and then produced by an RF generator. SWIFT excitation also allows to remove ions of predefined m/z ranges from the ICR cell. This results in storage of a small m/z range, or after repeated SWIFT pulsing of a single nominal mass out of a broad band spectrum. Those ions are then accessible for ultrahigh resolution and accurate mass measurements.

> **Note:** Although resolution and mass accuracy are already high in broadband spectra ($R = 3 \times 10^4$–10^5, $\Delta m < 10^{-3}$ u), the performance of FT-ICR instruments improves by at least one order of magnitude by storing only narrow m/z ranges, because fewer ions in the cell mean less distortion by coulombic interactions.

4.6.6 Detection in FT-ICR-MS

The ICR voltage signal strength at the detector plates is inversely proportional to ion mass if the monitoring circuit is predominantly resistive, and is independent of ion mass if the circuit is predominantly capacitive. [198] Image current detection at room temperature is typically less sensitive than ion counting techniques in beam instruments. However, for a modern FT-ICR instrument equipped with typical detection circuits, the detection limit to yield a signal-to-noise ratio of 3:1 corresponds to roughly 200 ions, provided these are excited to travel at half of the maximum cyclotron radius. [200]

An interesting correlation exists between the transient signal and the pressure in the ICR cell. [197,201] In a perfect vacuum, the orbiting motion would solely be

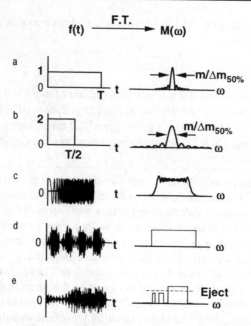

Fig. 4.52. Time-domain (left) and frequency-domain (right) excitation waveforms: (**a**), (**b**) rectangular pulses; (**c**) "chirp" excitation; (**d**), (**e**) SWIFT excitations, with (**e**) formed to eject a certain mass range from the cell. Reproduced from Ref. [201] by permission. © John Wiley & Sons, 1998.

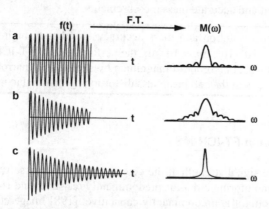

Fig. 4.53. Simulated time-domain ICR signals (left) and frequency-domain spectra (right) for (**a**) low pressure, (**b**) medium pressure, and (**c**) high pressure. Reproduced from Ref. [201] by permission. © John Wiley & Sons, 1998.

dampened by induction, whereas in the presence of residual gas collisions will finally slow down the ions to thermal energy. The simulated time-domain ICR signals at i) low pressure (almost no collisions during acquisition time), ii) medium pressure (single collision on the average), and iii) high pressure (several collisions during acquisition time) have been calculated, and then the expected frequency-domain peak shapes have been obtained by Fourier transformation (Fig. 4.53). As the resolution in FT-ICR-MS is also proportional to the acquisition time of the FID signal, ultrahigh vacuum is advantageous.

> **Note:** Besides from elongated detection intervals, mass accuracy of FT-ICR-MS also benefits by a factor of 3–10 from internal mass calibration compared to external calibration (Chap. 3.3.5). External calibration works the better, the closer the cyclotron radius and the number of calibrant ions fit the conditions of the intended analytical measurements. [200]

4.6.7 External Ion Sources for FT-ICR-MS

Ion traps, ICR cells as well as QITs, are best operated with the number of trapped ions close to their respective optimum, because otherwise ion trajectories are distorted by coulombic repulsion. Hence, external ion sources in combination with ion transfer optics capable of controlling the number of ions injected are ideally attached to ion traps. Currently, MALDI [207] and ESI (Fig. 4.54) [192-194,199,208] ion sources are predominating in FT-ICR work. The ion production may either be regulated by the source itself, or alternatively, by some device to collect and store the desired amount of ions from that source until injection into the ICR. For that purpose, linear RF multipole ion traps are often employed (Chap. 4.4.6), [118,209] but other systems are also in use. [195] RF-only multipoles are commonly used to transfer the ions into the cell (Chap. 4.4.4). For the injection, it is important to adjust the conditions so that the ions have low kinetic energy in z-direction in order not to overcome the shallow trapping potential.

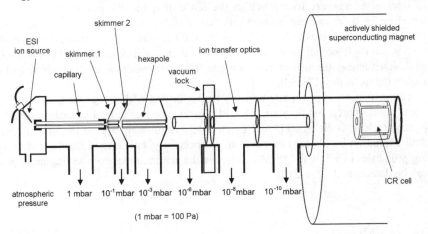

Fig. 4.54. Ion transfer optics and differential pumping stages to adapt an ESI source to an FT-ICR instrument. Only the ICR cell is inside the superconducting magnet. By courtesy of Bruker Daltonik, Bremen.

While some buffer gas is beneficial in case of QITs, ICR cells are preferably operated at the lowest pressure available. The typical path from an external ion source into the ICR cell is therefore characterized by multistep differential pumping to achieve some 10^{-8}–10^{-7} Pa in the cell.

Note: The need for almost perfect vacuum, i.e., extremely long mean free paths, in FT-ICR mass spectrometers arises from the combination of high ion velocities of several 10^3 m s^{-1}, observation intervals in the order of seconds, and the effect of collisions on peak shape.

4.6.8 Tandem MS with FT-ICR Instruments

FT-ICR mass spectrometers belong to the *tandem-in-time* category of instruments. The stage of precursor ion selection (MS1) is accomplished by selectively storing the ions of interest, whereas all others are ejected by means of a suitably tailored excitation pulse, e.g., using the SWIFT technique. [206] FT-ICR mass spectrometers are also capable of MS^n.

Collision-induced dissociation (CID) requires sufficiently fast ions to collide with some gas atoms in order to cause dissociation. The collision gas is admitted to the ICR cell through a rapid pulsed valve. The increase in ion kinetic energy is effected by short (0.2–0.5 ms) irradiation at or close to the cyclotron frequency of the precursor ion. [210,211] Nowadays, *sustained off-resonance irradiation* (SORI) has established as the standard CID method in FT-ICR, [212,213] although there are others of comparable effect. [214,215] Irradiating slightly off the resonance frequency makes ions undergo acceleration–deceleration cycles throughout the duration of the RF pulse. As a consequence, only small increments of internal energy are transferred to the ion during the RF pulse. The ions can therefore be irradiated for a sustained period (0.5-1 s) without causing ejection, that otherwise at f_c was unavoidable. SORI-CID results in sequential activation of ions by multiple collisions of low (<10 eV) kinetic energy ions with the collision gas. Nonetheless, the use of collision gas is in contradiction to the high vaccum requirements of the ICR cell.

The activation step can alternatively be performed without gas by means of infrared multiphoton dissociation (IRMPD) or electron capture dissociation (ECD) (Chap. 2.12.2). Both IRMPD and ECD, solely require storage of the ions during their excitation by photons or electrons, respectively. It is one of the most charming properties of FT-ICR-MS/MS that even the accurate mass of the fragment ions can be determined. [216,217]

Fig. 4.55. Bruker Daltonics ApexIII FT-ICR-MS. By courtesy of Bruker Daltonik, Bremen.

4.7 Hybrid Instruments

As indicated in the section on linear ion traps (Chap. 4.4.6), mass spectrometers may be constructed by combining different types of mass analyzers in a single so-called *hybrid instrument*. [86,100] The driving force to do so is the desire to obtain mass spectrometers ideally combining the advantagous properties of each mass analyzer they are composed of in order to have powerful, but still affordable machines for tandem MS. The development of hybrid instruments started from magnetic sector-quadrupole hybrids, either BqQ, [218,219] EBqQ, [220] or BEqQ. [221-223] Numerous other systems such as magnetic sector-QIT, [224] magnetic sector-oaTOF, [225,226] QITTOF, [147,227,228] QqTOF, [47,229-231] QqLIT, [134] LITICR, and QqICR followed (Table 4.4).

The adaptation of different mass analyzers to each other demands for sophisticated interfaces, because of the largely differing requirements concerning ion kinetic energies. The magnetic sector-quadrupole or magnetic sector-QIT hybrids, for example, require keV ions exiting the sector to be decelerated to some 10 eV before entering the qQ section, or to be slowed down and pulsed into the QIT, respectively.

Geometries composing of an oaTOF as MS2 bear the advantage that advanced TOFs offer accurate mass measurements close the the accuracy of magnetic sector instruments. Currently, QqTOF systems can be regarded as the commercially most successful hybrid. While the linear quadrupole serves as MS1 in MS/MS experiments, it is operated in RF-only mode when tandem MS is not intended, because this allows to acquire full-range spectra at high resolution using the TOF analyzer (Fig. 4.56).

The next higher level of performance can be achieved by replacing the oaTOF MS2 with an FT-ICR analyzer while employing a linear ion trap (Thermo Electron LTQ-FT) or a quadrupole as MS1 (Bruker Daltonik APEX-Q).

> **Note:** Besides accommodating to their versatility, there is nothing new to understand with hybrid instruments. Though exotic at a first glance, hybrids are governing today's market of mass spectrometers.

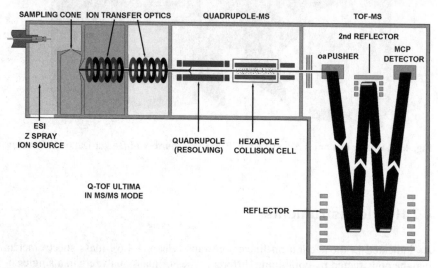

Fig. 4.56. Schematic of the Q-TOF Ultima with ESI ion source in MS/MS mode. The TOF analyzer has a double reflector for higher resolution. Courtesy of Waters Corporation, MS Technologies, Manchester, UK.

Table 4.4. Hybrid mass analyzers

MS1	Properties of MS1	MS2	Properties of MS2
BE or EB	LR and HR	Q	LR, low energy CID
BE or EB	LR and HR	QIT	LR, low energy CID, MS^n
EB or EBE	LR and HR	oaTOF	LR, low energy CID, high sensitivity
Q	LR, low energy CID	oaTOF	HR, high sensitivity
Q	LR, low energy CID	LIT	LR, higher sensitivity than QqQ
QIT	LR, low energy CID, MS^n	TOF	HR, high sensitivity
LIT	LR, low energy CID, MS^n	ICR	ultrahigh resolution and mass accuracy

4.8 Detectors

The simplest detector is a *Faraday cup*, i.e., an electrode where the ions deposit their charge. The electric current flowing away from that electrode results in a voltage when passing through a resistor of high impedance. Faraday cups are still in use to measure abundance ratios with highest accuracy in *isotope ratio mass spectrometry* (IR-MS). [232] In the era of Mattauch-Herzog type instruments, the *photographic plate* has been the standard detection system (Chap. 4.3.3). With the advent of scanning mass spectrometers, *secondary electron multipliers* (SEM) became predominant. [233] These rely on the emission of secondary electrons from surfaces upon impact of energetic ions. Progress has been made to employ *cryogenic detectors*, a rather special type of an ion counting detector for high-mass ions in TOF-MS. [234] The first commercial cryogenic detector TOF instrument has only recently been released (Comet Macromizer, Flamatt, Switzerland). Ion counting detectors are not used in FT-ICR-MS where *image current detection* yields superior results (Chap. 4.6.6).

> **Note:** Ion counting detectors also give signals upon impact of energetic neutrals, electrons, or photons. Therefore, care has to be taken, not to allow other particles than the mass-analyzed ions to hit the detector.

4.8.1 Discrete Dynode Electron Multipliers

When an energetic particle impinges on the surface of a metal or a semiconductor, secondary electrons are emitted from that surface. The ease of such an emission is determined by the *electron(ic) work function* of the respective material, e.g., BeCu alloy oxide ($w_e \approx 2.4$ eV). [235] The higher the velocity of the impacting particle [233,236] and the lower the electron work function of the surface, the larger the number of secondary electrons. If an electrode opposite to the location of emission is held at more positive potential, all emitted electrons will be accelerated towards and hit that surface where they in turn cause the release of several electrons each. The avalanche of electrons produced over 12–18 discrete dynode stages held at about 100 V more positive potential each causes an electric current large enough to be detected by a sensitive preamplifier. Such a detector is called *secondary electron multiplier* (SEM, Fig. 4.57). [237] The dynodes are normally cup-shaped, but stacks of Venetian blind-like dynodes have also been in use. Due to a certain air sensitivity of the emissive layer and in order to prevent arcing due to the high voltage, electron multipliers require operation in high vacuum.

The ion currents actually reaching the first dynode are chiefly in the picoampere range, but may span over a 10^{-18}–10^{-9} A range. Depending on the applied voltage, SEMs provide a gain of 10^6–10^8. [237] The resulting current at the electron trap is the input of a nearby preamplifier providing another 10^6–10^9 gain. Its output current is then converted to a voltage signal which finally can be translated to an intensity value by means of an analog-to-digital converter (ADC, Chap. 4.2.5).

Example: A singly-charged ion corresponds to a charge of 1.6×10^{-19} C, and 1 A is equal to 1 C s^{-1}. Thus, an ion current of 10^{-15} A $= 10^{-15}$ C s^{-1} is provided by about 6000 ions per second. If the detection of these ions during a scan in a GC-MS run taking 1 s over the m/z 40–400 range yields a mass spectrum consisting of some 30–60 peaks, this corresponds to 100–200 ions per peak. Such conditions define the detection limit of a scanning mass spectrometer.

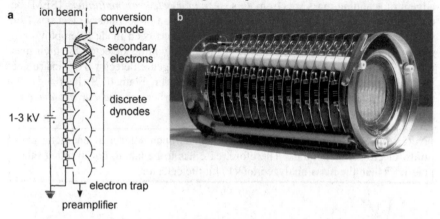

Fig. 4.57. Discrete dynode electron multipliers. (**a**) Schematic of a 14-stage SEM. (**b**) Photograph of an old-fashioned 16-stage Venetian blind-type SEM clearly showing the resistors and ceramics insulators between the stacking dynodes at its side. (**a**) Adapted from Ref. [238] by permission. © Springer-Verlag Heidelberg, 1991.

4.8.2 Channel Electron Multipliers

The cascade of secondary electrons can also be produced in a continuous tube. Such detectors, known as *channel electron multipliers* (CEM) or just *channeltrons*, are more compact and less expensive than discrete dynode SEMs. CEMs are preferably used in benchtop instruments. Their gain depends on the length-to-diameter ratio with an optimum ratio around 40–80. [239] In a CEM, the high voltage drops continuously from the ion entrance to the electron exit of the tube requiring a sufficiently high resistance of the semiconducting material to withstand high voltage of about 2 kV. This is accomplished by an emissive layer of silicon dioxide overlying a conductive layer of lead oxide on the supporting heavily lead-doped glas tube. [237,240] Straight CEMs are unstable at gains exceeding 10^4 because positive ions created inside by EI of residual gas are accelerated towards the input side of the tube where they randomly contribute to the signal causing spurious output pulses. [240,241] A curved design shortens the free path for ion acceleration thereby suppressing noise from this so-called *ion-feedback*. Curved CEMs provide gains up to 10^8 (Figs. 4.58, 4.59).

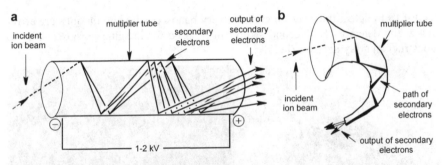

Fig. 4.58. Schematic of linear channel electron multiplier (**a**) and curved channel electron multiplier (**b**). By courtesy of Waters Corporation, MS Technologies, Manchester, UK..

Fig. 4.59. Photograph of a channeltron multiplier.

4.8.3 Microchannel Plates

An extreme reduction of the size of a linear channeltron tube to some micrometers in diameter can be achieved. The cross section of such a single tube is by far too small to be of any use, however, millions of these tubes put together in a "bundle" yield a *channel electron multiplier array*; more common terms are *microchannel plate* or *multichannel plate* (MCP, Fig. 4.60). To avoid that the ions enter the microchannels parallel to their axis, these are inclined by some degrees from the perpendicular to the plate's surface. The gain of an MCP is 10^3–10^4, i.e., much lower than that of a SEM or CEM. Instead of a single MCP, two MCPs are often sandwiched together in such a way that the small angles oppose each other (*Chevron plate*, Fig. 4.61) to obtain gains of 10^6–10^7. Occasionally, even three MCPs are stacked analogously (*z-stack*, gain up to 10^8). [241,242]

MCPs are produced as round plates of various sizes. Those of 2–5 cm in diameter are typically employed in mass spectrometers. An MCP may either be operated to give an integral output for all incident ions during a certain time interval, or the location of the impact may be conserved by connecting sectors of the MCP to individual registration channels. The first setup is more widespread, e.g., in

TOF-MS to detect the electron current every nanosecond while the ions are arriving at the detector. The second setup can be used for imaging purposes, e.g., to construct an array detector (below).

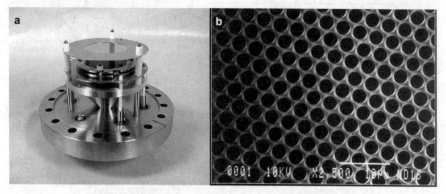

Fig. 4.60. MCP detector (shimmering surface) mounted on top of a flange (**a**) and SEM micrograph of a high resolving MCP showing channels of 2 μm diameter (**b**). By courtesy of (**a**) R.M. Jordan Company, Grass Valley, CA and of (**b**) Burle Industries, Baesweiler.

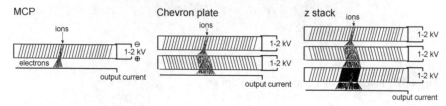

Fig. 4.61. Stacking of MCPs to increase gain. From left: single MCP, Chevron plate, and z-stack configuration. Note the loss of spacial resolution upon stacking.

4.8.4 Post-Acceleration and Conversion Dynode

Discrimination of slower ions as compared to faster ions is observed with SEMs, [243] CEMs, and MCPs as well. [239,244,245] This means a reduction in sensitivity upon reduction of the acceleration voltage of a mass spectrometer, and of course, with increasing ion mass (Eq. 4.3). In order to reduce such effects and especially to improve sensitivity for high-mass ions, *post-acceleration detectors* have been developed. [236,245] In post-acceleration detectors the ions are accelerated immediatly in front of the detector by a voltage of 10–30 kV before they hit the first dynode or the first MCP.

The electron output of ion counting detectors is usually measured with reference to ground potential, i.e., the first dynode is set to negative high voltage to achieve acceleration of the electrons down the dynode assembly. In case of sector or TOF instruments, the deceleration of keV negative ions by the detector voltage is of comparatively minor importance. However, the slow ions exiting from quadrupole mass analyzers would stop before they can reach the detector. Therefore, *conversion dynodes* are frequently placed in front of the SEM or CEM (Fig. 4.62). [237] These are robust electrodes set to high potential (5–20 kV) of a polarity suitable to attract the ions actually exiting from the mass analyzer. Their impact on the conversion dynode creates secondary ions or electrons that can be used for subsequent detection. A conversion dynode detector also serves as a post-acceleration detector and gives almost equal sensitivity for positive and negative ion detection. In addition, neutrals and photons cannot reach the detector if the conversion dynode is placed out of the line of sight.

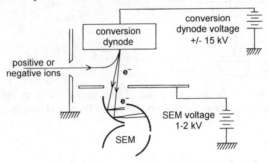

Fig. 4.62. Detector configuration with conversion dynode. By courtesy of JEOL, Tokyo.

4.8.5 Focal Plane Detectors

Magnetic sector instruments are scanning devices usually focusing ions of one m/z value after the other onto a point detector. However, the mass dispersive element of these instruments, i.e., the magnetic sector, is able to produce an image of several neighboring m/z values simultaneously. *Focal plane detectors* (FPD) or *array detectors*, as they are often termed, are able of detecting a small m/z range, i.e., ± 2–5 % of the center mass, at a time. [239,246,247] The ions impinging along the focal plane on the surface of an MCP (usually a Chevron plate) are converted to electrons. The electrons from the backside of the MCP stack are then converted into photons by a phosphor screen, and the light image is guided onto a photo-diode array or a CCD detector by means of a fiber optical device. Such a multi-channel electro-optical detection system typically improves sensitivity or signal-to-noise ratio, respectively, by a factor of 20–100, because less of the ion current is lost without being detected and fluctuations in the ion current are compensated. [248,249]

The resolution of an FPD is theoretically limited by the number of channels (512–2048). In practice, it is even less because the image suffers some broadening as it passes from the first MCP to the photodiode array (Fig. 4.63). Therefore, instruments with FPD can normally be switched from FPD to SEM detection, e.g., by vertical electrostatic deflection of the ion beam (Finnigan MAT900). Furthermore, quadrupole lenses or an inhomogeneous ESA behind the magnet are employed to achieve variable dispersion, i.e., to zoom the m/z range of simultaneous detection. Recent developments point towards fully integrated FPDs on a silicon chip. [250]

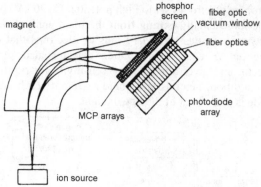

Fig. 4.63. Array detector in the focal plane of a magnetic sector to detect a small mass range simultanously. By courtesy of Thermo Electron (Bremen) GmbH.

4.9 Vacuum Technology

Vacuum systems are integral parts of any mass spectrometer, but vacuum technology definitely is a field of its own. [251-255] Thus, the discussion of mass spectrometer vacuum systems will be restricted to the very basics.

Table 4.5. Pressure ranges in vacuum technology

Pressure Range [Pa]	Pressure Range [mbar]	Pressure Range [mtorr]	Vacuum	Gas Flow
$10^5 - 10^2$	1 bar – 1 mbar	750 torr – 750 mtorr	rough vacuum (RV)	viscous flow
$10^2 - 10^{-1}$	$1 - 10^{-3}$	$750 - 0.75$	medium vacuum (MV)	Knudsen flow
$10^{-1} - 10^{-5}$	$10^{-3} - 10^{-7}$	$0.75 - 7.5 \cdot 10^{-5}$	high vacuum (HV)	molecular flow
$< 10^{-5}$	$< 10^{-7}$	$< 7.5 \cdot 10^{-5}$	ultrahigh vacuum (UHV)	molecular flow

4.9.1 Basic Mass Spectrometer Vacuum System

Generally, two pumping stages are employed to generate the high vacuum of a mass spectrometer. Usually, *rotary vane pumps* having *pumping speeds* of about $4-16 \text{ m}^3 \text{ h}^{-1}$ are used to generate a medium vacuum of several Pascal. They are then connected to high vacuum pumps in a way that their high pressure side exhales into the medium vacuum of the rotary vane pumps, i.e., they are operated as *backing pumps*. This way, each pumping stage contributes a compression of some 10^4–10^5 to the total factor of 10^9–10^{10} between atmospheric pressure and the high vacuum of 10^{-4}–10^{-5} Pa (Table 4.5). The high vacuum pumps can either be *turbo-molecular pumps*, *oil diffusion pumps*, or *cryopumps*. [256,257]

Example: The vacuum system of non-benchtop mass spectrometers consists of one to three rotary vane pumps and two or three turbo pumps. Rotary vane pumps are used for the inlet system(s) and as backing pumps for the turbo pumps. One turbo pump is mounted to the ion source housing, another one or two are operated at the analyzer. Thereby, a differentially pumped system is provided where local changes in pressure, e.g., from reagent gas in CI or collision gas in CID, do not have a noteworthy effect on the whole vacuum chamber.

4.9.2 High Vacuum Pumps

Turbomolecular pumps or *turbo pumps* having pumping speeds of 200–500 l s^{-1} are currently the standard high vacuum pumps in mass spectrometry. A high speed rotor (50,000–60,000 rpm) is employed to transport the molecules out of the *vacuum manifold (vacuum chamber)*. Turbo pumps can be switched on and off in minutes, have low power consumption (about 100 W), and thus can be operated either air- or water cooled. Furthermore, they provide clean, in particular oil-free, high vacuum, are compact and can be mounted either vertically below or horizontally at the sides of the vacuum manifold. Their disadvantages are the risk of sudden damage (similar to hard disk drives) and potential high frequency noise. Fortunately, modern turbo pumps run for many years and their noise is negligible.

Oil diffusion pumps offer high pumping speeds (600–2000 l s^{-1}) at the cost of high power consumption (0.4–2 kW) and the need for a strong cooling water flow. The oil (perfluoropolypropylene glycols; Fomblin™, Santovac™) is thermally evaporated and the oil vapor supplies transport of gas molecules that enter it by diffusion. The gas-loaded oil vapor is condensed and the gas is removed from the liquid by the action of the backing pump. As diffusion pumps have no moving mechanical parts they are highly reliable and extremely silent. However, elongated use as well as sudden venting causes severe oil contaminations of the vacuum manifold. Diffusion pumps have almost been abandoned from modern mass spectrometers.

Cryopumps adsorb (freeze) residual gas to a surface cooled to the temperature of liquid nitrogen. They are highly efficient and silent and provide clean vacuum, but cannot be operated without interruptions to recover the adsorber. Cryopumps are typically operated in combination with turbo pumps because they are only started after high vacuum conditions are reached. Otherwise, the adsorber would soon be saturated.

4.10 Buying an Instrument

To some of us it can unexpectedly happen that we are faced with the task of having to buy a mass spectrometer, eventually for ourselves or on behalf of someone else. References to certain commercial instruments made in this chapter were in no way intended to preclude such a 100,000–600,000 € decision. The below guide may be useful in selecting an instrument that meets your requirements best:

- Define the tasks the mass spectrometer is to be acquired for: i) Need for one or several ionization methods? ii) GC-MS or LC-MS required? iii) High resolution and accurate mass desirable? iv) Is tandem MS an issue? v) What sensitivity is needed? vi) Is quantitative work important?
- Check your budget. A powerful and versatile second-hand machine from a proven company can be better than a toy-like single-purpose benchtop system.
- Get into contact with sales persons of all manufacturers offering suitable systems. Compare prices and modalities of customer training and support.
- Use independent information from the published literature and ask currrent users of your top-choice mass spectrometers for the special strengths and shortcomings of those systems.
- Modern instruments form a unit with their data system. Does this offer the features you want? Does it allow for "manual" settings or corrections?
- Make appointments for demonstration of those instruments that are likely to be acquired, e.g., for the top three of your ranking. Do not use completely unknown samples for this purpose. Otherwise, problems associated with your analyte might erroneously be regarded as a lack of instrumental performance.

Reference List

1. Ligon, W.V., Jr. Molecular Analysis by Mass Spectrometry. *Science* **1979**, *205*, 151-159.
2. Brunnée, C. The Ideal Mass Analyzer: Fact or Fiction? *Int. J. Mass Spectrom. Ion Proc.* **1987**, *76*, 125-237.
3. Beynon, J.H. Instruments, in *Mass Spectrometry and its Applications to Organic Chemistry*, 1st ed.; Elsevier: Amsterdam, 1960; pp. 4-27.
4. Habfast, K.; Aulinger, F. Massenspektrometrische Apparate, in *Massenspektrometrie*, Kienitz, H., editor; Verlag Chemie: Weinheim, 1968; pp. 29-124.
5. Aulinger, F. Massenspektroskopische Geräte, in *Massenspektrometrie*, Kienitz, H., editor; Verlag Chemie: Weinheim, 1968; 125-154.
6. Brunnée, C. New Instrumentation in Mass Spectrometry. *Int. J. Mass Spectrom. Ion Phys.* **1982**, *45*, 51-86.
7. Brunnée, C. 50 Years of MAT in Bremen. *Rapid Commun. Mass Spectrom.* **1997**, *11*, 694-707.
8. Chapman, J.R.; Errock, G.A.; Race, J.A. Science and Technology in Manchester: the Nuture of Mass Spectrometry. *Rapid Commun. Mass Spectrom.* **1997**, *11*, 1575-1586.
9. McLuckey, S.A. Intrumentation for Mass Spectrometry: 1997, *Adv. Mass Spectrom.*, **1998**, *14*, 153-196.
10. Badman, E.R.; Cooks, R.G. Miniature Mass Analyzers. *J. Mass Spectrom.* **2000**, *35*, 659-671.
11. Baykut, G.; Franzen, J. Mobile Mass Spectrometry; a Decade of Field Applications. *Trends Anal. Chem.* **1994**, *13*, 267-275.
12. Prieto, M.C.; Kovtoun, V.V.; Cotter, R.J. Miniaturized Linear TOF Mass Spec-

trometer With Pulsed Extraction. *J. Mass Spectrom.* **2002**, *37*, 1158-1162.

13. Fenselau, C.; Caprioli, R. Mass Spectrometry in the Exploration of Mars. *J. Mass Spectrom.* **2003**, *38*, 1-10.

14. Arkin, C.R.; Griffin, T.P.; Ottens, A.K.; Diaz, J.A.; Follistein, D.W.; Adams, F.W.; Helms, W.R. Evaluation of Small Mass Spectrometer Systems for Permanent Gas Analysis. *J. Am. Soc. Mass Spectrom.* **2002**, *13*, 1004-1012.

15. Stephens, W.E. A Pulsed Mass Spectrometer With Time Dispersion. *Phys. Rev.* **1946**, *69*, 691.

16. Cameron, A.E.; Eggers, D.F. An Ion "Velocitron". *Rev. Sci. Instrum.* **1948**, *19*, 605-607.

17. Wolff, M.M.; Stephens, W.E. A Pulsed Mass Spectrometer With Time Dispersion. *Rev. Sci. Instrum.* **1953**, *24*, 616-617.

18. Wiley, W.C.; McLaren, I.H. Time-of-Flight Mass Spectrometer With Improved Resolution. *Rev. Sci. Instrum.* **1955**, *26*, 1150-1157.

19. Wiley, W.C.; McLaren, I.H. Reprint of: Time-of-Flight Mass Spectrometer With Improved Resolution. *J. Mass Spectrom.* **1997**, *32*, 4-11.

20. Harrington, D.B. The Time-of-Flight Mass Spectrometer, in *Advances in Mass Spectrometry*, Waldron, J.D., editor; Pergamon Press: Oxford, 1959; 249-265.

21. Gohlke, R.S.; McLafferty, F.W. Early Gas Chromatography/Mass Spectrometry. *J. Am. Soc. Mass Spectrom.* **1993**, *4*, 367-371.

22. Guilhaus, M. The Return of Time-of-Flight to Analytical Mass Spectrometry. *Adv. Mass Spectrom.* **1995**, *13*, 213-226.

23. Guilhaus, M.; Mlynski, V.; Selby, D. Perfect Timing: TOF-MS. *Rapid Commun. Mass Spectrom.* **1997**, *11*, 951-962.

24. Karas, M.; Hillenkamp, F. Laser Desorption Ionization of Proteins With Molecular Masses Exceeding 10,000 Daltons. *Anal. Chem.* **1988**, *60*, 2299-2301.

25. Weickhardt, C.; Moritz, F.; Grotemeyer, J. TOF-MS: State-of-the-Art in Chemical Analysis and Molecular Science. *Mass Spectrom. Rev.* **1997**, *15*, 139-162.

26. Cotter, R.J. *Time-of-Flight Mass Spectrometry: Instrumentation and Applications in Biological Research; ACS*: Washington, DC, 1997.

27. Enke, C.G. The Unique Capabilities of Time-of-Flight Mass Analyzers. *Adv. Mass Spectrom.* **1998**, *14*, 197-219.

28. Fuerstenau, S.D.; Benner, W.H. Molecular Weight Determination of Megadalton DNA Electrospray Ions Using Charge Detection Time-of-Flight Mass Spectrometry. *Rapid Commun. Mass Spectrom.* **1995**, *9*, 1528-1538.

29. Fuerstenau, S.D.; Benner, W.H.; Thomas, J.J.; Brugidou, C.; Bothner, B.; Suizdak, G. Mass Spectrometry of an Intact Virus. *Angew. Chem. , Int. Ed.* **2001**, *40*, 541-544.

30. Price, P. Standard Definitions of Terms Relating to Mass Spectrometry. A Report From the Committee on Measurements and Standards of the Amercian Society for Mass Spectrometry. *J. Am. Soc. Mass Spectrom.* **1991**, *2*, 336-348.

31. Todd, J.F.J. Recommendations for Nomenclature and Symbolism for Mass Spectroscopy Including an Appendix of Terms Used in Vacuum Technology. *Int. J. Mass Spectrom. Ion. Proc.* **1995**, *142*, 211-240.

32. Guilhaus, M. Principles and Instrumentation in TOF-MS. Physical and Instrumental Concepts. *J. Mass Spectrom.* **1995**, *30*, 1519-1532.

33. Ioanoviciu, D. Ion-Optical Solutions in Time-of-Flight Mass Spectrometry. *Rapid Commun. Mass Spectrom.* **1995**, *9*, 985-997.

34. Cotter, R.J. Time-of-Flight Mass Spectrometry for the Analysis of Biological Molecules. *Anal. Chem.* **1992**, *64*, 1027A-1039A.

35. Mamyrin, B.A. Laser Assisted Reflectron Time-of-Flight Mass Spectrometry. *Int. J. Mass Spectrom. Ion Proc.* **1994**, *131*, 1-19.

36. Vestal, M.L.; Juhasz, P.; Martin, S.A. Delayed Extraction MALDI-TOF-MS. *Rapid Commun. Mass Spectrom.* **1995**, *9*, 1044-1050.

37. Cordero, M.M.; Cornish, T.J.; Cotter, R.J.; Lys, I.A. Sequencing Peptides Without Scanning the Reflectron: Post-Source Decay With a Curved-Field Reflectron TOF Spectrometer. *Rapid Commun. Mass Spectrom.* **1995**, *9*, 1356-1361.

38. Cornish, T.J.; Cotter, R.J. A Curved-Field Reflectron for Improved Energy Focusing of Product Ions in TOF-MS.

Rapid Commun. Mass Spectrom. **1993,** *7,* 1037-1040.

39. Cornish, T.J.; Cotter, R.J. A Curved Field Reflection TOF Mass Spectrometer for the Simultaneous Focusing of Metastable Product Ions. *Rapid Commun. Mass Spectrom.* **1994,** *8,* 781-785.

40. Schuerch, S.; Schaer, M.; Boernsen, K.O.; Schlunegger, U.P. Enhanced Mass Resolution in MALDI Linear Time-of-Flight Mass Spectrometry. *Biol. Mass Spectrom.* **1994,** *23,* 695-700.

41. Brown, R.S.; Lennon, J.J. Mass Resolution Improvement by Incorporation of Pulsed Ion Extraction in a MALDI Linear Time-of-Flight Mass Spectrometer. *Anal. Chem.* **1995,** *67,* 1998-2003.

42. Colby, S.M.; King, T.B.; Reilly, J.P. Improving the Resolution of MALDI-TOF-MS by Exploiting the Correlation Between Ion Position and Velocity. *Rapid Commun. Mass Spectrom.* **1994,** *8,* 865-868.

43. Whittal, R.M.; Li, L. High-Resolution MALD in Linear TOF-MS. *Anal. Chem.* **1995,** *67,* 1950-1954.

44. Dawson, J.H.J.; Guilhaus, M. Orthogonal-Acceleration Time-of-Flight Mass Spectrometer. *Rapid Commun. Mass Spectrom.* **1989,** *3,* 155-159.

45. Mirgorodskaya, O.A.; Shevchenko, A.A.; Chernushevich, I.V.; Dodonov, A.F.; Miroshnikov, A.I. Electrospray-Ionization TOF-MS in Protein Chemistry. *Anal. Chem.* **1994,** *66,* 99-107.

46. Coles, J.; Guilhaus, M. Orthogonal Acceleration - a New Direction for TOF-MS: Fast, Sensitive Mass Analysis for Continuous Ion Sources. *Trends Anal. Chem.* **1993,** *12,* 203-213.

47. Guilhaus, M.; Selby, D.; Mlynski, V. Orthogonal Acceleration TOF-MS. *Mass Spectrom. Rev.* **2000,** *19,* 65-107.

48. Selby, D.S.; Mlynski, V.; Guilhaus, M. A 20 KV Orthogonal Acceleration TOF Mass Spectrometer for MALDI. *Int. J. Mass Spectrom.* **2001,** *210/211,* 89-100.

49. Selditz, U.; Nilsson, S.; Barnidge, D.; Markides, K.E. ESI/TOF-MS Detection for Microseparation Techniques. *Chimia* **1999,** *53,* 506-510.

50. Prazen, B.J.; Bruckner, C.A.; Synovec, R.E.; Kowalski, B.R. Enhanced Chemical Analysis Using Parallel Column Gas Chromatography With Single-Detector TOF-MS and Chemometric Analysis. *Anal. Chem.* **1999,** *71,* 1093-1099.

51. Hirsch, R.; Ternes, T.A.; Bobeldijk, I.; Weck, R.A. Determination of Environmentally Relevant Compounds Using Fast GC/TOF-MS. *Chimia* **2001,** *55,* 19-22.

52. Hsu, C.S.; Green, M. Fragment-Free Accurate Mass Measurement of Complex Mixture Components by Gas Chromatography/Field Ionization-oaTOF-MS: an Unprecedented Capability for Mixture Analysis. *Rapid Commun. Mass Spectrom.* **2001,** *15,* 236-239.

53. *Tandem Mass Spectrometry;* McLafferty, F.W., editor; John Wiley & Sons: New York, 1983.

54. Busch, K.L.; Glish, G.L.; McLuckey, S.A. *Mass Spectrometry/Mass Spectrometry;* Wiley VCH: New York, 1988.

55. Kaufmann, R.; Kirsch, D.; Spengler, B. Sequencing of Peptides in a TOF Mass Spectrometer: Evaluation of Postsource Decay Following MALDI. *Int. J. Mass Spectrom. Ion Proc.* **1994,** *131,* 355-385.

56. Brunelle, A.; Della-Negra, S.; Depauw, J.; Joret, H.; LeBeyec, Y. TOF-MS With a Compact Two-Stage Electrostatic Mirror: Metastable-Ion Studies With High Mass Resolution and Ion Emission From Thick Insulators. *Rapid Commun. Mass Spectrom.* **1991,** *5,* 40-43.

57. Schueler, B.; Beavis, R.; Ens, W.; Main, D.E.; Tang, X.; Standing, K.G. Unimolecular Decay Measurements of Secondary Ions From Organic Molecules by TOF-MS. *Int. J. Mass Spectrom. Ion Proc.* **1989,** *92,* 185-194.

58. Tang, X.; Ens, W.; Mayer, F.; Standing, K.G.; Westmore, J.B. Measurement of Unimolecular Decay in Peptides of Masses Greater Than 1200 Units by a Reflecting TOF-MS. *Rapid Commun. Mass Spectrom.* **1989,** *3,* 443-448.

59. Tang, X.; Ens, W.; Standing, K.G.; Westmore, J.B. Daughter Ion Mass Spectra From Cationized Molecules of Small Oligopeptides in a Reflecting Time-of-Flight Mass Spectrometer. *Anal. Chem.* **1988,** *60,* 1791-1799.

60. Boesl, U.; Weinkauf, R.; Schlag, E. Reflectron TOF-MS and Laser Excitation for the Analysis of Neutrals, Ionized Molecules and Secondary Fragments. *Int. J. Mass Spectrom. Ion Proc.* **1992,** *112,* 121-166.

61. Cornish, T.J.; Cotter, R.J. Collision-Induced Dissociation in a Tandem Time-of-Flight Mass Spectrometer With Two

Single-Stage Reflectrons. *Org. Mass Spectrom.* **1993**, *28*, 1129-1134.

62. Medzihradszky, K.F.; Campbell, J.M.; Baldwin, M.A.; Falick, A.M.; Juhasz, P.; Vestal, M.L.; Burlingame, A.L. The Characteristics of Peptide Collision-Induced Dissociation Using a High-Performance MALDI-TOF/TOF Tandem Mass Spectrometer. *Anal. Chem.* **2000**, *72*, 552-558.

63. Schnaible, V.; Wefing, S.; Resemann, A.; Suckau, D.; Bücker, A.; Wolf-Kümmeth, S.; Hoffmann, D. Screening for Disulfide Bonds in Proteins by MALDI in-Source Decay and LIFT-TOF/TOF-MS. *Anal. Chem.* **2002**, *74*, 4980-4988.

64. Yergey, A.L.; Coorssen, J.R.; Backlund, P.S.; Blank, P.S.; Humphrey, G.A.; Zimmerberg, J.; Campbell, J.M.; Vestal, M.L. De Novo Sequencing of Peptides Using MALDI/TOF-TOF. *J. Am. Soc. Mass Spectrom.* **2002**, *13*, 784-791.

65. Giannakopulos, A.E.; Thomas, B.; Colburn, A.W.; Reynolds, D.J.; Raptakis, E.N.; Makarov, A.A.; Derrick, P.J. Tandem Time-of-Flight Mass Spectrometer With a Quadratic-Field Ion Mirror. *Rev. Sci. Instrum.* **2002**, *73*, 2115-2123.

66. Nier, A.O. The Development of a High Resolution Mass Spectrometer: a Reminiscence. *J. Am. Soc. Mass Spectrom.* **1991**, *2*, 447-452.

67. Nier, A.O. Some Reminiscences of Mass Spectrometry and the Manhattan Project. *J. Chem. Educ.* **1989**, *66*, 385-388.

68. Nier, A.O. Some Reflections on the Early Days of Mass Spectrometry at the University of Minnesota. *Int. J. Mass Spectrom. Ion Proc.* **1990**, *100*, 1-13.

69. Duckworth, H.E.; Barber, R.C.; Venkatasubramanian, V.S. *Mass Spectroscopy;* 2nd ed.; Cambridge University Press: Cambridge, 1986.

70. Cooks, R.G.; Beynon, J.H.; Caprioli, R.M. Instrumentation, in *Metastable Ions,* Elsevier: Amsterdam, 1973; 5-18.

71. Morrison, J.D. Ion Focusing, Mass Analysis, and Detection, in *Gaseous Ion Chemistry and Mass Spectrometry*, Futrell, J.H., editor; John Wiley & Sons: New York, 1986; pp. 107-125.

72. Dempster, A.J. A New Method of Positive Ray Analysis. *Phys. Rev.* **1918**, *11*, 316-325.

73. Mattauch, J.; Herzog, R. Über Einen Neuen Massenspektrographen. *Z. Phys.* **1934**, *89*, 786-795.

74. Bainbridge, K.T.; Jordan, E.B. Mass-Spectrum Analysis. 1. The Mass Spectrograph. 2. The Existence of Isobars of Adjacent Elements. *Phys. Rev.* **1936**, *50*, 282-296.

75. Johnson, E.G.; Nier, A.O. Angular Aberrations in Sector Shaped Electromagnetic Lenses for Focusing Beams of Charged Particles. *Phys. Rev.* **1953**, *91*, 10-17.

76. Morgan, R.P.; Beynon, J.H.; Bateman, R.H.; Green, B.N. The MM-ZAB-2F Double-Focussing Mass Spectrometer and MIKE Spectrometer. *Int. J. Mass Spectrom. Ion Phys.* **1978**, *28*, 171-191.

77. Hintenberger, H.; König, L.A. Über Massenspektrometer mit vollständiger Doppelfokussierung zweiter Ordnung. *Z. Naturforsch.* **1957**, *12A*, 443.

78. Guilhaus, M.; Boyd, R.K.; Brenton, A.G.; Beynon, J.H. Advantages of a Second Electric Sector on a Double-Focusing Mass Spectrometer of Reversed Configuration. *Int. J. Mass Spectrom. Ion Proc.* **1985**, *67*, 209-227.

79. Bill, J.C.; Green, B.N.; Lewis, I.A.S. A High Field Magnet With Fast Scanning Capabilities. *Int. J. Mass Spectrom. Ion Phys.* **1983**, *46*, 147-150.

80. Matsuda, H. High-Resolution High-Transmission Mass Spectrometer. *Int. J. Mass Spectrom. Ion Proc.* **1985**, *66*, 209-215.

81. Matsuda, H. Double-Focusing Mass Spectrometers of Short Path Length. *Int. J. Mass Spectrom. Ion Proc.* **1989**, *93*, 315-321.

82. Cooks, R.G.; Beynon, J.H.; Caprioli, R.M. *Metastable Ions;* Elsevier: Amsterdam, 1973.

83. Hipple, J.A. Detection of Metastable Ions With the Mass Spectrometer. *Phys. Rev.* **1945**, *68*, 54-55.

84. Hipple, J.A.; Fox, R.E.; Condon, E.U. Metastable Ions Formed by Electron Impact in Hydrocarbon Gases. *Phys. Rev.* **1946**, *69*, 347-356.

85. Beynon, J.H.; Cooks, R.G.; Amy, J.W.; Baitinger, W.E.; Ridley, T.Y. Design and Performance of a Mass-Analyzed Ion Kinetic Energy (MIKE) Spectrometer. *Anal. Chem.* **1973**, *45*, 1023A.

86. Amy, J.W.; Baitinger, W.E.; Cooks, R.G. Building Mass Spectrometers and a

Philosophy of Research. *J. Am. Soc. Mass Spectrom.* **1990**, *1*, 119-128.

87. Millington, D.S.; Smith, J.A. Fragmentation Patterns by Fast Linked Electric and Magnetic Field Scanning. *Org. Mass Spectrom.* **1977**, *12*, 264-265.

88. Morgan, R.P.; Porter, C.J.; Beynon, J.H. On the Formation of Artefact Peaks in Linked Scans of the Magnet and Electric Sector Fields in a Mass Spectrometer. *Org. Mass Spectrom.* **1977**, *12*, 735-738.

89. Lacey, M.J.; Macdonald, C.G. The Use of Two Linked Scanning Modes in Alternation to Analyze Metastable Peaks. *Org. Mass Spectrom.* **1980**, *15*, 134-137.

90. Lacey, M.J.; Macdonald, C.G. Interpreting Metastable Peaks From Double Focusing Mass Spectrometers. *Org. Mass Spectrom.* **1980**, *15*, 484-485.

91. Mouget, Y.; Bertrand, M.J. Graphical Method for Artefact Peak Interpretation, and Methods for Their Rejection, Using Double and Triple Sector Magnetic Mass Spectrometers. *Rapid Commun. Mass Spectrom.* **1995**, *9*, 387-396.

92. Evers, E.A.I.M.; Noest, A.J.; Akkerman, O.S. Deconvolution of Composite Metastable Peaks: a New Method for the Determination of Metastable Transitions Occurring in the First Field Free Region. *Org. Mass Spectrom.* **1977**, *12*, 419-420.

93. Boyd, R.K.; Porter, C.J.; Beynon, J.H. A New Linked Scan for Reversed Geometry Mass Spectrometers. *Org. Mass Spectrom.* **1981**, *16*, 490-494.

94. Lacey, M.J.; Macdonald, C.G. Constant Neutral Spectrum in Mass Spectrometry. *Anal. Chem.* **1979**, *51*, 691-695.

95. Boyd, R.K.; Beynon, J.H. Scanning of Sector Mass Spectrometers to Observe the Fragmentations of Metastable Ions. *Org. Mass Spectrom.* **1977**, *12*, 163-165.

96. Jennings, K. R. Scanning Methods for Double-Focusing Mass Spectrometers. Almoster Ferreira, M. A. [118], 7-21. 1984. Dordrecht, D. Reidel Publishing. NATO ASI Series C: Mathematical and Physical Sciences: Ionic Processes Gas Phase.

97. Fraefel, A.; Seibl, J. Selective Analysis of Metastable Ions. *Mass Spectrom. Rev.* **1984**, *4*, 151-221.

98. Boyd, R.K. Linked-Scan Techniques for MS/MS Using Tandem-in-Space Instruments. *Mass Spectrom. Rev.* **1994**, *13*, 359-410.

99. Walther, H.; Schlunegger, U.P.; Friedli, F. Quantitative Determination of Compounds in Mixtures by B^2E = Const. Linked Scans. *Org. Mass Spectrom.* **1983**, *18*, 572-575.

100. Futrell, J.H. Development of Tandem MS. One Perspective. *Int. J. Mass Spectrom.* **2000**, *200*, 495-508.

101. Fenselau, C. Tandem MS: the Competitive Edge for Pharmacology. *Annu. Rev. Pharmacol. Toxicol.* **1992**, *32*, 555-578.

102. Srinivas, R.; Sülzle, D.; Weiske, T.; Schwarz, H. Generation and Characterization of Neutral and Cationic 3-Silacyclopropenylidene in the Gas Phase. Description of a New BEBE Tandem Mass Spectrometer. *Int. J. Mass Spectrom. Ion Proc.* **1991**, *107*, 369-376.

103. Paul, W. Elektromagnetische Käfige für neutrale und geladene Teilchen. *Angew. Chem.* **1990**, *102*, 780-789.

104. Paul, W. Electromagnetic Traps for Charged and Neutral Particles, in *Nobel Prize Lectures in Physics 1981-1990*, World Scientific Publishing: Singapore, 1993; pp. 601-622.

105. Paul, W.; Steinwedel, H. A New Mass Spectrometer Without Magnetic Field. *Z. Naturforsch.* **1953**, *8A*, 448-450.

106. Paul, W.; Raether, M. Das elektrische Massenfilter. *Z. Phys.* **1955**, *140*, 262-273.

107. Lawson, G.; Todd, J.F.J. Radio-Frequency Quadrupole Mass Spectrometers. *Chem. Brit.* **1972**, *8*, 373-380.

108. Dawson, P.H. *Quadrupole Mass Spectrometry and Its Applications;* 1st ed.; Elsevier: New York, 1976.

109. Dawson, P.H. Quadrupole Mass Analyzers: Performance, Design and Some Recent Applications. *Mass Spectrom. Rev.* **1986**, *5*, 1-37.

110. Blaum, K.; Geppert, C.; Müller, P.; Nörtershäuser, W.; Otten, E.W.; Schmitt, A.; Trautmann, N.; Wendt, K.; Bushaw, B.A. Properties and Performance of a Quadrupole Mass Filter Used for Resonance Ionization Mass Spectrometry. *Int. J. Mass Spectrom.* **1998**, *181*, 67-87.

111. Amad, M.H.; Houk, R.S. High-Resolution Mass Spectrometry With a Multiple Pass Quadrupole Mass Analyzer. *Anal. Chem.* **1998**, *70*, 4885-4889.

112. Liyu, Y.; Amad, M.H.; Winnik, W.M.; Schoen, A.E.; Schweingruber, H.; Mylchreest, I.; Rudewicz, P.J. Investigation of an Enhanced Resolution Triple Quad-

rupole Mass Spectrometer for High-Throughput LC-MS/MS Assays. *Rapid Commun. Mass Spectrom.* **2002**, *16*, 2060-2066.

113. Denison, D.R. Operating Parameters of a Quadrupole in a Grounded Cylindrical Housing. *J. Vac. Sci. Technol.* **1971**, *8*, 266-269.

114. Dawson, P.H.; Whetten, N.R. Nonlinear Resonances in Quadrupole Mass Spectrometers Due to Imperfect Fields. II. Quadrupole Mass Filter and the Monopole Mass Spectrometer. *Int. J. Mass Spectrom. Ion Phys.* **1969**, *3*, 1-12.

115. Brubaker, W.M. Comparison of Quadrupole Mass Spectrometers With Round and Hyperbolic Rods. *J. Vac. Sci. Technol.* **1967**, *4*, 326.

116. Gibson, J.R.; Taylor, S. Prediction of Quadrupole Mass Filter Performance for Hyperbolic and Circular Cross Section Electrodes. *Rapid Commun. Mass Spectrom.* **2000**, *14*, 1669-1673.

117. Chen, W.; Collings, B.A.; Douglas, D.J. High-Resolution Mass Spectrometry With a Quadrupole Operated in the Fourth Stability Region. *Anal. Chem.* **2000**, *72*, 540-545.

118. Huang, Y.; Guan, S.; Kim, H.S.; Marshall, A.G. Ion Transport Through a Strong Magnetic Field Gradient by Radio Frequency-Only Octupole Ion Guides. *Int. J. Mass Spectrom. Ion Proc.* **1996**, *152*, 121-133.

119. Tolmachev, A.V.; Udseth, H.R.; Smith, R.D. Radial Stratification of Ions As a Function of Mass to Charge Ratio in Collisional Cooling Radio Frequency Multipoles Used As Ion Guides or Ion Traps. *Rapid Commun. Mass Spectrom.* **2000**, *14*, 1907-1913.

120. Thomson, B.A. 1997 McBryde Medal Award Lecture Radio Frequency Quadrupole Ion Guides in Modern Mass Spectrometry. *Can. J. Chem.* **1998**, *76*, 499-505.

121. Lock, C.M.; Dyer, E. Characterization of High Pressure Quadrupole Collision Cells Possessing Direct Current Axial Fields. *Rapid Commun. Mass Spectrom.* **1999**, *13*, 432-448.

122. Lock, C.M.; Dyer, E. Simulation of Ion Trajectories Through a High Pressure RF-only Quadrupole Collision Cell by SIMION 6.0. *Rapid Commun. Mass Spectrom.* **1999**, *13*, 422-431.

123. Adlhart, C.; Hinderling, C.; Baumann, H.; Chen, P. Mechanistic Studies of Olefin Metathesis by Ruthenium Carbene Complexes Using ESI-MS/MS. *J. Am. Chem. Soc.* **2000**, *122*, 8204-8214.

124. Yost, R.A.; Enke, C.G. Selected Ion Fragmentation With a Tandem Quadrupole Mass Spectrometer. *J. Am. Chem. Soc.* **1978**, *100*, 2274-2275.

125. Yost, R.A.; Enke, C.G.; McGilvery, D.C.; Smith, D.; Morrison, J.D. High Efficiency Collision-Induced Dissociation in an Rf-Only Quadrupole. *Int. J. Mass Spectrom. Ion Phys.* **1979**, *30*, 127-136.

126. Yost, R.A.; Enke, C.G. Triple Quadrupole Mass Spectrometry for Direct Mixture Analysis and Structure Elucidation. *Anal. Chem.* **1979**, *51*, 1251A-1262A.

127. Hunt, D.F.; Shabanowitz, J.; Giordani, A.B. Collision Activated Decompositions in Mixture Analysis With a Triple Quadrupole Mass Spectrometer. *Anal. Chem.* **1980**, *52*, 386-390.

128. Dawson, P.H.; French, J.B.; Buckley, J.A.; Douglas, D.J.; Simmons, D. The Use of Triple Quadrupoles for Sequential Mass Spectrometry: The Instrument Parameters. *Org. Mass Spectrom.* **1982**, *17*, 205-211.

129. Dawson, P.H.; French, J.B.; Buckley, J.A.; Douglas, D.J.; Simmons, D. The Use of Triple Quadrupoles for Sequential Mass Spectrometry: A Detailed Case Study. *Org. Mass Spectrom.* **1982**, *17*, 212-219.

130. Thomson, B.A.; Douglas, D.J.; Corr, J.J.; Hager, J.W.; Jolliffe, C.L. Improved CID Efficiency and Mass Resolution on a Triple Quadrupole Mass Spectrometer. *Anal. Chem.* **1995**, *67*, 1696-1704.

131. Mao, D.; Douglas, D.J. H/D Exchange of Gas Phase Bradykinin Ions in a Linear Quadrupole Ion Trap. *J. Am. Soc. Mass Spectrom.* **2003**, *14*, 85-94.

132. Hofstadler, S.A.; Sannes-Lowery, K.A.; Griffey, R.H. Enhanced Gas-Phase Hydrogen-Deuterium Exchange of Oligonucleotide and Protein Ions Stored in an External Multipole Ion Reservoir. *J. Mass Spectrom.* **2000**, *35*, 62-70.

133. Collings, B.A.; Campbell, J.M.; Mao, D.; Douglas, D.J. A Combined Linear Ion Trap TOF System With Improved Performance and MS^n Capabilities. *Rapid Commun. Mass Spectrom.* **2001**, *15*, 1777-1795.

134. Hager, J.W. A New Linear Ion Trap Mass Spectrometer. *Rapid Commun. Mass Spectrom.* **2002,** *16*, 512-526.

135. Schwartz, J.C.; Senko, M.W.; Syka, J.E.P. A Two-Dimensional Quadrupole Ion Trap Mass Spectrometer. *J. Am. Soc. Mass Spectrom.* **2002,** *13*, 659-669.

136. Collings, B.A.; Scott, W.R.; Londry, F.A. Resonant Excitation in a Low-Pressure Linear Ion Trap. *J. Am. Soc. Mass Spectrom.* **2003,** *14*, 622-634.

137. March, R.E.; Hughes, R.J. *Quadrupole Storage Mass Spectrometry;* John Wiley & Sons: Chichester, 1989.

138. March, R.E. Quadrupole Ion Trap Mass Spectrometry: Theory, Simulation, Recent Developments and Applications. *Rapid Commun. Mass Spectrom.* **1998,** *12*, 1543-1554.

139. March, R.E. Quadrupole Ion Trap Mass Spectrometry. A View at the Turn of the Century. *Int. J. Mass Spectrom.* **2000,** *200*, 285-312.

140. Stafford, G., Jr. Ion Trap Mass Spectrometry: a Personal Perspective. *J. Am. Soc. Mass Spectrom.* **2002,** *13*, 589-596.

141. *Practical Aspects of Ion Trap Mass Spectrometry;* March, R.E.; Todd, J.F.J., editors; CRC Press: Boca Raton, 1995; Vol. 1 - Fundamentals of Ion Trap Mass Spectrometry.

142. *Practical Aspects of Ion Trap Mass Spectrometry;* March, R.E.; Todd, J.F.J., editors; CRC Press: Boca Raton, 1995; Vol. 2 - Ion Trap Instrumentation.

143. *Practical Aspects of Ion Trap Mass Spectrometry;* March, R.E.; Todd, J.F.J., editors; CRC Press: Boca Raton, 1995; Vol. 3 - Chemical, Environmental, and Biomedical Applications.

144. Yoshinari, K. Theoretical and Numerical Analysis of the Behavior of Ions Injected into a Quadrupole Ion Trap Mass Spectrometer. *Rapid Commun. Mass Spectrom.* **2000,** *14*, 215-223.

145. Alheit, R.; Kleinadam, S.; Vedel, F.; Vedel, M.; Werth, G. Higher Order Non-Linear Resonances in a Paul Trap. *Int. J. Mass Spectrom. Ion Proc.* **1996,** *154*, 155-169.

146. Wuerker, R.F.; Shelton, H.; Langmuir, R.V. Electrodynamic Containment of Charged Particles. *J. Appl. Phys.* **1959,** *30*, 342-349.

147. Ehlers, M.; Schmidt, S.; Lee, B.J.; Grotemeyer, J. Design and Set-Up of an External Ion Source Coupled to a Quad-rupole-Ion-Trap Reflectron-Time-of-Flight Hybrid Instrument. *Eur. J. Mass Spectrom.* **2000,** *6*, 377-385.

148. Forbes, M.W.; Sharifi, M.; Croley, T.; Lausevic, Z.; March, R.E. Simulation of Ion Trajectories in a Quadrupole Ion Trap: a Comparison of Three Simulation Programs. *J. Mass Spectrom.* **1999,** *34*, 1219-1239.

149. Coon, J.J.; Steele, H.A.; Laipis, P.; Harrison, W.W. LD-APCI: a Novel Ion Source for the Direct Coupling of Gel Electrophoresis to Mass Spectrometry. *J. Mass Spectrom.* **2002,** *37*, 1163-1167.

150. Stafford, G.C., Jr.; Kelley, P.E.; Syka, J.E.P.; Reynolds, W.E.; Todd, J.F.J. Recent Improvements in and Analytical Applications of Advanced Ion Trap Technology. *Int. J. Mass Spectrom. Ion Proc.* **1984,** *60*, 85-98.

151. Dawson, P.H.; Hedman, J.W.; Whetten, N.R. Mass Spectrometer. *Rev. Sci. Instrum.* **1969,** *40*, 1444-1450.

152. Dawson, P.H.; Whetten, N.R. Miniature Mass Spectrometer. *Anal. Chem.* **1970,** *42*, 103A-108A.

153. Griffiths, I.W.; Heesterman, P.J.L. Quadrupole Ion Store (QUISTOR) Mass Spectrometry. *Int. J. Mass Spectrom. Ion Proc.* **1990,** *99*, 79-98.

154. Griffiths, I.W. Recent Advances in Ion-Trap Technology. *Rapid Commun. Mass Spectrom.* **1990,** *4*, 69-73.

155. Kelley, P. E.; Stafford, G. C., Jr.; Syka, J. E. P.; Reynolds, W. E.; Louris, J. N.; Todd, J. F. J. New advances in the operation of the ion trap mass spectrometer. *Adv. Mass Spectrom.* **1986,** *10B*, 869-870.

156. Splendore, M.; Lausevic, M.; Lausevic, Z.; March, R.E. Resonant Excitation and/or Ejection of Ions Subjected to DC and RF Fields in a Commercial Quadrupole Ion Trap. *Rapid Commun. Mass Spectrom.* **1997,** *11*, 228-233.

157. Creaser, C.S.; Stygall, J.W. A Comparison of Overtone and Fundamental Resonances for Mass Range Extension by Resonance Ejection in a Quadrupole Ion Trap Mass Spectrometer. *Int. J. Mass Spectrom.* **1999,** *190/191*, 145-151.

158. Williams, J.D.; Cox, K.A.; Cooks, R.G.; McLuckey, S.A.; Hart, K.J.; Goeringer, D.E. Resonance Ejection Ion Trap Mass Spectrometry and Nonlinear Field Contributions: The Effect of Scan Direction

on Mass Resolution. *Anal. Chem.* **1994,** *66,* 725-729.

159. Cooks, R. G.; Amy, J. W.; Bier, M.; Schwartz, J. C.; Schey, K. New mass spectrometers. Adv Mass Spectrom. **1989,** *11A,* 33-52.

160. Kaiser, R.E., Jr.; Louris, J.N.; Amy, J.W.; Cooks, R.G. Extending the Mass Range of the Quadrupole Ion Trap Using Axial Modulation. *Rapid Commun. Mass Spectrom.* **1989,** *3,* 225-229.

161. Weber-Grabau, M.; Kelley, P.; Bradshaw, S.; Hoekman, D.; Evans, S.; Bishop, P. Recent advances in ion-trap technology. Adv Mass Spectrom. **1989,** *11A.,* 152-153

162. Siethoff, C.; Wagner-Redeker, W.; Schäfer, M.; Linscheid, M. HPLC-MS With an Ion Trap Mass Spectrometer. *Chimia* **1999,** *53,* 484-491.

163. Eades, D.M.; Johnson, J.V.; Yost, R.A. Nonlinear Resonance Effects During Ion Storage in a Quadrupole Ion Trap. *J. Am. Soc. Mass Spectrom.* **1993,** *4,* 917-929.

164. Makarov, A.A. Resonance Ejection From the Paul Trap: A Theoretical Treatment Incorporating a Weak Octapole Field. *Anal. Chem.* **1996,** *68,* 4257-4263.

165. Doroshenko, V.M.; Cotter, R.J. Losses of Ions During Forward and Reverse Scans in a QIT Mass Spectrometer and How to Reduce Them. *J. Am. Soc. Mass Spectrom.* **1997,** *8,* 1141-1146.

166. von Busch, F.; Paul, W. Nonlinear Resonances in Electric Mass-Filters As a Consequence of Field Irregularities. *Z. Phys.* **1961,** *164,* 588-594.

167. Dawson, P.H.; Whetten, N.R. Nonlinear Resonances in Quadrupole Mass Spectrometers Due to Imperfect Fields. I. Quadrupole Ion Trap. *Int. J. Mass Spectrom. Ion Phys.* **1969,** *2,* 45-59.

168. Wang, Y.; Franzen, J. The Non-Linear Ion Trap. Part 3. Multipole Components in Three Types of Practical Ion Trap. *Int. J. Mass Spectrom. Ion Proc.* **1994,** *132,* 155-172.

169. Franzen, J. The Non-Linear Ion Trap. Part 5. Nature of Non-Linear Resonances and Resonant Ion Ejection. *Int. J. Mass Spectrom. Ion Proc.* **1994,** *130,* 15-40.

170. Brodbelt, J.S.; Louris, J.N.; Cooks, R.G. Chemical Ionization in an Ion Trap Mass Spectrometer. *Anal. Chem.* **1987,** *59,* 1278-1285.

171. Doroshenko, V.M.; Cotter, R.J. Injection of Externally Generated Ions into an Increasing Trapping Field of a QIT Mass Spectrometer. *J. Mass Spectrom.* **1997,** *31,* 602-615.

172. Van Berkel, G.J.; Glish, G.L.; McLuckey, S.A. ESI Combined With QIT Mass Spectrometry. *Anal. Chem.* **1990,** *62,* 1284-1295.

173. Wang, Y.; Schubert, M.; Ingendoh, A.; Franzen, J. Analysis of Non-Covalent Protein Complexes up to 290 kDa Using Electrospray Ionization and Ion Trap Mass Spectrometry. *Rapid Commun. Mass Spectrom.* **2000,** *14,* 12-17.

174. Schwartz, J.C.; Jardine, I. High-Resolution Parent-Ion Selection/-Isolation Using a QIT Mass Spectrometer. *Rapid Commun. Mass Spectrom.* **1992,** *6,* 313-317.

175. Louris, J.N.; Cooks, R.G.; Syka, J.E.P.; Kelley, P.E.; Stafford, G.C., Jr.; Todd, J.F.J. Instrumentation, Applications, and Energy Deposition in Quadrupole Ion-Trap Tandem Mass Spectrometry. *Anal. Chem.* **1987,** *59,* 1677-1685.

176. McLuckey, S.A.; Glish, G.L.; Kelley, P.E. CID of Negative Ions in an Ion Trap Mass Spectrometer. *Anal. Chem.* **1987,** *59,* 1670-1674.

177. Hashimoto, Y.; Hasegawa, H.; Yoshinari, K. Collision-Activated Infrared Multiphoton Dissociation in a Quadrupole Ion Trap Mass Spectrometer. *Anal. Chem.* **2003,** *75,* 420-425.

178. Kuzma, M.; Jegorov, A.; Kacer, P.; Havlícek, V. Sequencing of New Beauverolides by HPLC and MS. *J. Mass Spectrom.* **2001,** *36,* 1108-1115.

179. Lawrence, E.O.; Livingston, M.S. The Production of High-Speed Light Ions Without the Use of High Voltages. *Phys. Rev.* **1932,** *40,* 19-35.

180. Comisarow, M.B.; Marshall, A.G. The Early Development of Fourier Transform Ion Cyclotron Resonance (FT-ICR) Spectroscopy. *J. Mass Spectrom.* **1996,** *31,* 581-585.

181. Smith, L.G. New Magnetic Period Mass Spectrometer. *Rev. Sci. Instrum.* **1951,** *22,* 115-116.

182. Sommer, H.; Thomas, H.A.; Hipple, J.A. Measurement of E/M by Cyclotron Resonance. *Phys. Rev.* **1951,** *82,* 697-702.

183. Baldeschwieler, J.D. Ion Cyclotron Resonance Spectroscopy. *Science* **1968**, *159*, 263-273.

184. *Fourier Transform Ion Cyclotron Resonance Mass Spectrometry: Analytical Applications;* Asamoto, B., editor; John Wiley & Sons: New York, 1991.

185. Comisarow, M.B.; Marshall, A.G. Fourier Transform Ion Cyclotron Resonance Spectroscopy. *Chem. Phys. Lett.* **1974**, *25*, 282-283.

186. Comisarow, M.B.; Marshall, A.G. Frequency-Sweep FT-ICR Spectroscopy. *Chem. Phys. Lett.* **1974**, *26*, 489-490.

187. Wanczek, K.-P. ICR Spectrometry - A Review of New Developments in Theory, Instrumentation and Applications. I. 1983-1986. *Int. J. Mass Spectrom. Ion Proc.* **1989**, *95*, 1-38.

188. Marshall, A.G.; Grosshans, P.B. FT-ICR-MS: the Teenage Years. *Anal. Chem.* **1991**, *63*, 215A-229A.

189. Amster, I.J. Fourier Transform Mass Spectrometry. *J. Mass Spectrom.* **1996**, *31*, 1325-1337.

190. Dienes, T.; Salvador, J.P.; Schürch, S.; Scott, J.R.; Yao, J.; Cui, S.; Wilkins, C.L. FTMS - Advancing Years (1992-Mid. 1996). *Mass Spectrom. Rev.* **1996**, *15*, 163-211.

191. Marshall, A.G. Milestones in FT-ICR Technique Development. *Int. J. Mass Spectrom.* **2000**, *200*, 331-356.

192. Smith, R.D. Evolution of ESI-MS and FT-ICR for Proteomics and Other Biological Applications. *Int. J. Mass Spectrom.* **2000**, *200*, 509-544.

193. He, F.; Hendrickson, C.L.; Marshall, A.G. Baseline Mass Resolution of Peptide Isobars: A Record for Molecular Mass Resolution. *Anal. Chem.* **2001**, *73*, 647-650.

194. Bossio, R.E.; Marshall, A.G. Baseline Resolution of Isobaric Phosphorylated and Sulfated Peptides and Nucleotides by ESI-FT-ICR-MS: Another Step Toward MS-Based Proteomics. *Anal. Chem.* **2002**, *74*, 1674-1679.

195. White, F.M.; Marto, J.A.; Marshall, A.G. An External Source 7 T FT-ICR Mass Spectrometer With Electrostatic Ion Guide. *Rapid Commun. Mass Spectrom.* **1996**, *10*, 1845-1849.

196. Castleman Jr., A.W.; Futrell, J.H.; Lindinger, W.; Märk, T.D.; Morrison, J.D.; Shirts, R.B.; Smith, D.L.; Wahrhaftig, A.L. *Gaseous Ion Chemistry and Mass Spectrometry;* Futrell, J.H., editor; John Wiley and Sons: New York, 1986.

197. Comisarow, M.B.; Marshall, A.G. Theory of FT-ICR-MS. I. Fundamental Equations and Low-Pressure Line Shape. *J. Chem. Phys.* **1976**, *64*, 110-119.

198. Comisarow, M.B. Signal Modeling for Ion Cyclotron Resonance. *J. Chem. Phys.* **1978**, *69*, 4097-4104.

199. Hughey, C.A.; Rodgers, R.P.; Marshall, A.G. Resolution of 11,000 Compositionally Distinct Components in a Single ESI-FT-ICR Mass Spectrum of Crude Oil. *Anal. Chem.* **2002**, *74*, 4145-4149.

200. Marshall, A.G.; Hendrickson, C.L. FT-ICR Detection: Principles and Experimental Configurations. *Int. J. Mass Spectrom.* **2002**, *215*, 59-75.

201. Marshall, A.G.; Hendrickson, C.L.; Jackson, G.S. FT-ICR-MS: a Primer. *Mass Spectrom. Rev.* **1998**, *17*, 1-35.

202. Shi, S.D.H.; Drader, J.J.; Freitas, M.A.; Hendrickson, C.L.; Marshall, A.G. Comparison and Interconversion of the Two Most Common Frequency-to-Mass Calibration Functions for FT-ICR-MS. *Int. J. Mass Spectrom.* **2000**, *195/196*, 591-598.

203. Caravatti, P.; Allemann, M. The Infinity Cell: a New Trapped-Ion Cell With Radiofrequency Covered Trapping Electrodes for FT-ICR-MS. *Org. Mass Spectrom.* **1991**, *26*, 514-518.

204. Huang, Y.; Li, G.-Z.; Guan, S.; Marshall, A.G. A Combined Linear Ion Trap for Mass Spectrometry. *J. Am. Soc. Mass Spectrom.* **1997**, *8*, 962-969.

205. Guan, S.; Marshall, A.G. Ion Traps for FT-ICR-MS: Principles and Design of Geometric and Electric Configurations. *Int. J. Mass Spectrom. Ion Proc.* **1995**, *146/147*, 261-296.

206. Guan, S.; Marshall, A.G. Stored Waveform Inverse Fourier Transform (SWIFT) Ion Excitation in Trapped-Ion Mass Spectrometry: Theory and Applications. *Int. J. Mass Spectrom. Ion Proc.* **1996**, *157/158*, 5-37.

207. Solouki, T.; Emmet, M.R.; Guan, S.; Marshall, A.G. Detection, Number, and Sequence Location of Sulfur-Containing Amino Acids and Disulfide Bridges in Peptides by Ultrahigh-Resolution MALDI-FT-ICR-MS. *Anal. Chem.* **1997**, *69*, 1163-1168.

208. Wu, Z.; Hendrickson, C.L.; Rodgers, R.P.; Marshall, A.G. Composition of Explosives by ESI-FT-ICR-MS. *Anal. Chem.* **2002**, *74*, 1879-1883.

209. Wang, Y.; Shi, S.D.H.; Hendrickson, C.L.; Marshall, A.G. Mass-Selective Ion Accumulation and Fragmentation in a Linear Octopole Ion Trap External to a FT-ICR Mass Spectrometer. *Int. J. Mass Spectrom.* **2000**, *198*, 113-120.

210. Cody, R.B.; Freiser, B.S. CID in a FT-ICR Mass Spectrometer. *Int. J. Mass Spectrom. Ion Phys.* **1982**, *41*, 199-204.

211. Cody, R.B.; Burnier, R.C.; Freiser, B.S. Collision-Induced Dissociation With FTMS. *Anal. Chem.* **1982**, *54*, 96-101.

212. Gauthier, J.W.; Trautman, T.R.; Jacobson, D.B. Sustained Off-Resonance Irradiation for Collision-Activated Dissociation Involving FT-ICR-MS. CID Technique That Emulates Infrared Multiphoton Dissociation. *Anal. Chim. Acta* **1991**, *246*, 211-225.

213. Senko, M.W.; Speir, J.P.; McLafferty, F.W. Collisional Activation of Large Multiply Charged Ions Using FT-MS. *Anal. Chem.* **1994**, *66*, 2801-2808.

214. Boering, K.A.; Rolfe, J.; Brauman, J.I. Control of Ion Kinetic Energy in Ion Cyclotron Resonance Spectrometry: Very-Low-Energy CID. *Rapid Commun. Mass Spectrom.* **1992**, *6*, 303-305.

215. Lee, S.A.; Jiao, C.Q.; Huang, Y.; Freiser, B.S. Multiple Excitation Collisional Activation in FTMS. *Rapid Commun. Mass Spectrom.* **1993**, *7*, 819-821.

216. Cody, R.B.; Freiser, B.S. High-Resolution Detection of Collision-Induced Dissociation Fragments by FT-MS. *Anal. Chem.* **1982**, *54*, 1431-1433.

217. Cody, R.B. Accurate Mass Measurements on Daughter Ions From Collisional Activation in FTMS. *Anal. Chem.* **1988**, *60*, 917-923.

218. McLuckey, S.A.; Glish, G.L.; Cooks, R.G. Kinetic Energy Effects in Mass Spectrometry/Mass Spectrometry Using a Sector/Quadrupole Tandem Instrument. *Int. J. Mass Spectrom. Ion Phys.* **1981**, *39*, 219-230.

219. Glish, G.L.; McLuckey, S.A.; Ridley, T.Y.; Cooks, R.G. A New "Hybrid" Sector/Quadrupole Mass Spectrometer for MS/MS. *Int. J. Mass Spectrom. Ion Phys.* **1982**, *41*, 157-177.

220. Bradley, C.D.; Curtis, J.M.; Derrick, P.J.; Wright, B. Tandem Mass Spectrometry of Peptides Using a Magnetic Sector/Quadrupole Hybrid-the Case for Higher Collision Energy and Higher Radio-Frequency Power. *Anal. Chem.* **1992**, *64*, 2628-2635.

221. Schoen, A.E.; Amy, J.W.; Ciupek, J.D.; Cooks, R.G.; Dobberstein, P.; Jung, G. A Hybrid BEQQ Mass Spectrometer. *Int. J. Mass Spectrom. Ion Proc.* **1985**, *65*, 125-140.

222. Ciupek, J.D.; Amy, J.W.; Cooks, R.G.; Schoen, A.E. Performance of a Hybrid Mass Spectrometer. *Int. J. Mass Spectrom. Ion Proc.* **1985**, *65*, 141-157.

223. Louris, J.N.; Wright, L.G.; Cooks, R.G. New Scan Modes Accessed With a Hybrid Mass Spectrometer. *Anal. Chem.* **1985**, *57*, 2918-2924.

224. Loo, J.A.; Münster, H. Magnetic Sector-Ion Trap Mass Spectrometry With Electrospray Ionization for High Sensitivity Peptide Sequencing. *Rapid Commun. Mass Spectrom.* **1999**, *13*, 54-60.

225. Strobel, F.H.; Solouki, T.; White, M.A.; Russell, D.H. Detection of Femtomole and Sub-Femtomole Levels of Peptides by Tandem Magnetic Sector/Reflectron TOF Mass Spectrometry and MALDI. *J. Am. Soc. Mass Spectrom.* **1991**, *2*, 91-94.

226. Bateman, R.H.; Green, M.R.; Scott, G.; Clayton, E. A Combined Magnetic Sector-Time-of-Flight Mass Spectrometer for Structural Determination Studies by Tandem MS. *Rapid Commun. Mass Spectrom.* **1995**, *9*, 1227-1233.

227. Aicher, K.P.; Müller, M.; Wilhelm, U.; Grotemeyer, J. Design and Setup of an Ion Trap-Reflectron-TOF Mass Spectrometer. *Eur. Mass Spectrom.* **1995**, *1*, 331-340.

228. Wilhelm, U.; Aicher, K.P.; Grotemeyer, J. Ion Storage Combined With Reflectron TOF Mass Spectrometry: Ion Cloud Motions As a Result of Jet-Cooled Molecules. *Int. J. Mass Spectrom. Ion Proc.* **1996**, *152*, 111-120.

229. Morris, H.R.; Paxton, T.; Dell, A.; Langhorne, J.; Berg, M.; Bordoli, R.S.; Hoyes, J.; Bateman, R.H. High Sensitivity CID Tandem MS on a Novel Quadrupole-oaTOF Mass Spectrometer. *Rapid Commun. Mass Spectrom.* **1996**, *10*, 889-896.

230. Shevchenko, A.; Chernushevich, I.V.; Ens, W.; Standing, K.G.; Thompson, B.; Wilm, M.; Mann, M. Rapid 'De Novo' Peptide Sequencing by a Combination of

nanoESI, Isotopic Labeling and a Quadrupole-TOF Mass Spectrometer. *Rapid Commun. Mass Spectrom.* **1997**, *11*, 1015-1024.

231. Hopfgartner, G.; Chernushevich, I.V.; Covey, T.; Plomley, J.B.; Bonner, R. Exact Mass Measurement of Product Ions for the Structural Elucidation of Drug Metabolites With a Tandem Quadrupole oaTOF Mass Spectrometer. *J. Am. Soc. Mass Spectrom.* **1999**, *10*, 1305-1314.

232. Platzner, I.T.; Habfast, K.; Walder, A.J.; Goetz, A. *Modern Isotope Ratio Mass Spectrometry;* 1st ed.; John Wiley & Sons: Chichester, 1997.

233. Stanton, H.E.; Chupka, W.A.; Inghram, M.G. Electron Multipliers in Mass Spectrometry; Effect of Molecular Structure. *Rev. Sci. Instrum.* **1956**, *27*, 109.

234. Frank, M. Mass Spectrometry With Cryogenic Detectors. *Nucl. Instrum. Methods Phys. Res. ,A* **2000**, *444*, 375-384.

235. Allen, J.S. An Improved Electron-Multiplier Particle Counter. *Rev. Sci. Instrum.* **1947**, *18*, 739-749.

236. Wang, G.H.; Aberth, W.; Falick, A.M. Evidence Concerning the Identity of Secondary Particles in Post-Acceleration Detectors. *Int. J. Mass Spectrom. Ion Proc.* **1986**, *69*, 233-237.

237. Busch, K.L. The Electron Multiplier. *Spectroscopy* **2000**, *15*, 28-33.

238. Schröder, E. *Massenspektrometrie - Begriffe und Definitionen;* Springer-Verlag: Heidelberg, 1991.

239. Boerboom, A.J.H. Array Detection of Mass Spectra, a Comparison With Conventional Detection Methods. *Org. Mass Spectrom.* **1991**, *26*, 929-935.

240. Kurz, E.A. Channel Electron Multipliers. *Am. Laboratory* **1979**, *11*, 67-82.

241. Wiza, J.L. Microchannel Plate Detectors. *Nucl. Instrum. Methods* **1979**, *162*, 587-601.

242. Laprade, B.N.; Labich, R.J. Microchannel Plate-Based Detectors in Mass Spectrometry. *Spectroscopy* **1994**, *9*, 26-30.

243. Alexandrov, M.L.; Gall, L.N.; Krasnov, N.V.; Lokshin, L.R.; Chuprikov, A.V. Discrimination Effects in Inorganic Ion-Cluster Detection by SEM in Mass Spectrometry Experiments. *Rapid Commun. Mass Spectrom.* **1990**, *4*, 9-12.

244. Geno, P.W.; Macfarlane, R.D. Secondary Electron Emission Induced by Impact of Low-Velocity Molecular Ions on a MCP. *Int. J. Mass Spectrom. Ion Proc.* **1989**, *92*, 195-210.

245. Hedin, H.; Håkansson, K.; Sundqvist, B.U.R. On the Detection of Large Organic Ions by Secondary Electron Production. *Int. J. Mass Spectrom. Ion Proc.* **1987**, *75*, 275-289.

246. Hill, J.A.; Biller, J.E.; Martin, S.A.; Biemann, K.; Yoshidome, K.; Sato, K. Design Considerations, Calibration and Applications of an Array Detector for a Four-Sector Tandem Mass Spectrometer. *Int. J. Mass Spectrom. Ion Proc.* **1989**, *92*, 211-230.

247. Birkinshaw, K. Fundamentals of Focal Plane Detectors. *J. Mass Spectrom.* **1997**, *32*, 795-806.

248. Cottrell, J.S.; Evans, S. The Application of a Multichannel Electro-Optical Detection System to the Analysis of Large Molecules by FAB-MS. *Rapid Commun. Mass Spectrom.* **1987**, *1*, 1-2.

249. Cottrell, J.S.; Evans, S. Characteristics of a Multichannel Electrooptical Detection System and Its Application to the Analysis of Large Molecules by FAB-MS. *Anal. Chem.* **1987**, *59*, 1990-1995.

250. Birkinshaw, K.; Langstaff, D.P. The Ideal Detector. *Rapid Commun. Mass Spectrom.* **1996**, *10*, 1675-1677.

251. Hucknall, D.J. *Vacuum Technology and Applications;* 1st ed.; Butterworth-Heinemann: Oxford, 1991.

252. Pupp, W.; Hartmann, H.K. *Vakuum-Technik - Grundlagen und Anwendungen;* Fachbuchverlag Leipzig: Leipzig, 1991.

253. Wutz, M.; Adam, H.; Walcher, W. *Theory and Practice of Vacuum Technology;* 5th ed.; Vieweg: Braunschweig/-Wiesbaden, 1992.

254. *Foundations of Vacuum Science and Technology;* Lafferty, J.M., editor; John Wiley & Sons: New York, 1998.

255. *Leybold Vacuum Products and Reference Book;* Umrath, W., editor; Leybold Vacuum GmbH: Köln, 2001.

256. Busch, K.L. Vacuum in Mass Spectroscopy. Nothing Can Surprise You. *Spectroscopy* **2000**, *15*, 22-25.

257. Busch, K.L. High-Vacuum Pumps in Mass Spectrometers. *Spectroscopy* **2001**, *16*, 14-18.

5 Electron Ionization

The use of *electron ionization* (EI) [1] dates back to the infancy of mass spectrometry in the early 20th century. Ionization is effected by shooting energetic electrons onto a neutral that must have been transferred into the gas phase before. EI definitely is the classical approach to ionization in organic mass spectrometry, and only the production of ions in electrical discharges and by *thermal ionization* (TI) of inorganic salts have earlier been in use. [2] Nevertheless, EI still represents an important technique for the analysis of low- to medium-polarity, non-ionic organic compounds in the range of molecular weights up to $M_r \approx 1000$. Until recent years, EI has been termed *electron impact ionization* or simply *electron impact* (EI).

The physicochemical aspects of the ionization process in general, ion internal energy, and the principles determining the reaction pathways of excited ions have already been addressed (Chap. 2). After a brief repetition of some of these issues we will go more deeply into detail from the analytical point of view. Next, we will discuss technical and practical aspects concerning the construction of EI ion sources and sample introduction systems. Finally, this chapter directly leads over to the interpretation of EI mass spectra (Chap. 6).

> **Note:** Sample introduction systems such as reservoir inlets, chromatographs, and various types of direct probes (Chap. 5.3) are of equal importance to other ionization methods. The same holds valid for the concepts of sensitivity, detection limit, and signal-to-noise ratio (Chap. 5.2.4) and finally to all sorts of ion chromatograms (Chap. 5.4).

5.1 Behavior of Neutrals Upon Electron Impact

5.1.1 Formation of Ions

If an electron carrying several tens of electronvolts is impacting on a neutral molecule, some of this energy is transferred to the neutral. Even though part of this energy will contribute to translational or rotational energy of the molecule as a whole, the major fraction has to be stored in internal modes. Among the internal modes rotational excitation cannot store significant amounts of energy, whereas vibration and especially electronic excitation are capable of an uptake of several electronvolts each (Chap. 2.2 to 2.7). [3,4]

If the electron in terms of energy transfer collides effectively with the neutral, the energy transferred can exceed the *ionization energy* (IE) of the neutral. Then – from the mass spectrometric point of view – the most desirable process can occur: ionization by ejection of one electron generating a positive *molecular ion*

$$M + e^- \rightarrow M^{+\bullet} + 2e^- \tag{5.1}$$

Depending on the analyte and on the energy of the primary electrons, doubly charged and even triply charged ions may be observed

$$M + e^- \rightarrow M^{2+} + 3e^- \tag{5.2}$$

$$M + e^- \rightarrow M^{3+\bullet} + 4e^- \tag{5.3}$$

In general, multiply-charged ions are of very low abundance in EI mass spectra. [5] Nonetheless, they play a role where analytes can easily stabilize a second or even third charge, e.g., in case of large conjugated π-systems.

Example: The EI mass spectrum of a phenanthroperylene derivative shows a series of comparatively intensive doubly-charged ions, the one at *m/z* 178 being the doubly-charged molecular ion, the others representing doubly-charged fragment ions (Fig. 5.1). [6] The occurrence of doubly-charged ions of moderate abundance is quite common in the mass spectra of polycyclic aromatic hydrocarbons (PAHs). For another example of multiply-charged ions in EI spectra cf. Chap. 3.5., for the LDI spectrum cf. Chap. 10.4.

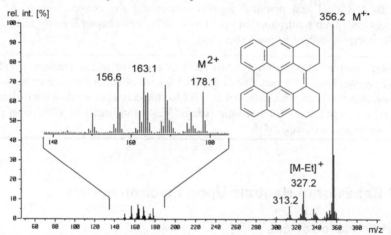

Fig. 5.1. EI mass spectrum of 1,2,3,4,5,6-hexahydrophenanthro[1,10,9,8-*opqra*]perylene. All signals in the expanded inset correspond to doubly-charged ions. Adapted from Ref. [6] with permission. © Elsevier Science, 2002.

5.1.2 Processes Accompanying Electron Ionization

In addition to the desired generation of molecular ions, several other events can result from electron-neutral interactions (Fig. 5.2). A less effective interaction brings the neutral into an electronically excited state without ionizing it. As the energy of the primary electrons increases, the abundance and variety of the ionized species will also increase, i.e., electron ionization may occur via different channels, each of which gives rise to characteristic ionized and neutral products. This includes the production of the following type of ions: molecular ions, fragment ions, multiply charged ions, metastable ions, rearrangement ions, and ion pairs. [7]

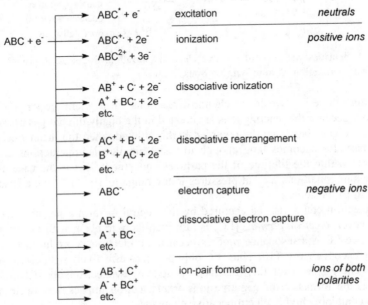

Fig. 5.2. Processes under electron ionization conditions. Under certain conditions, Penning ionization could also play a role (Chap. 2.2.1.). Adapted from Ref. [8] with permission. © Springer-Verlag Heidelberg, 1991.

Most of these processes are very fast. Ionization happens on the low femtosecond timescale, direct bond cleavages require between some picoseconds to several tens of nanoseconds, and rearrangement fragmentations usually proceed in much less than a microsecond (Fig. 5.3 and Chap. 2.7). Finally, some fragment ions may even be formed after the excited species has left the ion source giving rise to metastable ion dissociation (Chap. 2.7). The ion residence time within an electron ionization ion source is about 1 μs. [9]

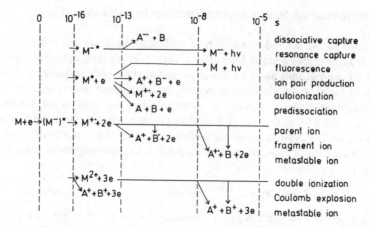

Fig. 5.3. Schematic time chart of possible electron ionization processes. Reprinted from Ref. [4] with permission. © John Wiley & Sons, 1986.

Although there is a wealth of chemical reactions, all of them are strictly unimolecular, because the reacting ions are created in the highly diluted gas phase. In practice, the gas phase may be regarded highly diluted when the mean free path for the particles becomes long enough to make bimolecular interactions almost impossible within the lifetime of the particles concerned. This is the case under high vacuum conditions, i.e., at pressures in the range of 10^{-5}–10^{-4} Pa, as usually realized in EI ion sources.

The electron could also be captured by the neutral to form a negative radical ion. However, *electron capture* (EC) is rather unlikely to occur with electrons of 70 eV since EC is a resonance process because no electron is produced to carry away the excess energy. [10] Thus, EC only proceeds effectively with electrons of very low energy, i.e., from thermal electrons up to a few electronvolts (Chap. 7.4). However, the formation of negative ions from electron impact may occur with analytes containing highly electronegative elements.

Exceptions: The formation of negative ions from tungsten hexafluoride, WF_6, has been studied under conventional EI conditions. At higher electron energies, $WF_6^{-\cdot}$, WF_5^-, and F^- ions have been found. [11] Organic analytes such as perfluorokerosene also yield negative ions, but sensitivity is lower by some orders of magnitude than for positive-ion EI.

5.1.3 Efficiency of Electron Ionization

The *ionization energy* (IE) defines the minimum energy required for ionization of the neutral concerned. Most molecules have IEs in the 7–15 eV range (Chap. 2.2, Tab. 2.1). If an impacting electron carrying just an amount of energy equal to IE would quantitatively transfer it when colliding with the neutral, ionization would take place. Such an event is of low probability and therefore, the *ionization efficiency* is close to zero with such electrons. However, a slight increase in electron energy brings about a steady increase in ionization efficiency.

Strictly speaking, every species has an ionization efficiency curve of its own. The overall efficiency of EI depends on the intrinsic properties of the ionization process as well as on the ionization cross section of the analyte (Chap. 2.4). Fortunately, the curves of ionization cross section versus electron energy are all of the same type, exhibiting a maximum at electron energies around 70 eV (Fig. 5.4). This explains why EI spectra are almost exclusively acquired at 70 eV.

Reasons for measuring EI spectra at 70 eV:
- There is no atom or molecule that cannot be ionized at 70 eV, whereas at 15 eV gases such as He, Ne, Ar, H_2, and N_2 would not.
- The plateau of the ionization efficiency curve around 70 eV makes small variations in electron energy negligible; in practice EI works equally well at 60–80 eV.
- This assures better reproducibility of spectra, and therefore allows comparison of spectra obtained from different mass spectrometers or from mass spectral databases (Chap 5.7).

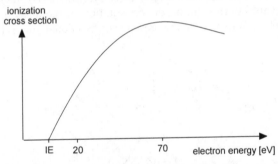

Fig. 5.4. Generalized ionization efficiency curve for EI. The onset of ionization is marked by the *IE* of the respective compound. The curve shows a plateau around 70 eV. Adapted from Ref. [8] with permission. © Springer-Verlag Heidelberg, 1991.

5.1.4 Practical Consequences of Internal Energy

Even comparatively small molecular ions can exhibit a substantial collection of first, second, and higher generation fragmentation pathways. The activation energy for those reactions is provided by their excess energy in the order of several electronvolts received upon 70 eV EI. Typically, ion dissociations are endothermic, and thus each fragmentation step consumes some ion internal energy. Highly excited ions may carry enough internal energy to allow for several consecutive cleavages, whereas others may just undergo one or even none. The latter reach the detector as still intact molecular ions. The fragmentation pathways of the same generation are competing with each other, their products being potential precursors for the next generation of fragmentation processes.

Example: Imagine the hypothetical molecular ion ABCD$^{+\bullet}$ undergoing three competing first generation fragmentations, two of them rearrangements, one a homolytic bond cleavage (Fig. 5.5). Next, three first generation fragment ions have several choices each, and thus the second generation fragment ions Y$^{+\bullet}$ and YZ^{+} are both formed on two different pathways. The formation of the same higher generation fragment ion on two or more different pathways is a widespread phenomenon. A closer look reveals that Y$^{+\bullet}$ and Y^{+} are different although having identical empirical formulas, because one of them is an odd-electron and the other an even-electron species. Nevertheless, they would both contribute to one common peak at the same nominal m/z in a mass spectrum of ABCD (Chap. 3.1.4).

In an EI mass spectrum of ABCD, all ionic species formed as a result of numerous competing and consecutive reactions would be detected, but there is no simple rule which of them should give rise to intensive peaks and which of them would be almost overlooked.

Obviously, the first generation fragment ions should be more closely related to the initial structure of ABCD than those of the second or even third generation. Fortunately, such higher generation (and therefore low-mass) fragment ions can also reveal relevant information on the constitution of the analyte. In particular, they yield reliable information on the presence of functional groups (Chap. 6).

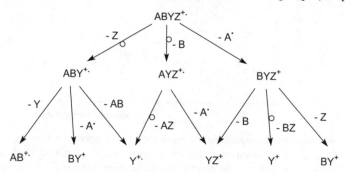

Fig. 5.5. Possible fragmentation pathways for hypothetical molecular ions ABCD$^{+\bullet}$ having internal energies in the range typical for EI.

5.1.5 Low-Energy Electron Ionization Mass Spectra

The excess energy deposited onto a molecular ion can obviously be decreased at low electron energy. The use of 12–15 eV electrons instead of the routinely employed 70 eV electrons still allows to ionize most analytes while reducing disadvantageous fragmentation, e.g., the EI mass spectra of large hydrocarbons benefit from such measures. [12] Especially in conjunction with low ion source temperature, e.g., 70 °C instead of 200 °C, the effect is convincing (Fig 5.6). The approach of recording *low-energy, low-temperature EI mass spectra* has been published for a long time, [13] and has been extensively investigated since then. [14–16]

Low-energy, low-temperature mass spectra are usually much easier to interpret than their conventional 70 eV counterparts, for three reasons: i) the relative intensity of the molecular ion peak is enhanced, thus permitting this important peak to be identified more readily, ii) the overall appearance of the spectrum is simpler because the extent to which fragmentation occurs is greatly reduced, and iii) the fragmentation pattern is dominated by a few characteristic primary fragmentations carrying the largest portion of structural information.

Unfortunately, there are also some disadvantages: i) the decreased ionization efficiency at 12–15 eV also means a significant loss of sensitivity (Fig. 5.4), ii) low ion source temperatures cause long-lasting "memory" of previous samples due to slow desorption from the surfaces that have been in contact with the sample vapor, and iii) a weak molecular ion peak may well be enhanced, however, a spectrum showing no molecular ion peak at 70 eV will not turn into a spectrum exhibiting a strong molecular ion peak at 12 eV.

Nonetheless, during the first decades of analytical mass spectrometry low energy EI spectra have been the only way to minimize fragmentation, and thereby to increase the relative intensity of a weak molecular ion peak. Nowadays, EI mass spectra are preferably complemented with spectra obtained from so-called *soft ionization methods* (Chaps. 7–11).

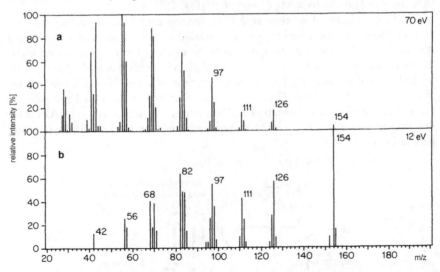

Fig. 5.6. Comparison of 70 eV (**a**) and 12 eV EI mass spectra (**b**) of undecan-1-ol, $C_{11}H_{24}O$, $M_r = 172$. The molecular ion peak is absent in both spectra, but the $[M–H_2O]^{+\bullet}$ peak and the primary fragments become more prominent at 12 eV. (The detailed fragmentation pathways of aliphatic alcohols are discussed in Chaps. 6.2.5, 6.10.3) Reproduced from Ref. [15] with permission. © John Wiley & Sons, 1988.

> **Note:** The detection of signals from previously measured samples in the mass spectrum of the actual analyte is usually termed *memory* or *memory effect*. It is caused by contaminations of ion source or sample introduction system. The best way to reduce memory effects is to use the lowest amount of sample necessary to produce good spectra, to keep ion source and ion source housing at elevated temperature, and to allow some minutes for pumping between subsequent measurements.

5.2 Electron Ionization Ion Sources

5.2.1 Layout of an Electron Ionization Ion Source

The basic layout of any ion source has already been introduced (Chap. 4.1). The design is arranged so that an ion generated within an electric field as realized between two oppositely charged plates, will be accelerated towards the plate of opposite charge sign. If the attracting plate has a hole or a slit, a beam of approximately monoenergetic ions is produced. For practical reasons, the attracting electrode is usually grounded and the location of ion generation is set to high potential. Doing so allows to keep the whole mass analyzer grounded, thereby contributing substantially to safety of operation (Fig. 5.7).

The beam of neutral gaseous analyte molecules enters the *ionization chamber* or *ion volume*, i.e., the actual region of ionization within the *ion source block*, in a line vertical to the paper plane and crosses the electron beam in the center. In order to reduce loss of ions by neutralizing collisions with the walls, the ions are pushed out immediately after generation by action of a low voltage applied to the *repeller* electrode. [17,18] They are then accelerated and focused towards the mass analyzer. Efficient ionization and ion extraction are of key importance for the construction of ion sources producing focusable ion currents in the nanoampere range. [19]

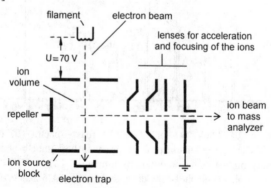

Fig. 5.7. Schematic layout of a typical EI ion source. Adapted from Ref. [8] with permission. © Springer-Verlag Heidelberg, 1991.

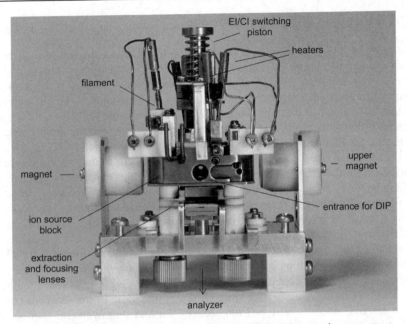

Fig. 5.8. EI/CI/FAB combination ion source of a JEOL JMS-700 sector instrument.

Modern ion sources are of compact design to simplify the handling during exchange and cleaning (Fig. 5.8). They are usually constructed as *combination ion sources*, that can be switched from EI to *chemical ionization* (CI), *fast atom bombardment* (FAB), or *field ionization/field desorption* (FI/FD). Frequently, EI/CI/FAB and EI/FI/FD [20,21] ion sources are offered. Some instruments are equipped with ion sources having ion volumes that are exchangeable via a (separate) vacuum lock to allow for their exchange without breaking the high vacuum, e.g., Micromass Autospec, Finnigan MAT 95 and MAT 900 series. Unfortunately, more recent developments of EI ion sources are mainly communicated in patents.

Note: For a high vacuum system and for the ion source in particular *any sample* introduction also means introduction of *contamination*. Even though ion sources are operated at 150–250 °C, decomposition products of low volatility cannot be removed by pumping. Instead, they form semiconducting layers on the surfaces which in turn cause defocusing due to badly defined potentials. Therefore, cleaning of an ion source is necessary on a regular basis. This includes disassembling, use of fine abrasives to polish the lenses, baking of the insulating ceramics, and thorough washing of the parts.

5.2.2 Generation of Primary Electrons

The beam of ionizing electrons is produced by thermionic emission from a resistively heated metal wire or *filament* typically made of rhenium or tungsten. The filament reaches up to 2000 °C during operation. Some reduction of the working temperature without loss of electron emission (1–10 mA mm^{-2}) can be achieved by use of thoriated iridium or thoriated rhenium filaments. [22] There is a wide variety of filaments available from different manufactures working almost equally well, e.g., the filament can be a straight wire, a ribbon, or a small coil (Fig. 5.9).

The potential for acceleration of the primary electrons has to be carefully shielded from the acceleration voltage for the ions. Otherwise, badly defined ion kinetic energies cause substantial loss of resolution and mass accuracy. In addition, a pair of permanent magnets with their field aligned in parallel to the electron beam is attached above the filament and below the electron trap, thus preventing the electrons from spreading all over the ion volume (Fig. 5.8). This is achieved by action of the Lorentz force on divergent electrons which are forced to travel on helical paths; the resulting electron beam is about 1 mm in diameter. These features developed by Bleakney, [23] and refined by Nier [24] were destined to direct the design of EI ion sources ever since [2,18,25] (the basic EI ion source is sometimes referred to as "Nier-type" ion source). Further improvements can be obtained by i) more efficient shielding of the ion volume from the filament, a measure also reducing thermal decomposition of the analyte, and ii) by use of stronger ion source magnets. [17] Nevertheless, for quadrupole mass analyzers with a tolerance of several electronvolts in energy spread, EI ion sources without magnets have also been constructed successfully. [26]

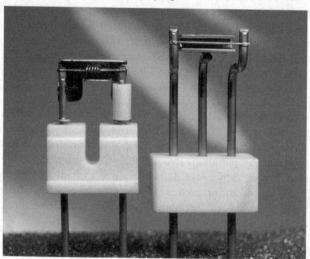

Fig. 5.9. Filaments for EI/CI ion sources. A coiled filament of the VG ZAB-2F (left) and a straight wire filament of the JEOL JMS-700 (right). The shields behind the filament are at the same potential as the wire itself and the white parts are made of ceramics for insulation.

To achieve a more stable mode of operation, in modern instruments the heating current for the filament is *emission-controlled*, i.e., the current of the electron trap is used to keep emission comparatively independent from actual ion source conditions. Typical emission currents are in the range of 50–400 μA.

> **Note:** The lifetime of a filament is several weeks under proper operating conditions. However, harsh conditions such as aggressive analytes or reagent gases in chemical ionization, too high emission current, and in particular sudden breakdown of the high vacuum have devastating consequences.

5.2.3 Overall Efficiency of an Electron Ionization Ion Source

The overall efficiency of an EI ion source depends on the intrinsic properties of the ionization process, the ion source design, and the actual operation parameters. However, the most astonishing fact is the extremely low ionization probability for a neutral entering the EI ion source. Only a very minor fraction of the sample introduced becomes ionized, whereas the vast majority gets lost via the vacuum pumps. This poor ionization probability is caused by the disadvantageous combination of long mean free paths for ions and electrons and the low collision cross section of the electron itself. Remember, the path through the effective ionization volume is merely about 1 mm. Even though, EI provides high sensitivity as compared to other ionization methods and does not really suffer from such low figures.

5.2.3.1 Sensitivity

The term *sensitivity* specifies the overall response of any analytical system for a certain analyte when run under well-defined conditions of operation.

> **Definition:** The *sensitivity* is the slope of a plot of analyte amount versus signal strength. In mass spectrometry, sensitivity is reported as ionic charge of a specified *m/z* reaching the detector per mass of analyte used. The sensitivity is given in units of C μg^{-1} for solids. [27] For gaseous analytes, it can be specified as the ratio of ion current to analyte partial pressure in units of A Pa^{-1}. [28,29]

According to the above definition, sensitivity does not only depend on the ionization efficiency of EI or any other ionization method. Also relevant are the extraction of ions from the ion source, the mass range acquired during the experiment, and the transmission of the mass analyzer. Therefore, the complete experimental conditions have to be stated with sensitivity data.

Example: Modern magnetic sector instruments are specified to have a sensitivity of about 4×10^{-7} C μg^{-1} for the molecular ion of methylstearate, *m/z* 298, at $R = 1000$ in 70 eV EI mode. The charge of 4×10^{-7} C corresponds to 2.5×10^{12} electron charges. One microgram of methylstearate is equivalent to 3.4×10^{-9} mol or 2.0×10^{15} molecules. For a molecule entering the ionization volume one therefore calculates a chance of 1 : 800 to become ionized and to reach the detector as a molecular ion. The fraction of the neutrals that become ionized is definitely larger, maybe 1 : 100, because a significant fraction of them will dissociate to form fragment ions of other *m/z* values. Even then, a ratio of about 1 : 100 indicates an almost perfect ion source design as compared to $1 : 10^4-10^5$ in the 1970s.

5.2.3.2 Detection Limit

The term *detection limit* is almost self-explanatory, yet it is often confused with sensitivity.

Definition: The detection limit defines the smallest flow or the lowest amount of analyte necessary to obtain a signal that can be distinguished from the background noise. The detection limit is valid for one well-specified analyte upon treatment according to a certain analytical protocol. [27-29]

Of course, the sensitivity of an instrumental setup is of key importance to low detection limits; nevertheless, the detection limit is a clearly different quantity. The detection limit may either be stated as a relative measure in trace analysis, e.g., 1 ppb of dioxin in waste oil samples (equivalent to $1 \mu g \ kg^{-1}$ of sample), or as an absolute measure, e.g., 10 femtomol of substance P with a certain MALDI instrument.

5.2.3.3 Signal-to-Noise Ratio

The *signal-to-noise ratio* (S/N) describes the uncertainty of an intensity measurement and provides a quantitative measure of a signal's quality.

Definition: The *signal-to-noise ratio* (S/N) quantifies the ratio of the intensity of a signal relative to noise.

Noise results from the electronics of an instrument, and thus noise is not only present *between* the signals but also *on* the signals. Consequently, intensity measurements are influenced by noise. Real and very numerous background signals of various origin, e.g., FAB or MALDI matrices, GC column bleed and impurities, can appear as if they were electronic noise, so-called "chemical" noise. In the strict sense, this should be separated from electronic noise and should be reported as *signal-to-background ratio* (S/B) [29]. In practice, doing so can be difficult.

Noise is statistical in nature, and therefore can be reduced by elongated data acquisition and subsequent summing or averaging of the spectra, respectively. Accordingly, an intensive peak has a better S/N than a low intensity peak within the same spectrum. The reduction of noise is proportional to the square root of acquisition time or number of single spectra that are averaged, [30] e.g., the noise is reduced by a factor of 3.16 by averaging 10 spectra or by a factor of 10 by averaging 100 spectra, respectively.

Example: A signal may be regarded to be clearly visible at S/N ≥ 10, a value often stated with detection limits. A mass spectrometer in good condition yields S/N > 10^4 which means in turn that even isotopic peaks of low relative intensity can be reliably measured, provided there is no interference with background signals (Fig. 5.10).

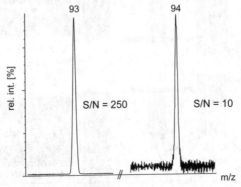

Fig. 5.10. Illustration of the signal-to-noise ratio. The first isotopic peak of toluene molecular ion, *m/z* 93, (70 eV EI, *R* = 4000) shows about S/N = 250, whereas the second isotopic peak at *m/z* 94 has S/N = 10. The theoretical ratio of intensities of *m/z* 92 : 93 : 94 = 100 : 7.7 : 0.3, i.e., the ratio of intensities also reflects the S/N ratio.

5.2.4 Optimization of Ion Beam Geometry

The extraction of ions and the shape of the ion beam can largely be improved if the acceleration voltage is applied in two or more successive steps instead of a single step. In addition, by dividing the plates into an upper and a lower half or a left and a right one, the ion beam can be directed up and down or adjusted from right to left, respectively, by applying slightly different voltages to the corresponding halves (Figs. 5.7 and 5.8). This also allows for focusing of the ion beam.

In practice, the ion optical properties of an ion source are optimized by means of ion trajectory calculations. [31] The standard tool for this task is the SIMION software suite, [32-35] while there are others, too. [36] Thus, the optimum number, positions, voltages, and eventually shapes of the plates are determined (Fig. 5.11). In order to compensate for slight mechanical deviations from theory and for effects exerted by contamination of the plates during elongated use, the voltages can be adjusted to yield optimum conditions.

> **Note:** Usually, some *tuning* of the instrument is performed before actually starting measurements. Such a tuning procedure mainly means adjusting the voltages applied to the ion acceleration lenses of the ion source and eventually of some additional components of the ion optical system. Modern instruments have automatic tuning routines that are often faster and even more thorough than routine manual tuning would be. In any case, a reasonable mass spectrometer data system will also allow for manual corrections.

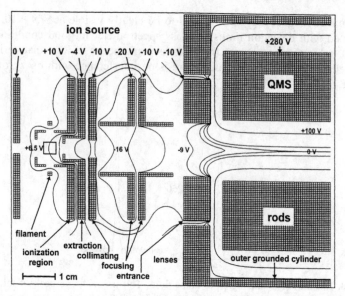

Fig. 5.11. Schematic of the EI ion source of a quadrupole mass spectrometer with equipotentials as calculated by the computer simulation program SIMION three-dimensional version 6.0. Reproduced from Ref. [33] with permission. © Elsevier Science, 1998.

5.3 Sample Introduction

For the purpose of sample introduction, any *sample introduction system* (also *sample inlet system* or *inlet*) suitable for the respective compound can be employed. Hence, *direct probes*, *reservoir inlets*, *gas chromatographs* and even *liquid chromatographs* can be attached to an EI ion source. Which of these inlet systems is to be preferred depends on the type of sample going to be analyzed. Whatever type the inlet system may be, it has to manage the same basic task, i.e., the transfer of the analyte from atmospheric conditions into the high vacuum of the EI ion source; Table 5.1 provides an overview.

5.3.1 Direct Insertion Probe

5.3.1.1 Design of Direct Insertion Probes

The oldest but obviously old-fashioned way to introduce solids into a mass spectrometer ion source is by directly placing them inside an ion source which is mounted to the instrument thereafter and heated. [37] The use of a micro *sample vial* or *crucible* containing some 0.1–2 µg of the analyte that can be directly

Table 5.1. Sample introduction systems for EI-MS

Inlet System	Principle	Analytes
direct insertion probe, DIP	sample in heated/cooled glass/metal vial as particles or film of analyte	solids, waxes or high boiling liquids
direct exposure probe, DEP	sample particles or film of analyte on resistively heated metal filament	solids of extremely low volatility, especially if thermally labile
reservoir/reference inlet	heated reservoir with sample vapor	low to medium boiling liquids
gas chromatograph, GC	elutes directly into ion source	volatile components of mixtures
liquid chromatograph, LC	connected via particle beam interface	analytes suitable for EI that cannot be separated by GC due to high polarity

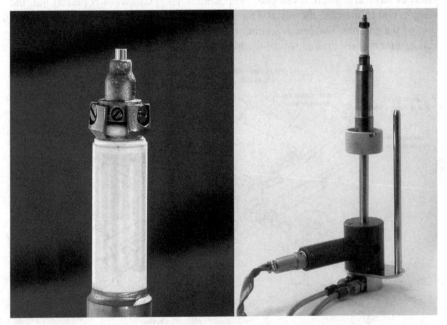

Fig. 5.12. DIP of a JEOL JMS-700 sector instrument for use with EI, chemical ionization (CI) and field ionization (FI). The copper probe tip holds the glass sample vial and is fitted to a temperature-controlled heater (left). The heater, a thermocouple, and circulation water cooling are provided inside. The (white) ceramics insulator protects the operator from the high voltage of the ion source.

brought into the proximity of the ion source by means of a *direct insertion probe* (DIP) is much more convenient (Fig. 5.12). DIPs have been described since the late 1950s and slowly became widespread in use in the mid-1960s. [38-40]

A DIP basically consists of a stainless steel shaft with a tip suitable to insert the sample vial. The shaft is polished to provide a proper vacuum seal against the O-rings of a vacuum lock. Being transferred into the high vacuum of the ion source housing through the *vacuum lock*, [40] the probe is pushed in until it contacts the ion source block (Fig. 5.13). Spectra are then acquired while the analyte evaporates or sublimes from the vial directly into the ionization volume of the ion source. Typically, heating of the sample holder up to about 500 °C can be applied to enforce evaporation of the sample, which sometimes is accompanied by decomposition prior to evaporation. Special high-temperature probes reaching up to 1000 °C are also available. Modern direct probes have *temperature programed heaters* [41] allowing to set rates of 5–150 °C min^{-1} or are equipped with *ion current-controlled heaters*. [42]

To prevent more volatile samples from sudden evaporation a circulation water cooling is often incorporated in the DIP, and refrigerated probes for more volatile samples have also been developed. [43] Sometimes, glass wool is placed into the sample vial to increase the surface for adsorption, and thus to slow down evaporation of the sample. Sudden evaporation causes distorted spectra and may even result in a temporary breakdown of the high vacuum.

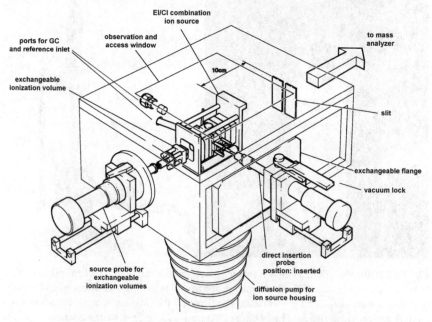

Fig. 5.13. Ion source housing of a Autospec magnetic sector instrument. The ion source can be accessed from several directions to allow for simultaneous connection to DIP/DEP, GC and reference inlet system. By courtesy of Waters Corporation, MS Technologies, Manchester, UK.

5.3.1.2 Sample Vials

Sample vials for use with DIPs usually are about 2 mm in outer diameter and 10–20 mm in length. They are made from borosilicate glass or aluminum (Fig. 5.14). In general, the vials are disposed after use. However, although bearing the risk of memory effects or adsorption of analyte to residual pyrolysis products from previous samples, reusable quartz vials have also been employed in the past. The analyte may either be loaded in solution into the vial or as a tiny piece if it is of solid or waxy consistence (Fig. 5.15). Using solutions of known concentration allows for more reproducible loading of the vials which is usually not a prerequisite. The solvent should then be evaporated before insertion of the probe into the vacuum lock, but it may also be removed by carefully opening the rough evacuation valve.

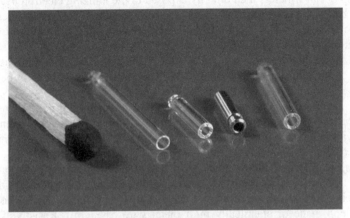

Fig. 5.14. Sample vials for different DIPs. From left: VG ZAB-2F, Finnigan TSQ700 glass and aluminum version, and JEOL JMS-700. The match illustrates the scale.

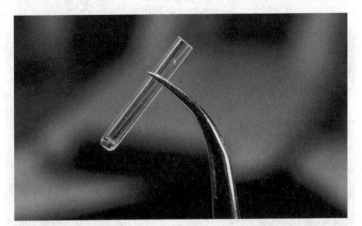

Fig. 5.15. Sample vial filled with analyte. The bright spot halfway between the tip of the tweezers and upper rim of the vial is the piece of solid material to be analyzed. Use of more sample does not have any advantage; it only causes ion source contamination.

> **Note:** Disposable vials are free of memory effects. Glass vials bear the additional advantage of showing how much sample is placed at which position in the vial. Afterwards, it is also possible whether the analyte decomposed during the measurement to yield some black residue.

5.3.2 Direct Exposure Probe

Employing a *direct exposure probe* (DEP) may be helpful in case of analytes that cannot be evaporated from a sample vial without complete decomposition. [44] Here, the analyte is applied from solution or suspension to the outside of a thin wire loop or pin which is then directly exposed to the ionizing electron beam. This method has also been termed *in-beam* electron ionization. Early work describing the direct exposure of a sample to the electron beam came from Ohashi, [45,46] Constantin, [47] and Traldi. [48,49]

The idea behind the rapid heating of a DEP is to achieve evaporation faster than thermal degradation of the sample. [50,51] This principle is realized in perfection with *energy-sudden* methods (Chaps. 9, 10).

If the analyte is exposed to energetic electrons the method is called *direct electron ionization* (DEI) or *desorption electron ionization* (DEI), and accordingly it is termed *direct* or *desorption chemical ionization* (DCI) if the analyte is immersed into the reagent gas under conditions of chemical ionization (Chap. 7).

There are different types of DEPs in that some of them rely on conductive heating from the ion source block and/or the heated tip of a modified DIP, [49] and others – now widespread in use – that are capable of rapid resistive heating of a little loop made of chemically inert metal wire (rhenium). Resistively heated probes allow rates of several hundred $°C\ s^{-1}$ and temperatures up to about 1500 °C (Fig. 5.16). As a consequence of rapid heating, fast scanning, e.g., 1 s per scan over the *m/z* range of interest, is required to follow the evaporation of the analyte.

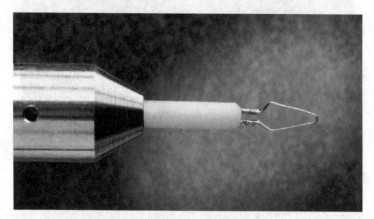

Fig. 5.16. Tip of a direct exposure probe of a GC-oaTOF mass spectrometer for EI, CI, FI). By courtesy of Waters Corporation, MS Technologies, Manchester, UK.

In either case, the use of a DEP allows to extend the temperature range for evaporation. In addition, it reduces thermal degradation as a result of heating the analyte faster than its thermal decomposition usually proceeds, and therefore expands the range of applications for EI and CI to some extent. Whatsoever, employing direct exposure probes is by far no replacement of real desorption ionization methods. [52,53]

Example: The DEI technique has been applied to obtain mass spectra of four underivatized amino acids. [49] The method allowed for the observation of molecular ions and some primary fragment ions in contrast to conventional EI conditions which do not yield molecular ions. However, these additional signals are comparatively weak (Fig. 5.17).

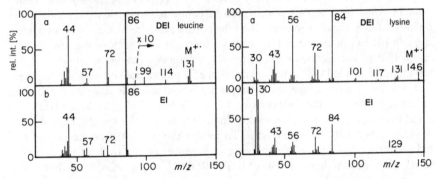

Fig. 5.17. Comparison of DEI (upper) and EI spectra (lower) of the amino acids leucine and lysine. In the leucine DEI spectrum the intensities above *m/z* 90 are shown in 10fold expansion. Adapted from Ref. [49] with permission. © John Wiley & Sons, 1982.

5.3.2.1 Pyrolysis Mass Spectrometry

The enormous temperatures attained on resistively heated sample holders can also be used to intentionally enforce the decomposition of non-volatile samples, thereby yielding characteristic pyrolysis products. *Pyrolysis mass spectrometry* (Py-MS) can be applied to synthetic polymers, [54] fossil biomaterial, [55] food [56] and soil [57] analysis and even to characterize whole bacteria. [58]

In polymer analysis, for example, Py-MS does of course not yield molecular weight distributions, but the type of polymer and the monomer units it is based on can usually be identified by Py-MS. [54] A detailed treatment of this branch of mass spectrometry is beyond the scope of the present book.

5.3.3 Reference Inlet System

Highly volatile samples cannot be introduced into the ion source by means of a direct insertion probe even when cooling is applied. A *reference inlet system* or *reservoir inlet system* is better suited for that purpose. [59] The name of this type of inlet has been coined by the fact that reference inlets have been intended – and often are used for – the introduction of a mass calibrant (or *reference*) independent of the analyte in order to achieve internal mass calibration in accurate mass measurements (Chap. 3.3). Fluorocarbons such as perfluorotributylamine (FC43) or perfluorokerosene (PFK) can be admitted to the ion source in this way.

Reference inlets serve equally well for the analysis of gases, solvents, and similar volatile samples. They are especially convenient when a continuous signal is desired for instrument tuning or long-lasting MS/MS experiments in ion chemistry. In addition, the components of a mixture are admitted to the ion source without fractionation, i.e., without affecting their partial pressures. This property of reservoir inlets has extensively been used in the petroleum industry.

A reference inlet system basically consists of a vessel heated to 80–200 °C that i) can be filled via a septum port, ii) can be connected to the ion source via a toggle valve and a needle valve to adjust the partial pressure in the ion source housing, and iii) can be connected to some vacuum pump to remove the sample after completion of the measurement. Typically, reference inlets have volumes of 30–100 ml and are filled with a few microliters of liquid sample by means of a microliter syringe. Heating is applied to suppress adsorption of the analytes to the walls and thus, to speed up the final removal of sample (Fig. 5.17). As the analyte is exposed to elongated heating, the application of reference inlet systems should be restricted to thermally stable analytes. To reduce the risks of catalytic degradation and selective adsorption of components of mixtures on the walls of the reservoir, an *all-glass heated inlet system* (AGHIS) [60,61] and a teflon-coated reservoir

Fig. 5.18. Reservoir inlet of a JEOL JMS-700 sector instrument with the septum injection port opened. The "operation valve" switches between evacuation, isolation and admission of the sample; a needle valve allows regulation of the sample flow. The GC transfer line crosses in the upper background from the GC (left) to the ion source housing (upper right).

inlet systems have been commercially available as an alternative to stainless steel systems. To this end, *dynamic batch inlet systems* (DBIS) have been developed, where the analyte is transferred from the reservoir into the ion source by means of a capillary and where a carrier gas (H_2, He) can be used to reduce fractionation of mixtures having an extremely wide range of boiling points. [61]

The *liquid introduction system* represents another variation of reservoir inlets. Here, the ion source housing serves as the reservoir into which a few microliters of liquid are introduced by means of a kind of "micro DIP". [62]

5.3.4 Gas Chromatograph

A *gas chromatograph* (GC) can be used for the chromatographic separation of volatile analytes in complex mixtures prior to mass spectrometric analysis. This becomes especially advantageous if the GC elutes directly ("online") into the ion source of a mass spectrometer, so-called *GC-MS coupling*. [63-65] Packed GC columns with a high flow can be connected via a jet-separator, but these are almost out of use at present. [66] Capillary columns provide flow rates in the order of a few milliliters per minute, therefore their back end can be connected directly at the entrance of the ion volume.

EI is well suited as an ionization method for GC-MS applications. The specific properties of GC-MS are discussed later (Chap. 12). For the moment, it is sufficient to notice that proper coupling of a GC to an EI ion source neither exerts substantial effects on the EI process nor does it alter the fragmentation pathways of the ions.

5.3.5 Liquid Chromatograph

A *liquid chromatograph* (LC) can be used for the chromatographic separation of polar analytes of low volatility prior to mass spectrometric analysis. Usually, atmospheric pressure ionization methods are employed when direct coupling of a liquid chromatograph to a mass spectrometer is required, so-called *LC-MS coupling*. However, it can be desirable to obtain EI spectra of LC-separated analytes or of dissolved analytes in general, e.g., where the analyte is only accessible by EI or where EI spectra are required for searching spectral databases. For this purpose, a *particle beam interface* (PBI) may be employed. [67,68] A particle beam interface provides removal of the solvent by nebulization into an evacuated desolvation chamber, and thereby produces a beam of microscopic sample particles that are introduced into an EI ion source. Designs different from PBI have also been developed. [69] (Chap. 12).

5.4 Ion Chromatograms

It is obvious from the working principle of gas and liquid chromatographs that different components of a mixture are eluted at different *retention time* from the chromatographic column. As the mass spectrometer serves as the chromatographic detector in GC-MS and LC-MS experiments, its output must somehow represent the chromatogram. In contrast to the chromatograms obtained with simple chromatographic detectors (FID, WLD, UV), the chromatogram as produced by the mass spectrometer is built up from a large *set of consecutively acquired mass spectra*, each of them containing qualitative information on the eluting species. Thus, continuously changing mass spectra are obtained during the chromatographic elution, as one component after the other yields a mass spectrum of its own, i.e., the number of final mass spectra extracted at least equals the number of components separated. Because mass spectral chromatograms represent ionic abundances as a function of retention time, these are termed *ion chromatograms*.

Fractionation to a certain degree is also observed during evaporation of mixtures from a DIP. The separation can by far not be compared to that of chromatographic systems, nonetheless it often reveals valuable information on impurities accompanying the main product, e.g., remaining solvents, plasticizers, vacuum grease, or synthetic by-products.

> **Note:** The terms *ion profile* and *ion pherogram* have been suggested to replace *ion chromatogram* when DIPs or electrophoresis are used, respectively, as these cannot produce chromatograms in the strict sense. [29] However, this would add yet more terms to one of the rather vague areas of our terminology which is even insufficiently covered by official recommendations. [27-29]

5.4.1 Total Ion Current

The *total ion current* (TIC) can either be measured by a *hardware TIC monitor* before mass analysis, or it can be reconstructed by the data system from the spectra after mass analysis. [27] Thus, the TIC represents a measure of the overall intensity of ion production or of mass spectral output as a function of time, respectively. The TIC obtained by means of *data reduction*, [28] i.e., by mathematical construction from the mass spectra as successively acquired while the sample evaporates, is also termed *total ion chromatogram* (TIC). For this purpose, the sum of all ion intensities belonging to each of the spectra is plotted as a function of time or scan number, respectively.

> **Note:** Modern instruments usually do not support hardware TIC measurements, but including the mass spectrometers of the 1970s, there used to be a hardware TIC monitor, i.e., an ampere meter on the electronics panel. The TIC was obtained by measuring the ion current caused by those ions hitting the ion source exit plate instead of passing through its slit.

5.4.2 Reconstructed Ion Chromatogram

The term *reconstructed ion chromatogram* (RIC) is used to describe the intensity of a given *m/z* or *m/z* range plotted as a function of time or scan number. Plotting RICs is especially useful to identify a target compound of known *m/z* from complex DIP measurements or more often from GC-MS or LC-MS data. In other words, the RIC allows to extract at what time during a measurement the target compound is eluted. RICs can also be used to uncover the relationship of certain *m/z* values to different mass spectra obtained from the measurement of a single (impure) sample (Fig. 5.19).

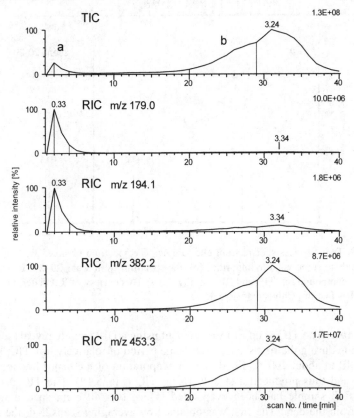

Fig. 5.19. Ion chromatograms from a DIP measurement. The TIC (upper trace) shows a bump during early scans that is also reflected in the RICs of *m/z* 179.0 and 194.1. Spectrum (**a**) belongs to volatile impurities, whereas the RICs of *m/z* 382.2 and 453.3 are related to the target compound (**b**) that evaporates upon heating (cf. Fig. 5.20). By courtesy of R. Gleiter, Organisch-Chemisches Institut, Universität Heidelberg.

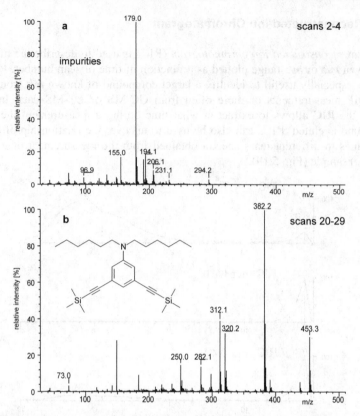

Fig. 5.20. Two EI mass spectra from one sample. The spectrum obtained from scans 2–4 corresponds to some volatile impurities (**a**), the spectrum from scans 20–29 (**b**) represents the target compound (cf. TIC and RICs in Fig. 5.19). By courtesy of R. Gleiter, Organisch-Chemisches Institut, Universität Heidelberg.

Example: The TIC of an EI measurement using a DIP already reveals some ion production during the first scans. After some period of relatively low TIC, heating of the DIP to about 200 °C finally led to evaporation of the major fraction of the sample, and thus produced a steeply rising TIC (Fig. 5.19). The TIC decreased again as all sample had been evaporated. Thus, two different spectra were extracted from the data. The first was obtained by averaging scans 2–4, the second by averaging scans 20–29 (Fig. 5.20). Spectrum (**a**) corresponds to some impurities, whereas spectrum (**b**) represents the target compound, $C_{28}H_{47}NSi_2$.

Using RICs, the ions belonging to the same component can readily be identified from their paralleling dependence in time (Fig. 5.19). Rising of the m/z 194 chromatogram during late scans results from the presence of a minor m/z 194 signal in the mass spectrum of the target compound. (More examples for the use of RICs are presented in Chap. 12.)

5.5 Mass Analyzers for EI

In case of EI, there is no restriction to a certain type of mass analyzer (Chap. 4). As EI is an old method, magnetic sector instruments have been dominating EI applications in the early days of mass spectrometry. With the increasing importance of GC-MS, single quadrupole instruments were gaining importance due to their significantly higher scan speeds. The demand for MS/MS capability to improve analytical selectivity made triple quadrupole instruments and quadrupole ion traps become widespread in use. In recent years, the value of accurate mass measurements has been rediscovered, thereby donating a significant advantage to oaTOF and magnetic sector analyzers in contrast to quadrupole systems of either type. ICR analyzers may also be used in conjunction with EI, but they rarely are because of their high cost.

5.6 Analytes for EI

Classical organic chemistry provides a wide variety of potential analytes for electron ionization, the only limitation being that the analyte should be accessible to evaporation or sublimation without significant thermal decomposition. These requirements are usually met by saturated and unsaturated aliphatic and aromatic hydrocarbons and their derivatives such as halides, ethers, acids, esters, amines, amides etc. Heterocycles generally yield useful EI spectra, and flavones, steroids, terpenes and comparable compounds can successfully be analyzed by EI, too. Therefore, EI represents *the standard method* for such kind of samples.

GC-EI-MS permits the direct analysis of mixtures, e.g., to analyze synthetic byproducts; an advantage that made GC-EI-MS benchtop instruments become widespread in modern synthetic laboratories. The GC-EI-MS combination is especially successful in monitoring environmental pollutants such as polycyclic aromatic hydrocarbons (PAHs), polychlorinated biphenyls (PCBs), polychlorinated dibenzodioxins (PCDDs), polychlorinated dibenzofuranes (PCDFs), or other volatile organic compounds (VOCs).

Whereas low and medium polarity analytes are usually well suited for EI, highly polar or even ionic compounds, e.g., diols or polyalcohols, amino acids, nucleosides, peptides, sugars, and organic salts should not be subject to EI unless properly derivatized prior to EI-MS. [70-75]

There is – as with any other ionization method – no strict upper limit for molecular mass, nevertheless a range of up to 800–1000 u is a realistic estimate. Exceptions up to 1300 u are observed, if the analyte is extremely unpolar, e.g., from numerous fluoroalkyl or trialkylsilyl groups which also significantly contribute to molecular mass.

5.7 Mass Spectral Databases for EI

Provided EI spectra have been measured under some sort of standard conditions (70 eV, ion source at 150–250 °C, pressure in the order of 10^{-4} Pa), they exhibit very good reproducibility. This is not only the case for repeated measurements on the same instrument, but also between mass spectrometers having different types of mass analyzers, and/or coming from different manufacturers. This property soon led to the collection of large EI mass spectral libraries, either printed [76-78] or computerized. [79] The best established EI mass spectral databases are the NIST/EPA/NIH Mass Spectral Database and the Wiley/NBS Mass Spectral Database, each of them giving access to about 120,000 evaluated spectra. [80-83]

Often, EI spectral databases are (optionally) included in the software package of new mass spectrometers. Thus, the EI spectra measured are directly searchable in the database by similarity search typically yielding a list of several hits. Alternatively, stand-alone solutions allow sophisticated search algorithms and strategies [84-86] to be used. Although the coverage of these databases is enormous, and an easy-to-use user interface is provided, one should be aware of potential pitfalls: even the highest score hit can be wrong, perhaps simply because the spectrum of the actual compound is still not included. Even though the unknown can generally not be perfectly identified by simple comparison with a library spectrum, mass spectral databases are highly useful, e.g., to find spectra of similar compounds or isomers for comparison, to get an idea of the spectra of an expected compound class, to yield examples for teaching etc.

> **Note:** Although similarity searches generally provide useful search results, it is always necessary to cross check the hits from the database. Even the highest score hit can be wrong. Some interpretational skills will be advantageous ...

Reference List

1.   Field, F.H.; Franklin, J.L. *Electron Impact Phenomena and the Properties of Gaseous Ions;* 1st ed.; Academic Press: New York, 1957.

2.   Nier, A.O. Some Reflections on the Early Days of Mass Spectrometry at the University of Minnesota. *Int. J. Mass Spectrom. Ion Proc.* **1990,** *100,* 1-13.

3.   Märk, T.D. Fundamental Aspects of Electron Impact Ionization. *Int. J. Mass Spectrom. Ion Phys.* **1982,** *45,* 125-145.

4.   Märk, T.D. Electron Impact Ionization, in *Gaseous ion Chemistry and Mass Spectrometry,* Futrell, J.H., editor; John Wiley and Sons: New York, 1986; pp. 61-93.

5.   Meyerson, S.; Van der Haar, R.W. Multiply Charged Organic Ions in Mass Spectra. *J. Chem. Phys.* **1962,** *37,* 2458-2462.

6.   Wolkenstein, K.; Gross, J.H.; Oeser, T.; Schöler, H.F. Spectroscopic Characterization and Crystal Structure of the 1,2,3,4,5,6-Hexahydrophenanthro[1,10,9,8-*Opqra*]Perylene. *Tetrahedron Lett.* **2002,** *43,* 1653-1655.

7.   Selby, D.S.; Mlynski, V.; Guilhaus, M. A 20 KV Orthogonal Acceleration Time-of-Flight Mass Spectrometer for Matrix-Assisted Laser Desorption/Ionization. *Int. J. Mass Spectrom.* **2001,** *210/211,* 89-100.

8. Schröder, E. *Massenspektrometrie - Be-griffe und Definitionen;* 1st ed.; Springer-Verlag: Heidelberg, 1991.

9. Meier, K.; Seibl, J. Measurement of Ion Residence Times in a Commercial Electron Impact Ion Source. *Int. J. Mass Spectrom. Ion Phys.* **1974,** *14,* 99-106.

10. Harrison, A.G. Fundamentals of Gas Phase Ion Chemistry, in *Chemical Ionization Mass Spectrometry,* 2nd ed.; CRC Press: Boca Raton, 1992; Chapter 2, pp. 26.

11. De Wall, R.; Neuert, H. The Formation of Negative Ions From Electron Impact With Tungsten Hexafluoride. *Z. Naturforsch. ,A.* **1977,** *32A,* 968-971.

12. Ludányi, K.; Dallos, A.; Kühn, Z.; Vékey, D. Mass Spectrometry of Very Large Saturated Hydrocarbons. *J. Mass Spectrom.* **1999,** *34,* 264-267.

13. Remberg, G.; Remberg, E.; Spiteller-Friedmann, M.; Spiteller, G. Massenspektren schwach angeregter Moleküle. 4. Mitteilung. *Org. Mass Spectrom.* **1968,** *1,* 87-113.

14. Bowen, R.D.; Maccoll, A. Low-Energy, Low-Temperature Mass Spectra. I. Selected Derivatives of Octane. *Org. Mass Spectrom.* **1983,** *18,* 576-581.

15. Brophy, J.J.; Maccoll, A. Low-Energy, Low-Temperature Mass Spectra. 9. The Linear Undecanols. *Org. Mass Spectrom.* **1988,** *23,* 659-662.

16. Melaku, A.; Maccoll, A.; Bowen, R.D. Low-Energy, Low-Temperature Mass Spectra. Part 17: Selected Aliphatic Amides. *Eur. Mass Spectrom.* **1997,** *3,* 197-208.

17. Schaeffer, O.A. An Improved Mass Spectrometer Ion Source. *Rev. Sci. Instrum.* **1954,** *25,* 660-662.

18. Fock, W. Design of a Mass Spectrometer Ion Source Based on Computed Ion Trajectories. *Int. J. Mass Spectrom. Ion Phys.* **1969,** *3,* 285-291.

19. Koontz, S.L.; Denton, M.B. A Very High Yield Electron Impact Ion Source for Analytical Mass Spectrometry. *Int. J. Mass Spectrom. Ion Phys.* **1981,** *37,* 227-239.

20. Hogg, A.M.; Payzant, J.D. Design of a Field Ionization/Field Desorption/Electron Impact Ion Source and its Performance on a Modified AEIMS9 Mass Spectrometer. *Int. J. Mass Spectrom. Ion Phys.* **1978,** *27,* 291-303.

21. Brunnée, C. A Combined Field Ionisation-Electron Impact Ion Source for High Molecular Weight Samples of Low Volatility. *Z. Naturforsch. ,B* **1967,** *22* , 121-123.

22. Habfast, K. Massenspektrometrische Funktionselemente: Ionenquellen, in *Massenspektrometrie,* 1st ed.; Kienitz, H., editor; Verlag Chemie: Weinheim, 1968; Chapter B 1.2, pp. 43-74.

23. Bleakney, W. A New Method of Positive-Ray Analysis and its Application to the Measurement of Ionization Potentials in Mercury Vapor. *Physical Review* **1929,** *34,* 157-160.

24. Nier, A.O. Mass Spectrometer for Isotope and Gas Analysis. *Rev. Sci. Instrum.* **1947,** *18,* 398-411.

25. Nier, A.O. The Development of a High Resolution Mass Spectrometer: a Reminiscence. *J. Am. Soc. Mass Spectrom.* **1991,** *2,* 447-452.

26. Swingler, D.L. Mass Spectrometer Ion Source with High Yield. *J. Appl. Phys.* **1970,** *41,* 1496-1499.

27. Price, P. Standard Definitions of Terms Relating to Mass Spectrometry. A Report From the Committee on Measurements and Standards of the Amercian Society for Mass Spectrometry. *J. Am. Chem. Soc. Mass Spectrom.* **1991,** *2,* 336-348.

28. Todd, J.F.J. Recommendations for Nomenclature and Symbolism for Mass Spectroscopy Including an Appendix of Terms Used in Vacuum Technology. *Int. J. Mass Spectrom. Ion Proc.* **1995,** *142,* 211-240.

29. Sparkman, O.D. *Mass Spec Desk Reference;* 1st ed.; Global View Publishing: Pittsburgh, 2000.

30. Kilburn, K.D.; Lewis, P.H.; Underwood, J.G.; Evans, S.; Holmes, J.; Dean, M. Quality of Mass and Intensity Measurements From a High Performance Mass Spectrometer. *Anal. Chem.* **1979,** *51,* 1420-1425.

31. Morrison, J.D. Ion Focusing, Mass Analysis, and Detection, in *Gaseous Ion Chemistry and Mass Spectrometry,* Futrell, J.H., editor; John Wiley & Sons: New York, 1986; pp. 107-125.

32. Dahl, D.A.; Delmore, J.E.; Appelhans, A.D. SIMION PC/PS2 Electrostatic Lens Design Program. *Rev. Sci. Instrum.* **1990,** *61,* 607-609.

33. Blaum, K.; Geppert, C.; Müller, P.; Nörtershäuser, W.; Otten, E.W.; Schmitt, A.; Trautmann, N.; Wendt, K.; Bushaw,

B.A. Properties and Performance of a Quadrupole Mass Filter Used for Resonance Ionization Mass Spectrometry. *Int. J. Mass Spectrom.* **1998**, *181*, 67-87.

34. Ehlers, M.; Schmidt, S.; Lee, B.J.; Grotemeyer, J. Design and Set-Up of an External Ion Source Coupled to a Quadrupole-Ion-Trap Reflectron-Time-of-Flight Hybrid Instrument. *Eur. J. Mass Spectrom.* **2000**, *6*, 377-385.

35. Dahl, D.A. SIMION for the Personal Computer in Reflection. *Int. J. Mass Spectrom.* **2000**, *200*, 3-25.

36. Forbes, M.W.; Sharifi, M.; Croley, T.; Lausevic, Z.; March, R.E. Simulation of Ion Trajectories in a Quadrupole Ion Trap: a Comparison of Three Simulation Programs. *J. Mass Spectrom.* **1999**, *34*, 1219-1239.

37. Cameron, A.E. Electron-Bombardment Ion Source for Mass Spectrometry of Solids. *Rev. Sci. Instrum.* **1954**, *25*, 1154-1156.

38. Reed, R.I. Electron Impact and Molecular Dissociation. Part I. Some Steroids and Triterpenoids. *J. Chem. Soc.* **1958**, 3432-3436.

39. Gohlke, R.S. Obtaining the Mass Spectra of Non-Volatile or Thermally Unstable Compounds. *Chem. Industry* **1963**, 946-948.

40. Junk, G.A.; Svec, H.J. A Vacuum Lock for the Direct Insertion of Samples into a Mass Spectrometer. *Anal. Chem.* **1965**, *37*, 1629-1630.

41. Kankare, J.J. Simple Temperature Programmer for a Mass Spectrometer Direct Insertion Probe. *Anal. Chem.* **1974**, *46*, 966-967.

42. Franzen, J.; Küper, H.; Riepe, W.; Henneberg, D. Automatic Ion Current Control of a Direct Inlet System. *Int. J. Mass Spectrom. Ion Phys.* **1973**, *10*, 353-357.

43. Sawdo, R.M.; Blumer, M. Refrigerated Direct Insertion Probe for Mass Spectrometry. *Anal. Chem.* **1976**, *48*, 790-791.

44. Cotter, R.J. Mass Spectrometry of Nonvolatile Compounds by Desorption From Extended Probes. *Anal. Chem.* **1980**, *52*, 1589A-1602A.

45. Ohashi, M.; Nakayama, N. In-Beam Electron Impact Mass Spectrometry of Aliphatic Alkohols. *Org. Mass Spectrom.* **1978**, *13*, 642-645.

46. Ohashi, M.; Tsujimoto, K.; Funakura, S.; Harada, K.; Suzuki, M. Detection of Pseudomolecular Ions of Tetra- and Pentasaccharides by in-Beam Electron Ionization Mass Spectrometry. *Spectroscopy Int. J.* **1983**, *2*, 260-266.

47. Constantin, E.; Nakatini, Y.; Ourisson, G.; Hueber, R.; Teller, G. Spectres De Masse De Phospholipides Et Polypeptides Non Proteges. Une Méthode Simple D'Obtention Du Spectre Complet. *Tetrahedron Lett.* **1980**, *21*, 4745-4746.

48. Traldi, P.; Vettori, U.; Dragoni, F. Instrument Parameterization for Optimum Use of Commercial Direct Inlet Systems. *Org. Mass Spectrom.* **1982**, *17*, 587-592.

49. Traldi, P. Direct Electron Impact - a New Ionization Technique? *Org. Mass Spectrom.* **1982**, *17*, 245-246.

50. Udseth, H.R.; Friedman, L. Analysis of Styrene Polymers by Mass Spectrometry With Filament-Heated Evaporation. *Anal. Chem.* **1981**, *53*, 29-33.

51. Daves, G.D., Jr. Mass Spectrometry of Involatile and Thermally Unstable Molecules. *Accounts of Chemical Research* **1979**, *12*, 359-365.

52. Peltier, J.M.; MacLean, D.B.; Szarek, W.A. Determination of the Glycosidic Linkage in Peracetylated Disaccharides Comprised of D-Glucopyranose Units by Use of Desorption Electron-Ionization Mass Spectrometry. *Rapid Commun. Mass Spectrom.* **1991**, *5*, 446-449.

53. Kurlansik, L.; Williams, T.J.; Strong, J.M.; Anderson, L.W.; Campana, J.E. Desorption Ionization Mass Spectrometry of Synthetic Porphyrins. *Biomed. Mass Spectrom.* **1984**, *11*, 475-481.

54. Qian, K.; Killinger, W.E.; Casey, M.; Nicol, G.R. Rapid Polymer Identification by In-Source Direct Pyrolysis Mass Spectrometry and Library Searching Techniques. *Anal. Chem.* **1996**, *68*, 1019-1027.

55. Meuzelaar, H.L.C.; Haverkamp, J.; Hileman, F.D. *Pyrolysis Mass Spectrometry of Recent and Fossil Biomaterials;* 1st ed.; Elsevier: Amsterdam, 1982.

56. Guillo, C.; Lipp, M.; Radovic, B.; Reniero, F.; Schmidt, M.; Anklam, E. Use of Pyrolysis-Mass Spectrometry in Food Analysis: Applications in the Food Analysis Laboratory of the European Commissions' Joint Research Center. *J. Anal. Appl. Pyrolysis* **1999**, *49*, 329-335.

57. Schulten, H.-R.; Leinweber, P. Characterization of Humic and Soil Particles by Analytical Pyrolysis and Computer Modeling. *J. Anal. Appl. Pyrolysis* **1996**, *38*, 1-53.

58. Basile, F.; Beverly, M.B.; Voorhees, K.J. Pathogenic Bacteria: Their Detection and Differentiation by Rapid Lipid Profiling With Pyrolysis Mass Spectrometry. *Trends Anal. Chem.* **1998**, *17*, 95-109.

59. Caldecourt, V.J. Heated Sample Inlet System for Mass Spectrometry. *Anal. Chem.* **1955**, *27*, 1670.

60. Peterson, L. Mass Spectrometer All-Glass Heated Inlet. *Anal. Chem.* **1962**, *34*, 1850-1851.

61. Roussis, S.G.; Cameron, A.S. Simplified Hydrocarbon Compound Type Analysis Using a Dynamic Batch Inlet System Coupled to a Mass Spectrometer. *Energy & Fuels* **1997**, *11*, 879-886.

62. Pattillo, A.D.; Young, H.A. Liquid Sample Introduction System for a Mass Spectrometer. *Anal. Chem.* **1963**, *35*, 1768.

63. Message, G.M. *Practical Aspects of Gas Chromatography/Mass Spectrometry;* 1st ed.; John Wiley & Sons: New York, 1984.

64. Hübschmann, H.-J. *Handbuch Der GC-MS - Grundlagen Und Anwendungen;* 1st ed.; Verlag Chemie: Weinheim, 1996.

65. Budde, W.L. *Analytical Mass Spectrometry;* 1st ed.; ACS and Oxford University Press: Washington, D.C. and Oxford, 2001.

66. Gohlke, R.S.; McLafferty, F.W. Early Gas Chromatography/Mass Spectrometry. *J. Am. Soc. Mass Spectrom.* **1993**, *4*, 367-371.

67. Willoughby, R.C.; Browner, R.F. Monodisperse Aerosol Generation Interface for Combining Liquid Chromatography with Mass Spectroscopy. *Anal. Chem.* **1984**, *56*, 2625-2631.

68. Winkler, P.C.; Perkins, D.D.; Williams, D.K.; Browner, R.F. Performance of an Improved Monodisperse Aerosol Generation Interface for Liquid Chromatography/Mass Spectrometry. *Anal. Chem.* **1988**, *60*, 489-493.

69. Brauers, F.; von Bünau, G. Mass Spectrometry of Solutions: a New Simple Interface for the Direct Introduction of Liquid Samples. *Int. J. Mass Spectrom. Ion Proc.* **1990**, *99*, 249-262.

70. *Handbook of Derivates for Chromatography;* 1st ed.; Blau, G.; King, G.S., editors; Heyden & Son: London, 1977.

71. Poole, C.F. Recent Advances in the Silylation of Organic Compounds for Gas Chromatography, in *Handbook of derivates for chromatography*, 1st ed.; Blau, G.; King, G.S., editors; Heyden & Son: London, 1977; Chapter 4, pp. 152-200.

72. Svendsen, J.S.; Sydnes, L.K.; Whist, J.E. Mass Spectrometric Study of Dimethyl Esters of Trimethylsilyl Ether Derivatives of Some 3-Hydroxy Dicarboxylic Acids. *Org. Mass Spectrom.* **1987**, *22*, 421-429.

73. Svendsen, J.S.; Whist, J.E.; Sydnes, L.K. A Mass Spectrometric Study of the Dimethyl Ester Trimethylsilyl Enol Ether Derivatives of Some 3-Oxodicarboxylic Acids. *Org. Mass Spectrom.* **1987**, *22*, 486-492.

74. Scribe, P.; Guezennec, J.; Dagaut, J.; Pepe, C.; Saliot, A. Identification of the Position and the Stereochemistry of the Double Bond in Monounsaturated Fatty Acid Methyl Esters by Gas Chromatography/Mass Spectrometry of Dimethyl Disulfide Derivatives. *Anal. Chem.* **1988**, *60*, 928-931.

75. Pepe, C.; Sayer, H.; Dagaut, J.; Couffignal, R. Determination of Double Bond Positions in Triunsaturated Compounds by Means of Gas Chromatography/Mass Spectrometry of Dimethyl Disulfide Derivatives. *Rapid Commun. Mass Spectrom.* **1997**, *11*, 919-921.

76. Abrahamsson, S.; Stenhagen, E.; McLafferty, F.W. *Atlas of Mass Spectral Data;* 1st ed.; John Wiley & Sons: New York, 1969; Vol. 1-3.

77. *Eight Peak Index of Mass Spectra;* 3rd ed.; Royal Society of Chemistry: London, 1983; Vol. 1-3.

78. McLafferty, F.W.; Stauffer, D.B. *The Wiley/NBS Registry of Mass Spectral Data;* 2nd ed.; Wiley-Interscience: New York, 1989; Vol. 1-7.

79. McLafferty, F.W.; Gohlke, R.S. Mass-Spectrometric Analysis: Spectral-Data File Utilizing Machine Filing and Manual Searching. *Anal. Chem.* **1959**, *31*, 1160-1163.

80. Stein, S.E.; Ausloos, P.; Lias, S.G. Comparative Evaluations of Mass Spectral Databases. *J. Am. Soc. Mass Spectrom.* **1991**, *2*, 441-443.

81. McLafferty, F.W.; Stauffer, D.B.; Twiss-Brooks, A.B.; Loh, S.Y. An Enlarged Data Base of Electron-Ionization Mass Spectra. *J. Am. Soc. Mass Spectrom.* **1991**, *2*, 432-437.

82. McLafferty, F.W.; Stauffer, D.B.; Loh, S.Y. Comparative Evaluations of Mass Spectral Data Bases. *J. Am. Soc. Mass Spectrom.* **1991**, *2*, 438-440.

83. Henneberg, D.; Weimann, B.; Zalfen, U. Computer-Aided Interpretation of Mass Spectra Using Databases with Spectra and Structures. I. Structure Searches. *Org. Mass Spectrom.* **1993**, *28*, 198-206.

84. Stein, S.; Scott, D.R. Optimization and Testing of Mass Spectral Library Search Algorithms for Compound Identification. *J. Am. Soc. Mass Spectrom.* **1994**, *5*, 859-866.

85. Stein, S.E. Estimating Probabilities of Correct Identification From Results of Mass Spectral Library Searches. *J. Am. Soc. Mass Spectrom.* **1994**, *5*, 316-323.

86. Lebedev, K.S.; Cabrol-Bass, D. New Computer Aided Methods for Revealing Structural Features of Unknown Compounds Using Low Resolution Mass Spectra. *J. Chem. Inf. Comput. Sci.* **1998**, *38*, 410-419.

6 Fragmentation of Organic Ions and Interpretation of EI Mass Spectra

The following chapter introduces one of the key disciplines of organic mass spectrometry: the common fragmentation pathways of organic ions and the resulting methodology for the interpretation of *electron ionization* (EI) mass spectra. Of course, a single chapter cannot be comprehensive and thus, further reading may be desirable. [1-6] Applications of mass spectrometry to organic stereochemistry are treated in a monograph, [7] and in addition, there is a vast number of original publications dealing with classes of compounds or fragmentation pathways: these should be consulted to solve a particular problem. This chapter is an attempt to present a systematic introduction to the topic rather by emphasizing the most important fragmentation pathways than by dwelling on countless compounds. It is an attempt to teach the basic skills and to provide a guideline for further "learning by doing". Throughout the chapter we will keep an eye on the relationship between fragmentation patterns and gas phase ion chemistry. In order to successfully work through these pages, some knowledge of the general concept of mass spectrometry (Chap. 1) and of the basics of electron ionization (Chap. 5) are prerequisite. In addition, you should be familiar with the fundamentals of gas phase ion chemistry (Chap. 2) as well as with isotopic mass and isotopic distributions (Chap. 3).

> **Note:** Although the discussion of common fragmentation pathways of organic ions is embedded here in the context of EI mass spectrometry, their occurrence is not restricted to this technique. The reactions of isolated gaseous ions do not directly depend on the ionization method, but are almost exclusively governed by intrinsic properties of the respective ion and by its internal energy (Chap. 2).

6.1 Cleavage of a Sigma-Bond

6.1.1 Writing Conventions for Molecular Ions

Electron ionization mainly creates singly charged positive ions by ejection of one electron out of the neutral. If the precursor was a molecule, M, it will have an even number of electrons, i.e., an *even-electron* or *closed-shell* species. The molecular ion formed upon EI must then be a positive *radical ion*, $M^{+\bullet}$, (*odd-electron* or *open-shell*) ion.

Definition: The *molecular ion* has the same empirical formula as the corresponding neutral molecule. The neutral and its molecular ion only differ by one (or more) electron(s). A singly charged molecular ion can either be a positive radical ion, $M^{+\bullet}$, or a negative radical ion, $M^{-\bullet}$ (not in case of EI). The mass of this ion corresponds to the sum of the masses of the most abundant isotopes of the various atoms that make up the molecule (with a correction for the electron(s) lost or gained, Chap. 3.1.4). [8,9]

Table 6.1. Symbolism for indication of charge and radical state

State of Species	Symbol	Examples
even-electron ion	$+, -$	CH_3^+, NH_4^+, EtO^-, CF_3^-
odd-electron ion	$+\bullet, -\bullet$	$CH_4^{+\bullet}$, $C_{60}^{+\bullet}$, $C_{60}^{-\bullet}$, $CCl_4^{-\bullet}$
radical, no charge	$\bullet$	$CH_3^\bullet$, $OH^\bullet$, $H^\bullet$, $Br^\bullet$

Note: The symbolism $M^{+\bullet}$ does *not mean one added* electron. The radical symbol is added to the molecular ion only to *indicate a remaining unpaired* electron after ionization. Addition of one electron to a neutral would make it a negative radical ion, $M^{-\bullet}$ (electron capture).

When initially formulating the ionization process, we did not consider where the charge would reside (Chap. 2.2.1), e.g., for methane we wrote:

$$CH_4 + e^- \rightarrow CH_4^{+\bullet} + 2e^- \tag{6.1}$$

The tremendous changes in electronic structure and bonding that arise with the loss of an electron are by far better visualized by a formula representation including the σ-bonding electrons of the C–H bond. In small non-functionalized molecules such as methane, it is not possible to avoid the loss of a valence electron from a σ-bond upon electron ionization. The same is true for all other saturated hydrocarbons or a hydrogen molecule, for example. The resulting methane molecular ion may *formally* be represented either as a species resembling a combination of a methyl cation plus a hydrogen radical or a methyl radical plus a proton. Nevertheless, one may still assume that it resembles more the intact molecule than the evolving fragments – a statement being also in accordance with the basic assumptions of the QET (Chap. 2.1). Otherwise, fragmentation would proceed almost spontaneously, thereby excluding the detection of the molecular ion.

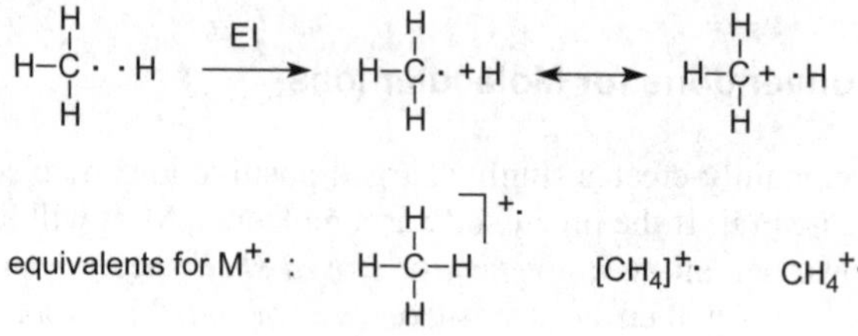

Scheme 6.1.

Free electron pairs or π-orbitals are the preferred sites of a molecule to deliver an electron. The ease of doing so is directly reflected by their effect on ionization energy (Chap. 2.2.2). [10] Therefore, it is common practice to write the charge of a molecular ion as if localized at the position of lowest ionization energy. This is of course not in full agreement with the real charge distributions in ions, but it has been established as a useful first step when writing down a fragmentation scheme. Thus, the formula representations of molecular ions of acetone, *N,N*-dimethyl-propylamine, tetrahydrothiophene, benzene, and 2-methyl-1-pentene are:

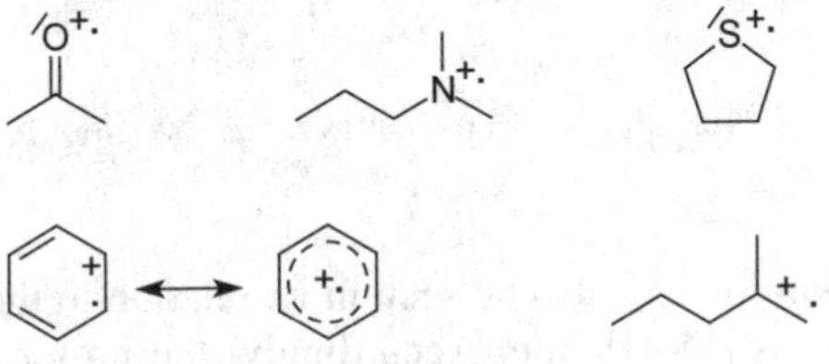

Scheme 6.2.

6.1.2 σ-Bond Cleavage in Small Non-Functionalized Molecules

The most intensive peaks in the EI mass spectrum of methane are the molecular ion, *m/z* 16 and the fragment ion at *m/z* 15 (Fig. 6.1). Explicitly writing the electrons helps to understand the subsequent dissociations of $CH_4^{+\bullet}$ to yield CH_3^+, *m/z* 15, by H˙ loss (σ_1) or H^+ by $CH_3^{\bullet}$ loss (σ_2), respectively. In general, it is more convenient to write the molecular ion in one of the equivalent forms. The charge and radical state are then attached to the brackets (often abbreviated as ⌉) enclosing the molecule.

> **Note:** Ions are detected *at* a certain *m/z* value, and therefore only ionic species may be correlated with peaks. Neutral losses are exclusively indicated by the mass difference between peaks which is given in units of u.

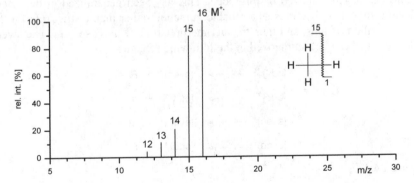

Fig. 6.1. EI mass spectrum of methane. Used by permission of NIST. © NIST 2002.

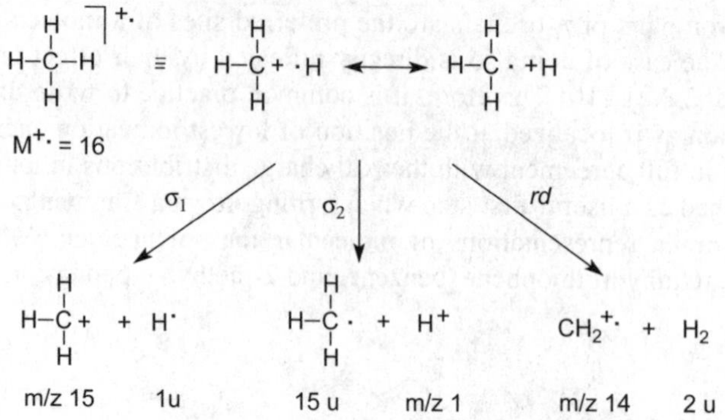

Scheme 6.3.

The CH_3^+ fragment ion can also be seen in its relation to the molecular ion, i.e., it may be described as $[M–H]^+$ ion. Accordingly, the proton could be written as $[M–CH_3]^+$. The $[M–H_2]^{+\bullet}$ ion at *m/z* 14, i.e., carbene molecular ion this case, results from a rearrangement fragmentation (*rd*). Rearrangements are discussed later in this chapter.

The σ-bond cleavage represents a simple but extremely widespread type of fragmentation. Its occurrence does neither require particular functional groups nor heteroatoms in a molecular ion. The σ-bond cleavage proceeds via a loose transition state and therefore can become very rapid, provided sufficient ion internal energy is available (Chap. 2.6.4). Even if additional and perhaps energetically more favorable fragmentation pathways are gaining importance, σ-bond cleavages will not vanish completely.

6.1.3 'Even-Electron Rule'

The *dissociation* or *fragmentation* of ions as it is usually referred to in mass spectrometry, yields a fragment ion and a neutral. Depending on whether the fragmenting ion is an even-electron or an odd-electron species, certain fragmentation pathways may be allowed or forbidden. This has been generalized in the 'even-electron rule'. [11] It serves as a reliable guideline rather than a strict rule for the fragmentation of ions, and thus for the interpretation of mass spectra. The 'even-electron rule' can be summarized in the following scheme:

$$[\text{odd}]^{+\bullet} \longrightarrow [\text{even}]^{+} + R^{\bullet}$$

$$[\text{odd}]^{+\bullet} \longrightarrow [\text{odd}]^{+\bullet} + n$$

$$[\text{even}]^{+} \longrightarrow [\text{even}]^{+} + n$$

$$[\text{even}]^{+} \not\longrightarrow [\text{odd}]^{+\bullet} + R^{\bullet}$$

Scheme 6.4.

'**Even-electron rule**': Odd-electron ions (such as molecular ions and fragment ions formed by rearrangements) may eliminate either a radical or an even-electron neutral species, but even electron ions (such as protonated molecules or fragments formed by a single bond cleavage) will not usually lose a radical to form an odd-electron cation. In other words, the successive loss of radicals is forbidden. [12]

Example: According to the 'even-electron rule' the molecular ion of methane should undergo the following dissociations:

$$CH_4^{+\bullet} \rightarrow CH_3^+ + H^\bullet \qquad (6.2)$$

$$CH_4^{+\bullet} \rightarrow CH_2^{+\bullet} + H_2 \qquad (6.3)$$

Despite that it might initially be generated in a different electronic state and/or conformation, the $CH_2^{+\bullet}$ fragment ion can be expected to decompose further in a way identical to the molecular ion of carbene:

$$CH_2^{+\bullet} \rightarrow C^{+\bullet} + H_2 \qquad (6.4)$$

$$CH_2^{+\bullet} \rightarrow CH^+ + H^\bullet \qquad (6.5)$$

Loss of a radical from a radical ion creates an even-electron fragment ion, CH_3^+ in this case, which preferably may undergo subsequent loss of a molecule:

$$CH_3^+ \rightarrow CH^+ + H_2 \qquad (6.6)$$

Subsequent loss of $H^\bullet$ from CH_3^+ should not occur. It is noteworthy that even such a simple fragmentation scheme offers two independent pathways for the generation of CH^+ (reactions 6.5 and 6.6).

Note: It is important to assign the correct charge and radical state to all species encountered and to carefully track them through a fragmentation scheme. Otherwise, "impossible" fragmentation pathways may be formulated, thereby misleading the assignment of elemental composition and molecular constitution.

Numerous exceptions of the even-electron rule have been described. [12-15] Violations tend to occur where the loss of a radical from an even-electron ion leads to the formation of an exceptionally stable ionic species or where highly excited ions are involved. Representatives of highly excited ions are small molecular ions having only a few degrees of freedom to randomize internal energy [13] or ions being subject to high-energy collisional activation [15,16] (Chap. 12). The rule and some exceptions will be mentioned where appropriate.

Note: In chemistry, there usually is no strict "either ... or"; instead the ions behave like "more ... than ...".

6.1.4 σ-Bond Cleavage in Small Functionalized Molecules

The introduction of a charge-localizing heteroatom into a molecule is accompanied by obvious changes in the appearance of the mass spectrum. Withdrawal of an electron upon EI must not necessarily affect a σ-bond, because the electron can be supplied from one of the free electron pairs of the heteroatom.

In the EI mass spectrum of iodomethane the molecular ion, $CH_3–I^{+\bullet}$, is detected as the base peak at m/z 142 (Fig. 6.2). However, the most important characteristic of the spectrum is the $[M–CH_3]^+$ peak at m/z 127. This signal is due to formation of I^+ by a process that can also be classified as a σ-bond cleavage. Here, a single electron shifts from the intact C–I bond to the radical-site, thereby rupturing the bond. The peak is accompanied by a much less intense one at m/z 128 due to formation of $HI^{+\bullet}$ by rearrangement. (The m/z 128 peak must not be interpreted as an isotopic peak of m/z 127, because iodine is monoisotopic!) The remaining peaks can be explained as for methane before, i.e., peaks can be assigned as $[M–H]^+$, m/z 141, $[M–H_2]^{+\bullet}$, m/z 140, and $[M–H–H_2]^+$, m/z 139. Analogous to $H^\bullet$ loss, $I^\bullet$ loss also occurs generating the CH_3^+ ion, m/z 15.

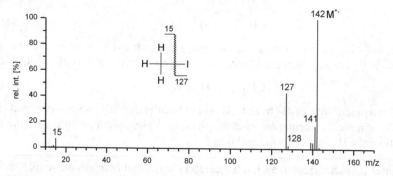

Fig. 6.2. EI mass spectrum of iodomethane. The molecular ion mainly decomposes to yield ionic fragments representing its major constituent groups, i.e., I^+, m/z 127 and CH_3^+, m/z 15. Spectrum used by permission of NIST. © NIST 2002.

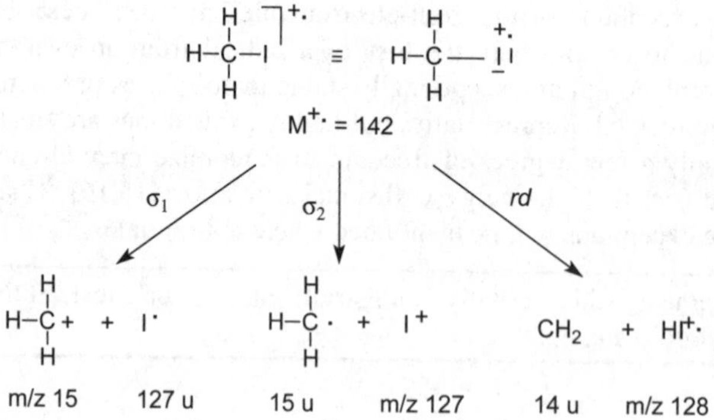

Scheme 6.5.

Note: The detection of the positive halogen ion and a less intensive peak due to the hydrogen halogenide is a generally observed characteristic of halogenated compounds. In accordance with the relative electronegativities of the halogens their intensities follow the order $I^+ > Br^+ > Cl^+ > F^+$. In case of Br and Cl the isotopic patterns give additional evidence for the presence of the respective halogen (Chap. 6.2.6). [17]

6.2 Alpha-Cleavage

6.2.1 α-Cleavage of Acetone Molecular Ion

The EI mass spectrum of acetone is comparatively simple. It basically shows three important peaks at m/z 58, 43, and 15. According to the formula C_3H_6O, the peak at m/z 58 corresponds to the molecular ion. The base peak at m/z 43 is related to this signal by a difference of 15 u, a neutral loss which can almost always be assigned to loss of a methyl radical, $CH_3{}^{\bullet}$. The m/z 15 peak may then be expected to correspond to the ionic counterpart of the methyl radical, i.e., to the $CH_3{}^+$ carbenium ion (Fig. 6.3). The question remains, as to whether this mass spectrum can be rationalized in terms of ion chemistry. Let us therefore consider the steps of electron ionization and subsequent fragmentation in greater detail.

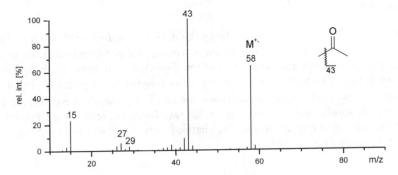

Fig. 6.3. EI mass spectrum of acetone. Used by permission of NIST. © NIST 2002.

6.2.1.1 α-Cleavage – A Radical-Site Initiated Fragmentation

The acetone molecule possesses two non-bonding electron pairs at the oxygen that will at least formally be the preferred source of the ejected electron (Chap.2.2.2). The excited molecular ion may then cleave off a methyl radical by simply shifting *one electron* (single-barbed arrow or "fishhook") from the CO–CH₃ bond to the radical site at the oxygen atom allowing the products to drift apart (Scheme 6.6). This homolytic bond cleavage is a *radical-site initiated* process with *charge retention*, i.e., the ionic charge resides within the moiety where it was initially located. The process is also known as *α-cleavage*. The neutral fragment, $CH_3^{\cdot}$, is not detected by the mass spectrometer, whereas the charged fragment, $C_2H_3O^+$, gives rise to the base peak m/z 43:

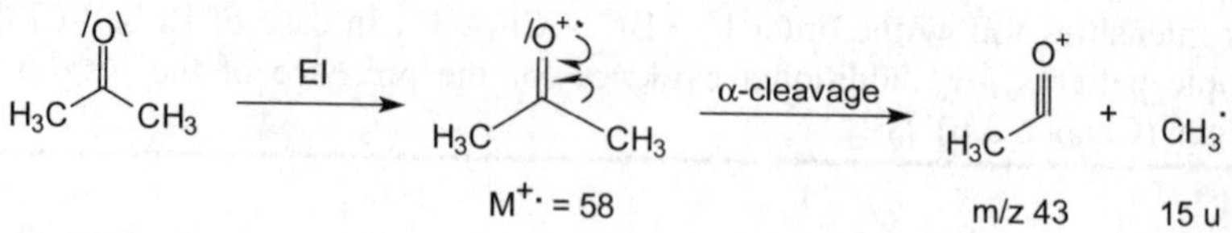

$$M^{+\cdot} = 58 \qquad\qquad m/z\ 43 \qquad 15\ u$$

Scheme 6.6.

> **Note:** The term *α-cleavage* for this widespread *radical-site initiated* process with *charge retention* can be misleading, because the bond cleaved is not directly attached to the radical site, but to the next neighboring atom.

Including the ionization process, the free electron pairs, and single-barbed arrows for each moving electron in the scheme is not necessary but presents a valuable aid. Alternatively, the α-cleavage may be indicated in a simplified manner:

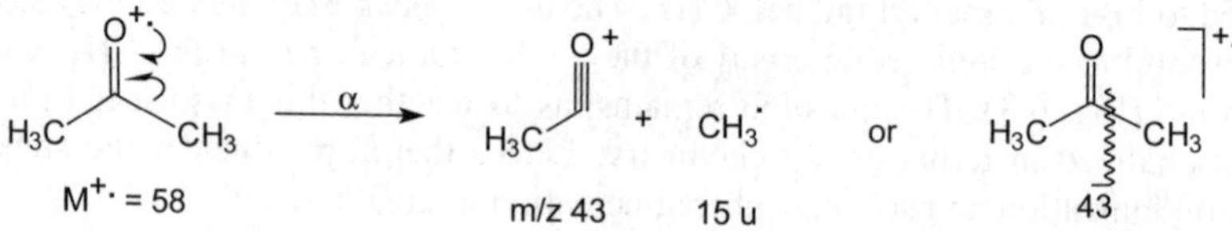

$$M^{+\cdot} = 58 \qquad m/z\ 43 \qquad 15\ u \qquad\qquad 43$$

Scheme 6.7.

The first abbreviated form only shows the decisive electron shift, but it still explicitly gives the structures of the products. The second is useful to indicate which bond will be cleaved and what m/z the ionic product will have. Such a writing convention can of course be used for any other fragmentation pathway. The ionic product of the α-cleavage, an *acyclium ion*, will not exactly have the angular structure as shown, however drawing it this way helps to identify the fragment as a part of the initial molecular ion. The charge in acylium ions can be substantially resonance-stabilized:

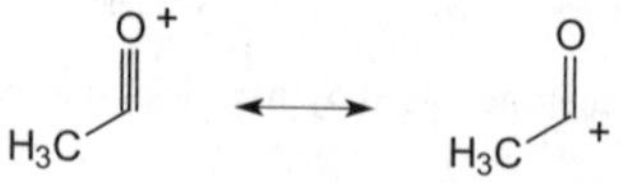

Scheme 6.8.

6.2.2 Stevenson's Rule

The origin of the m/z 43 peak in the EI mass spectrum of acetone should be quite clear now and we may examine the formation of the CH_3^+ ion, m/z 15, next. In principle the ionic charge may reside on either fragment, the acylium or the alkyl. In case of acetone, the formation of the acylium ion, CH_3CO^+, m/z 43, is preferred over the formation of the small carbenium ion, CH_3^+, m/z 15. The question which of the two incipient fragments will preferably keep the charge can be answered by means of Stevenson's Rule.

Stevenson's rule: When a fragmentation takes place, the positive charge remains on the fragment with the lowest ionization energy.

This criterion was originally established for the fragmentation of alkanes by Stevenson [18] and was later demonstrated to be generally valid. [19,20] The rule can be rationalized on the basis of some ion thermochemical considerations (Fig. 6.4). Assuming no reverse activation barrier, the difference in thermodynamic stability as expressed in terms of the difference of heats of formation of the respective products determines the preferred dissociation pathway:

$$\Delta H_{f(B+;A\bullet)} - \Delta H_{f(A+;B\bullet)} = \Delta\Delta H_{f(Prod)} = \Delta E_0 \qquad (6.7)$$

This can also be expressed in relation to the ionization potentials of the radicals formed by both dissociations:

$$\Delta E_0 = IE_{B\bullet} - IE_{A\bullet} = E_{02} - E_{01} \qquad (6.8)$$

Therefore, the radical having the lower ionization energy will dominate among the products. [19,21] For example, in case of the α-cleavage of acetone, $CH_3C^\bullet{=}O$ has a by 2.8 eV lower IE than $^\bullet CH_3$ (Tab. 6.2). As ion internal energies are subject to a wide distribution, there is a substantial fraction of ions that may also dissociate to form the energetically more demanding pair of fragments. This is the reason for the "soft" character of the rule.

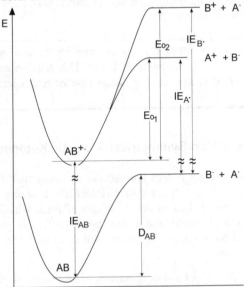

Fig. 6.4. Thermochemical description of Stevenson's rule; D_{AB} homolytic bond dissociation energy of bond A–B, IE ionization energy.

> **Note:** The homolytic dissociation of a C–C bond always proceeds to yield both product pairs, their relative abundances being basically governed by Stevenson's rule.

Without having the thermochemistry data at hand, it is not trivial to decide which pair of products will be preferred over the other. In general, the formation of the higher substituted and/or larger carbenium ion is preferred, because it can more easily stabilize a charge. However, the tendency is the same for the radicals and one may expect loss of ethyl to be favored over loss of methyl, for example. Thus, the formation of both the ionic and the radical fragments are of decisive influence on the final distribution of products.

Table 6.2. Ionization energies of some radicals.[a]

Radical	IE[b] [eV]	Radical	IE[b] [eV]
H$^\bullet$	13.6	CH$_3$O$^\bullet$	10.7
$^\bullet$CH$_3$	9.8	$^\bullet$CH$_2$OH	7.6
$^\bullet$C$_2$H$_5$	8.4	CH$_3$C$^\bullet$=O	7.0
n-$^\bullet$C$_3$H$_7$	8.2	C$_2$H$_5$C$^\bullet$=O	5.7
i-$^\bullet$C$_3$H$_7$	7.6	$^\bullet$CH$_2$Cl	8.8
n-$^\bullet$C$_4$H$_9$	8.0	$^\bullet$CCl$_3$	8.1
i-$^\bullet$C$_4$H$_9$	7.9	C$_6$H$_5$$^\bullet$	8.3
s-$^\bullet$C$_4$H$_9$	7.3	C$_6$H$_5$CH$_2$$^\bullet$	7.2
t-$^\bullet$C$_4$H$_9$	6.8	$^\bullet$CH$_2$NH$_2$	6.3

[a] IE data extracted from Ref. [22] with permission. © NIST 2002.
[b] All values rounded to one decimal place.

> **Note:** The validity of Stevenson's rule requires no reverse activation energy barrier to exist for the fragmentation pathway. This requirement is usually fulfilled for simple bond cleavages, but not in case of rearrangement fragmentations.

6.2.3 α-Cleavage of Non-Symmetrical Aliphatic Ketones

When a ketone grows larger it does not necessarily imply that it has two identical alkyl groups at the carbonyl. In case of different alkyls at the carbonyl, Stevenson's rule may also be applied to decide which of them will dominantly be detected as part of the acylium ion and which should preferably give rise to a carbenium ion. Overall, a nonsymmetrical ketone will yield four primary fragment ions in its EI mass spectrum.

Example: The 70 eV EI mass spectrum of butanone shows an ethyl loss, m/z 43, which is largely preferred over methyl loss, m/z 57 (Fig. 6.5). Furthermore, the C$_2$H$_5$$^+$ ion, m/z 29, is more abundant than the less stable CH$_3$$^+$ ion, *m/z* 15. If the alkyls become larger than ethyl, another pathway of ion dissociation will occur in addition (Chap. 6.7).

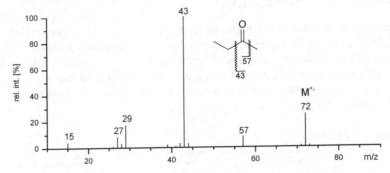

Fig. 6.5. EI mass spectrum of butanone. Used by permission of NIST. © NIST 2002.

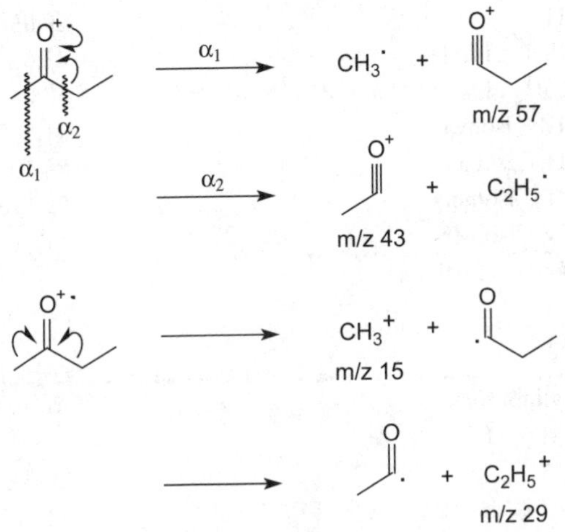

Scheme 6.9.

Different from ketones [23] the mass spectra of aliphatic aldehydes [24,25] reveal that α-cleavage is one of the less important primary fragmentations. This is because α-cleavage of aldehyde molecular ions leads to the formation of energetically unfavorable products, i.e., loss of H⋅ or formation of the formyl ion, CHO^+, *m/z* 29. It was demonstrated by means of high-resolution mass spectrometry (Chap. 3.3), that only a minor portion of the *m/z* 29 peak originates from CHO^+ ions, but the major fraction of about 65–75 % is due to $C_2H_5^+$ ions from σ-cleavage. [26]

6.2.4 Acylium Ions and Carbenium Ions

6.2.4.1 Series of Homologous Ions

Upon extrapolation to larger ketones one can expect to observe larger acylium and alkyl fragments. The occurrence of series of homologous ions is a feature that can be very helpful to deduce structural information from mass spectra. Ions such as the acylium ion series and the carbenium ion series are also known as *characteristic ions*. Learning the nominal masses of the first members of each series by heart is useful (Tables 6.2 and 6.3).

Table 6.3. Carbenium ions

m/z	Carbenium Ions $[C_nH_{2n+1}]^+$	Accurate Mass [u][a]
15	CH_3^+	15.0229
29	$C_2H_5^+$	29.0386
43	$n\text{-}C_3H_7^+$, $i\text{-}C_3H_7^+$	43.0542
57	$n\text{-}C_4H_9^+$, $i\text{-}C_4H_9^+$, $sec\text{-}C_4H_9^+$, $tert\text{-}C_4H_9^+$	57.0699
71	$C_5H_{11}^+$ isomers	71.0855
85	$C_6H_{13}^+$ isomers	85.1011
99	$C_7H_{15}^+$ isomers	99.1168
113	$C_8H_{17}^+$ isomers	113.1325

a Values rounded to four decimal places.

Table 6.4. Acylium ions

m/z	Acylium Ions $[C_nH_{2n-1}O]^+$	Accurate Mass [u][a]
29	HCO^+	29.0022
43	CH_3CO^+	43.0178
57	$C_2H_5CO^+$	57.0335
71	$C_3H_7CO^+$	71.0491
85	$C_4H_9CO^+$	85.0648
99	$C_5H_{11}CO^+$ isomers	99.0804
113	$C_6H_{13}CO^+$ isomers	113.0961
127	$C_7H_{15}CO^+$ isomers	127.1117

a Values rounded to four decimal places.

> **Note:** The acylium ions and the saturated carbenium ions are isobaric, i.e., they have the same nominal mass. However, their exact masses are different, i.e., CH_3CO^+, m/z 43.0178, and $C_3H_7^+$, m/z 43.0542.

6.2.4.2 Differentiation Between Carbenium Ions and Acylium Ions

Besides having different accurate mass, carbenium ions and acylium ions fortunately exhibit different dissociation pathways allowing their differentiation from the appearance of a mass spectrum. Acylium ions undergo CO loss (–28 u), thereby forming carbenium ions, but they do not eliminate hydrogen molecules:

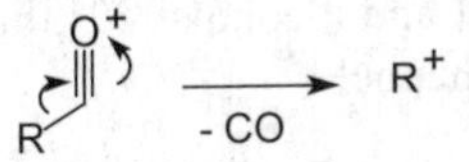

$$R-C{\equiv}O^+ \xrightarrow[-\ CO]{} R^+$$

Scheme 6.10.

Example: The dissociation $CH_3CO^+ \rightarrow CH_3^+ + CO$ presents a second pathway for the generation of the m/z 15 peak in the mass spectra of acetone (Fig. 6.3) and butanone (Fig. 6.5), and accordingly the process $CH_3CH_2CO^+ \rightarrow C_2H_5^+ + CO$ also yields an m/z 29 ion. Thus, this pathway competes with the one-step formation of carbenium ions from ketone molecular ions by charge migration.

Carbenium ions, especially from ethyl to butyl show remarkable dehydrogenation that gives rise to a characteristic accompanying pattern of peaks at $m/z-2$ and $m/z-4$, i.e., at the low-mass side of the corresponding peak:

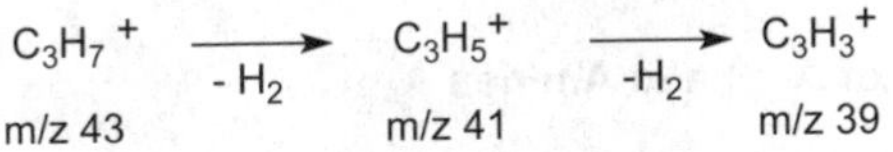

$$C_3H_7^+ \xrightarrow[-\ H_2]{} C_3H_5^+ \xrightarrow[-H_2]{} C_3H_3^+$$
$$\ \ m/z\ 43 \qquad\quad m/z\ 41 \qquad\quad m/z\ 39$$

Scheme 6.11.

Example: The mass spectra of both acetone and butanone show typical acylium ion peaks at m/z 43, whereas the signals in the spectra of isopropyl ethyl thioether (Fig. 6.9), of 1-bromo-octane, (Fig. 6.10), and of isomeric decanes (Fig. 6.18) may serve as examples for carbenium ion signals. The superimposition of both classes of ions causes signals representing an average pattern. The properties of larger carbenium ions are discussed in the section on alkanes (Chaps. 6.6.1 and 6.6.3).

6.2.5 α-Cleavage of Amines, Ethers, and Alcohols

The charge-localizing heteroatom can also be part of the aliphatic chain as it is the case with amines, ethers, and alcohols. The mechanism of the α-cleavage remains unaffected by this change, i.e., still the bond second to the charge site is cleaved. Nevertheless, the structure of the charged fragments is quite changed, and consequently their further fragmentation pathways are also different.

In case of amine molecular ions the α-cleavage is the absolutely dominating primary fragmentation pathway. The strong charge-stabilizing properties of the nitrogen atom keeps the fraction of charge migration fragments very low. The product ions are termed *immonium ions*, because these can formally be obtained by electrophilic addition of a proton or carbenium ion to imines. Immonium ions represent the most stable $[C_nH_{2n+2}N]^+$ isomers [27] and preferably undergo further decomposition on two widespread and thus highly important pathways for alkene loss [28,29] (Chap. 6.11). *Oxonium ions* are the oxygen analogs to immonium ions. As oxygen is somewhat less charge-stabilizing than nitrogen, the competing formation of carbenium ions is more pronounced among the primary fragmentations of aliphatic ethers [30,31] and alcohols [32], but still the α-cleavage represents the favored dissociation channel.

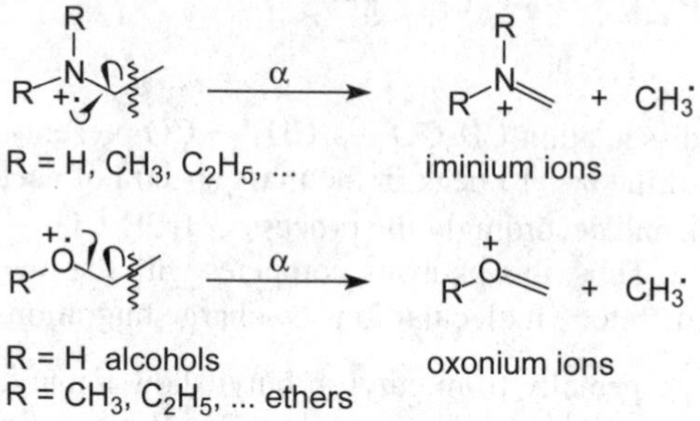

Scheme 6.12.

6.2.5.1 α-Cleavage of Aliphatic Amines

The intensity of the molecular ion of aliphatic amines decreases regularly with increasing molecular weight. [33] Analogous general behavior has also early been noted for other compound types such as the aliphatic ethers [31], hydrocarbons [34] and aldehydes [24], and others. The reason for the low stability of amine molecular ions is their predisposition to α-cleavage.

A comparison of the relative strength of functional groups to cause α-cleavage is summarized in Table 6.5. [6] This also corresponds to a rough measure of relative charge-stabilizing capability of the respective substituent, e.g., the ratio $H_2C=OH^+/H_2C=NH_2^+$ from 2-amino ethanol molecular ion is 2.3/100 and the ratio of $H_2C=OH^+/H_2C=SH^+$ from 2-thio ethanol molecular ion is 42/70. [20]

The mass spectrum of *N*-ethyl-*N*-methyl-propanamine, $C_6H_{15}N$, shows the molecular ion peak at *m/z* 101 (Fig. 6.6). The primary fragmentations of the molecular ion may well be explained in terms of the α-cleavage and accordingly, the peaks at *m/z* 72 and 86 can be assigned as immonium fragment ions due to ethyl and methyl loss, respectively. However, there are five instead of two channels for the α-cleavage possible! In order to identify the missing three dissociation channels, we have to write down the fragmentation scheme carefully checking every possibility. And of course, we have to examine the mass spectrum with sufficient care in order not to overlook an indicative peak.

Table 6.5. Relative strength of functional groups to cause α-cleavage; from Ref. [6] with permission. © Georg Thieme Verlag, 2002.

Functional Group	Relative Value
–COOH	1
–Cl, –OH	8
–Br	13
–COOMe	20
–CO– (ketones)	43
–OMe	100
–I , –SMe	≈ 110
–NHCOCH$_3$	128
–NH$_2$	990
acetals	1600
–NMe$_2$	2100

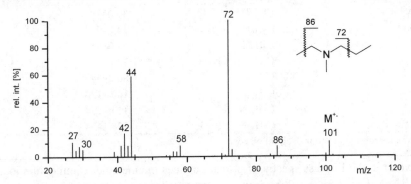

Fig. 6.6. EI mass spectrum of *N*-ethyl-*N*-methyl-propanamine. The molecular ion is detected at odd *m/z*, whereas the immonium fragment ions all have even *m/z* values. Spectrum used by permission of NIST. © NIST 2002.

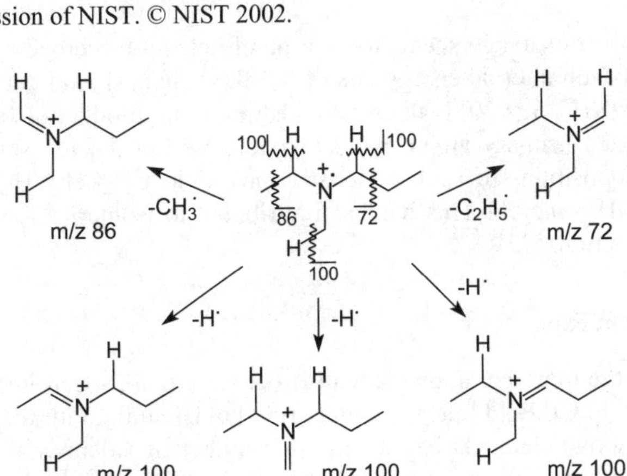

Scheme 6.13.

The peak at m/z 100 belongs to H$^{\bullet}$ loss and is also due to α-cleavage as can be easily recognized (Scheme 6.13). There are three different positions to cleave off the radical, and even seven almost equivalent hydrogens are available in total (for clarity only one them has been shown at any position in the scheme). Despite this multiple chance, the peak at m/z 100 is very weak, the reason for this being the unfavorable thermodynamics of H$^{\bullet}$ loss as compared to methyl loss (Table 2.2).

As with acylium ions and carbenium ions before, the series of homologous immonium ions is part of the mass spectrometrist's tool box. They can easily be recognized in the mass spectra and have even-numbered m/z values (Tab. 6.6). In the EI spectrum of N-ethyl-N-methyl-propanamine the series is completely present from m/z 30 up to m/z 100.

Table 6.6. Aliphatic immonium ions

m/z (16 + 14n)	Immonium Ions $[C_nH_{2n+2}N]^+$	Accurate Mass [u][a]
30	CH_4N^+	30.0338
44	$C_2H_6N^+$	44.0495
58	$C_3H_8N^+$	58.0651
72	$C_4H_{10}N^+$	72.0808
86	$C_5H_{12}N^+$	86.0964
100	$C_6H_{14}N^+$	100.1120
114	$C_7H_{16}N^+$	114.1277
128	$C_8H_{18}N^+$	128.1434

a Values rounded to four decimal places.

> **Note:** In the EI mass spectra of primary amines the methylene immonium ion, $CH_2=NH^+$, m/z 30, resulting from α-cleavage either represents the base peak or at least is the by far most abundant of the immonium ion series.

Even though α-cleavage seems to be a purely electronic process at first sight, the influence of ion internal energy and of the alkyl chain should not be neglected. Whereas $CH_2=NH^+$, m/z 30, is the predominating ionic product of fragmentations of highly excited primary amine molecular ions within the ion source, the metastable decompositions of such molecular ions yield $CH_2CH^+NH_2$, m/z 44, and $CH_2CH_2CH^+NH_2$, m/z 58. This has been attributed to isomerization via H$^{\bullet}$ shifts prior to dissociation. [35]

6.2.5.2 Nitrogen Rule

Restricting to the more common elements in organic mass spectrometry (H, B, C, N, O, Si, S, P, F, Cl, Br, I, etc.), a simple rule holds valid: With the exception of N, all of the above elements having an odd number of valences also possess an odd mass number and those having an even number of valences have even mass numbers. This adds up to molecular masses fulfilling the *nitrogen rule* (Tab. 6.7).

Nitrogen rule: If a compound contains an even number of nitrogen atoms (0, 2, 4, ...), its monoisotopic molecular ion will be detected at an even-numbered m/z (integer value). Vice versa, an odd number of nitrogens (1, 3, 5, ...) is indicated by an odd-numbered m/z.

Table 6.7. Demonstration of the nitrogen rule

Number of Nitrogens	Examples	$M^{+\bullet}$ at m/z
0	methane, CH_4	16
0	acetone, C_3H_6O	58
0	chloroform, $CHCl_3$	118
0	[60]fullerene, C_{60}	720
1	ammonia, NH_3	17
1	acetonitrile, C_2H_3N	41
1	pyridine, C_5H_5N	79
1	N-ethyl-N-methyl-propanamine, $C_6H_{15}N$	101
2	urea, CH_4N_2O	60
2	pyridazine, $C_4H_4N_2$	80
3	triazole, $C_2H_3N_3$	69
3	hexamethylphosphoric triamide, HMPTA, $C_6H_{18}N_3OP$	179

The rule may also be extended for use with fragment ions. This makes a practical tool to distinguish even-electron from odd-electron fragment ions and thus simple bond cleavages from rearrangements. However, some additional care has to be taken when applying the extended rule, because nitrogen might also be contained in the neutral, e.g., loss of NH_3, 17 u, or of $CONH_2^{\bullet}$, 44 u.

Rule: Cleaving off a radical (that contains no nitrogen) from any ion changes the integer m/z value from odd to even or vice versa. Loss of a molecule (that contains no nitrogen) from an ion produces even mass fragments from even mass ions and odd mass fragments from odd mass ions.

Examples: To rationalize the mass spectrum of methane, reactions 6.2–6.6 were proposed. They all obey the rule. You should check the mass spectra and fragmentation schemes throughout this chapter for additional examples of the nitrogen rule:

$$CH_4^{+\bullet} \rightarrow CH_3^+ + H^{\bullet}; \ m/z \ 16 \rightarrow m/z \ 15 \qquad (6.2)$$

$$CH_4^{+\bullet} \rightarrow CH_2^{+\bullet} + H_2; \ m/z \ 16 \rightarrow m/z \ 14 \qquad (6.3)$$

$$CH_2^{+\bullet} \rightarrow C^{+\bullet} + H_2; \ m/z \ 14 \rightarrow m/z \ 12 \qquad (6.4)$$

$$CH_2^{+\bullet} \rightarrow CH^+ + H^{\bullet}; \ m/z \ 14 \rightarrow m/z \ 13 \qquad (6.5)$$

$$CH_3^+ \rightarrow CH^+ + H_2; \ m/z \ 15 \rightarrow m/z \ 13 \qquad (6.6)$$

6.2.5.3 α-Cleavage of Aliphatic Ethers and Alcohols

The molecular ions of aliphatic ethers do not behave much different from those of amines. [30,31] In constrast, oxygen is not directing their primary dissociations as strongly to the α-cleavage as nitrogen does in amines. Although α-cleavage is still dominant, there is a stronger tendency towards formation of carbenium ion fragments by charge migration. As can be expected, the structure of the alkyls also exerts a significant influence on the selection of a cleavage pathway.

Example: The 70 eV EI mass spectra of methyl propyl ether and diethyl ether are shown below (Fig. 6.7). Although being isomers, their spectra are clearly different. In case of methyl propyl ether, α-cleavage can occur by $H^{\bullet}$ loss, m/z 73, and definitely preferable by $C_2H_5^{\bullet}$ loss, m/z 45, which gives rise to the base peak. The advantage of $C_2H_5^{\bullet}$ loss over $CH_3^{\bullet}$ loss becomes evident from the diethyl

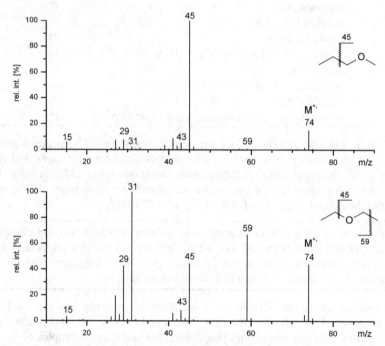

Fig. 6.7. EI mass spectra of methylpropylether and diethyl ether. Spectra used by permission of NIST. © NIST 2002.

ether spectrum. In accordance with Stevenson's rule, even two ethyl groups cause much less $CH_3^{\bullet}$ loss. It is tempting, although not trivial, to quantify the relative "ease" of fragmentation from the ratio of peak intensities. However, the primary fragment will undergo further fragmentation at an unknown rate, e.g., ethene loss from the m/z 59 oxonium ion to yield the m/z 31 oxonium ion. Generally, oxonium ions are highly indicative for aliphatic ethers and alcohols. In the EI spectrum of diethyl ether the series from m/z 31 to m/z 73 is present (Table 6.8).

In case of alkanols, the methylene oxonium ion, $CH_2=OH^+$, m/z 31, deserves special attention. Resulting from α-cleavage, it undoubtedly marks spectra of primary alkanols, where it either represents the base peak or at least is the by far most abundant of the oxonium ion series (Fig. 6.8). [32] The second important fragmentation route of aliphatic alcohols, loss of H_2O, is discussed in Chap. 6.10.

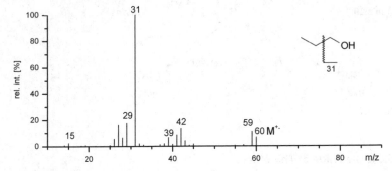

Fig. 6.8. EI mass spectrum of 1-propanol. Used by permission of NIST. © NIST 2002.

Table 6.8. Aliphatic oxonium ions

m/z (17 + 14n)	Oxonium Ions $[C_nH_{2n+1}O]^+$	Accurate Mass [u][a]
31	CH_3O^+	31.0178
45	$C_2H_5O^+$	45.0335
59	$C_3H_7O^+$	59.0491
73	$C_4H_9O^+$	73.0648
87	$C_5H_{11}O^+$	87.0804
101	$C_6H_{13}O^+$	101.0961
115	$C_7H_{15}O^+$	115.1117
129	$C_8H_{17}O^+$	129.1274

a Values rounded to four decimal places.

6.2.5.4 Charge Retention at the Heteroatom

The peak at m/z 45 in the spectrum of diethyl ether can neither be explained by α-cleavage nor by α-cleavage plus a subsequent cleavage, because the latter would demand carbene loss to occur, which is almost never observed (one of the rare examples is presented by formation of $HI^{+\bullet}$ from $CH_3I^{+\bullet}$). Obviously, the C–O bond can access a direct pathway for cleavage. Although some C–N bond cleavage can be observed in EI spectra of aliphatic amines, the cleavage of the C–O bond is gaining importance for the molecular ions of aliphatic ethers. It is a simple σ-bond cleavage as discussed for methane molecular ions, the only difference being the fact that one of the atoms connected by the σ-bond is not a carbon atom. As the heteroatom can stabilize the charge better than a primary carbon, the RX^+ fragment is preferably formed in such a case (σ_1). The product ion may then rearrange by 1,2-hydride shift to form an oxonium ion. [36] The formation of $C_2H_5^+$ fragment ions, m/z 29, directly competes with σ_1.

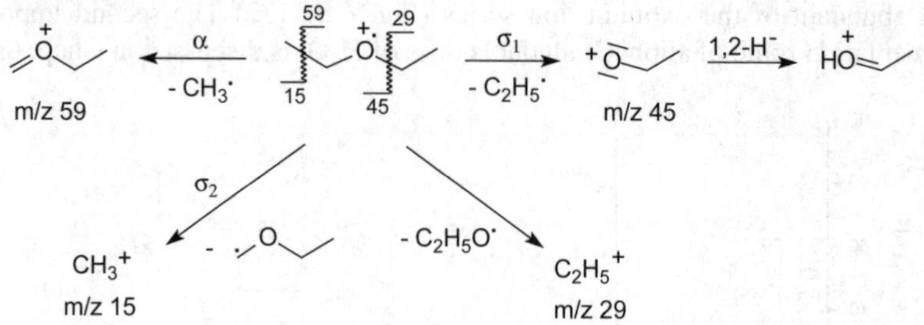

Scheme 6.14.

6.2.5.5 α-Cleavage of Thioethers

The EI mass spectra of thiols and thioethers also show a series of onium ions generated by α-cleavage of the molecular ion (Table 6.9). *Sulfonium ions* can easily be recognized from the isotopic pattern of sulfur (Fig. 6.9). The fragmentation patterns of thioethers will be discussed in greater detail later (Chap. 6.5.2 and 6.12.4).

Table 6.9. Aliphatic sulfonium ions

m/z (33 + 14n)	Sulfonium Ions $[C_nH_{2n+1}S]^+$	Accurate Mass [u][a]
47	CH_3S^+	46.9950
61	$C_2H_5S^+$	61.0106
75	$C_3H_7S^+$	75.0263
89	$C_4H_9S^+$	89.0419
103	$C_5H_{11}S^+$	103.0576
117	$C_6H_{13}S^+$	117.0732
131	$C_7H_{15}S^+$	131.0889
145	$C_8H_{17}S^+$	145.1045

a Values rounded to four decimal places.

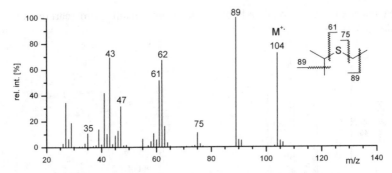

Fig. 6.9. EI mass spectrum of ethylisopropylthioether. Due to the isotopic composition of sulfur, the intensity of the [M+2]$^{+\cdot}$ ion is increased. The corresponding contributions can also be recognized for the sulfonium ions at *m/z* 47, 61, 75, and 89. The *m/z* 35 peak is due to H$_3$S$^+$ ions; also compare the spectrum of ethylpropylthioether (Fig. 6.45). Spectrum used by permission of NIST. © NIST 2002.

6.2.6 α-Cleavage of Halogenated Hydrocarbons

Aliphatic halogenated hydrocarbons do not show abundant fragment ions due to α-cleavage. [17] As was found with aliphatic amines and ethers before, the molecular ion peak decreases in intensity as the molecular weight and branching of the alkyl chain increase. In general, the relative intensity of the molecular ion peak falls in the order I > Br > Cl > F (Chap. 6.1.4). For bromine the molecular ions are almost of the same relative intensity as for hydrogen, i.e., for the corresponding hydrocarbon. This corresponds to the inverse order of halogen electronegativities. Higher electronegativity of the halogen also causes a higher ionization energy of the RX molecule and a rise of α-cleavage products.

Example: The 70 eV EI mass spectrum of 1-bromo-octane incorporates the characteristic fragmentations of aliphatic halogenated hydrocarbons (Fig. 6.10). All bromine-containing fragments are readily recognized from their bromine isotopic pattern (Chap. 3.1, 3.2). The product of the α-cleavage, [CH$_2$Br]$^+$, *m/z* 91, 93, is of minor intensity, whereas a [M–57]$^+$ ion, *m/z* 135, 137, dominates the spectrum. Obviously, there is preference to form this specific fragment ion, but not its homologues. This preference for the [M–57]$^+$ ion, [C$_4$H$_8$Br]$^+$, has been attributed to the possibility to yield a five-membered cyclic bromonium ion of low ring strain. This reaction differs from the preceding bond cleavages in that it is a displacement of an alkyl group, but it shares the common property of being a one-step process. [37] Analogous behavior is observed for the chlorohydrocarbons from hexyl through octadecyl that show the [C$_4$H$_8$Cl]$^+$ ion, *m/z* 91, 93. Fortunately, the isotopic patterns of Br and Cl are clearly different, thereby avoiding confusion of the isobaric [C$_4$H$_8$Cl]$^+$ and [CH$_2$Br]$^+$ ions. Iodo- and fluorohydrocarbons do not exhibit such cyclic halonium ions in their EI spectra. The remaining fragments can be rationalized by σ-cleavages leading to carbenium ions (Table 6.2)

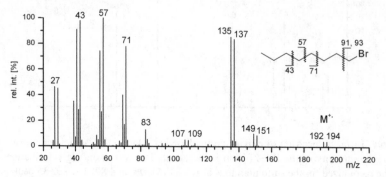

Fig. 6.10. EI mass spectrum of 1-bromo-octane. The product of the α-cleavage, *m/z* 91, 93 is of minor intensity, whereas the cyclic bromonium ion, *m/z* 135 and 137, dominates the spectrum. Spectrum used by permission of NIST. © NIST 2002.

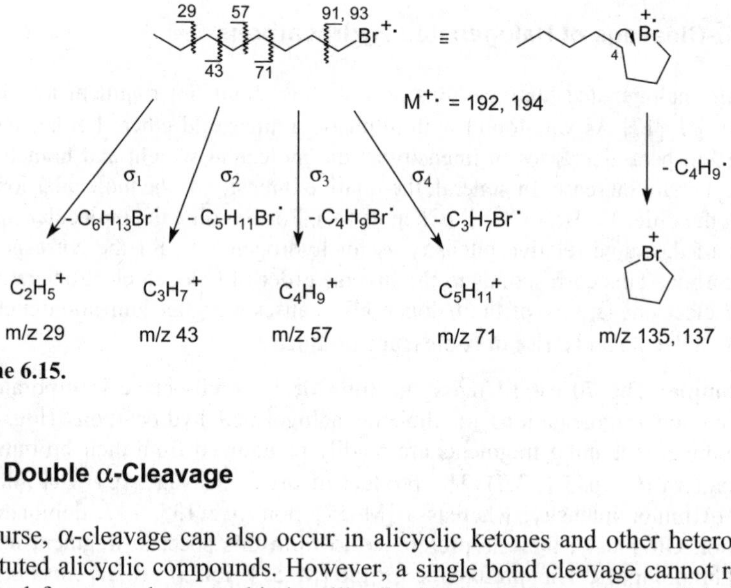

Scheme 6.15.

6.2.7 Double α-Cleavage

Of course, α-cleavage can also occur in alicyclic ketones and other heteroatom-substituted alicyclic compounds. However, a single bond cleavage cannot release a neutral fragment, because this is still adhering to another valence of the functional group:

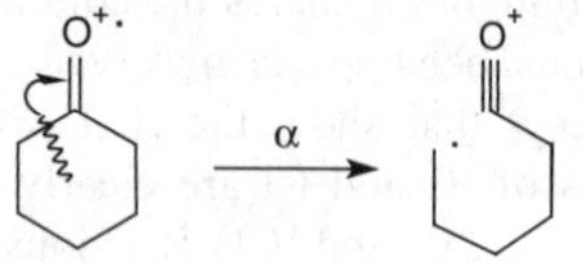

Scheme 6.16.

6.2.7.1 Propyl Loss of Cyclohexanone Molecular Ion

The mass spectrum of cyclohexanone has been examined by deuterium-labeling to reveal the mechanism effective for propyl loss, $[M–43]^+$, m/z 55, from the molecular ion, $M^{+\bullet} = 98$. [38,39] The corresponding signal represents the base peak of the spectrum (Fig. 6.11). Obviously, one deuterium atom is incorporated in the fragment ion that is shifted to m/z 56 in case of the $[2,2,6,6-D_4]$isotopomer. These findings are consistent with a three-step mechanism for propyl loss, i.e., with a double α-cleavage and an intermediate 1,5-H$^\bullet$ shift.

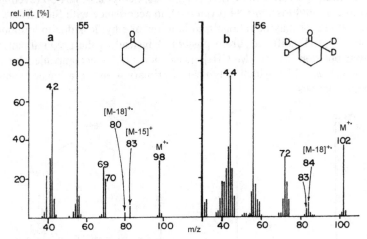

Fig. 6.11. 70 eV EI mass spectra of cyclohexanone (**a**) and its deuterated isotopomer $[2,2,6,6-D_4]$ (**b**). Adapted from Ref. [38] with permission. © Springer-Verlag Wien, 1964.

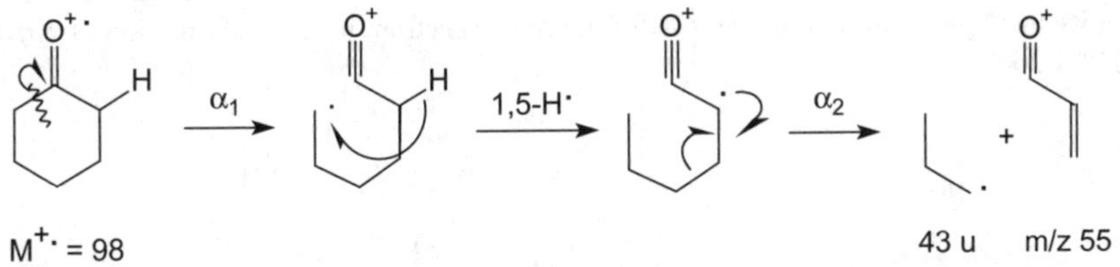

Scheme 6.17.

Note: Care should be taken when trying to formulate complex fragmentation processes in a concerted manner as it often was the case in early publications. There are strong arguments that gas phase ionic reactions are generally step-wise processes; i.e., it is more probable that such a fragmentation consists of several discrete steps instead of one that demands simultaneous breaking and making of several bonds. [37]

6.2.7.2 Double α-Cleavage for the Identification of Regioisomers

Following the logic of the deuterium labeling, regioisomers of cyclohexanones, cyclohexylamines, cyclohexylalcohols and others can be identified by strict application of the above mechanism. Regardless of some limitations of the method, e.g., that it is impossible to distinguish a 2,3-dimethyl from a 2-ethyl or 3-ethyl derivative, the method provides a valuable aid in structure elucidation.

Example: Propyl and pentyl loss from 2-ethyl-cyclohexylamine molecular ion, $M^{+\cdot} = 127$ (odd m/z) are competitive (Fig. 6.12). Pentyl loss, m/z 56 (even m/z), is favored over propyl loss, m/z 84 (even m/z), in accordance with Stevenson's Rule. The peak at m/z 98 may be rationalized in terms of ethyl loss due to a minor contribution of 1,4-H$^{\cdot}$ shift, i.e., from position 3 instead of the predominant 1,5-H$^{\cdot}$ shift from position 2. The $[M-CH_3]^+$ peak, m/z 112, is accompanied by a $[M-NH_3]^{+\cdot}$ signal, m/z 110, which is typical of primary – and to a lower extend also secondary – amines.

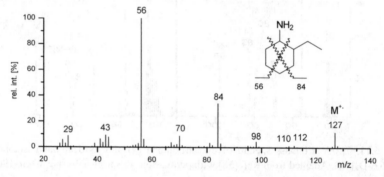

Fig. 6.12. EI mass spectrum of 2-ethyl-cyclohexylamine. Double α-cleavage allows to identify the C_2-substitution on one side of the ring. Spectrum used by permission of NIST. © NIST 2002.

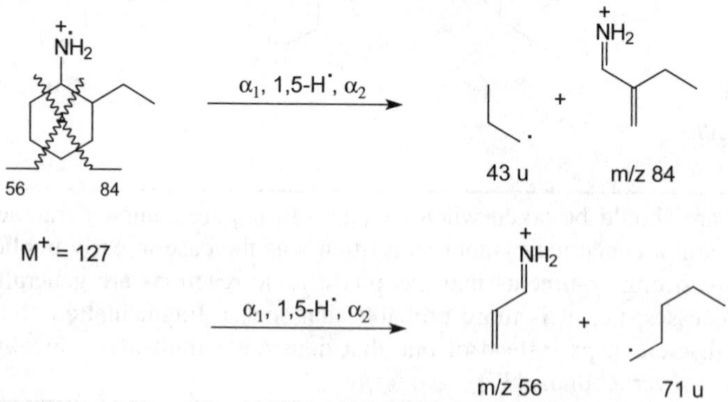

Scheme 6.18.

6.3 Distonic Ions

6.3.1 Definition of Distonic Ions

In the intermediate generated by the first α-cleavage of cyclohexyl compounds charge and radical are not located at the same atom as is the case with molecular ions, but at distant positions. The term *distonic ion* was derived from the Greek word for 'separate' to describe such ionic species. [40] Distonic ions represent an ionic class of their own. [40-42]

> **Definition:** A *distonic ion* is a positive radical ion, which would formally arise by ionization of a zwitterion or a diradical, by isomerization or fragmentation of a classical molecular ion, or by ion-molecule reactions. Consequently, distonic ions have charge and radical at separate atoms in a conventional valence bond description. [42,43]

However, bearing charge and radical at separate sites is not a sufficient condition for an ion to be denoted distonic, e.g., the ethylene molecular ion may be written as such, but the corresponding neutral is *not* best represented as zwitterion and therefore the ethylene molecular ion by definition is *not* a distonic ion:

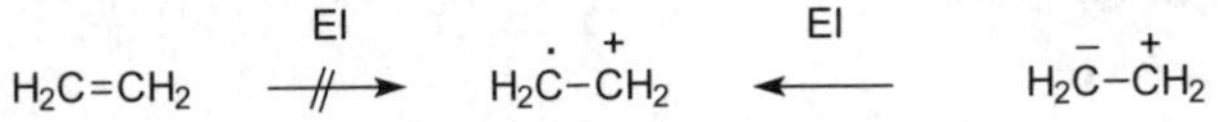

$$H_2C=CH_2 \quad \xrightarrow{\text{EI}} \quad H_2\overset{.}{C}-\overset{+}{C}H_2 \quad \xleftarrow{\text{EI}} \quad H_2\overset{-}{C}-\overset{+}{C}H_2$$

Scheme 6.19.

The expressions *nonclassical* and *hypervalent* ion have also been used by some authors to describe distonic ions, but these are incorrect and thus should no longer be used. The term *ylidion* is limited to species where charge and radical are at adjacent positions. Thus, to describe the distance between charge and radical site, the terms α- (1,2-) *distonic ion*, β- (1,3-) *distonic* ion, γ- (1,4-) *distonic ion*, and so forth are now in use: [42,43]

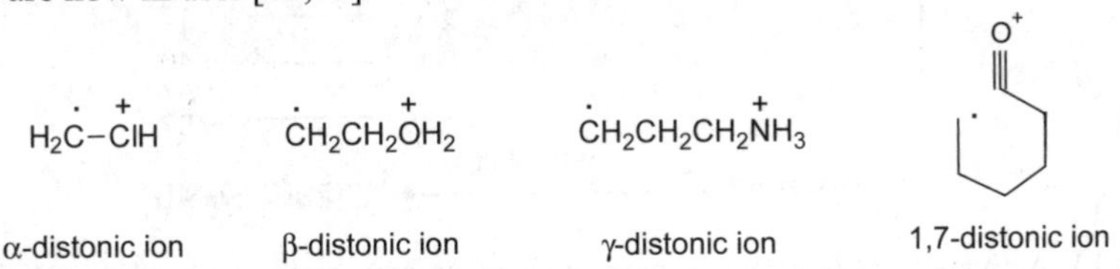

$H_2\overset{.}{C}-\overset{+}{C}lH$	$\overset{.}{C}H_2CH_2\overset{+}{O}H_2$	$\overset{.}{C}H_2CH_2CH_2\overset{+}{N}H_3$	
α-distonic ion	β-distonic ion	γ-distonic ion	1,7-distonic ion

Scheme 6.20.

6.3.2 Formation and Properties of Distonic Ions

Cleavage of a bond without immediate dissociation of the precursor radical ion is one way for the generating distonic ions. Isomerization of molecular ions by hydrogen radical shift frequently leads to distonic ions prior to fragmentation and, moreover, the distonic isomers are often thermodynamically more stable than their "classical" counterparts: [40,44]

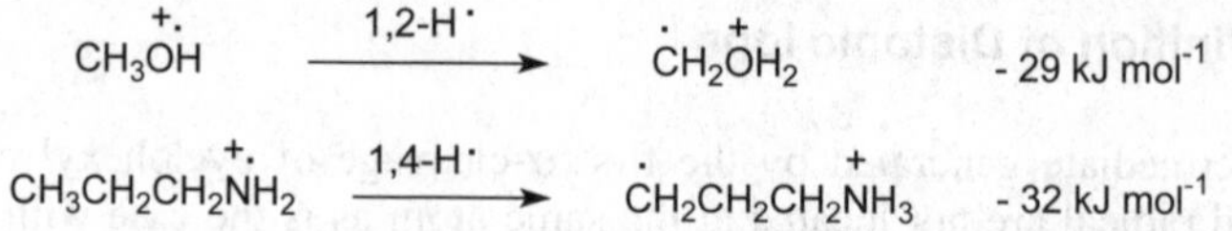

CH3OH
1,2-H·
CH2OH2
- 29 kJ mol-1
CH3CH2CH2NH2
1,4-H·
CH2CH2CH2NH3
- 32 kJ mol-1

Scheme 6.21.

The mutual interconversion of "classical" and distonic ions is not a fast process, because the isomers are separated by a comparatively high energy barrier. The activation energy for 1,2-H$^•$ shifts is substantially larger than for longer distance shifts such as 1,4- or 1,5-H$^•$ shifts. The activation energies and the heats of reaction have been determined for H$^•$ shifts in primary amine molecular ions (Fig. 6.13). [35,45,46] It has been argued that with increasing number of atoms between the positions the ring strain in the respective C–H$^•$–N transition states is reduced and that this is the reason for a significant decrease in activation energy. [46] In addition to the influence on transition states, the longer distance from charge to radical site also lowers the heats of formation of the distonic ions with respect to the "classical" ion precursors, i.e., any of these isomerizations is exothermic.

energy [kJ mol-1]
1,2-H·
n = 0
175
1,3-H·
n = 1
144
1,4-H·
n = 2
68
1,5-H·
n = 3
18
0
- 17
- 40
- 47
- 57
CH3(CH2)nNH2+·
1,x-H· shift
·CH2(CH2)nNH3+

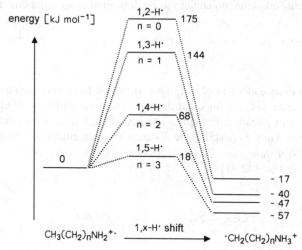

Fig. 6.13. Activation energies for isomerization of primary amine molecular ions to distonic isomers with the heats of formation of the precursor $M^{+•}$ ions normalized to zero. [46]

6.3.3 Distonic Ions as Intermediates

Although amine-derived distonic ions have been mainly presented up to here, the occurrence of distonic ions is by far not restricted to this compound class. Instead, distonic ions are of high relevance as these ions play an important role as the central intermediates and products in the dissociation reactions of many ionized molecules. Further, it seems now likely that the long-lived molecular ions of many organic compounds may exist as their distonic forms. [43] It has been shown, for example, that the long-lived radical ions of simple organophosphates spontaneously isomerize to distonic isomers, [47] and that the ring-opening product of cyclopropane molecular ion is also distonic. [48] The next distonic ions we are going to learn about are the intermediates of the McLafferty rearrangement (Chap. 6.7).

6.4 Benzylic Bond Cleavage

The α-cleavage in molecular ions of ketones, amines, ethers and similar functionalized compounds yields specific cleavage products of high importance for structure elucidation. Analogous behavior is observed in the mass spectra of phenylalkanes. [49]

6.4.1 Cleavage of the Benzylic Bond in Phenylalkanes

Molecular ions of phenylalkanes are comparatively stable due to the good charge stabilizing properties of the aromatic ring and thus, they normally give rise to intense peaks. Those molecular ions, possessing a benzylic bond preferably show

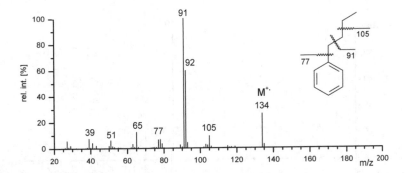

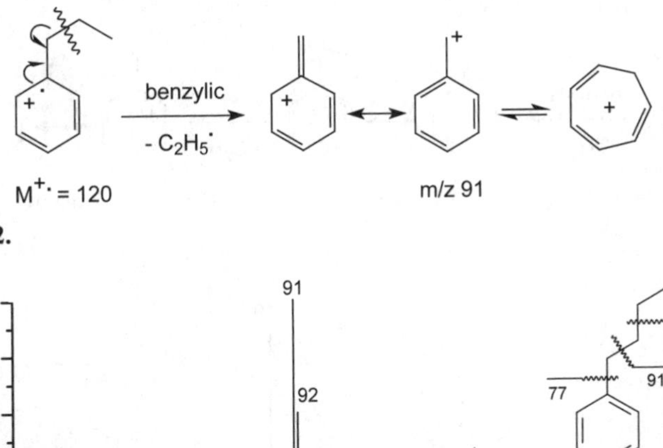

Scheme 6.22.

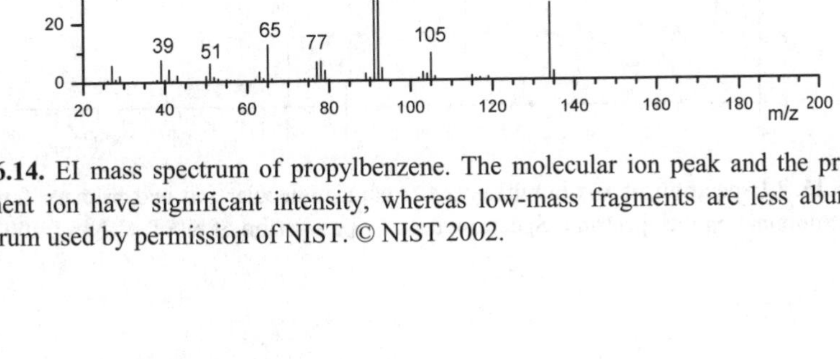

Fig. 6.14. EI mass spectrum of propylbenzene. The molecular ion peak and the primary fragment ion have significant intensity, whereas low-mass fragments are less abundant. Spectrum used by permission of NIST. © NIST 2002.

cleavage of that bond as compared to the phenylic or homobenzylic position. As with the α-cleavage, the process is radical-initiated and follows the same basic scheme, i.e., dissociation of the bond second to the radical-site after transfer of a single electron towards the radical-site. The mass spectrum of *n*-propylbenzene is an example (Fig. 6.14). The cleavages of the phenylic and homobenzylic bond are much less emphasized as can be seen from the minor *m/z* 77 and *m/z* 105 peak in the mass spectrum of propylbenzene. Nevertheless, phenylic bond cleavage does also occur and gives rise to further fragments. Due to its high thermodynamic stability, the $[C_7H_7]^+$ ion is not only formed by benzylic cleavage, but it is also obtained from many other fragmentation processes of phenylalkanes. Almost any mass spectrum of such compounds clearly exhibits the corresponding peak at *m/z* 91.

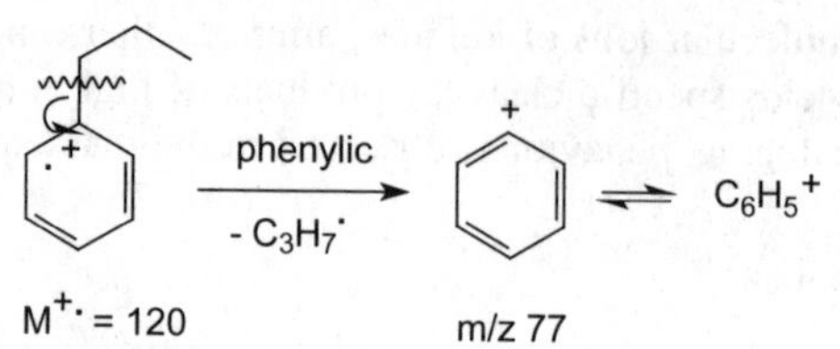

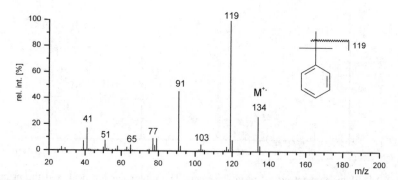

$$\xrightarrow[- C_3H_7^{\bullet}]{\text{phenylic}}$$

$$\rightleftharpoons C_6H_5^+$$

$$M^{+\bullet} = 120 \qquad\qquad m/z\ 77$$

Scheme 6.23.

Example: In the mass spectrum of *tert*-butylbenzene, $C_{10}H_{14}$, $M_r = 134$, the $[C_7H_7]^+$ ion is formed by C_2H_4 loss from the product of benzylic cleavage, *m/z* 119, despite such a rearrangement would not be expected at first sight (Fig. 6.15).

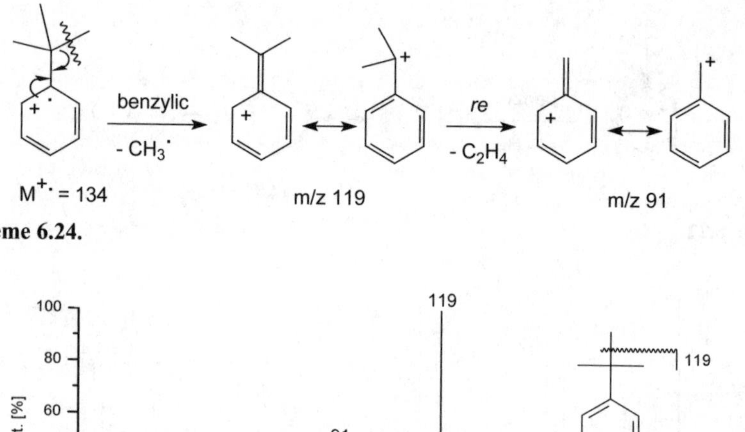

$$M^{+\bullet} = 134 \xrightarrow[- CH_3^{\bullet}]{\text{benzylic}} \quad m/z\ 119 \xrightarrow[- C_2H_4]{re} \quad m/z\ 91$$

Scheme 6.24.

Fig. 6.15. EI spectrum of *tert*-butylbenzene. Again, molecular ion and primary fragment ion are dominating the spectrum. Spectrum used by permission of NIST. © NIST 2002.

Note: The mere occurrence of the $[C_7H_7]^+$ ion – especially when it is of low intensity – is *not sufficient* to prove that the analyte belongs to the phenylalkanes. This ion and its characteristic fragments may also be observed *if there is some way at all* to generate a $[C_7H_7]^+$ fragment ion.

6.4.2 The Further Fate of $[C_6H_5]^+$ and $[C_7H_7]^+$

As already indicated in the preceding schemes, the resonance-stabilized benzyl ion, $[C_7H_7]^+$, initially formed by benzylic bond cleavage reversibly isomerizes to the tropylium and tolyl ion isomers. The isomerization of $[C_7H_7]^+$ ions has been the subject of numerous studies, [50] revealing the tropylium ion as the thermodynamically most stable isomer. [51,52]

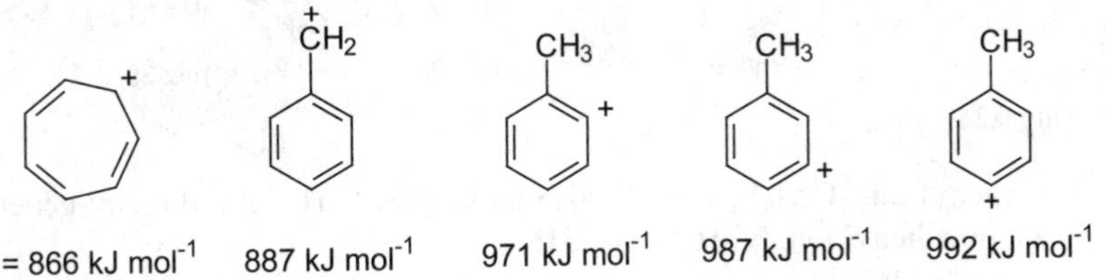

$\Delta H_f = 866 \text{ kJ mol}^{-1}$ 887 kJ mol^{-1} 971 kJ mol^{-1} 987 kJ mol^{-1} 992 kJ mol^{-1}

Scheme 6.25.

Provided sufficient lifetime, the process of benzyl/tropylium isomerization can be reversible ($E_0 = 167 \text{ kJ mol}^{-1}$ [50]), and then it necessarily goes along with a *scrambling* of hydrogens and of carbons as can be demonstrated by use of isotopically labeled compounds. [53] The degree of scrambling also depends on ion internal energy and on ion lifetime. In addition, it has been shown that the benzyl-to-tropylium ratio depends on the structure of the precursor molecular ion. Thus, $[C_7H_7]^+$ ions from benzyl chloride yield 57 % benzyl and 43 % tropylium upon 70 eV EI, but basically remain benzyl at low ion internal energies as realized by 12 eV EI (Table 6.10).

Table 6.10. Benzyl-to-tropylium ratios from various precursors

Precursor	Conditions	% Benzyl of $[C_7H_7]^+$
benzyl fluoride	12 eV EI	100
benzyl fluoride	methane CI	64
benzyl chloride	12 eV EI	92
benzyl chloride	70 eV EI	57
1,2-diphenylethane	11.5 eV EI	> 90
benzyl acetate	isobutane CI	90

> **Note:** The pictorial term *scrambling* is used in mass spectrometry to describe rapid processes of (intramolecular) positional interchange of atoms. Scrambling may occur with hydrogens or may involve the complete carbon skeleton of an ion. Aryl radical ions and protonated aryl compounds are well known for their numerous scrambling processes. [54,55]

6.4.2.1 Formation of Characteristic Fragments from $[C_7H_7]^+$ and $[C_6H_5]^+$ Ions

Whatever the initial or preferred structure, the $[C_7H_7]^+$ ions dissociate by loss of ethine, C_2H_2, to yield the cyclopentadienyl ion, $[C_5H_5]^+$, m/z 51, which fragments to yield $[C_3H_3]^+$, m/z 39, upon repeated ethine loss:

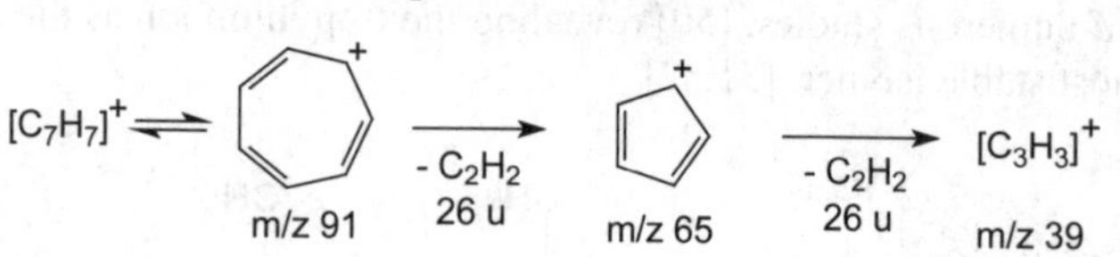

Scheme 6.26.

The phenyl ion $[C_6H_5]^+$, m/z 77, also undergoes C_2H_2 loss, thereby generating the cyclobutadienyl ion, $[C_4H_3]^+$, m/z 51:

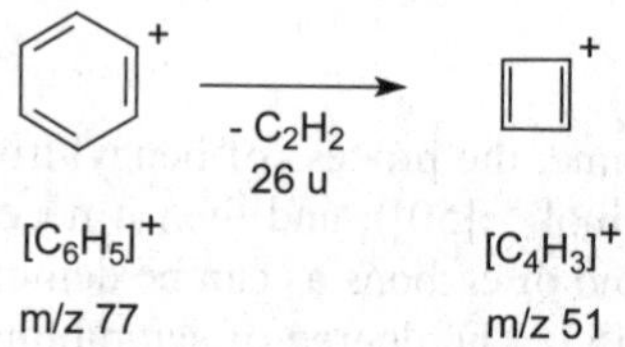

Scheme 6.27.

> **Note:** Overall, this gives the typical appearance of the ion series m/z 39, 51, 65, 77, 91 in the EI mass spectra of phenylalkanes. However, one needs to keep aware of the fact, that there are two competing reaction sequences leading to this series: (i) $M^{+\bullet} \rightarrow 91 \rightarrow 65 \rightarrow 39$ and (ii) $M^{+\bullet} \rightarrow 77 \rightarrow 51$; the latter usually being of lower intensity.

6.4.3 Isomerization of $[C_7H_8]^{+\bullet}$ and $[C_8H_8]^{+\bullet}$ Ions

The molecular ions of toluene and cycloheptatriene are capable of mutual isomerization prior to dissociation. [49,53,56] It has been known for long that hydrogen loss from toluene molecular ions takes place as if all eight positions within these ions were equivalent, and this has been interpreted in terms of complete hydrogen scrambling prior to dissociation. [57] The extent of isomerization becomes apparent from the close similarity of the toluene and cycloheptatriene EI mass spectra

(Fig. 6.16). Loss of H· from $[C_7H_8]^{+\cdot}$ leads to the formation of $[C_7H_7]^+$ ions, preferably of tropylium structure, [54] and thus, to the typical fragment ion series described above.

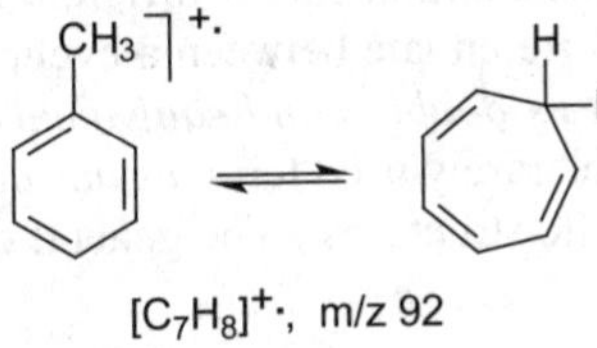

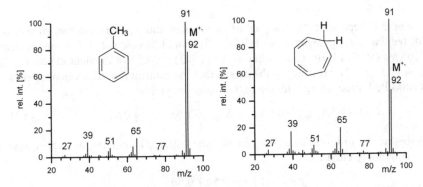

Scheme 6.28.

Fig. 6.16. Close similarity of the 70 eV EI mass spectra of toluene (left) and cycloheptatriene (right). Spectra used by permission of NIST. © NIST 2002.

Extensive isomerizations prior to dissociation are also observed for other ions having similar structural features. Deuterium labeling of some bicyclic aromatic systems has established that complete or partial randomization of the aromatic hydrogens prior to dissociation can occur upon electron impact. [58] In detail, such observations were made in molecular ions of i) biphenyl prior to loss of $CH_3^·$, C_2H_2, or $C_3H_3^·$, ii) 1- and 2-cyanonaphthalenes before HCN expulsion, and iii) benzothiophene before C_2H_2 expulsion. Cyclooctatetraene and styrene molecular ions, $[C_8H_8]^{+\cdot}$, m/z 104, interconvert prior to loss of ethine to yield benzene molecular ion, m/z 78. Because of the substantially lower overall activation energy, most of the $[C_6H_6]^{+\cdot}$ ions formed result from dissociation of cyclooctatetraene molecular ion although this has the higher heat of formation. [59]

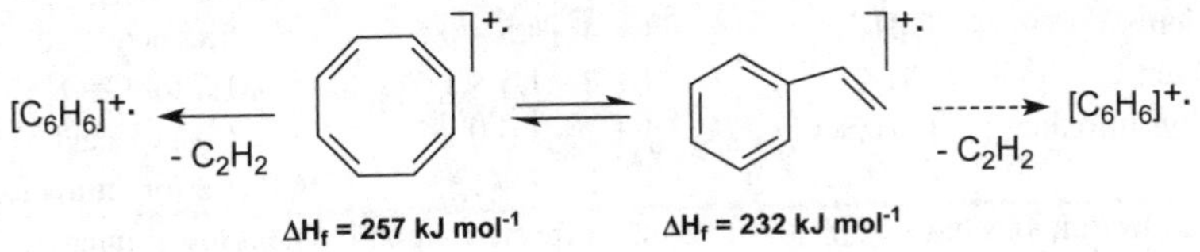

Scheme 6.29.

6.4.4 Rings Plus Double Bonds

Based on valence rules, a general algorithm can be derived to distinguish between ions of different degrees of unsaturation and/or different number of rings. As this algorithm does not allow to differentiate between a cyclic substructure and a double bond, the result is known as *double bond equivalents* (DBE) or simply *rings plus double bonds* ($r + d$). (The rarely used term *unsaturation* is not recommended because it does not imply cyclic structures.) The general equation for the determination of $r + d$ is:

$$r + d = 1 + 0.5 \cdot \sum_i^{i_{max}} N_i \, (V_i - 2) \qquad (6.9)$$

where V_i represents the valence of an element and N_i is the number of atoms. For regular use, the expressions $0.5 \cdot (V_i - 2)$ should be calculated. For monovalent elements (H, F, Cl, Br, I) one obtains $0.5 \cdot (1 - 2) = -0.5$, for divalent elements (O, S, Se) $0.5 \cdot (2 - 2) = 0$ which is the reason that the contribution of divalent elements is cancelled. Proceeding with the higher valences yields:

$$r + d = 1 - 0.5 N_{mono} + 0.5 N_{tri} + N_{tetra} + 1.5 N_{penta} + 2 N_{hexa} + ... \qquad (6.10)$$

Restriction to formulas of the general type $C_cH_hN_nO_o$ reduces the expression to the commonly cited form:

$$r + d = 1 + c - 0.5h + 0.5n \qquad (6.11)$$

Here, other monovalent elements than hydrogen (F, Cl, Br, I) are counted "as hydrogens", other trivalent elements such as phosphorus are counted "as nitrogen" and tetravalent elements (Si, Ge) are handled the same way as carbons.

The $r + d$ algorithm produces integers for odd-electron ions and molecules, but non-integers for even-electron ions that have to be rounded to the next lower integer, thereby allowing to distinguish even- from odd-electron species.

Table 6.11. Examples for the application of the $r + d$ algorithm according to Eq. 6.11

Name	Formula	$r + d$	Comment
decane	$C_{10}H_{22}{}^{+\bullet}$	$1 + 10 - 11 = 0$	
cyclohexanone	$C_6H_{10}O^{+\bullet}$	$1 + 6 - 5 = 2$	$1\,r + 1\,d$
benzene	$C_6H_6{}^{+\bullet}$	$1 + 6 - 3 = 4$	$1\,r + 3\,d$
benzonitrile	$C_7H_5N^{+\bullet}$	$1 + 7 - 2.5 + 0.5 = 6$	$1\,r + 5\,d$
butyl ion	$C_4H_9{}^{+}$	$1 + 4 - 4.5 = 0.5^a$	fragment^a
acetyl ion	CH_3CO^+	$1 + 2 - 1.5 = 1.5^a$	1 d for C=O^b
dimethylsulfoxide	$C_2H_6SO^{+\bullet}$	$1 + 2 - 3 = 0$	1 d expected^c
		$1 + 3 - 3 = 1^c$	1 d for tetravalent S

a Even-electron ions have 0.5 $r + d$ more than expected. Round to next lower integer.
b Values are correct for resonance structure with charge at carbon.
c Attention! Sulfur is tetravalent in DMSO and should be handled as carbon.

Care has to be taken when elements of changing valence are encountered, e.g., S in sulfoxides and sulfones or P in phosphates. In such a case, Eqs. 6.9 or 6.10 have to be used, whereas Eq. 6.11 yields erroneous results.

> **Note:** It is good practice to apply the $r + d$ algorithm as soon as the empirical formula of a molecular ion or a fragment ion seems to be apparent. This yields valuable information on possible substructures.

6.5 Allylic Bond Cleavage

6.5.1 Cleavage of the Allylic Bond in Aliphatic Alkenes

The mass spectra of alkenes are governed by the alkyl part of the molecule and by the specific reactions characteristic for the double bond. Molecular ions are generally of low intensity – as is the case with all other purely aliphatic compounds discussed so far – and may even be fully absent. [2,4,30] There is some preference for the cleavage of the allylic bond in the alkene molecular ions which can be treated analogously to α-cleavage and benzylic cleavage, but the double bond is the weakest cleavage-directing functional group:

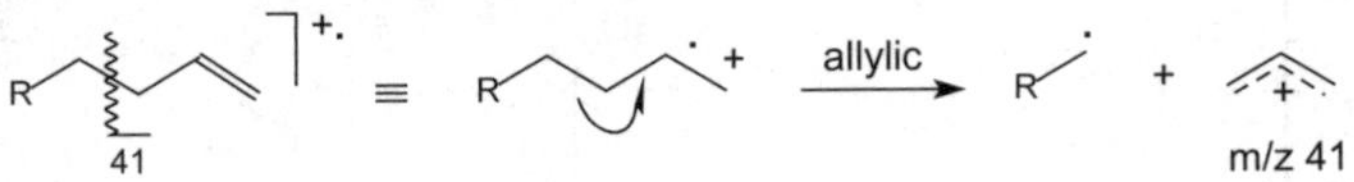

Scheme 6.30.

6.5.1.1 Isomerization Prior to Allylic Bond Cleavage

As hydrogen rearrangements prior to dissociation are prevalent in alkenes, the radical site migrates along the chain, thereby obscuring the location of the double bond. [60]

The radical or non-radical character of alkene molecular and fragment ions, respectively, determines the extent to which these ions equilibrate to a mixture of interconverting structures prior to decomposition. Odd-electron ions have lower thresholds for decomposition and higher barriers for isomerization than even-electron ions, thus explaining the reduced tendency for isomerization of $[C_nH_{2n}]^{+\bullet}$ molecular ions as compared to $[C_nH_{2n-1}]^+$ fragment ions. [61] Moreover, the molecular size seems to be a major factor for the isomerization of unsaturated hydrocarbon ions. Decreasing molecular size leads to more isomerization prior to decomposition. [62]

^{2}H- and ^{13}C-labeling revealed that the decomposing molecular ions of 1-, 2-, 3-, and 4-octene isomerize to a mixture of interconverting structures within 10^{-9} s after ionization. The equilibration of double bond isomers is mainly due to radical-site migration accompanied by hydrogen rearrangements, which do not involve the terminal hydrogens to a larger extent. [63]

> **Note:** Extensive isomerization prior to or between single steps of fragmentation makes the mass spectra of isomeric alkenes become very similar.

Example: The spectra of 1-decene, (*E*)-5-decene, and (*Z*)-5-decene are shown below (Fig. 6.17). Whereas in the spectrum of 1-decene the basepeak at *m/z* 41 can easily be explained as a result of allylic cleavage, the base peak of both

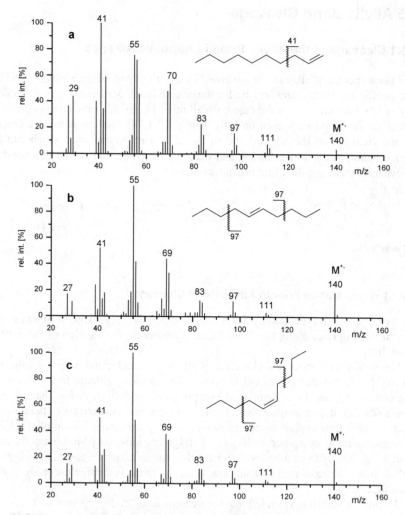

Fig. 6.17. EI mass spectra of isomeric decenes. The differences between 1-decene (**a**), 5(*E*)-decene (**b**), and 5(*Z*)-decene (**c**) are visible, but without reference spectra the isomers cannot be identified. Spectra used by permission of NIST. © NIST 2002.

stereoisomeric 5-decenes is at m/z 55 and therefore, cannot be the result of direct allylic cleavage. Nonetheless, without having reference spectra at hand, it is not possible to find clear evidence for a certain isomer. At m/z 97, there is only a minor signal from allylic cleavage, and unfortunately such a peak is also observed in the spectrum of 1-decene. Consecutive fragmentations of the primary cleavage product give one reason for this close similarity, another being the influence of H$^{\cdot}$ migrations prior to dissociation of the alkene molecular ions. The differences between the stereoisomers are negligible and even comparison to reference spectra does not permit their proper identification.

6.5.2 Methods for the Localization of the Double Bond

The localization of a double bond is an important step in structure elucidation and therefore, it is not astonishing that numerous approaches have been made to overcome the above limitations. The methods "to freeze" isomerization include i) epoxidation [64], ii) iron and copper ion chemical ionization [65,66], iii) field ionization [67], iv) collision-induced dissociation [60], v) formation of thioether derivatives, [68,69] and others.

Example: The oxidative addition of dimethyl disulfide (DMDS) transforms the double bond to its 1,2-bis-thiomethyl derivative (**a**). Induced by charge localization at either sulfur atom, the molecular ions of DMDS adducts are prone to α-cleavage at the former double bond position (**b**). This gives rise to sulfonium ions that are readily identified from the mass spectrum (Chap. 6.2.5). The method can be extended to dienes, trienes, and alkynes. [70,71] (For the mass spectral fragmentation of thioethers cf. Chap. 6.12.4).

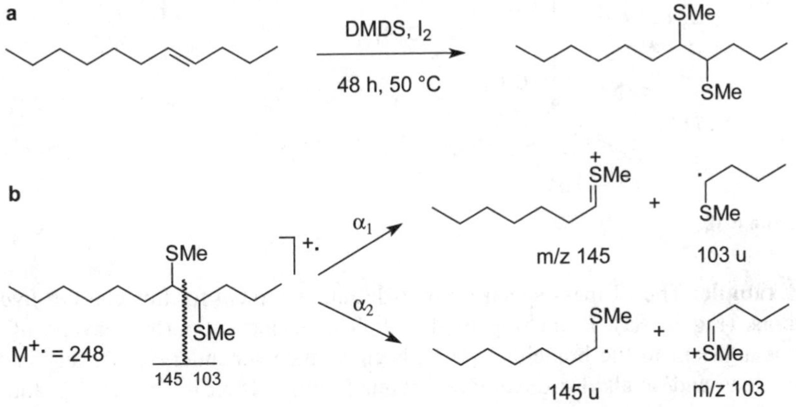

Scheme 6.31.

6.6. Cleavage of Non-Activated Bonds

6.6.1 Saturated Hydrocarbons

In a molecular ion of a straight-chain n-alkane there is no preferred position to localize the charge. Consequently, as in the case of methane, longer chain alkanes also suffer from substantial weakening of bonds upon electron ionization. The α-cleavage is not observed anymore. As expected, the primary step of dissociation is unspecific σ-bond cleavage at any but C_1–C_2 bonds, provided there are others. Methyl loss is negligible due to the unfavorable thermodynamics of this process. The resulting carbenium ions subsequently dissociate by loss of alkene molecules, preferably propene to pentene. [72] This process can occur repeatedly as long as the chain length of the fragment ion suffices. [73,74] As a result of these competing and consecutive ion dissociations, low-mass carbenium ions, typically propyl, m/z 43, or butyl, m/z 57 give rise to the base peak of aliphatic hydrocarbon spectra (Table 6.2). As with the alkenes before, it has been shown that saturated alkyl fragments $[C_nH_{2n+1}]^+$ ($n = 3$–7) isomerize to common structures before decomposition, whereas the molecular ions retain their structure. [75] The molecular ion peak is usually visible, but never as strong as in spectra of aromatic compounds. [34] For highly branched hydrocarbons it may even be absent:

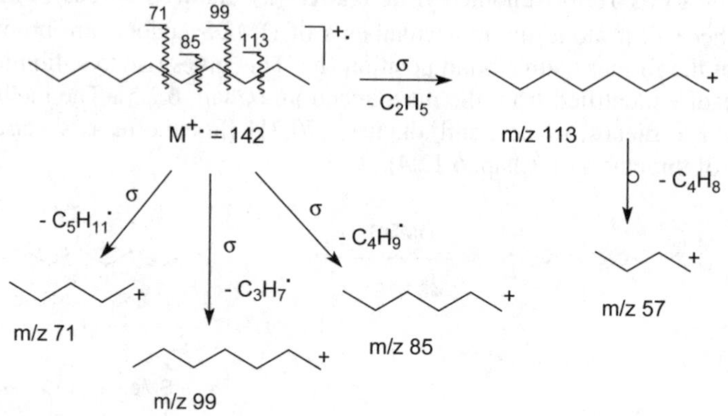

$$71 \quad 99$$
$$85 \quad 113$$
$$M^{+\cdot} = 142 \qquad \xrightarrow[\;-C_2H_5^{\cdot}\;]{\sigma} \qquad m/z\ 113$$

- $-C_5H_{11}^{\cdot}$ (σ) → m/z 71
- σ, $-C_3H_7^{\cdot}$ → m/z 99
- σ, $-C_4H_9^{\cdot}$ → m/z 85
- m/z 113 $\xrightarrow[\;-C_4H_8\;]{}$ m/z 57

Scheme 6.32.

Example: The EI mass spectrum of n-decane is typical for this class of hydrocarbons (Fig. 6.18a). Branching of the aliphatic chain supports cleavage of the bonds adjacent to the branching point, because then secondary or tertiary carbenium ions and/or alkyl radicals are obtained (Fig. 6.18b,c). This allows for the identification of isomers to a certain degree. Unfortunately, hydrocarbon molecular ions may undergo skeletal rearrangements prior to dissociations, thereby obscuring structural information.

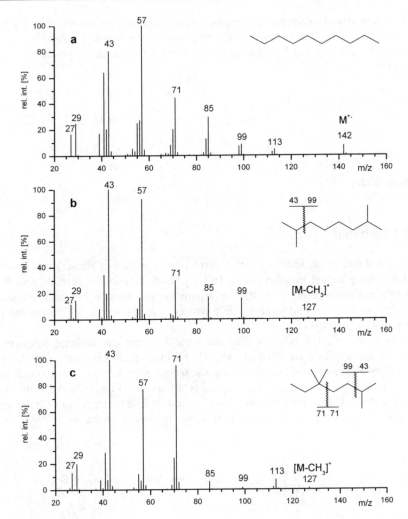

Fig. 6.18. EI mass spectra of *n*-decane (**a**), 2,7-dimethyloctane (**b**), and 2,5,5-trimethyl-heptane (**c**). Spectra used by permission of NIST. © NIST 2002.

A closer look at the spectrum of *n*-decane also reveals fragment ions at *m/z* 84, 98, and 112, i.e., rearrangement ions at even mass number. The origin by loss of H· from the accompanying carbenium ions at *m/z* 85, 99, and 113, respectively, can be excluded by application of the even-electron rule. Instead, alkane molecular ions may undergo alkane loss, [76] e.g.,

$$M^{+\cdot} = 142 \quad \xrightarrow[- C_3H_8]{rH} \quad m/z \ 112$$

Scheme 6.33.

Theoretical and experimental studies revealed a common intermediate for the elimination of $CH_3^{\cdot}$ and CH_4 from isobutane and other alkanes of similar size. The additional hydrogen needed may originate from either terminal position of the propyl moiety: [77,78]

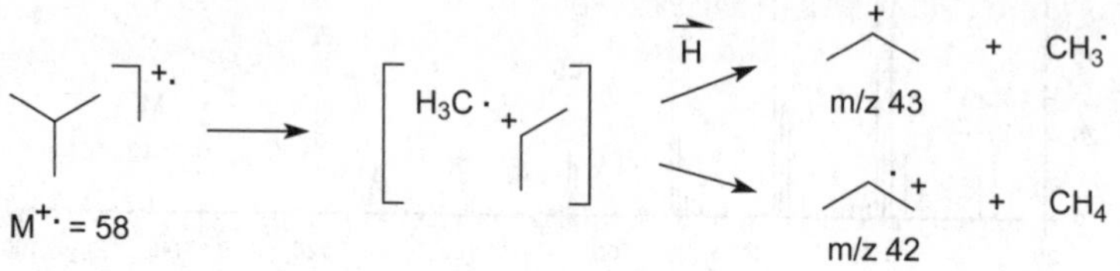

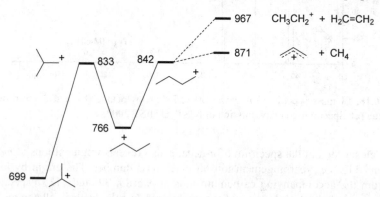

Scheme 6.34.

6.6.2 Carbenium Ions

As pointed out, carbenium ions make up the largest portion of alkane mass spectra and dissociate further by alkene loss. This type of fragmentation is rather specific, nevertheless isomerizations of the carbenium ion can happen before. Generally, such processes will yield a more stable isomer, e.g., the *tert*-butyl ion instead of other $[C_4H_9]^+$ isomers (Fig. 6.19).

Among the $[C_4H_9]^+$ isomers, only the *n*-butyl isomer can undergo fragmentations. It either yields an allyl ion, *m/z* 41, by loss of methane or an ethyl ion, *m/z* 29, by elimination of ethene. This is the reason why *tert*-butyl compounds exhibit strong *m/z* 41 and *m/z* 57 signals in their EI spectra (Figs. 6.20, 6.23). In case of metastable $[C_4H_9]^+$ ions, methane loss is about 20fold stronger than ethane loss because of the by 1 eV higher activation energy in case of ethene loss. [79]

Fig. 6.19. Energetics of the isomerization and fragmentation pathways of $[C_4H_9]^+$ ions. Heat of formations are in kJ mol^{-1}. Redrawn from Ref. [79] with permission. © American Chemical Society, 1977.

Example: In the EI mass spectrum of *tert*-butylchloride the molecular ion peak is absent and the highest *m/z* signal is due to [M-CH$_3$]$^+$; note the Cl isotopic pattern of this signal (Fig. 6.20). The remaining portion of the spectrum is determined by the *tert*-butyl group, causing the base peak at *m/z* 57. The *tert*-butyl ion directly originates from σ-cleavage of the C–Cl bond. The ratio of the [C$_3$H$_5$]$^+$, *m/z* 41, and [C$_2$H$_5$]$^+$, *m/z* 29, fragments correlates well with the energetically favorable CH$_4$ loss from [C$_4$H$_9$]$^+$ ions (Fig. 6.19).

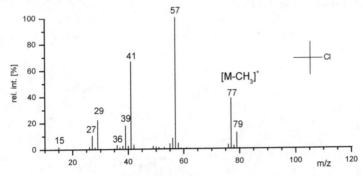

Fig. 6.20. EI mass spectrum of *tert*-butylchloride. Spectrum used by permission of NIST. © NIST 2002.

Other typical reactions of carbenium ions are alkene loss, provided sufficient chain length is available (Chap. 6.6.1), and dehydrogenation in case of the smaller ions such as ethyl, propyl, or butyl ion (Chap. 6.2.4.).

Carbenium ions, especially from ethyl to butyl show remarkable dehydrogenation that gives rise to a characteristic accompanying pattern of peaks at *m/z*–2 and *m/z*–4, i.e., at the low-mass side of the corresponding peak:

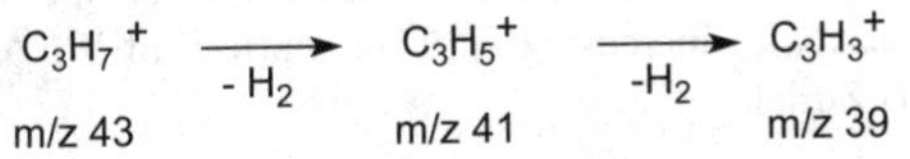

$$C_3H_7^+ \xrightarrow{-H_2} C_3H_5^+ \xrightarrow{-H_2} C_3H_3^+$$
$$\text{m/z 43} \qquad\qquad \text{m/z 41} \qquad\qquad \text{m/z 39}$$

Scheme 6.35.

Diarylmethyl and triarylmethyl ions (trityl ions) are even more stable than the *tert*-butyl ion which is impressively demonstrated by the commercial availability of solid [Ph$_3$C]$^+$[BF$_4$]$^-$ and similar salts. Triphenylchloromethane dissociates in polar, inert solvents such as SO$_2$, and therefore, it is not surprising that EI spectra of triphenylmethyl compounds almost exclusively exhibit this ion together with some of its fragments, whereas the molecular ion peak is usually absent. Field desorption allows to circumvent this problem (Chap. 8.5).

6.6.3 Very Large Hydrocarbons

The mass spectra of very large hydrocarbons obey the same rules as their lower-mass counterparts. However, their peculiarities are worth mentioning. In case of linear chains, the EI spectrum shows an evenly expanding series of carbenium ion fragments of ever decreasing intensity. Given a clean background, the molecular ion (< 1 % rel. int.) can well be recognized at even m/z followed by the first fragment ion, $[M-29]^+$ (Fig. 6.21).

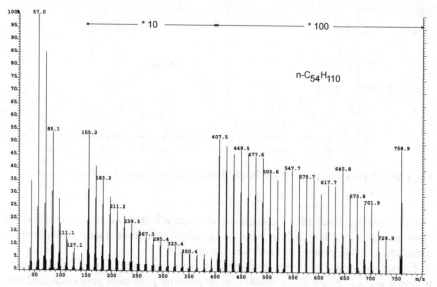

Fig. 6.21. DEI mass spectrum of tetrapentacontane, n-$C_{54}H_{110}$. The intensity scale is shown 10fold from m/z 150–400 and 100fold above m/z 400. Note how the first decimal of the m/z values continuously rises with increasing m/z. By courtesy of W. Amrein, Labor für Organische Chemie, ETH Zürich.

> **Note:** The first decimal of the m/z values of carbenium ions continuously rises with increasing m/z as a result of mass sufficiency of hydrogen (Chap. 3.3). Theoretically from $C_{32}H_{65}^+$ onwards (m/z 449.5081), rounding of the m/z value causes a shift to the next integer m/z, and thus result in confusion due to the nitrogen rule. Therefore, integer m/z values should not be used above m/z 400; instead use of the decimal is recommended for LR-MS.

Very large branched alkanes, such as 24,24-diethyl-19,29-dioctadecylheptatetracontane, $C_{87}H_{176}$, for example, pose difficulties to obtaining useful mass spectra and even 15 eV EI does not anymore allow for the detection of their molecular ions. [80] Beyond C_{40} alkanes, especially in case of mixtures such as hydrocarbon waxes or polyethylenes of low molecular weight, field desorption and matrix-assisted laser desorption/ionization are the ionization methods of choice (Chaps. 8, 10).

6.6.4 Recognition of the Molecular Ion Peak

The molecular ion peak directly provides valuable information on the analyte. Provided the peak being of sufficient intensity, in addition to mere molecular mass, the accurate mass can reveal the molecular formula of the analyte, and the isotopic pattern may be used to derive limits of elemental composition (Chaps. 3.2 and 3.3). Unfortunately, the peak of highest m/z in a mass spectrum must not necessarily represent the molecular ion of the analyte. This is often the case with EI spectra either as a result of rapidly fragmenting molecular ions or due to thermal decomposition of the sample (Chaps. 6.9 and 6.10.3)

6.6.4.1 Improving the Detection of the Molecular Ion

In general, the stability of the molecular ion increases if π-bonding electrons for the delocalization of the charge are available and it decreases in the presence of preferred sites for bond cleavage, e.g., by α-cleavage.

> **Rule of thumb:** The stability of molecular ions roughly decreases in the following order: aromatic compounds > conjugated alkenes > alkenes > alicyclic compounds > carbonyl compounds > linear alkanes > ethers > esters > amines > carboxylic acids > alcohols > branched alkanes. [81]

In EI mass spectrometry, the molecular ion peak can be increased to a certain degree by application of reduced electron energy and lower ion source temperature (Chap. 5.1.5). However, there are compounds that thermally decompose prior to evaporation or where a stable molecular ion does not exist. The use of soft ionization methods is often the best way to cope with these problems.

> **Note:** Even for the softest ionization method, there is no guarantee that the highest m/z ion must correspond to molecular mass.

6.6.4.2. Criteria for the Identification of the Molecular Ion

In order to derive reliable analytical information, it is therefore important to have some criteria at hand to identify the molecular ion.
1. The molecular ion must be the ion of highest m/z in the mass spectrum (besides the corresponding isotopic peaks).
2. It has to be an odd-electron ion, $M^{+\bullet}$.
3. The peaks at the next lowest m/z must be explicable in terms of reasonable losses, i.e., of common radicals or molecules. Signals at M–5 to M–14 and at M–21 to M–25 point towards a different origin of the presumed $M^{+\bullet}$.
4. Fragment ions may not show isotopic patterns due to elements that are not present in the presumed molecular ion.
5. No fragment ion may contain a larger number of atoms of any particular element than the molecular ion does.

Example: The EI mass spectrum of 2,5,5-trimethyl-heptane (Fig. 6.18) shows no molecular ion peak. The highest m/z peak is at m/z 113, demanding a nitrogen to be present; otherwise it must be interpreted as a fragment ion. The remaining mass spectrum does not reveal any hint for nitrogen-containing ions, e.g., immonium ions are not present. Instead, the spectrum suggest that there are carbenium ions and their accompanying fragments only. Thus, m/z 113 should be a fragment ion, most probably created by loss of C_xH_y. Here, it is explained as $[M–Et]^+$. However, $[M–Me]^+$ and other alkyl losses should also be taken into account.

Table 6.12. Commonly observed neutral losses from molecular ions

$[M–X]^+$	Radicals	$[M–XY]^{+\bullet}$	Molecules
–1	H$^\bullet$	–2	H_2
–15	$CH_3^\bullet$	–4	$2 \cdot H_2$
–17	OH$^\bullet$	–18	H_2O
–19	F$^\bullet$	–20	HF
–29	$C_2H_5^\bullet$	–27	HCN
–31	$OCH_3^\bullet$	–28	$CO, C_2H_4, (N_2)$
–33	SH$^\bullet$	–30	$H_2C=O, NO$
–35	Cl$^\bullet$	–32	CH_3OH
–43	$C_3H_7^\bullet$, $CH_3CO^\bullet$	–34	H_2S
–45	$OC_2H_5^\bullet$, $COOH^\bullet$	–36	HCl
–57	$C_4H_9^\bullet$	–42	$C_3H_6, H_2C=C=O$
–79	Br$^\bullet$	–44	CO_2
–91	$C_7H_7^\bullet$	–46	C_2H_5OH, NO_2
–127	I$^\bullet$	–60	CH_3COOH

6.7 McLafferty Rearrangement

Up to here, we have mainly discussed mass spectral fragmentations along the guideline of radical loss by simple cleavages, occasionally making some steps aside to learn about compound-specific reactions where appropriate. Next, we shall examine some pathways of alkene loss from molecular ions. Any loss of an intact molecule belongs to the rearrangement type of fragmentations. Among these, the *γ-H shift with β-cleavage*, commonly known as *McLafferty rearrangement*, surely is the most prominent.

6.7.1 McLafferty Rearrangement of Aldehydes and Ketones

The loss of alkenes from molecular ions of carbonyl compounds has early been noted. [23,82] Soon, a mechanism involving γ-H shift and β-cleavage has been proposed and studied in detail. [24-26,83,84] Strictly speaking, the term *McLafferty rearrangement* only describes an alkene loss from molecular ions of saturated aliphatic aldehydes, ketones, and carboxylic acids, the mechanism of which is analogous to the Norrish type-II photo-fragmentation in condensed-phase chemistry. However, a more generous use of the term to include all alkene losses essentially following this mechanism is useful for the recognition of analogies from a wide variety of molecular ions. Below, any fragmentation that can be described as a transfer of a γ-hydrogen to a double-bonded atom through a six-membered transition state with β-bond cleavage is regarded as McLafferty rearrangement (*McL*): [85,86]

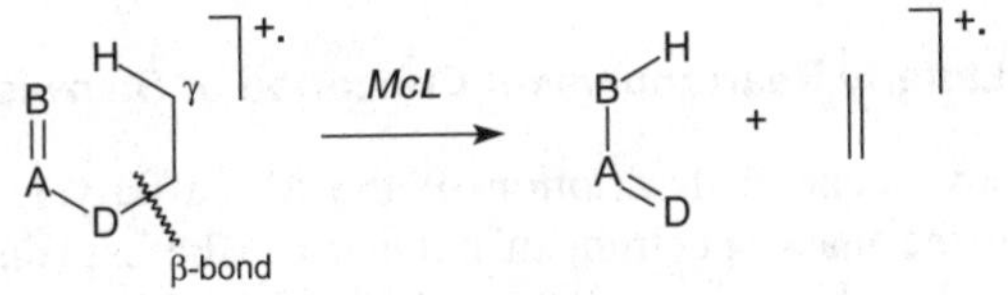

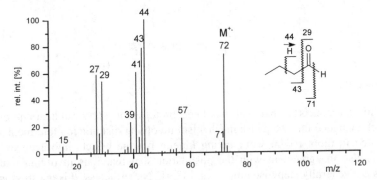

Scheme 6.36.

Example: The EI mass spectrum of butanal mainly shows carbenium fragment ions due to simple bond cleavage that can be easily recognized from their odd-numbered *m/z* values. However, the base peak is represented by a [M–28]$^{+\bullet}$ ion, *m/z* 44, obviously resulting from rearrangement. A closer look reveals that the charge-migration product, $C_2H_4^{+\bullet}$, *m/z* 28, is also present in the spectrum.

Fig. 6.22. 70 eV EI mass spectrum of butanal. The base peak at *m/z* 44 is caused by the product of McLafferty rearrangement. Note that here the peak at *m/z* 57 cannot result from $[C_4H_9]^+$, but is due to [M–CH₃]$^+$. Spectrum used by permission of NIST. © NIST 2002.

Requirements for the McLafferty rearrangement in a broad sense: i) the atoms A, B, and D can be carbons or heteroatoms, ii) A and B must be connected by a double bond, iii) at least one γ-hydrogen is available that iv) is selectively transferred to B via a six-membered transition state, v) causing alkene loss upon cleavage of the β-bond.

In addition to these general requirements, the distance between the γ-hydrogen and the double-bonded atom must be less than 1.8×10^{-10} m [87,88] and the C_γ–H bond must be in plane with the acceptor group. [89]

The McLafferty rearrangement itself proceeds via charge retention, i.e., as alkene loss from the molecular ion, but depending on the relative ionization energies of the respective enol and alkene products, the charge migration product, i.e., the corresponding alkene molecular ion is also observed. This is in accordance with Stevenson's rule (Chap. 6.2.2).

6.7.1.1 Is the McLafferty Rearrangement Concerted or Stepwise?

Following the above general description of the McLafferty rearrangement, the peak at m/z 44 in the mass spectrum of butanal can be explained by C_2H_4 loss from the molecular ion. The process may either be formulated in a concerted manner (**a**) or as a stepwise process (**b**):

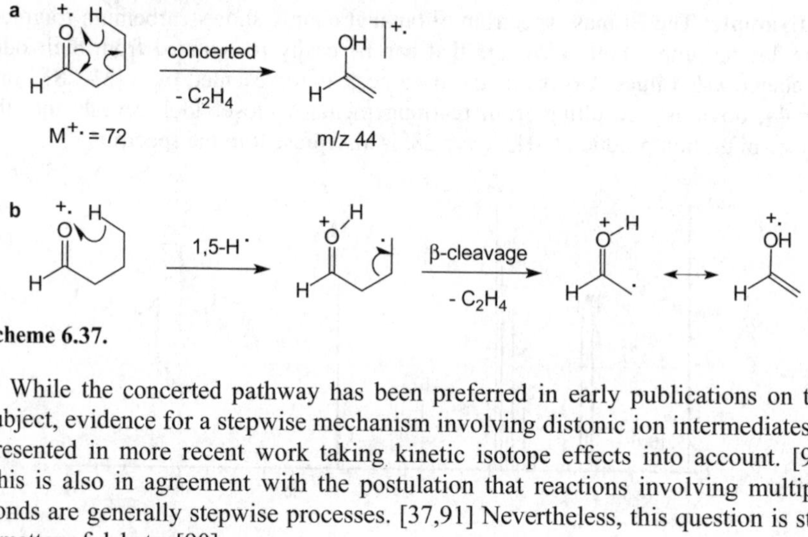

Scheme 6.37.

While the concerted pathway has been preferred in early publications on the subject, evidence for a stepwise mechanism involving distonic ion intermediates is presented in more recent work taking kinetic isotope effects into account. [90] This is also in agreement with the postulation that reactions involving multiple bonds are generally stepwise processes. [37,91] Nevertheless, this question is still a matter of debate. [90]

In principle, the enolic fragment ion may or may not tautomerize to the keto form before further fragmentation takes place:

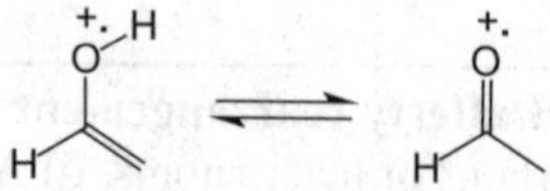

Scheme 6.38.

The gas-phase heats of formation of several enol positive ions of aliphatic aldehydes, ketones, acids, and esters were measured and compared with those of the corresponding keto ions. The enolic ions were found to be thermodynamically more stable by 58–129 kJ mol^{-1}. This is in marked contrast to the neutral tautomers, in which the keto forms are generally more stable. [92] The experimental findings also are in good agreement with MNDO calculations, [93] and support the hypothesis that reketonization does not play a major role for further fragmentation. Anyway, it is helpful to remind the possible tautomerization when seeking for subsequent decomposition pathways.

6.7.1.2 The Role of the γ-Hydrogen for the McLafferty Rearrangement

The occurrence of the McLafferty rearrangement is strictly limited to molecular ions possessing at least one γ-hydrogen for transfer to the terminal atom at the double bond. Thus, blocking the γ-position, e.g., by introduction of alkyl or halogen substituents, effectively hinders this dissociation pathway.

Example: In the EI mass spectrum of 3,3-dimethyl-2-butanone no fragment ion due to McLafferty rearrangement can be observed, because there is no γ-hydrogen available (Fig. 6.23). Instead, the products are exclusively formed by simple cleavages as evident from their odd-numbered m/z values. The highly stable *tert*-butyl ion, m/z 57, predominates over the acylium ion at m/z 43 (Chap. 6.6.2).

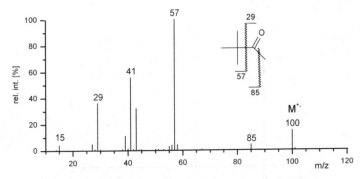

Fig. 6.23. EI spectrum of 3,3-dimethyl-2-butanone. The McLafferty rearrangement issuppressed, because there is no γ-hydrogen. Spectrum by permission of NIST. © NIST 2002.

6.7.2 Fragmentation of Carboxylic Acids and Their Derivatives

The mass spectra of carboxylic acids and their derivatives are governed by both α-cleavage and McLafferty rearrangement. As expected, α-cleavage may occur at either side of the carbonyl group causing OH• loss, $[M–17]^+$, or alternatively alkyl loss. Whereas the α-cleavages can even be observed for formic acid where they are leading to $[M–OH]^{+•}$, m/z 29, $[M–H]^{+•}$, m/z 45, and the respective charge migration products (Fig. 6.24), the McLafferty rearrangement can only occur from butanoic acid and its derivatives onwards. [82,84] Analogous to aliphatic aldehydes, the same fragment ions are obtained for a homologous series of carboxylic acids, provided they are not branched at the α-carbon. Thus, highly characteristic fragment ions make their recognition straightforward.

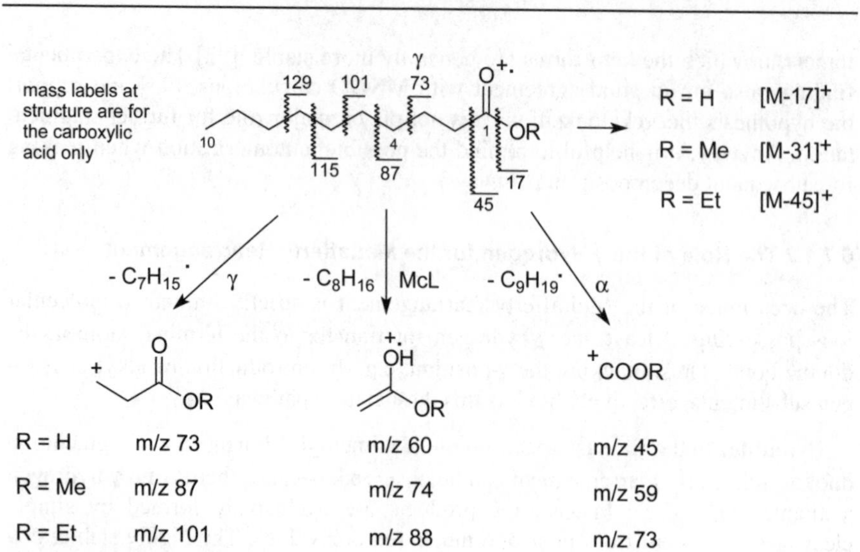

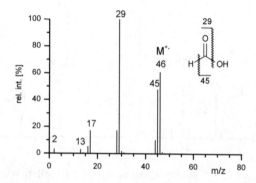

Scheme 6.39.

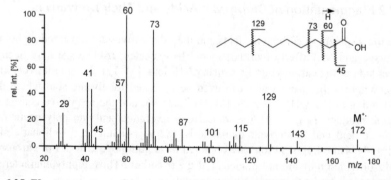

Fig. 6.24. EI mass spectrum of formic acid. Used by permission of NIST. © NIST 2002.

Fig. 6.25. EI mass spectrum of decanoic acid. Used by permission of NIST. © NIST 2002.

Example: The molecular ion of decanoic acid, $C_{10}H_{20}O_2^{+\cdot}$, preferably exhibits octene loss via McLafferty rearrangement yielding $C_2H_4O_2^{+\cdot}$, m/z 60, as the base peak of the spectrum (Fig. 6.25). This process is accompanied by several σ-bond cleavages among which γ-cleavage is clearly favored. Resulting from α-cleavage, OH$^\cdot$ loss can be expected, but is often represented by a very minor peak in the spectra of carboxylic acids. The other product, $[M–R]^+$, COOH$^+$, m/z 45, is almost always observed, however together with the complementary $[M–45]^+$ ion it normally also belongs to the ions of low abundance.

In aliphatic esters the fragments typical for carboxylic acids are shifted by 14 u to higher mass upon transition from the free acid to the methyl ester and by further 14 u for the ethyl ester. [85,86,94,95]

Examples: The mass spectrum of methyl heptanoate perfectly meets the standard, and thus all peaks may be explained by following Scheme 6.39 (Fig. 6.26a). In principle, the same is true for the isomeric methyl 2-methylhexanoate, but here care has to be taken not to interpret the mass spectrum as belonging to an ethyl ester. This is because the 2-methyl substituent resides on the ionic products of McLafferty rearrangement and γ-cleavage, shifting both of them 14 u upwards. Nevertheless, a look at the α-cleavage products, $[M–OMe]^+$, and $[COOMe]^+$, m/z 59, reveals the methyl ester.

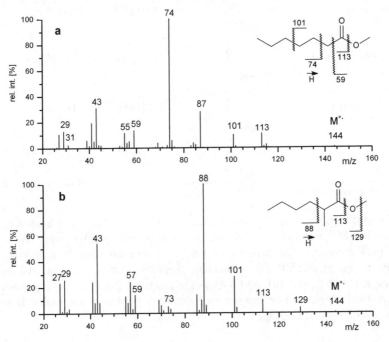

Fig. 6.26. EI mass spectra of methyl heptanoate (**a**) and methyl 2-methylhexanoate (**b**). Spectra used by permission of NIST. © NIST 2002.

For carboxylic acid ethyl and longer aliphatic chain esters the McLafferty rearrangement can also occur on the alkoxy branch (R^2) of the molecular ion. It then competes as a second alkene loss with the reaction at R^1:

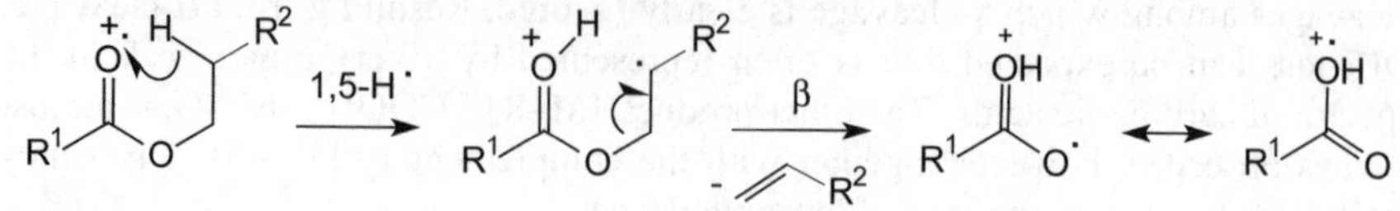

Scheme 6.40.

Table 6.13. Frequent product ions of the McLafferty rearrangement

Precursor	Product Structure	Formula	Accurate Mass [u][a]
aldehyde	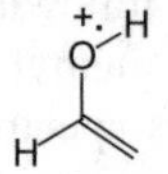	$C_2H_4O^{+\bullet}$	44.0257
alkyl methyl ketone	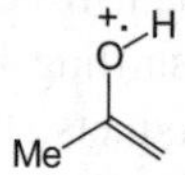	$C_3H_6O^{+\bullet}$	58.0413
carboxylic acid	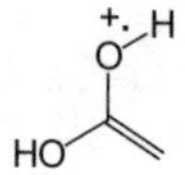	$C_2H_4O_2^{+\bullet}$	60.0206
carboxylic acid amide	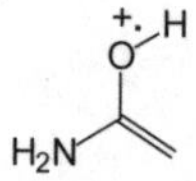	$C_2H_5NO^{+\bullet}$	59.0366
methyl carboxylates	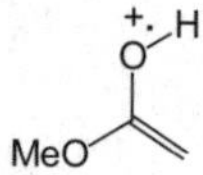	$C_3H_6O_2^{+\bullet}$	74.0362
ethyl carboxylates	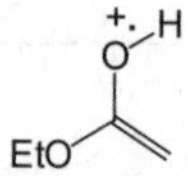	$C_4H_8O_2^{+\bullet}$	88.0519

a Values rounded to four decimal places.

Example: In the case of ethyl hexanoate, butene is eliminated from the molecular ion, $M^{+\bullet}$, m/z 144, via McLafferty rearrangement thereby giving rise to the base peak at m/z 88. This product ion may then undergo ethene loss to yield the fragment ion at m/z 60. The remaining fragments can be rationalized by γ-cleavage, $[M–Pr]^+$, m/z 101, and α-cleavage products, $[M–OEt]^+$, m/z 99, respectively. The carbenium ions representing the alkyl portion of the molecule result from several σ-bond cleavages.

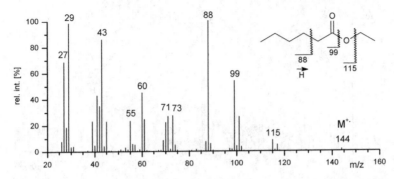

Fig. 6.27. EI spectrum of ethyl hexanoate. Used by permission of NIST. © NIST 2002.

6.7.3 McLafferty Rearrangement of Aromatic Hydrocarbons

In addition to the already described benzylic and phenylic cleavages (Chap. 6.4), phenylalkanes may undergo alkene loss by a mechanism that is perfectly analogous to the "true" McLafferty rearrangement, provided the alkyl substituent fulfills all requirements. The γ-hydrogen is transferred to the *ortho*-position where the aromatic ring serves as the accepting double bond:

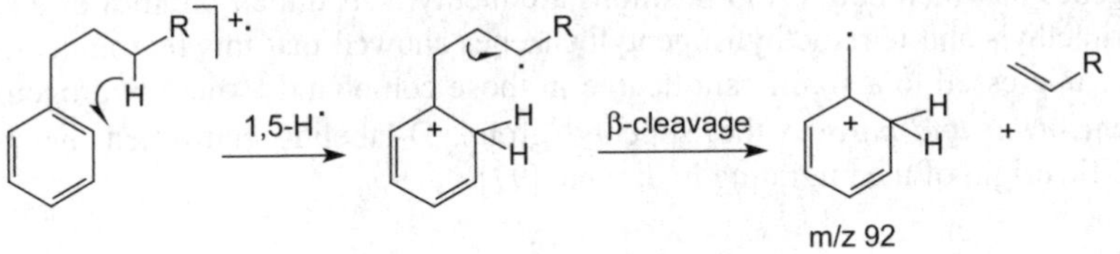

Scheme 6.41.

Independent of the alkyl substituent, $[C_7H_8]^{+\cdot}$, *m/z* 92, is obtained as the product ion, provided there are no other substituents at the ring. The product is an isomer of toluene molecular ion, and as such it readily stabilizes by H· loss to yield the even-electron $[C_7H_7]^+$ species, *m/z* 91, which then gives rise to the well-known characteristic fragments (*m/z* 65, 39). Provided that there is no prior isomerization of the molecular ion, this dissociation is prohibited if both *ortho*-positions are substituted and/or if there is no γ-hydrogen in the alkyl group.

Example: The base peak in the EI mass spectrum of (3-methylpentyl)-benzene is formed by McLafferty rearrangement of the molecular ion (Fig. 6.28). As long as pentene loss may occur, there is not much difference to spectra of other isomers such as 2-methylpentyl-, 4-methylpentyl, or *n*-hexyl. Reference spectra are needed to distinguish those isomers, because differences are mainly due to peak intensities, whereas only minor peaks might appear or vanish depending on the isomer.

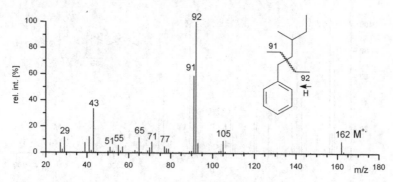

Fig. 6.28. EI mass spectrum of (3-methylpentyl)-benzene. McLafferty rearrangement and benzylic cleavage are clearly dominating. In the low-mass range carbenium ions and the "aromatic fragments" are present. Spectrum used by permission of NIST. © NIST 2002.

In case of alkyl benzylethers, aldehyde loss can occur following the same reaction pathway. For example, acetaldehyde is eliminated from the molecular ions of benzylethylether, thus producing $[C_7H_8]^{+\bullet}$ fragment ions. Again, evidence has been presented for a stepwise mechanism involving a distonic intermediate. [96]

From the above mechanism, non-hydrogen substitutents at the *ortho* positions of the aryl are expected to suppress the McLafferty rearrangement from alkylbenzenes. However, the site-specific γ-H rearrangement is even observed in alkylbenzenes in which both *ortho* positions are methyl substituted. Studies of a series of trimethyl- and tetramethylisopentylbenzenes showed that this rearrangement is only suppressed to a significant degree in those compounds where hydrogens are absent *ortho and para* to the isopentyl group. D-labeling confirmed the γ-site-specific origin of the migrating hydrogen. [97]

6.7.4 McLafferty Rearrangement with Double Hydrogen Transfer

Alkene loss via McLafferty rearrangement at the alkoxy group of aliphatic and aromatic carboxylic acid esters competes with yet another reaction path, where two hydrogens instead of one as in the "normal McLafferty product" are transferred to the charge site. This second pathway leading to alkenyl loss has early been noticed [94] and became known as *McLafferty rearrangement with double hydrogen transfer (r2H)*:

$$[R^1COOR^2]^{+\bullet} \xrightarrow{\ r2H\ } [R^1C(OH)_2]^+ \ + \ [R^2-2H]^{\bullet}$$

Scheme 6.42.

Labeling studies indicate that this process is by far not as site specific as its better known counterpart, however even this pathway lacks the high γ-specificity when occurring at the alkoxy group [95,98,99] Interestingly, both McLafferty rearrangement with single and with double hydrogen transfer remain unaffected even if the ester molecule serves as ligand in a transition-metal complex such as η^6-benzoic acid *n*-propyl ester tricarbonylchromium, for example. [99]

Based on thermochemical data, evidence has been presented for a two-step mechanism that finally yields a protonated carboxylic acid and the radical. However, while the final products are well described, the second step has not fully been elucidated: [100,101]

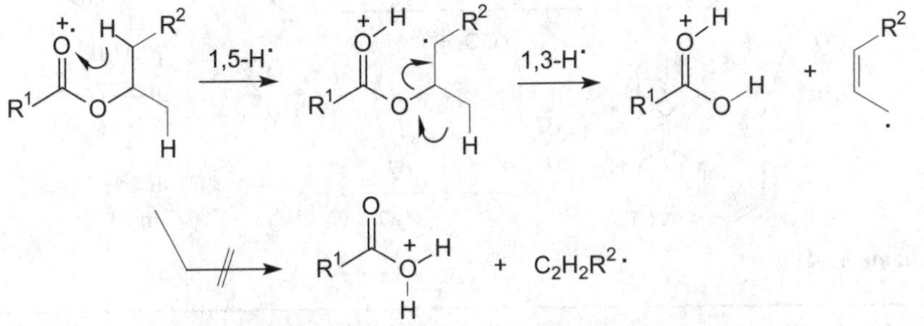

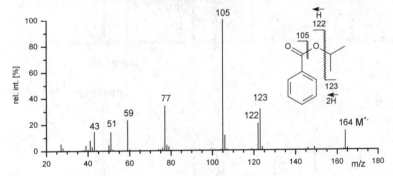

Scheme 6.43.

Studying the competition of McLafferty rearrangement either with charge retention or charge migration and double hydrogen transfer has revealed that ion-neutral complex intermediates (Chap. 6.12) can also play a role for the latter two processes. [102]

Example: The competition of alkene and alkenyl loss is demonstrated by the presence of peaks at *m/z* 122 (*McL*) and 123 (*r2H*) in the EI mass spectrum of iso-propyl benzoate (Fig. 6.29). [101] Of course, both primary fragmentations are competing with the predominating α-cleavage at the carbonyl group yielding the highly characteristic *benzoyl ion*, $[C_7H_5O]^+$, *m/z* 105. [103] The benzoyl ion, essentially an aromatic acylium ion, dissociates by CO loss to yield the phenyl ion, *m/z* 77, and finally by C_2H_2 loss the cyclobutadienyl ion, *m/z* 51 (Chap. 6.4.2).

Fig. 6.29. EI spectrum of isopropyl benzoate. Used by permission of NIST. © NIST 2002.

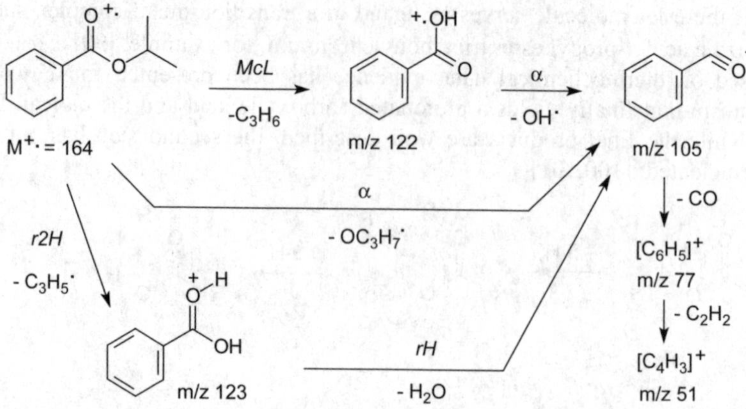

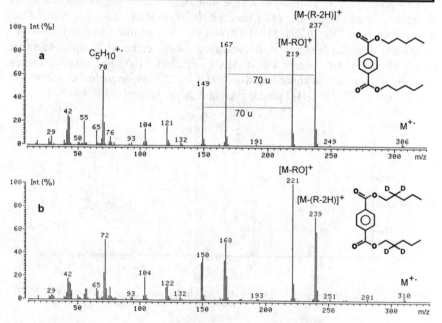

Scheme 6.44.

Note: The ion series m/z 51, 77, 105 is a reliable indicator for benzoyl substructures, e.g., from benzaldehyde, benzoic acid and its derivatives, acetophenone, benzophenone etc. Different from benzylic compounds, the peaks at m/z 39, 65, and 91 are almost absent. If a peak at m/z 105 and the complete series m/z 39, 51, 65, 77, 91 are present, this strongly points towards the composition $[C_8H_9]^+$, and thus to phenylalkanes. In case of doubt, HR-MS is the method of choice for their differentiation.

Fig. 6.30. EI mass spectra of dipentyl terephthalate (**a**) and a deuterated isotopomer (**b**). Adapted from Ref. [86] with permission. © John Wiley & Sons, 1985.

Example: The molecular ion of dipentyl terephthalate either fragments by α-cleavage to yield the $[M–RO]^+$ fragment ion, m/z 219, or by double hydrogen transfer causing the m/z 237 peak (Fig. 6.30). It is worth noting that there is no signal due to McLafferty rearrangement as might be expected at first sight. Even more interestingly, both of the above fragments are capable of pentene loss which is detected at m/z 149 and 167, respectively. This indicates that they may undergo McLafferty rearrangement although being even-electron ions. [86] The spectrum of the deuterium-labeled isotopomer reveals that there is no hydrogen scrambling preceding the α-cleavage, because the peak is completely shifted by 2 u to higher mass, m/z 221. The rearrangement, on the other hand, causes two peaks, one corresponding to $r2H$, m/z 239, and one corresponding to rHD, m/z 240.

6.7.4.1 Double Hydrogen Transfer of Phthalates

The fragment ions at m/z 149, $[C_8H_5O_3]^+$, and 167, $[C_8H_7O_4]^+$, are especially prominent in the EI spectra of phthalates. The formation of the $[C_8H_5O_3]^+$ ion has initially been attributed to a McLafferty rearrangement followed by loss of an alkoxy radical and final stabilization to a cyclic oxonium ion. [104] However, it has been revealed that four other pathways in total lead to its formation excluding the above one. [105,106] The two most prominent fragmentation pathways are:

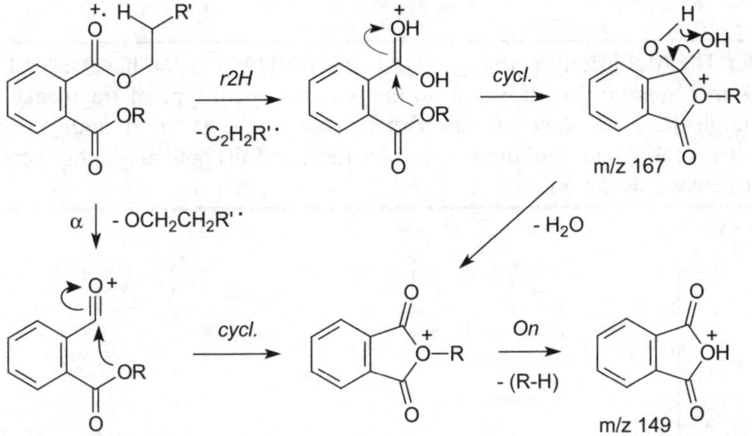

Scheme 6.45.

> **Note:** Phthalates are commonly used plasticizers in synthetic polymers. Especially di-2-ethylhexyl phthalate (also known as dioctyl phthalate, DOP) accounts for 45 % of all plasticizer usage in Western Europe. Unfortunately, they are extracted from the polymer by elongated exposure to solvents such as dichloromethane, chloroform or toluene, e.g., from syringes, tubing, vials etc. Therefore, they are often detected as impurities. They are easily recognized from their typical peaks at m/z 149 (often base peak), 167 and $[M–(R–2H)]^+$ (m/z 279 for DOP). The molecular ion is often absent in their EI spectra.

6.8 Retro-Diels-Alder Reaction

6.8.1 Properties of the Retro-Diels-Alder Reaction

Molecular ions containing a cyclohexene unit may fragment to form conjugated di-olefinic (*diene*) and mono-olefinic (*ene*) products. This fragmentation pathway was first recognized by Biemann as formally analogous to the retro-Diels-Alder (RDA) reaction of neutrals in the condensed phase. [81] The fragmentation can in principle proceed in a concerted manner (**a**) or stepwise, i.e., it can be regarded as a double α-cleavage with the first being an allylic cleavage (**b**):

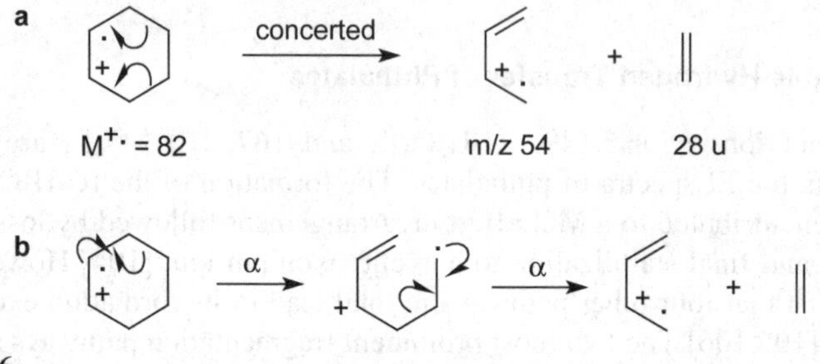

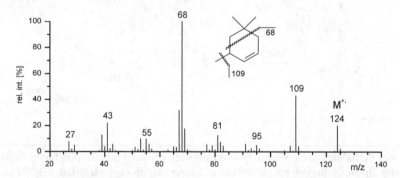

Scheme 6.46.

> **Note:** The McLafferty rearrangement and the RDA reaction have several features in common: i) both belong to the rearrangement type of fragmentations, although the name conceals this fact in case of the latter, ii) both represent pathways for alkene loss from molecular ions, and iii) both are highly versatile in structure elucidation.

Fig. 6.31. EI mass spectrum of 3,5,5-trimethylcyclohexene. The molecular ion undergoes isobutene loss, 56 u, via RDA reaction yielding the base peak at *m/z* 68, which is also the only significant peak at even-numbered *m/z*. The charge migration product at *m/z* 42 is of low intensity. Spectrum used by permission of NIST. © NIST 2002.

As most ion dissociations, the RDA reaction is endothermic, e.g., by 185 kJ mol^{-1} in case of ethene loss from cyclohexene molecular ion. [107] The charge usually resides at the diene fragment, but the ene fragment is also frequently observed in the mass spectra (Fig. 6.31). The fraction of charge retention and charge migration products has been studied. [108] As may be expected, competition for the charge is greatly influenced by the substitution pattern of the cyclohexene derivative and by the presence or absence of heteroatoms in the respective fragments.

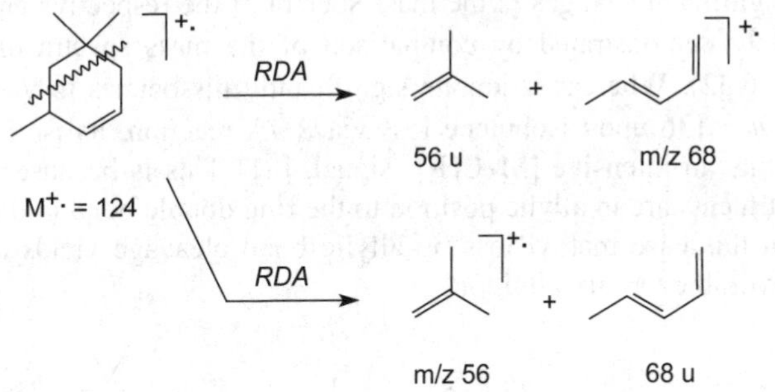

Scheme 6.47.

Provided the double bond does not migrate prior to RDA reaction, the alkene loss is perfectly regiospecific, and it does not really suffer from extensive substitution at the cyclohexene unit. Thus, substitution patterns can be revealed from the mass spectrum (**a**). The RDA reaction also proceeds independent of whether the six-membered ring contains heteroatoms or not (**b**):

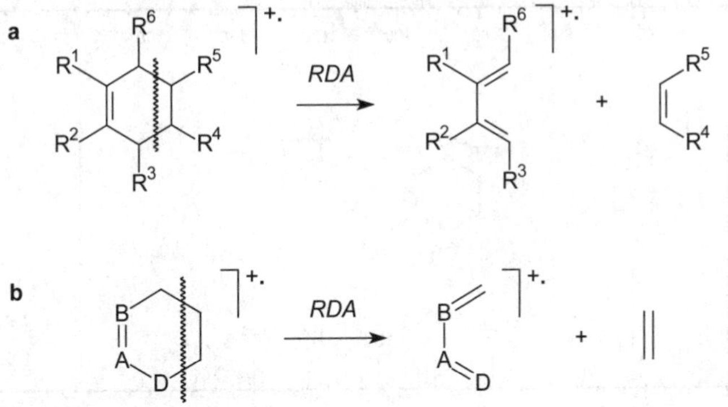

Scheme 6.48.

Obviously, the RDA reaction has the potential needed for a widespread mass spectral fragmentation, and consequently, its analytical and mechanistical aspects have repeatedly been reviewed. [107,109,110]

> **Note:** Almost any molecular ion having a six-membered ring that contains one double bond can undergo the RDA reaction to eliminate a (substituted) alkene or a corresponding heteroanalog.

6.8.2 Influence of Positional Isomerism on the RDA Reaction

As often observed in mass spectrometry, seemingly small changes in ion structure may cause significant changes in the mass spectra of the respective analytes. This is impressively demonstrated by comparison of the mass spectra of α- and β-ionone (Fig. 6.32). Whereas α-ionone significantly dissociates to yield the fragment ion at m/z 136 upon isobutene loss via RDA reaction, its isomer β-ionone mainly exhibits an intensive $[M–CH_3]^+$ signal. [81] This is because the geminal methyl substituents are in allylic position to the ring double bond in the latter, and moreover, in this case methyl loss by allylic bond cleavage yields a thermodynamically favorable tertiary allyl ion.

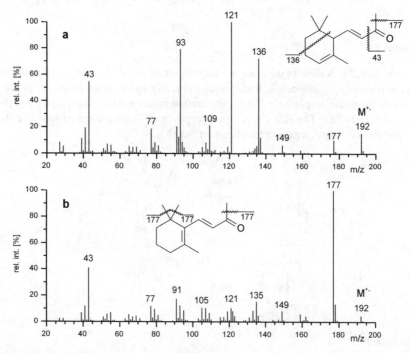

Fig. 6.32. EI mass spectra of α- (a) and β-ionone (b). RDA reaction proceeds via loss of isobutene, causing the m/z 136 peak in case of α-ionone, whereas ethene loss is almost absent in the β-ionone spectrum. Spectra used by permission of NIST. © NIST 2002.

6.8.3 Is the RDA Reaction Stepwise or Concerted?

It has been shown that certain unsaturated cyclic diketones readily decompose via RDA reaction if they are *cis* at the central ring junction, but not if *trans* (Scheme 6.49a) It was pointed out that this behavior would be expected if the mass spectral RDA reaction is concerted and follows symmetry rules analogous to those established for thermal reactions. [111]

On the other hand, the extent to which RDA reactions occur among various stereoisomeric bicyclic Δ^2-alkenes does not much depend on the stereochemistry of the ring juncture (Scheme 6.49b). These results were interpreted in terms of RDA reactions that do not follow orbital symmetry rules established for thermal reactions, and therefore rather proceed in a step-wise fashion. [112]

In summary, the actual structure of the molecular ion seems to exert strong effects on the RDA process in that it may be rather stepwise than concerted or vice versa. A detailed discussion of this issue taking thermochemical and theoretical chemistry data into account has been published by Tureček and Hanuš. [107]

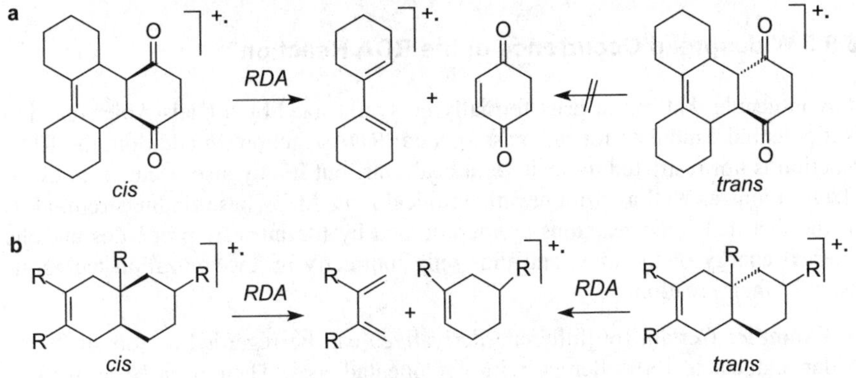

Scheme 6.49.

6.8.4 RDA Reaction in Natural Products

The RDA reaction is often observed from steroid molecular ions, and it can be very indicative of steroidal structure. [107,110,113,114] The extent of the RDA reaction depends on whether the central ring junction is *cis* or *trans*. The mass spectra of Δ^7-steroidal olefins, for example, showed a marked dependence upon the stereochemistry of the A/B ring juncture, in accordance with orbital symmetry rules for a thermal concerted process. In the *trans* isomer the RDA is much reduced as compared to the *cis* isomer. The effect was shown to increase at 12 eV, and as typical for a rearrangement, the RDA reaction became more pronounced, whereas simple cleavages almost vanished. This represented the first example of such apparent symmetry control in olefinic hydrocarbons. [114].

The RDA reaction is also a typical process of flavones, naphthoflavones, and methoxynaphthoflavones. [115] In many cases it provides intact A- and B-ring fragments, and therefore, it is of high relevance for structure elucidation. The intensity ratio of A- to B-ring fragments was found to be strongly influenced by the substituent position, i.e., to be very sensitive to the charge distribution within the molecular ion. [116]

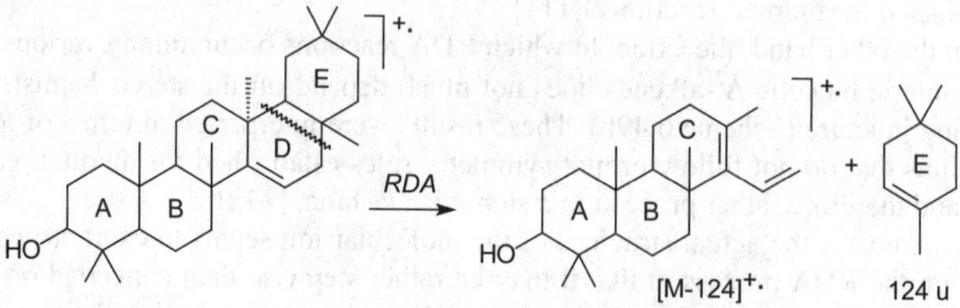

Scheme 6.50.

6.8.5 Widespread Occurrence of the RDA Reaction

Any molecule that can at least formally be synthesized by a Diels-Alder reaction is a potential candidate for the mass spectral RDA reaction. In addition, the RDA reaction is not restricted to positive radical ions, but it may also occur from even-electron ions as well as from negative radical ions, $M^{-\bullet}$. These findings remind us of the fact that ionic reactions are determined by the intrinsic properties and the internal energy of the ions, and thus only indirectly by the ionization technique used for their creation.

Example: Certain [60]fullerene derivatives can be regarded as formal Diels-Alder adducts of [60]fullerene with cyclopentadienes. Their positive as well as negative radical ions undergo RDA reaction at extremely high rates making the detection of a molecular ion very difficult. Instead, $C_{60}^{+\bullet}$ and $C_{60}^{-\bullet}$ are detected, respectively. [117,118] However, using negative-ion FAB or MALDI under care-

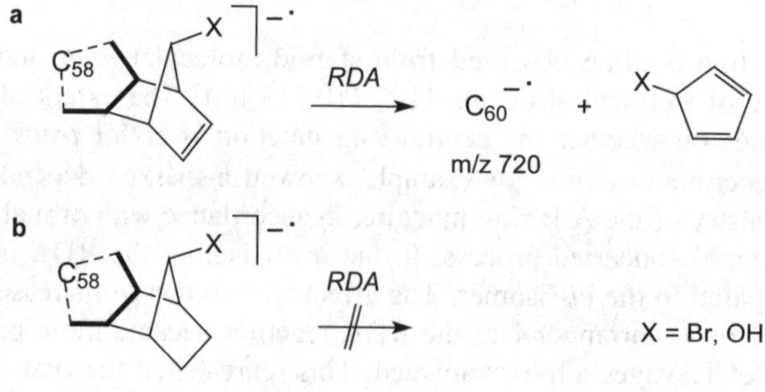

Scheme 6.51.

fully controlled conditions allows for the detection of M$^{-\bullet}$ and the RDA product (**a**). Removal of the double bond by hydration causes this fragmentation pathway to vanish (**b**) and allows for the detection of intense molecular anions. [117]

6.9 Elimination of Carbon Monoxide

As we know by now, the elimination of small stable molecules represents a frequent fragmentation route of odd- and even-electron ions. Second to alkene loss, we shall consider some of the numerous pathways for loss of carbon monoxide, CO. Here, the situation is quite different from that in the preceding sections, because there is no single mechanism for the elimination of CO. Instead, a wide variety of molecular and fragment ions can undergo loss of CO. Therefore, the following section is rather a compound-specific view in the general context of CO loss than a collection of compounds exhibiting a certain reaction mechanism in common. In retrospect, we have already encountered some cases of CO loss (Chap. 6.2.4.2 and 6.7.4).

6.9.1 CO Loss from Phenols

Phenols exhibit a strong molecular ion peak which often represents the base peak in their spectra. The most characteristic fragment ions of phenols are caused by loss of carbon monoxide, CO, from the molecular ion, [119] and subsequent H$^{\bullet}$ loss, thereby giving rise to [M–28]$^{+\bullet}$ and [M–29]$^{+}$ ions, respectively (Fig. 6.33). To these ions the compositions [M–CO]$^{+\bullet}$ and [M–CHO]$^{+}$ have been assigned by HR-MS. [120] This at first sight unexpected fragmentation proceeds via ketonization of the molecular ion prior to elimination of CO. The mechanism that can be rationalized in analogy to the behavior of neutral phenol, e.g., in the Reimer-Tiemann condensation, [121] has been verified by D-labeling. These experiments also revealed that only about one third of the H$^{\bullet}$ cleaved off from the cyclopentadiene ion, m/z 66, originates from the former OH, whereas the majority comes from the ring. [122]

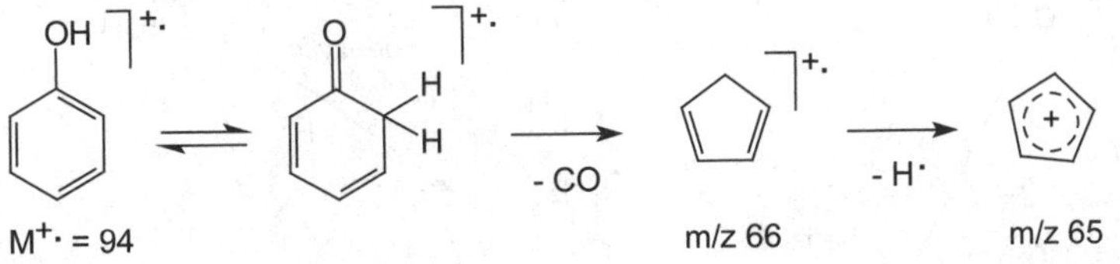

Scheme 6.52.

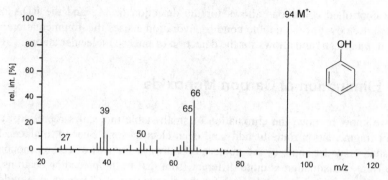

Fig. 6.33. EI mass spectrum of phenol showing an intensive peak due to CO loss at m/z 66. Spectrum used by permission of NIST. © NIST 2002.

Note: The above mechanism of CO loss from phenols is perfectly analogous to HCN loss from aniline and other aminoarenes (Chap. 6.14.2).

6.9.1.1 Substituted Phenols

Alkylphenol molecular ions preferably dissociate by benzylic bond cleavage, e.g., $[M–H]^+$ ions are observed in case of methylphenols. Provided a γ-hydrogen is available in alkyl substituents of sufficient chain length, the benzyclic bond cleavage competes with the McLafferty rearrangement, [123] the product ion of this being an isomer of methylphenol molecular ion. Loss of H• from the latter yields a $[C_7H_7O]^+$ ion, m/z 107, that may dissociate further by CO loss, thus forming a protonated benzene ion, $[C_6H_7]^+$, m/z 79. This tells us that CO loss occurs equally well from even-electron phenolic ions.

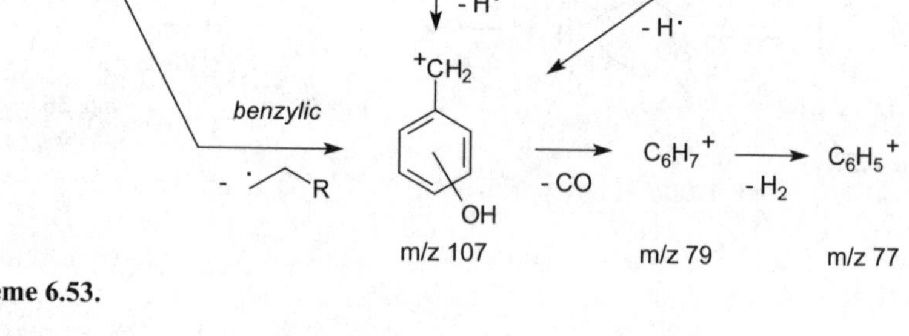

Scheme 6.53.

Example: Ethyl loss clearly predominates methyl loss in the EI mass spectrum of 2-(1-methylpropyl)-phenol. It proceeds via benzylic bond cleavage, the products of which are detected as the base peak at m/z 121 and m/z 135 (3 %), respectively (Fig. 6.34a). The McLafferty rearrangement does not play a role, as the peak at m/z 122 (8.8 %) is completely due to the ^{13}C isotopic contribution to the peak at m/z 121. From the HR-EI spectrum (Fig. 6.34b) the alternative pathway for the formation of a $[M–29]^+$ peak, i.e., $[M–CO–H]^+$, can be excluded, because the measured accurate mass of this singlet peak indicates $C_8H_9O^+$. HR-MS data also reveal that the peak at m/z 107 corresponds to $[M–CH_3–CO]^+$ and that the one at m/z 103 corresponds to $[M–C_2H_5–H_2O]^+$. Although perhaps unexpected, the loss of H_2O from phenolic fragment ions is not unusual.

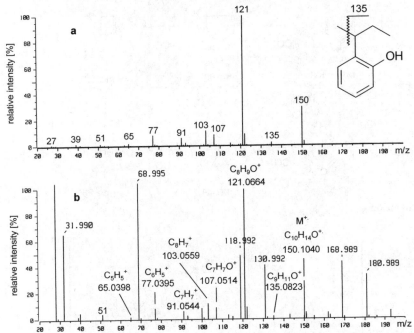

Fig. 6.34. LR- (**a**) and HR-EI (**b**) mass spectra of 2-(1-methylpropyl)-phenol. The elemental compositions as obtained from accurate mass measurement are directly attached to the corresponding peaks. Peaks with small-lettered labels belong to PFK and residual air used for internal mass calibration (Chap. 3.3).

6.9.2 CO and C_2H_2 Loss from Quinones

Quinones [124] and aromatic ketones such as flavones, [116] fluorenone, anthraquinone, and similar compounds [121] dissociate by competing and consecutive losses of CO and C_2H_2. Multiple CO losses may also occur subsequent to the RDA reaction of flavones. [116,125] As these molecules all have large π-electron systems to stabilize the charge, they all show intensive molecular ion peaks that often represent the base peak in those spectra. Typically any available carbonyl group is expelled as CO in the course of complete dissociation of the molecular ion; sometimes CO first, sometimes alternating with C_2H_2 loss. This gives rise to characteristic fragment ions resulting from sequences such as $[M-28-26-28]^{+\cdot}$ or $[M-28-26-26]^{+\cdot}$.

Example: 1,4-Benzoquinone represents the perfect prototype of this fragmentation pattern. The subsequent eliminations of intact molecules causes a series comprising of odd-electron fragment ions only (Fig. 6.35).

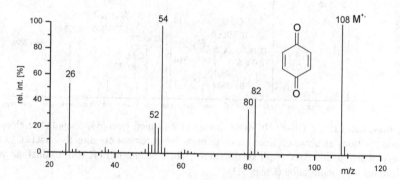

Scheme 6.54.

Fig. 6.35. EI mass spectrum of 1,4-benzoquinone. Spectrum used by permission of NIST. © NIST 2002.

6.9.3 Fragmentation of Arylalkylethers

6.9.3.1 CO and H2CO Loss from Arylmethylethers

The molecular ions of arylmethylethers preferably dissociate by loss of a form-aldehyde molecule or by loss of the alkyl group, [96,126] thereby yielding an even-electron phenolic ion, $C_6H_5O^+$, m/z 93, that readily expells CO (Fig. 6.36): [127,128]

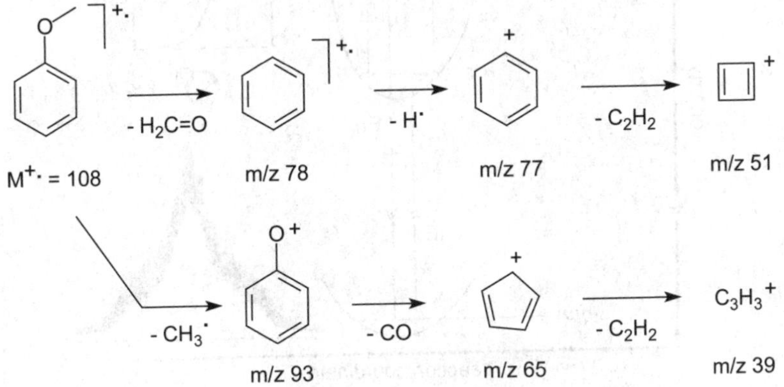

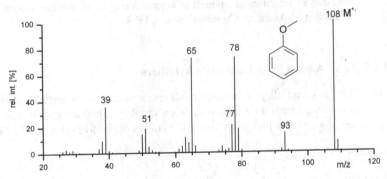

Scheme 6.55.

Fig. 6.36. EI mass spectrum of anisole exhibiting a strong $[M–H_2CO]^{+\bullet}$ peak at m/z 78. Spectrum used by permission of NIST. © NIST 2002.

Astonishingly, the study of the mechanism of formaldehyde loss from anisole revealed two different pathways for this process, one involving a four- and one a five-membered cyclic transition state (Fig. 6.37). [129] The four-membered transition state conserves aromaticity in the ionic product, which therefore has the lower heat of formation. Prompted by the observation of a composite metastable peak, this rather unusual behavior could be uncovered by deconvolution of two different values of kinetic energy release with the help of metastable peak shape analysis (Chap. 2.8).

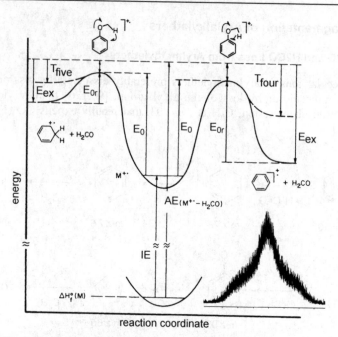

Fig. 6.37. Energetics of formaldehyde loss from anisole. The inset shows the composite metastable peak due to two different amounts of kinetic energy release. Adapted from Ref. [129] with permission. © American Chemical Society, 1973.

6.9.3.1 CO and Alkene Loss from Arylalkylethers

In case of longer chain alkyl substituents, alkene loss can occur analogous to the McLafferty rearrangement as is observed from suitably substituted phenylalkanes (Chap. 6.7.3), e.g., ethene loss from phenetole and its derivatives (Fig. 6.38). [126]

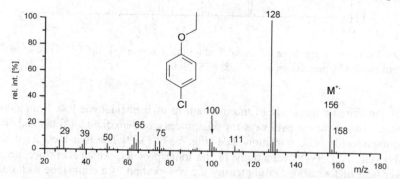

Fig. 6.38. EI mass spectrum of 4-chlorophenetole. The chlorine isotopic pattern is found in the signals corresponding to M$^{+\cdot}$, m/z 156, [M–C$_2$H$_4$]$^{+\cdot}$, m/z 128, [M–OEt]$^+$, m/z 111, and [M–C$_2$H$_4$–CO]$^{+\cdot}$, m/z 100. Spectrum used by permission of NIST. © NIST 2002.

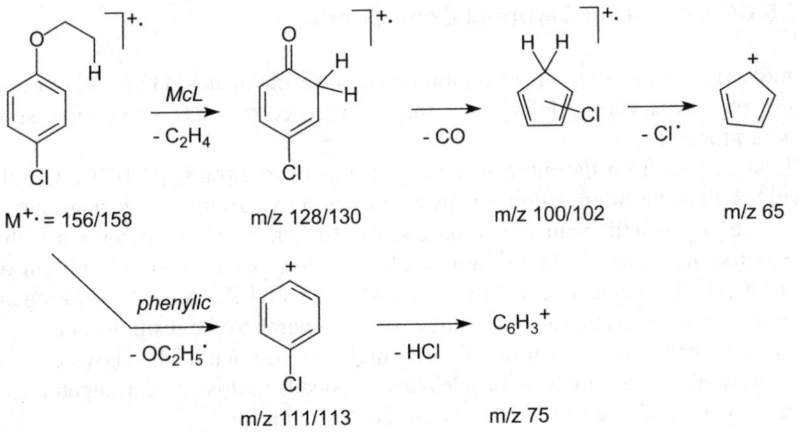

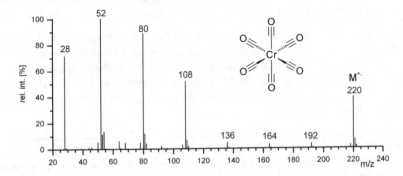

Scheme 6.56.

6.9.4 CO Loss from Transition Metal Carbonyl Complexes

Transition metal carbonyl complexes successively eliminate all CO ligands upon electron ionization until the bare metal remains. Pure carbonyl complexes as well as many other complexes with carbonyl ligands can therefore be readily identified from their characteristic CO losses which are – dependent on the metal – often observed in combination with the eye-catching isotopic pattern of the respective metal (Fig. 6.39). According to ^{13}C-labeling experiments, extensive isomerization may precede the dissociation of the carbonyl complexes. [130] Furthermore, the different bond strength of the M–CO bond in heterometallic transition-metal dimers can be deduced from the mass spectral fragmentation pattern of the carbonyl complexes, e.g., selective ^{12}CO loss from Mn in $[Mn(CO)_5Re(^{13}CO)_5]$ has been interpreted in terms of stronger Re–CO bonds. [131]

Fig. 6.39. EI mass spectrum of hexacarbonylchromium. All six CO ligands are eliminated until the bare metal ion, m/z 52, remains. The isotopic pattern of chromium can well be recognized from the more intensive signals. Used by permission by NIST. © NIST 2002.

6.9.5 CO Loss from Carbonyl Compounds

A molecule of CO may also be eliminated from malonates, [132] β-keto esters, [133] phenoxyacetates, [134] and many other compounds containing similar structural features.

Loss of CO from the molecular ions of anisoyl fluorides, $[CH_3OC_6H_4COF]^{+\bullet}$, involves fluorine atom migration from the carbonyl group to the benzene ring. [135,136] In o- and p-anisoyl fluorides, the fluorine atom migrates via a three-membered transition state to form molecular ions of o- and p-fluoroanisoles, $[CH_3OC_6H_4F]^{+\bullet}$, respectively. In the case of m-anisoyl fluoride, the fluorine atom migrates to the benzene ring via a three- or four-membered transition state.

Whatever the mechanism of CO loss might be, in none of the above cases CO loss proceeds by a simple bond cleavage. Instead, multistep rearrangements are necessary to "cut" the CO moiety out of the precursor ion.

Note: The detection of CO loss from molecular or fragment ions usually indicates the presence of carbonyl groups. However, it is less indicative of molecular structure than the highly specific reactions discussed before, because a multitude of rearrangement processes can be effective. These might even lead to CO loss in cases where no carbonyl group exists, e.g., from phenols.

6.9.6 Differentiation Between Loss of CO, N₂, and C₂H₄

Carbon monoxide, CO, 27.9949 u, is a nominal isobar of ethene, C_2H_4, 28.0313 u, and nitrogen, N_2, 28.0061 u, and the separation of these gases themselves by means of HR-MS has already been described (Chaps. 3.3.1 and 3.3.4). Accordingly, the difference in accurate mass for a given pair of peaks can be used to decide which of these neutrals has been eliminated, e.g., we employed HR-MS to uniquely identify the $[M-29]^+$ peak from 2-(1-methylpropyl)-phenol as $[M-C_2H_5]^+$ (Chap. 6.9.1). The elimination of N_2, CO, and/or C_2H_4 can even occur on consecutive and competing fragmentation pathways from the very same analyte. [137]

Example: The molecular ion of ethyl-*trans*-3-methyltetrazole-5-acrylate, $[C_7H_{10}O_2N_4]^{+\bullet}$, m/z 182.0798, exhibits two fragmentation sequences involving loss of N_2, C_2H_4, and/or CO, in different order. This way, isobaric fragment ions at m/z 98 and 126 are formed that could not be distinguished by LR-MS. HR-MS reveals both signals are doublets and allows the deconvolution of the two competing pathways (Table 6.14). [137]

Note: In case of doubt which neutral loss(es) are effective, it is highly recommended to obtain HR-MS data to avoid ambiguities. Moreover, the differences in accurate mass for a given pair of peaks can be used to identify the neutral loss even though the composition of the ions themselves might still be unknown.

Table 6.14. Example of the differentiation of isobaric neutral losses

Peak	Sequence a)	Sequence b)
	$M^{+\bullet} = 182.0798$	
	$\downarrow - N_2$	
singlet		154.0737
	$\downarrow - N_2$	$\downarrow - C_2H_4$
doublet	126.0675 (10 %)	126.0424 (90 %)
	$\downarrow - C_2H_4$	$\downarrow - CO$
doublet	98.0362 (61 %)	98.0475 (39 %)

6.10 Thermal Degradation Versus Ion Fragmentation

Processes such as decarbonylation, decarboxylation, elimination of water, and several other reactions may also occur prior to ionization, i.e., as non-mass spectral reactions, typically as a result of thermal degradation upon heating of the sample to enforce evaporation. In such a case, the mass spectrum obtained is not that of the analyte itself, but of its decomposition product(s). Sometimes, those thermal reactions are difficult to recognize, because the same neutral loss may also occur by a true mass spectral fragmentation of the corresponding molecular ion.

In even more disadvantageous circumstances, the thermal decomposition does not yield a single defined product, but a complex mixture that results in almost useless spectra, e.g., in case of highly polar natural products such as saccharides, nucleotides, and peptides or in case of ionic compounds such as organic salts or metal complexes.

6.10.1 Decarbonylation and Decarboxylation

Thermal degradation prior to ionization can cause decarbonylation or decarboxylation of the analyte. Decarbonylation, for example, is observed from α-ketocarboxylic acids and α-ketocarboxylic acid esters, whereas decarboxylation is typical behavior of β-oxocarboxylic acids such as malonic acid and its derivatives and di-, tri-, or polycarboxylic acids.

6.10.2 Retro-Diels-Alder Reaction

The Diels-Alder reaction is reversible at elevated temperature, and therefore its products can decompose prior to evaporation by RDA reaction of the neutral in the condensed phase. The mass spectral RDA reaction has already been discussed in detail (Chap. 6.8).

6.10.3 Loss of H_2O from Alkanols

6.10.3.1 Thermal H_2O Loss

Aliphatic alcohols show a strong tendency to thermally eliminate a water molecule. This is of special relevance if volatile alkanols are introduced via the reference inlet system or by means of a gas chromatograph. Then, the mass spectra correspond to the respective alkenes rather than to the alkanols that were intended to
be analyzed. The water is often not detected, simply because mass spectra are frequently acquired starting from m/z 40 to omit background from residual air.

6.10.3.2 Loss of H_2O from Alkanol Molecular Ions

Interestingly, the molecular ions of alkanols also exhibit a strong tendency to H_2O
loss, [32,138-140] which is their second important fragmentation route in addition
to α-cleavage (Chap. 6.2.5). It has been demonstrated that the large majority of
H_2O loss from the molecular ion proceeds by 1,4-elimination, whereas only some
percent of H_2O loss are due to other 1,x-eliminations. Following the labeling results, a cyclic intermediate can be assumed that decomposes further by loss of ethene. [139] These findings are not only supported by deuterium labeling studies but
also by the fact that H_2O loss does play a less important role in the mass spectra of
methanol, ethanol, and the propanols:

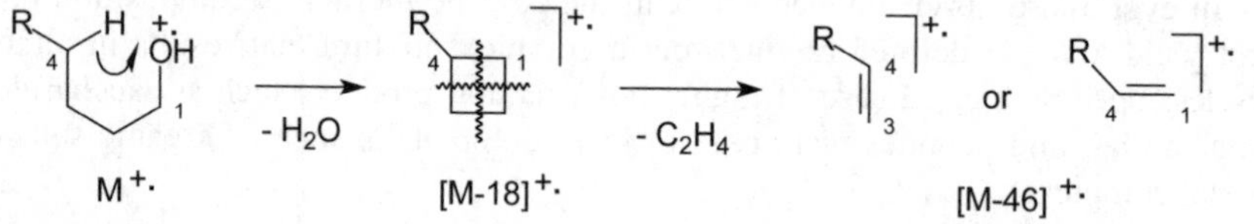

Scheme 6.57.

Example: The EI mass spectra of 1-hexanol, $M_r = 102$, and 1-hexene, $M_r = 84$,
show close similarity because the molecular ion peak is absent in the mass spectrum of hexanol (Fig. 6.40). However, a more careful examination of the hexanol
spectrum reveals peaks at m/z 18, 19, 31, and 45 that are absent in the hexene
spectrum. These are due to $H_2O^{+\bullet}$, H_3O^+, and to oxonium ions ($H_2C=OH^+$ and
$H_3CCH=OH^+$ in this case) which are reliable indicators of aliphatic alcohols and
ethers (Table 6.8).

In addition to the observation of oxonium ions, alkanols may occasionally be
identified from the occurrence of a seemingly [M–3] peak while the molecular ion
is absent (Fig. 6.41). The unusual difference of 3 u results from neighboring [M–
$CH_3]^+$ and [M–$H_2O]^{+\bullet}$ peaks. In these cases, a [M–33] peak indicates consecutive
losses of $CH_3^\bullet$ and H_2O in either order, i.e., [M-H_2O-$CH_3]^+$ and [M-CH_3-$H_2O]^+$.
(Chap. 6.6.4). Alternatively, the sequence of [M–$H_2O]^{+\bullet}$ and [M–H_2O–$C_2H_4]^{+\bullet}$
may occur.

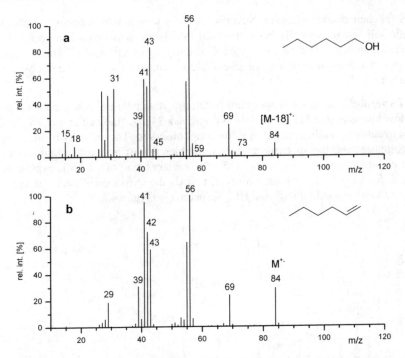

Fig. 6.40. Comparison of the EI mass spectra of 1-hexanol, $M_r = 102$ **(a)** and 1-hexene, $M_r = 84$ **(b)**. Spectra used with permission by NIST. © NIST 2002.

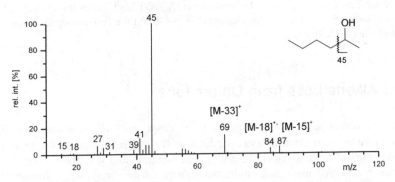

Fig. 6.41. EI mass spectrum of 2-hexanol, $M_r = 102$. Spectrum used with permission by NIST. © NIST 2002.

6.10.4 EI Mass Spectra of Organic Salts

Ammonium, phosphonium, oxonium salts and the like cannot be evaporated without substantial or complete decomposition, and thus it is not useful to employ EI-

MS for their characterization. Nevertheless, if a sample that happens to be an organic salt has incidentally been analyzed by EI-MS, it is helpful to know about their recognition. The EI mass spectra usually appear to result from the corresponding amines, phosphines, or ethers, but fortunately, the anion cannot hide perfectly from us.

Example: The EI mass spectrum of tetrabutylammonium iodide shows a peak of low intensity (0.6 %) for the $[Bu_4N]^+$ ion, m/z 242. A "molecular ion" of the salt that eventually might occur at m/z 369 is not observed. The majority of the sample decomposes and the spectrum closely resembles that of pure tributylamine showing its molecular ion peak at m/z 185 (the fragmentation of which is explained in Chap. 6.11.1). A closer look, however, reveals the existence of peaks at m/z 127 and 128 corresponding to I^+ and $HI^{+\bullet}$, respectively (Fig. 6.42).

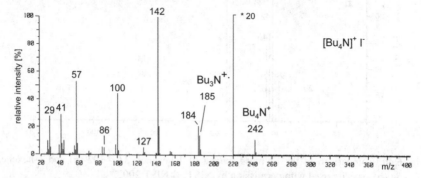

Fig. 6.42. EI mass spectrum of tetrabutylammonium iodide. The intensity scale is 20fold above m/z 220, i.e., disregarding the peaks at m/z 127 and 128 this spectrum is very similar to the mass spectrum of pure tributylamine (Fig. 6.43b; for the field desorption mass spectrum cf. Chap. 8.5.3).

6.11 Alkene Loss from Onium Ions

Aliphatic *onium ions* such as *immonium, oxonium,* and *sulfonium ions* have been introduced as even-electron ionic products of the α-cleavage occurring from molecular ions of amines, alcohols, and ethers or thiols and thioethers, respectively (Chap. 6.2.5). All these and analogous onium ions are capable of further fragmentation reactions, the majority of which are alkene losses [141] yielding fragments of high relevance for structure elucidation.

The first of these reactions requires a γ-hydrogen, and thus at least one C_3-alkyl substitutent. It may either be regarded as the *even-electron analogy of the McLafferty rearrangement* (*McL*, Chap. 6.7) as it will be treated here, [85,86,142] or alternatively, as a *retro-ene reaction.* [143,144]

The second alkene loss can occur from any onium ion bearing at least one C_2-alkyl moiety, which obviously is the least demanding prerequisite for an alkene loss (ethene) to proceed. By this process, the whole substituent is cleaved off the heteroatom with concomitant non-specific hydrogen transfer from the leaving group to the heteroatom. [30,33] In accordance with its occurrence from onium ions, this alkene loss is sometimes termed *onium reaction* (*On*). [4,6,145-148] Thus, we obtain the following general fragmentation scheme for alkene losses from suitably substituted onium ions:

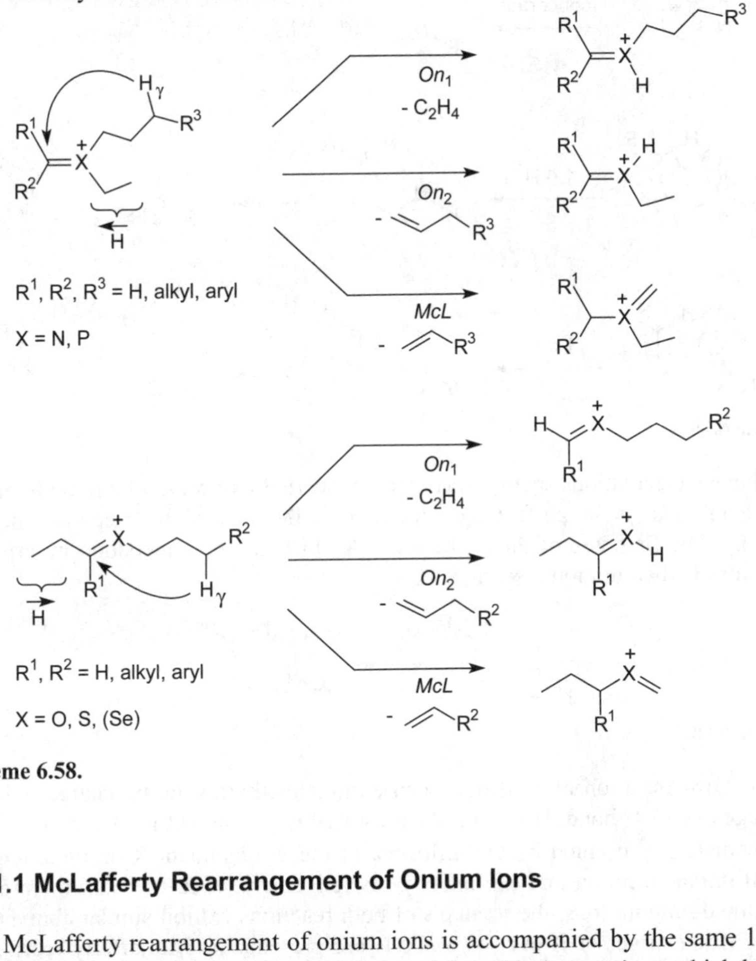

Scheme 6.58.

6.11.1 McLafferty Rearrangement of Onium Ions

The McLafferty rearrangement of onium ions is accompanied by the same 100 % regioselectivity for γ-H transfer as is observed for odd-electron ions which has repeatedly been demonstrated by deuterium labeling experiments. [142,145,147,149] This γ-selectivity also supports the general findings that immonium ions neither undergo random hydrogen nor carbon skeleton rearrangements prior to fragmentation. [27,142,150] The generally observed inertness of onium ions towards isomerization which is in contrast to $[C_nH_{2n+1}]^+$ (Chap. 6.6.1) and $[C_2H_{2n-1}]^+$ ions (Chap. 6.5.1) can be attributed to the preferred charge localization at the heteroatom. [151]

Although the origin of the moving hydrogen and the structures of the product ions of the McLafferty rearrangement are well known, it is still a matter of debate whether it is concerted (**a**) [144] or heterolytic stepwise (**b**) [142,145,149,152-154] or even might be stepwise involving homolytic bond cleavages (**c**): [146]

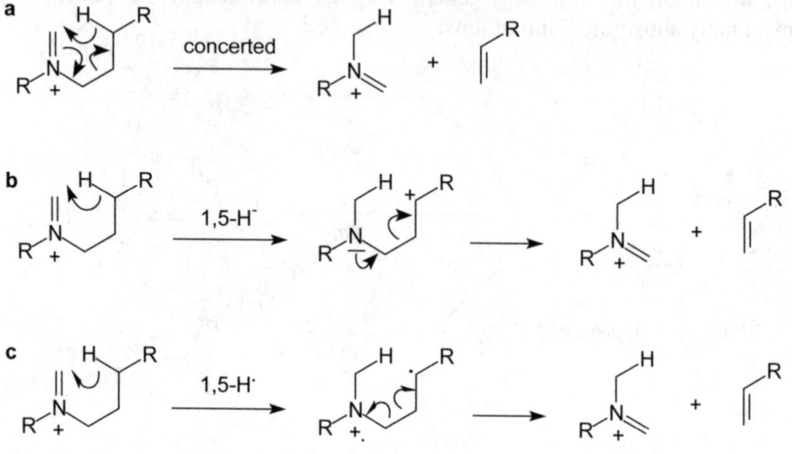

Scheme 6.59.

Whereas calculations seem to favor the concerted pathway, a large body of experimental data is in perfect agreement with the heterolytic stepwise mechanism (**b**). The first step of this, a 1,5-hydride shift, can be understood in terms of the onium-carbenium ion resonance:

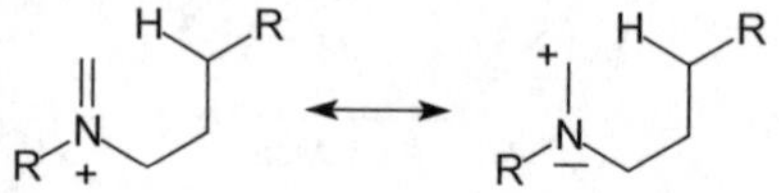

Scheme 6.60.

The carbenium ion intermediate then eliminates the alkene by charge-induced cleavage of a C–C bond. However, the most striking argument for a carbenium ion intermediate is presented by the influence of the γ-substituent R on the competition of onium reaction and McLafferty rearrangement. If R = H, i.e., for propyl-substituted iminum ions, the products of both reactions exhibit similar abundance. If R = Me or larger or if even two alkyls are present, the McLafferty rearrangement becomes extremely dominant, because then its intermediate is a secondary or tertiary carbenium ion, respectively, in contrast to a primary carbenium ion intermediate in case of R = H. The importance of relative carbenium ion stability for onium ion fragmentations (Chap. 6.11.2) will become more apparent when dealing with the mechanism of the onium reaction.

> **Note:** Regardless of whatever the exact mechanism and regardless of whatever the correct name, the McLafferty rearrangement of onium ions is one of those processes allowing to reliably track ionic structures through a mass spectrum.

Example: The comparison of the EI mass spectra of tripropylamine (Fig. 6.43a) and tributylamine [146] (Fig. 6.43b) clearly shows that the latter basically can be explained by two reactions: α-cleavage of the molecular ion at m/z 185 forming a dibutylmethyleneimmonium ion, m/z 142, undergoing double propene loss (42 u) via consecutive McLafferty rearrangements. The propyl substitutents of the primary iminum ion fragment from tripropylamine, m/z 114, on the other side allow for both fragmentations equally well, i.e., ethene loss via McLafferty rearrangement and propene loss via onium reaction, yielding peaks at m/z 86 and 72, respectively.

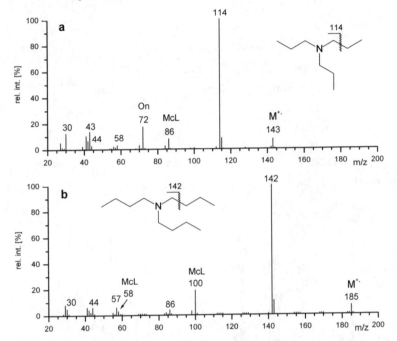

Fig. 6.43. EI mass spectra of tripropylamine (**a**) and tributylamine (**b**). For both compounds, the immonium ion series are completely present from m/z 30 onwards. Spectra used by permission of NIST. © NIST 2002.

Example: The McLafferty rearrangement of onium ions is not necessarily as obvious as in case of the immonium ions from aliphatic amines above. Especially, when other dissociation pathways are effectively competing, the corresponding signals can be of comparatively low intensity. The EI mass spectrum of butylisopropylether represents such a case: only the primary fragment ion at m/z 101 is able to undergo McLafferty rearrangement to form the oxonium ion at m/z 59 (Fig. 6.44 and Scheme 6.65). The peak at m/z 56 is due to $[C_4H_8]^{+\cdot}$ ions by loss of isopropanol from the molecular ion. [155,156] Eliminations of ROH from ionized ethers are usually of low abundance, but can gain importance in case of branching at the α-carbon.

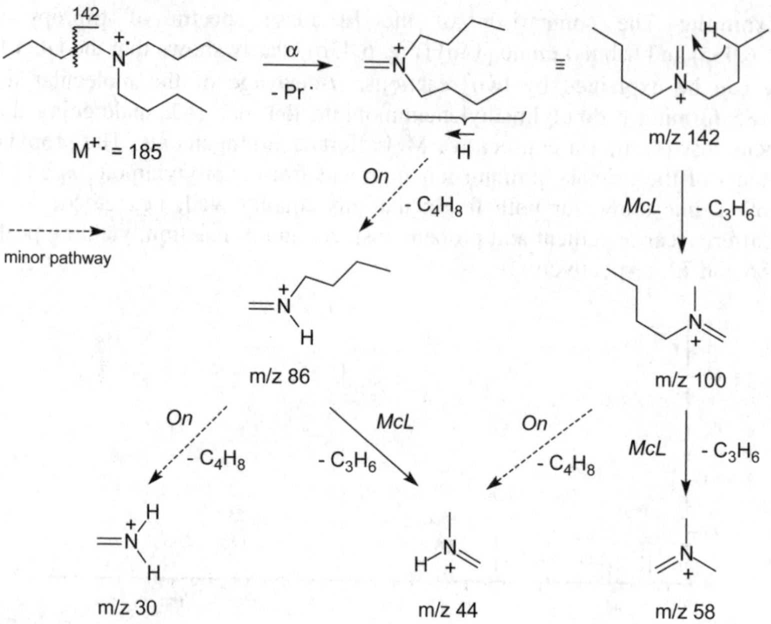

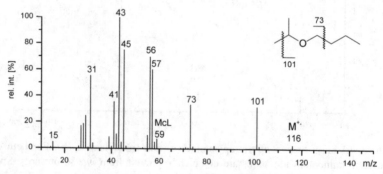

Scheme 6.61.

Fig. 6.44. EI mass spectrum of butylisopropylether. Spectrum used by permission of NIST. © NIST 2002.

6.11.2 Onium Reaction

The loss of alkenes from aliphatic onium ions via onium reaction comprises scission of the C–X bond and concomitant transfer of a hydrogen from the leaving alkyl moiety to the heteroatom, and a merely phenomenological description of this reaction has already been included in the preceding schemes.

Initially, the onium reaction was described to occur via β-hydrogen transfer from the leaving moiety to the nitrogen through a four-membered cyclic transition state. This assumption was mainly based on the reasoning that it could be observed for ethyl and larger substitutents. [33] Later, it was demonstrated by means of deuterium labeling that in case of pentyl-substituted oxonium ions this hydrogen does not exclusively originate from the β-position, but also comes from γ-, δ-, and to lesser extent from α-, and even ε-position. These results were interpreted in terms of competing reaction pathways through cyclic transition states of various ring sizes. [30]

6.11.2.1 Mechanism of the Onium Reaction

The simple concepts of cyclic transition states cannot appropriately describe the onium reaction. Instead, this appearingly simple reaction has to be broken up into a multi step process that involves *ion-neutral complex* (INC) intermediates. The properties of ion-neutral complexes are discussed in detail below (Chap. 6.12). Here, we restrict ourselves to the application of this concept to achieve a full and consistent explanation of the behavior of onium ions. The onium reaction of immonium ions [145,147,148,152,154,157-159] and oxonium ions [36,143,160-163] has been studied exhaustively. Imagine an isopropyl-propylmethyleneimmonium ion, m/z 114, where the reaction is initiated by heterolytic elongation of the C–N bond (Scheme 6.62). Having passed the transition state, the reaction would normally be expected to proceed by direct dissociation to yield a $C_3H_7^+$ ion, m/z 43 (imine loss). Alternatively, the incipient propyl ion might stay with the imine, thus forming an ion-neutral complex INC_1. The decision which of these channels will be followed depends on the ion internal energy, i.e., the less energetic ions will prefer INC formation, whereas the highly excited ones are prone to dissociation. An INC can be regarded as the gas phase analog to solvation of ions in the condensed phase. Here, the lone pair of the imine acts as a donor for the carbenium ion. As ion and neutral are mutually free and mainly held together by coulombic attraction, isomerization of the incipient 1-propyl ion to a 2-propyl ion via 1,2-hydride shift can take place. This goes along with a stabilization of about 60 kJ mol^{-1} (Table 2.3). The stabilization energy must of course be stored in internal degrees of freedom, thereby contributing to some further excitation of INC_1. Direct dissociation of INC_1 is also detected as imine loss. By considering the propyl ion as a protonated propene molecule it becomes reasonable that within the INC, a proton may be attracted by the imine having a by 100–140 kJ mol^{-1} higher proton affinity (PA, Chap. 2.11) than the leaving alkene. [142,164-166] Therefore, the proton is almost quantitatively transferred to the imine, and the heat of reaction of 140 kJ mol^{-1} causes additional excitation. Thus, INC_2, also called *proton-bridged complex* (PBC), finally decomposes to give an immonium ion, m/z 72, and propene, 42 u.

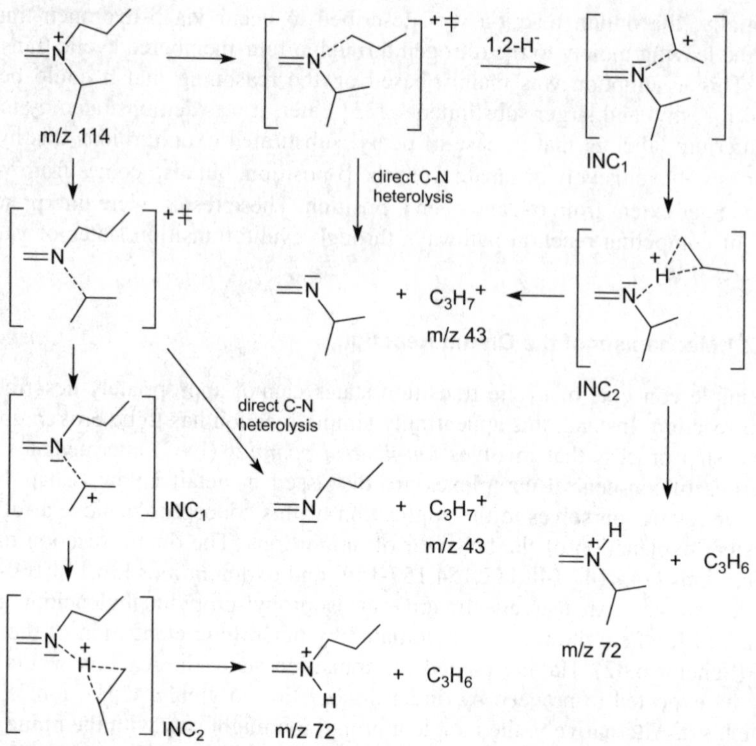

Scheme 6.62.

Example: The spectra of *N*-ethyl-*N*-methyl-propylamine (Fig. 6.6), tripropyl-amine, and tributylamine (Fig. 6.43) also exemplify alkene loss from immonium ions via onium reaction. Oxonium ions are involved in the fragmentation of diethylether and methylpropylether (Fig. 6.7) and many others have been published. [36,143,160-163] Below, the fragmentation of butyl-isopropylether is shown:

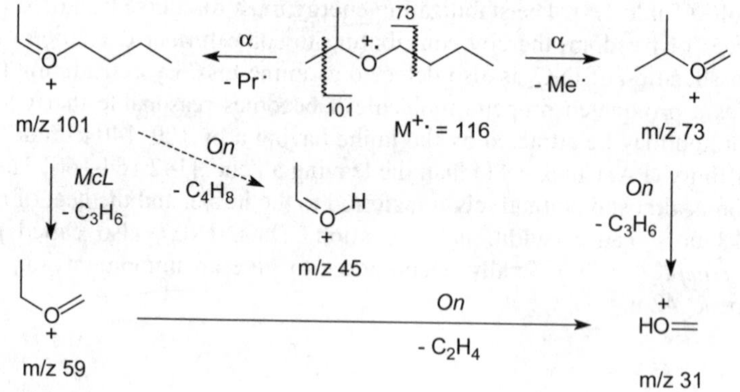

Scheme 6.63.

6.11.2.2 Competition of Onium Reaction and McLafferty Rearrangement

The onium reaction and the McLafferty rearrangement of onium ions are closely related to each other in so far that they often occur from the very same ion. However, the knowledge of the above mechanisms reveals significant differences in the way they proceed and it sheds light on the competition of onium reaction and McLafferty rearrangement: During the onium reaction alkyl substituents with no branch at C_α (adjacent to the heteroatom) have to be cleaved off via a transition state resembling a primary carbenium ion. Branching at C_α will effect dramatic change on the ease of bond heterolysis. As pointed out, analogous effects are observed for the McLafferty rearrangement in case of substitution at C_γ (Chap. 6.11.1 and Fig. 6.43). It becomes clear that the reaction pathway passing through the higher branched carbenium ion intermediate will be observed as the predominating process.

6.11.2.3 Aldehyde Loss from Oxonium Ions

In case of immonium ion fragmentations, the large difference of proton affinities, ΔPA, between imine and alkene clearly favors the formation of immonium ion plus neutral alkene, whereas imine loss is restricted to highly energetic precursors.

In decomposing oxonium ions the situation is quite different, i.e., the preference for alkene loss is much less emphasized and aldehyde (ketone) loss is gaining importance. The observed changes are in good agreement with the postulated mechanism of the onium reaction. [143,167] The alternative pairs of oxonium ion plus alkene and aldehyde plus carbenium ion may be formed with a preference for the first one, because ΔPA is comparatively small (20–60 kJ mol^{-1}) or even zero, e.g., for the acetone/isobutene pair: [143,166,167]

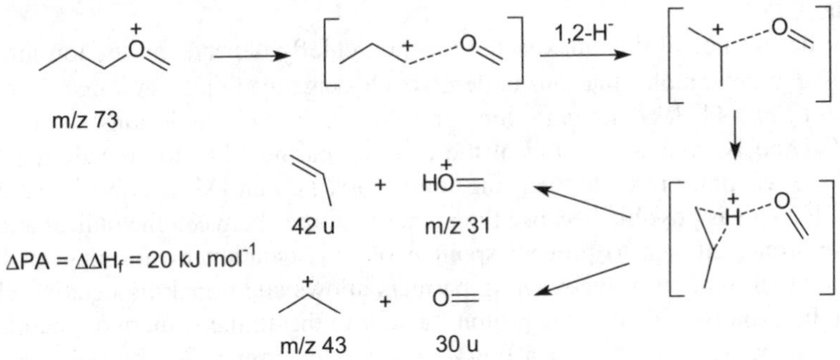

Scheme 6.64.

6.12 Ion-Neutral Complexes

During the above discussion of the mechanism of the onium reaction we encountered another type of reactive intermediates of unimolecular ion fragmentations: *ion-neutral complexes* (INC). [29,41,141,168-171] Different from *distonic ions* (Chap. 6.3) ion-neutral complex intermediates add some bimolecular reaction characteristics to fragmentation pathways of isolated ions. This is effected by allowing an incipient neutral fragment to remain some time with the ionic part, both derived from the same precursor ion, thus enabling processes that otherwise could only occur from bimolecular ion-molecule reactions. The final products of such fragmentations are governed by the properties of the species involved, e.g., by the relative proton affinities of the partners as was the case with the onium reaction.

6.12.1 Evidence for the Existence of Ion-Neutral Complexes

Perhaps the first suggestion of INCs came from Rylander and Meyerson. [172,173] The concept that the decomposition of oxonium and immonium ions involve INCs (Chap. 6.11.2) was successfully put forth by Bowen and Williams, [143,157,158,167,174] and the analogies to solvolysis were described by Morton. [168] Nonetheless, mass spectrometrist were too much used to strictly unimolecular reactions to assimilate such a concept without stringent proof.

Conclusive evidence for INCs was first presented by Longevialle and Botter [175,176] who demonstrated the transfer of a proton from an imine fragment to an amino group located on the opposite side of the rigid skeleton of bifunctional steroidal amines. [176] Such a proton transfer cannot proceed without the intermediacy of an INC; otherwise, a conventional hydrogen shift would have bridged the distance of about 10^{-9} m between the respective groups which is about five times too far.

The occurrence of this unexpected reaction chiefly depends on the ion internal energy: energetic molecular ions undergo α-cleavage leading to an immonium ion, $[C_2H_6N]^+$, *m/z* 44, less energetic ions and especially metastable ions show imine loss. Obviously, mutual rotation of the reacting partners has to precede the final acid-base reaction preferably causing imine loss, i.e., an $[M-C_2H_5N]^{+\bullet}$ fragment, $[M-43]^{+\bullet}$. This is possible because the attractive forces between the radical and the ion are strong enough to prevent spontaneous dissociation by forming an INC. Now, mutual rotation of the reacting partners allows additional dissociation channels to be explored. Finally, the proton transfer to the amine is thermodynamically favorable, because the proton affinity of ethylideneimine, 891 kJ mol^{-1}, is presumably lower than that of the steroidal amine radical (as estimated from the proton affinity of cyclohexylamine, [164] this is about 920 kJ mol^{-1}, i.e., $\Delta PA \approx 29$ kJ mol^{-1}).

Proton transfer will be prevented when the initial fragments separate too rapidly for the partners to rotate into a suitable configuration. Here, the lower limit for the intermediate's lifetime can be estimated to be 10^{-11} s. The competition between the α-cleavage and the INC-mediated reaction is therefore governed by the amount of ion internal energy.

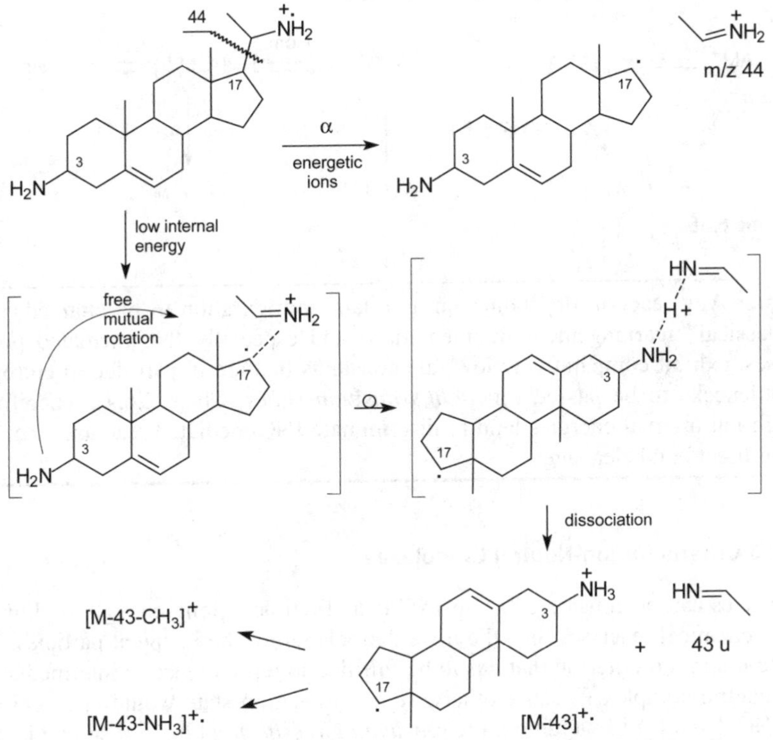

Scheme 6.65.

6.12.2 Attractive Forces in Ion-Neutral Complexes

The term ion-neutral complex is applied to species in which an ion and a neutral are held in association mainly by electrostatic attraction. [171] Instead, by action of covalent bonds, the INC must be held together by the attraction between a positive or negative charge and an electron-donating or -accepting group, respectively, e.g., by ion-dipole and ion-induced dipole interactions. The formation of an INC can be compared to solvolysis reactions in the condensed phase [168] with subsequent mutual solvation of the partners. Typically, stabilization energy V_r is in the range of 20–50 kJ mol^{-1}.

Consider the dissociation of an ion AM$^+$ that may either dissociate to form the fragments A$^+$ and M or the INC [A$^+$, M] allowing free mutual rotation and thus reorientation of the particles. Within the INC, A$^+$ and M can recombine only if they attain a well-defined mutual orientation, i.e., the system has to freeze rotational degrees of freedom. Such a configuration allowing covalent bonds to be formed is termed *locked-rotor critical configuration* [169-171,177] and any reaction channel, e.g., proton transfer or recombination, has to pass this *entropic bottleneck*. The INC [A⁺, M] can in principle undergo multiple isomerizations and may dissociate at any of these isomeric states:

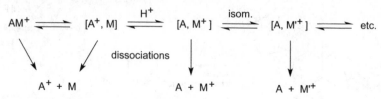

Scheme 6.66.

> **Note:** Any reaction depending on a certain configuration to be attained, i.e., "classical" rearrangement fragmentations and especially INC-mediated processes, exhibit comparatively low rate constants that are in part due to entropic bottlenecks to be passed (cf. *tight transition states*, Chap. 2.6.4). Therefore, high ion internal energies tend to discriminate INC-mediated reactions, but favor direct bond cleavages.

6.12.3 Criteria for Ion-Neutral Complexes

A species can be considered as an INC if its lifetime is long enough to allow for other chemical reactions than the mere dissociation of the incipient particles. This is the minimum criterion that has to be fulfilled to term a reactive intermediate an ion-neutral complex, because otherwise any transition state would also represent an INC. [171,178] In addition, the *reorientation criterion* [29,169] should be met, i.e., free reorientation of the particles involved must be possible. Although still regarded provisional by the authors, McAdoo and Hudson have provided a useful collection of additional characteristics of ion-neutral complex-mediated decompositions: [179,180]

1. Complete dissociation by the bond cleavage that forms the INC will be observed and will increase in importance relative to the INC-mediated pathway as internal energy increases.
2. INC-mediated processes will be among the lowest energy reactions of an ion.
3. Reactions within and between partners may occur below the dissociation thresholds for simple dissociations.
4. Alternative mechanisms consistent with the observations would require much higher energy transition states – transition states in which multiple bond breakings and makings occur simultaneously, or even transition states of impossible geometries.
5. The observed kinetic isotope effects are usually large (because of low excess energy in the transition states).
6. The kinetic energy releases in corresponding metastable decompositions may be very small.

6.12.4 Ion-Neutral Complexes of Radical Ions

The intermediacy of ion-neutral complexes is neither restricted to even-electron fragmentations nor to complexes that consist of a neutral molecule and an ion. In addition, radical-ion complexes and radical ion-neutral complexes occur that may dissociate to yield the respective fragments or can even *reversibly interconvert* by hydride, proton or hydrogen radical shifts. Many examples are known from aliphatic alcohols, [180-183] alkylphenylethers, [184-187] and thioethers. [188]

Example: The unimolecular reactions of the molecular ion of ethyl-propyl-thioether (Fig. 6.45, Scheme 6.67) [188] chiefly are i) loss of a methyl radical, $CH_3^{\bullet}$, from the ethyl or the propyl group, ii) loss of an ethyl radical, $C_2H_5^{\bullet}$, from the propyl group, both leading to *sulfonium ions* [189] (Chap. 6.2.5), and iii) elimination of propene or an allyl radical by transfer of one or two hydrogen atoms, respectively, from the propyl group to the sulfur. On the microsecond time scale, the loss of $CH_3^{\bullet}$ involves only the propyl entity and is preceded by an isomerization of this group. Partial loss of the positional identity of the hydrogen atoms of the propyl group occurs, but incorporation of hydrogen or carbon atoms from the ethyl group into the formed neutral species does not occur. Cleavage of a C–S bond assisted by a 1,2-hydride shift in the incipient carbenium ion leads to an INC of a thioethoxy radical and a 2-propyl ion. The INC may recombine to form the molecular ion of ethylisopropylthioether (cf. Fig. 6.9) prior to $CH_3^{\bullet}$ loss, or react by proton transfer to give another INC, which may dissociate or undergo hydrogen atom transfer, followed by elimination of an allyl radical. The partial loss of positional identity of the hydrogen atoms during the decomposition of the metastable ions is mainly a result of reversible proton transfer between the constituents, which competes favorably with 1,2-hydride shifts within the carbenium ion entity of the complex.

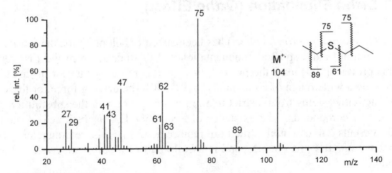

Fig. 6.45. EI mass spectrum of ethylpropylthioether. (Also compare to the spectrum of ethylisopropylthioether, Fig. 6.9. The isotopic pattern is discussed in Chap. 3.2.6) Spectrum used by permission of NIST. © NIST 2002.

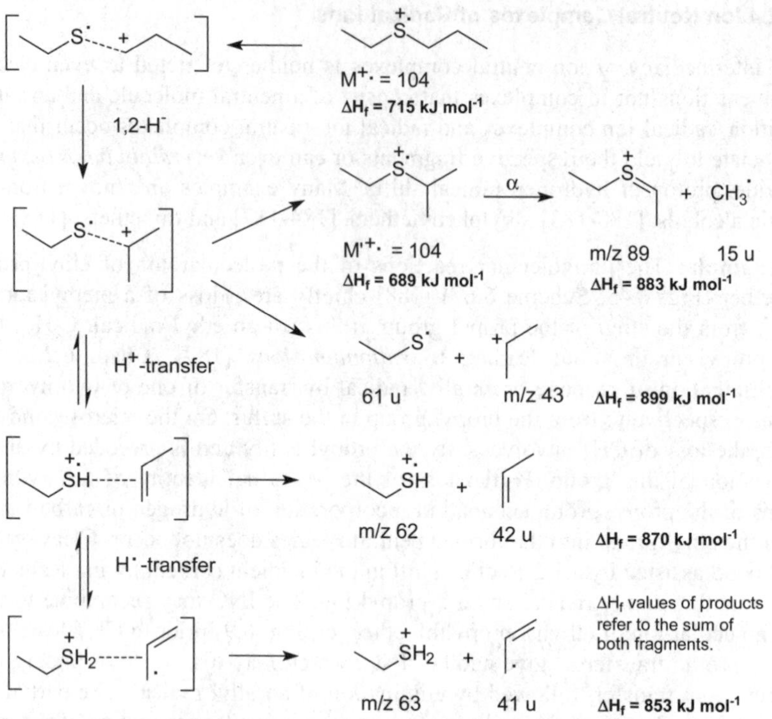

Scheme 6.67.

6.13 *Ortho* Elimination (*Ortho* Effect)

The *ortho elimination* (*ortho effect*) has been stated to belong to the diagnostically most important mass spectral fragmentations. [190] Indeed, from the first report on this process [104] until the present, there is a continuing interest in this topic concerning a wide range of compounds. [191-200] The driving force for these research activities stems from a strict relation of this process to the substitution pattern of aryl compounds. The existence of a suitably 1,2-disubstituted *cis*-double bond presents an essential structural requirement for this rearrangement-type fragmentation, and this double bond is usually part of an aromatic system, hence the name.

> **Note:** Although not being correct to describe a reaction, the term *ortho effect* is well established, and another occasionally used term, *retro-1,4-addition*, is redundant; *1,4-elimination* would describe the same. Unfortunately, this might be confused with 1,4-H_2O-eliminations from aliphatic alcohols, for example. Therefore, the term *ortho elimination* is suggested and used here.

6.13.1 *Ortho* Elimination from Molecular Ions

Commonly, *ortho* elimination refers to a hydrogen transfer via a six-membered transition state at *ortho*-disubstituted aromatic compounds. In practice, the reacting entities are almost in position to form this six-membered transition state. The general mechanism of the *ortho* elimination is as follows:

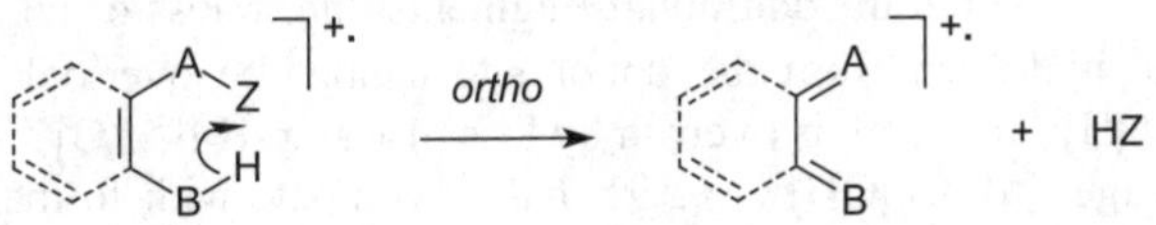

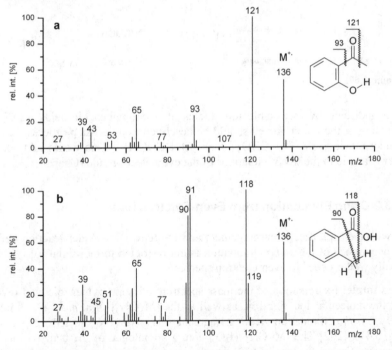

Scheme 6.68.

The charge can reside on either product, but charge retention at the diene product is generally predominant. As the 1,2-disubstituted *cis*-double bond must not necessarily belong to an aryl group, the optional part of the molecule has been drawn in dashed style in the above scheme. In addition to the substitution pattern, the *ortho* elimination requires substituent A to bear a hydrogen-accepting leaving group, Z, and substituent B to be a hydrogen donor, e.g., hydroxyl, amino, thiol, or even alkyl. [201]

Fig. 6.46. EI mass spectra of 2-hydroxy-acetophenone (**a**) and 2-methyl-benzoic acid (**b**). Spectra used by permission of NIST. © NIST 2002.

Example: The leaving methyl of 2-hydroxy-acetophenone (Fig. 6.46a) is not a suitable hydrogen acceptor, and therefore homolytic cleavages predominate. Here, these are loss of a methyl by α-cleavage, $[M–CH_3]^+$, m/z 121, and loss of an acyl by phenylic cleavage, $[M–COCH_3]^+$, m/z 93. On the other hand, all requirements for the *ortho* elimination are met in the 2-methyl-benzoic acid molecular ion (Fig. 6.46b). [104] As usual, homolytic cleavages are still competing with the *ortho* elimination, [193] but the additional fragments due to loss of intact molecules containing one hydrogen from the donor site cannot be overlooked. Thus, α-cleavage, $[M–OH]^+$, m/z 119, is overrun by loss of water, $[M–H_2O]^{+\bullet}$, m/z 118, and phenylic cleavage, $[M–COOH]^+$, m/z 91, has to compete with formal loss of formic acid, $[M–HCOOH]^{+\bullet}$, m/z 90. Consequently, such neutral losses are absent in the mass spectra of the *meta* and *para* isomers.

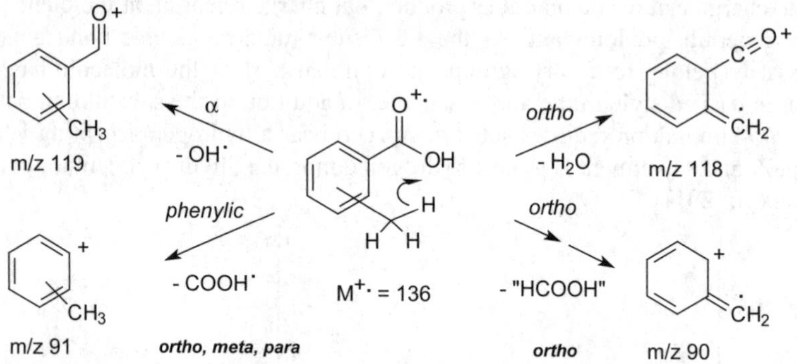

Scheme 6.69.

As indicated by metastable ion studies of isopropylbenzoic acids, [202] the formation of the ion at m/z 90 should be described as a two-step process, i.e., the product rather is $[M–OH–CHO]^{+\bullet}$ than $[M–HCOOH]^{+\bullet}$. This would add yet another example to the list of violations of the even-electron rule (Chap. 6.1.3).

6.13.2 *Ortho* Elimination from Even-Electron Ions

As with the McLafferty rearrangement and the retro-Diels-Alder reaction before, the occurrence of the *ortho* elimination is not restricted to molecular ions. It may equally well proceed in even-electron species.

Example: Examination of the mass spectrum of isopropylbenzoic acid reveals that the molecular ion, m/z 164, as well as the $[M–CH_3]^+$ ion, m/z 149, eliminate H_2O via *ortho* elimination yielding fragment ions at m/z 146 and 131, respectively (Fig. 6.47). [202] The $[M–CH_3–H_2O]^+$ ion, a homologue of the benzoyl ion, decomposes further by loss of CO, thereby creating a fragment that overall corresponds to $[M–CH_3–HCOOH]^+$, m/z 103.

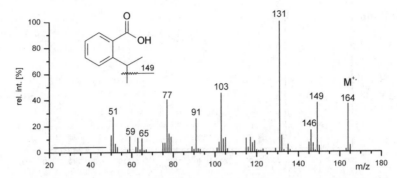

Fig. 6.47. EI mass spectrum of isopropylbenzoic acid. Spectrum used by permission of NIST. © NIST 2002.

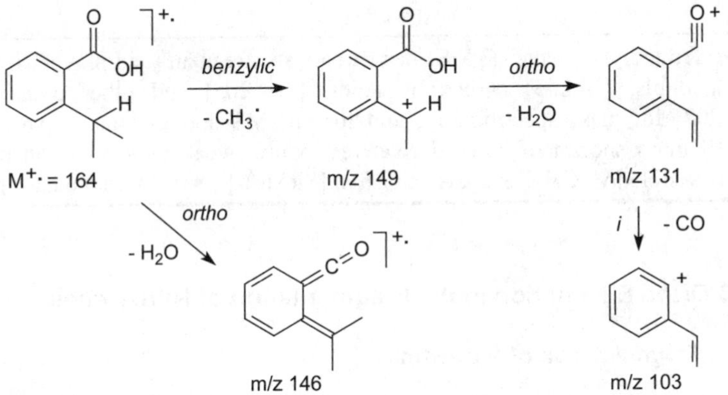

Scheme 6.70.

Example: The molecular ion of 1,2-bis(trimethylsiloxy)benzene, *m/z* 254, undergoes methyl loss by Si–C bond cleavage as typically observed for silanes (Fig. 6.48). Rearrangement of the [M–CH$_3$]$^+$ ion then yields [Si(Me)$_3$]$^+$, *m/z* 73 (base peak). This is not an *ortho* elimination with concomitant H$^•$ transfer as defined in the strict sense, but the observed reaction is still specific for the *ortho*-isomer. [190,203] In the spectra of the *meta*- and *para*-isomers the [Si(Me)$_3$]$^+$ ion is of lower abundance, the [M–CH$_3$]$^+$ ion representing the base peak in their spectra. Moreover, the *m/z* 73 ion is then generated directly from the molecular ion which is clearly different from the two-step pathway of the *ortho*-isomer.

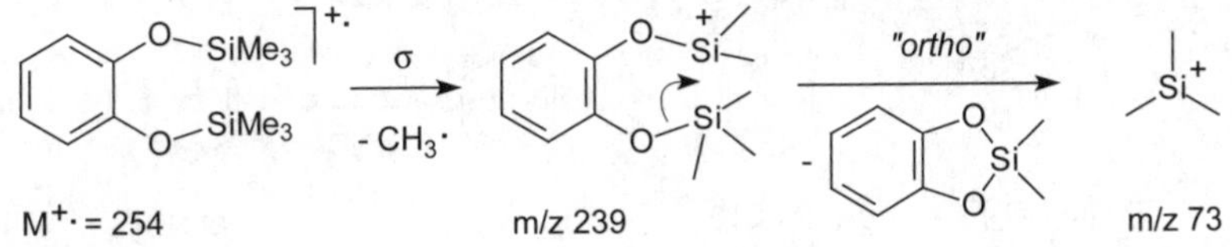

M$^{+•}$ = 254   *m/z* 239   *m/z* 73

Scheme 6.71.

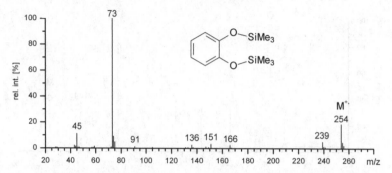

Fig. 6.48. EI mass spectrum of 1,2-bis(trimethylsiloxy)benzene. The isotopic pattern of silicon is clearly visible in the signals at m/z 73, 239, and 254 (Chap. 3.2.6). Spectrum used by permission of NIST. © NIST 2002.

Note: Trimethylsilylether (TMS) derivatives are frequently employed to volatize alcohols, [14,203] carboxylic acids, [204,205] and other compounds [206,207] for mass spectrometry, and for GC-MS applications in particular. The EI mass spectra of TMS derivatives exhibit weak molecular ion peaks, clearly visible [M–CH$_3$]$^+$ signals and often [Si(Me)$_3$]$^+$, m/z 73, as the base peak.

6.13.3 *Ortho* Elimination in the Fragmentation of Nitroarenes

6.13.3.1 Fragmentation of Nitroarenes

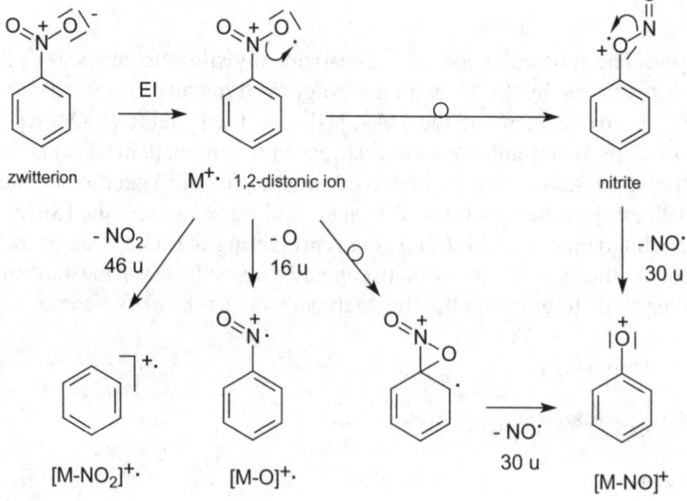

Scheme 6.72.

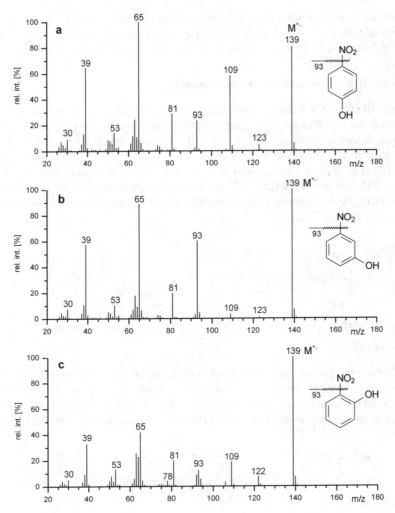

Fig. 6.49. EI mass spectra of isomeric nitrophenols. The fragment ion [M–OH]$^+$, *m/z* 122, instead of [M–O]$^{+\cdot}$, *m/z* 123, is only observed for the *ortho* isomer. Spectra used by permission of NIST. © NIST 2002.

Nitroarenes are recognized from their characteristic neutral losses due to the NO$_2$ substituent. Normally, all theoretically possible fragment ions, the plausible [M–NO$_2$]$^{+\cdot}$ and [M–O]$^{+\cdot}$ ions as well as the unexpected [M–NO]$^+$ ion, are observed. It is worth noting that molecular ions are 1,2-distonic by definition, because nitroarene molecules are best represented as zwitterion (Chap. 6.3). The molecular ion may either dissociate directly by loss of an oxygen atom or a NO$_2$ molecule or it may rearrange prior to loss of NO$^\cdot$. For the latter process, two reaction pathways have been uncovered, one of them involving intermediate formation of a nitrite, and the other proceeding via a three-membered cyclic intermediate. [208] Fragmentation of the nitrite then yields a phenolic-type ion that can decompose of the nitrite then yields a phenolic-type ion that can decompose further by elimination of CO. Thus, the characteristic series of $[M-16]^{+\bullet}$ (generally weak signal), $[M-30]^+$ and $[M-46]^{+\bullet}$ ions is obtained (Fig. 6.49a,b), e.g., for nitrobenzene:

6.13.3.2 *Ortho* Elimination from Nitroarenes

Ortho-substituted nitroarenes can be distinguished from *meta*- or *para*-substituted isomers due to a characteristic change in their mass spectra. In precence of a hydrogen-donating *ortho* subtituent, $[M-OH]^+$ replaces the $[M-O]^{+\bullet}$ the fragment ion, e.g., in case of the above nitrophenols, the ion at m/z 123 is replaced by one at m/z 122 (Fig. 6.49c). The mechanism of this OH$^\bullet$ loss is not quite clear, [209] because different proposals [210-212] are in good accordance with the experimental data. Here, the mechanism which resembles the previously discussed *ortho* eliminations is shown: [212]

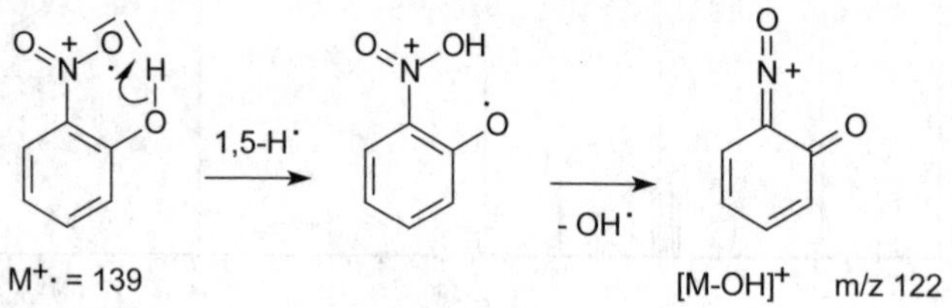

Scheme 6.73.

Ortho eliminations find widespread application in the structure elucidation of aromatic nitro compounds, e.g., nitroanilines, [200] dinitrophenols, [213] trinitroaromatic explosives, [214] and nitrophenyl-methanesulfonamides. [199] (Scheme 6.75 reproduced from Ref. [199] with permission. © IM Publications 1997):

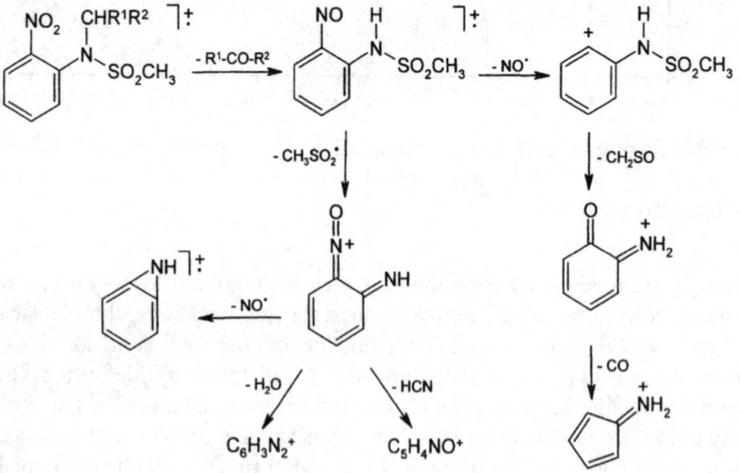

Scheme 6.74.

6.14 Heterocyclic Compounds

As mentioned in the beginning of this chapter, there is by far no chance for a comprehensive treatment of the fragmentations of organic ions in this context. Nevertheless, we should not finish without having briefly addressed the mass spectrometry of heterocyclic compounds, an issue filling a book of its own. [215]

6.14.1 Saturated Heterocyclic Compounds

The molecular ions of small saturated heterocyclic compounds exhibit a strong tendency for *transannular cleavages* that often give rise to the base peak. These transannular cleavages can formally be regarded as clean cuts across the ring. Other fragmentation pathways including ring-opening bond scissions followed by consecutive cleavages and α-cleavage also play a role and start to compete effectively with the characteristic ring cleavage as the number of substituents at the ring increases. [216-218]

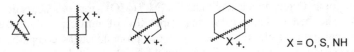

Scheme 6.75.

6.14.1.1 Three- and Four-Membered Saturated Heterocycles

The methyloxirane molecular ion, m/z 58, easily fragments by α-cleavage preferably yielding a $[M–CH_3]^+$ ion, m/z 43, and a $[M–H]^+$ ion, m/z 57 (Fig. 6.50). The more characteristic transannular ring cleavage leads to loss of formaldehyde, thus explaining the base peak at m/z 28 as a $C_2H_4^{+\bullet}$ ion. Ring opening and consecutive fragmentation give rise to fragment ions such as $[CH_2OH]^+$, m/z 31, $[CHO]^+$, m/z 29, and $[CH_3]^+$, m/z 15:

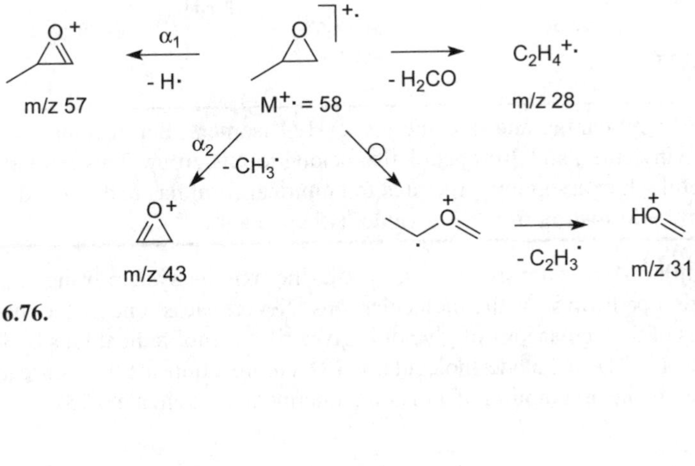

$$m/z\ 57 \quad \xleftarrow[-\,\text{H}\cdot]{\alpha_1} \quad M^{+\bullet} = 58 \quad \xrightarrow[-\,\text{H}_2\text{CO}]{} \quad C_2H_4^{+\bullet} \quad m/z\ 28$$

$$\alpha_2 \searrow {-\,\text{CH}_3^{\bullet}}$$

$$m/z\ 43 \qquad\qquad \xrightarrow[-\,\text{C}_2\text{H}_3^{\bullet}]{} \quad HO\quad m/z\ 31$$

Scheme 6.76.

100
80
60
40
20
0
rel. int. [%]
15
26
28
29
31
39
43
M+.
58
20
40
60
80
m/z
28
43

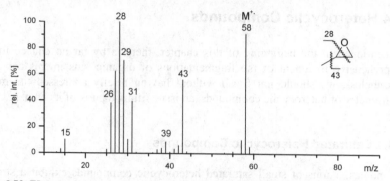

Fig. 6.50. EI mass spectrum of methyloxirane. Spectrum used by permission of NIST.

Example: Comparing the spectra of methyloxirane and its isomer oxetane reveals a clear difference in that the $[M–CH_3]^+$ ion, *m/z* 43, is almost absent because of the missing methyl group, whereas the $[M–H]^+$ ion, *m/z* 57, remains unaffected. (Fig. 6.51a). The transannular ring cleavage, however, benefits from this reduced number of competing channels and thus, the $[M–H_2CO]^{+\bullet}$ ion, *m/z* 28, is definitely dominant. The better charge-localizing capability of nitrogen as in azetidine makes radical-induced cleavages become more pronounced as compared to the oxetane spectrum, i.e., a $[M–CH_3]^+$ ion, *m/z* 42, is observed due to ring opening and the $[M–H]^+$ ion, *m/z* 56, from α-cleavage (Chap. 6.2) is also more prominent (Fig. 6.51b). Here, transannular ring cleavage of the *N*-heterocycle effects loss of methyleneimine, and the $[M–HCNH]^{+\bullet}$ ion gives rise to the base peak:

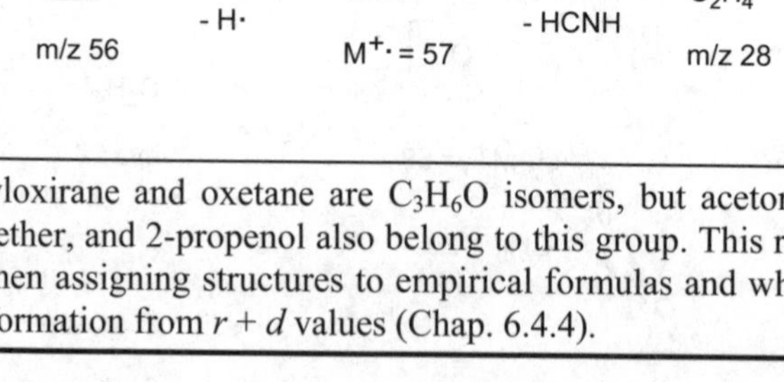

a
m/z 57
α
- H·
M+· = 58
- H2CO
C2H4+·
m/z 28
b
m/z 56
α
- H·
M+· = 57
- HCNH
C2H4+·
m/z 28

Scheme 6.77.

> **Note:** Methyloxirane and oxetane are C_3H_6O isomers, but acetone, propanal, methyl vinylether, and 2-propenol also belong to this group. This reminds us to be careful when assigning structures to empirical formulas and when deducing structural information from *r* + *d* values (Chap. 6.4.4).

Example: More extensive substitution at the oxirane system brings additional dissociation pathways for the molecular ions. Nevertheless, one of the main reaction paths of molecular ions of glycidols gives rise to enol radical ions by loss of a aldehyde (R = H) or ketone molecule. [218] The reaction mechanism can be rationalized by the assumption of a distonic intermediate (Scheme 6.78):

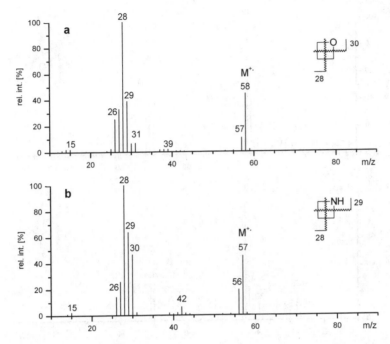

Fig. 6.51. EI mass spectra of oxetane (**a**) and azetidine (**b**). Spectra used by permission of NIST. © NIST 2002.

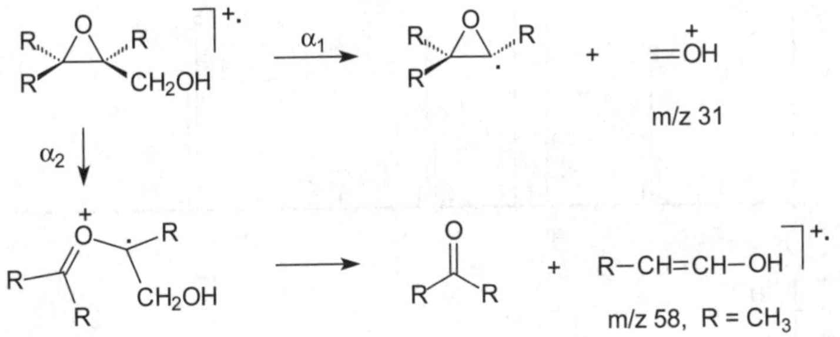

Scheme 6.78.

6.14.1.2 Five- and Six-Membered Saturated Heterocycles

Loss of formaldehyde is not only among the low critical energy processes of oxirane and oxetane molecular ions, but also of larger cyclic ethers such as tetrahydrofurane and tetrahydropyran. [219] Again, imine loss from *N*-heterocycles [220] behaves analogously. The mass spectra of tetrahydrofuran, pyrrolidine, tetrahydropyran, and piperidine are compared below (Fig. 6.52).

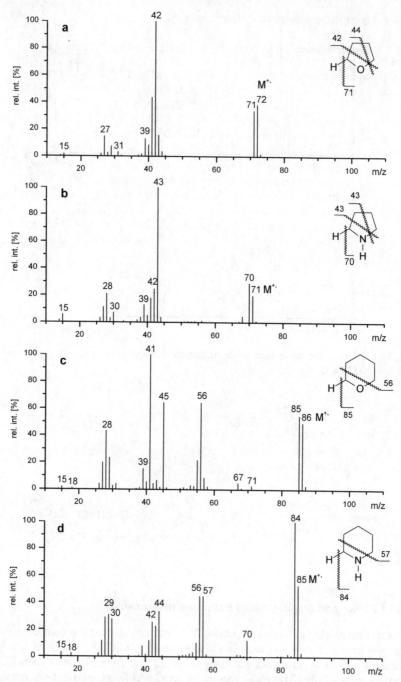

Fig. 6.52. EI mass spectra of tetrahydrofuran (**a**), pyrrolidine (**b**), tetrahydropyran (**c**), and piperidine (**d**). Spectra used by permission of NIST. © NIST 2002.

6.14.2 Aromatic Heterocyclic Compounds

6.14.2.1 Loss of Hydrogen Cyanide from *N*-Heterocycles

Pyridine and numerous other aromatic *N*-heterocycles eliminate a molecule of hydrogen cyanide, HCN, 27 u, from their molecular ions. It has been demonstrated that the molecule eliminated from pyridine molecular ions, *m/z* 79, definitely is HCN and not its isomer hydrogen isocyanide, HNC. [221,222] The fragmentation of pyridine molecular ions to form $[C_4H_4]^{+\bullet}$, *m/z* 52, and HCN was reported to proceed via a tight transition state, [223] which is in some disagreement with the small kinetic energy release of 43 meV. [224] The fragmentation threshold of 12.1 eV for this process is well beyond the ionization energy of 9.3 eV, [22] thus demonstrating comparatively high energy requirements. [223] HCN loss from aromatic *N*-heterocycles is the equivalent to C_2H_2 loss from aromatic hydrocarbons. As anticipated, the peak at *m/z* 53 in the spectrum of pyridine reveals that (after subtraction of the isotopic contribution of ^{13}C) a minor fraction of the pyridine molecular ions elimintates C_2H_2, 26 u (Fig. 6.53).

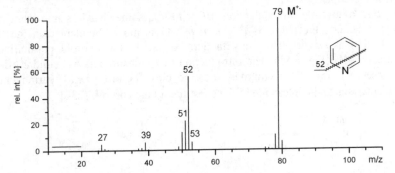

Fig. 6.53. EI mass spectrum of pyridine. HCN loss represents the most important primary fragmentation of the molecular ion. Spectrum used by permission of NIST. © NIST 2002.

Example: Indole molecular ions, *m/z* 117, preferably dissociate by loss of HCN (Fig. 6.54). [225] The $[C_7H_6]^{+\bullet}$ fragment ion, *m/z* 90, then stabilizes by H$^\bullet$ loss to form an even-electron species, $[C_7H_5]^+$, *m/z* 89, which decomposes further by loss of acetylene:

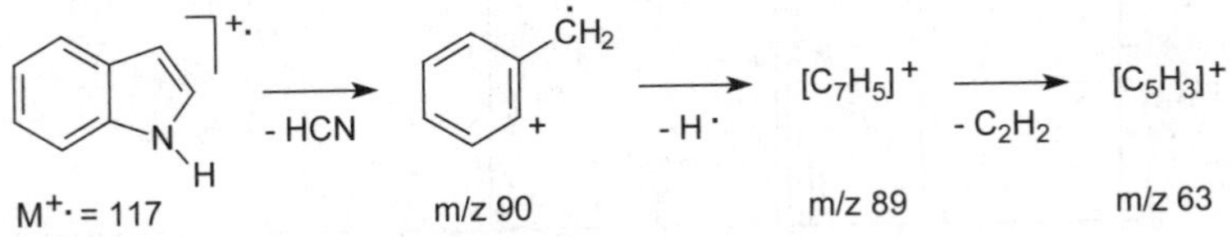

M$^{+\bullet}$ = 117   - HCN   m/z 90   + - H$^\bullet$ $[C_7H_5]^+$ m/z 89   - C$_2$H$_2$ $[C_5H_3]^+$ m/z 63

Scheme 6.79.

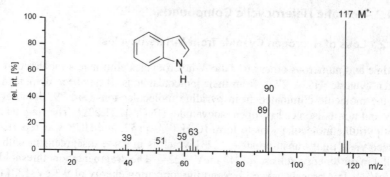

Fig. 6.54. EI mass spectrum of indole. Used by permission of NIST. © NIST 2002.

Indole derivatives are widespread natural compounds, and mass spectrometry of indole derivates, [225,226] especially of indole alkaloids, has been a topic of interest ever since. [227-229]

In case of pyrrole molecular ion, HCN loss is somewhat less important than with indole, i.e., the $[M-C_2H_2]^{+\bullet}$ ion at m/z 41 is more abundant than the $[M-HCN]^{+\bullet}$ ion at m/z 40. This may in part be due to the twofold possibility to eliminate C_2H_2 (Fig. 6.55). The introduction of N-substituents has similar effects as observed for the saturated heterocycles before, i.e., rearrangement fragmentations and α-cleavage of the substituent take control: [230]

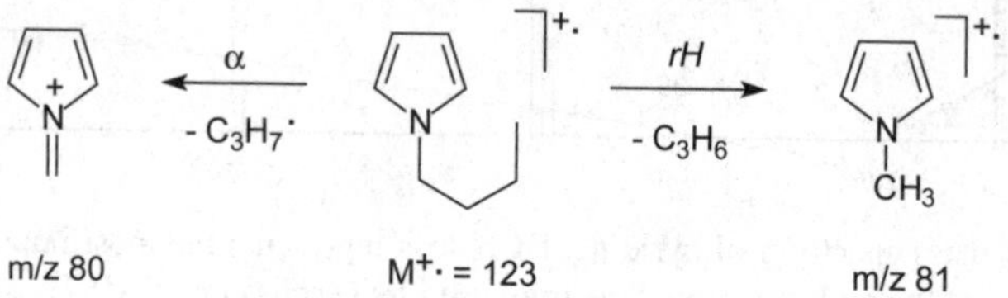

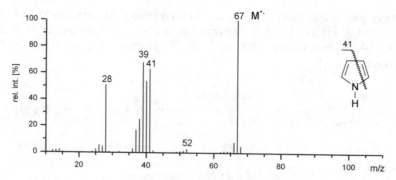

Scheme 6.80.

Fig. 6.55. EI mass spectrum of pyrrole. Elimination of C_2H_2 slightly predominates over loss of HCN. Spectrum used by permission of NIST. © NIST 2002.

6.14.2.2 Loss of Hydrogen Isocyanide from Aromatic Amines

The mass spectrum of aniline has been known since the early days of mass spectrometry. [122] Initially, the observed [M–27]⁺˙ ion has been interpreted in terms of HCN loss (Fig. 6.56a). The mechanism for loss of the elements of [H, N, C] from aminoarenes is perfectly analogous to CO loss from phenols (Chap. 6.9.1). [231] More recently, it could be demonstrated that loss of hydrogen isocyanide, HNC, occurs rather than losing the more stable neutral species HCN, a behavior typical of ionized pyridine. [222]

Interestingly, the three isomeric aminopyridine molecular ions display ion chemistry similar to aniline molecular ions, i.e., metastable HNC loss, [222] instead of HCN loss which should also be possible due to the pyridine core of the molecule (Fig. 6.56b).

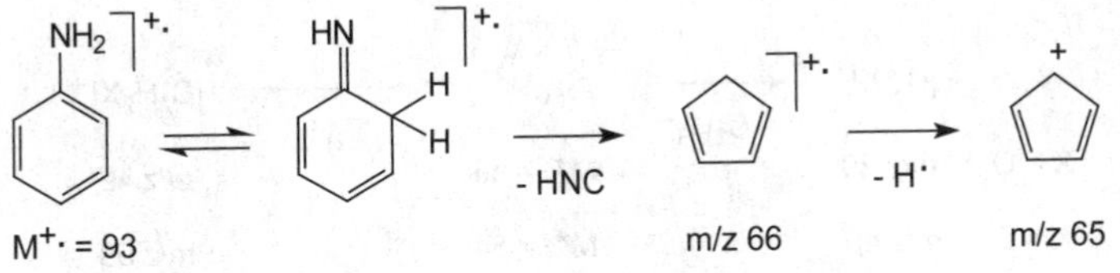

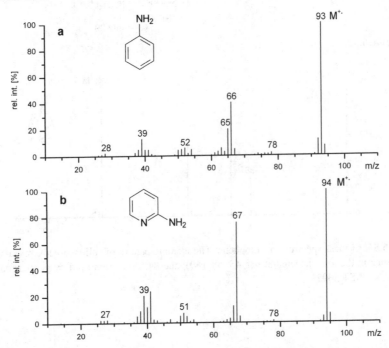

Scheme 6.81.

Fig. 6.56. EI mass spectra of aniline (**a**) and 2-aminopyridine (**b**). Different in mechanism from the N-heterocycles, but very similar in appearance: the aromatic amine molecular ions eliminate HNC. Spectra used by permission of NIST. © NIST 2002.

6.14.2.3 Fragmentation of Furane and Thiophene

The mass spectra of furanes are governed by a strong $[M–HCO]^+$ signal and the corresponding, but weaker peak at m/z 29 belonging to the formyl ion. [232] Analogous behavior is observed for thiophene, i.e., the spectrum shows a $[M–HCS]^+$ peak, m/z 39, and the thioformyl ion, $[HCS]^+$, at m/z 45 (Fig. 6.57). Mass spectra of [2-^{13}C]thiophene and of [2-D]thiophene showed that the thioformyl ion is generated after carbon skeleton rearrangement, whereas hydrogen scrambling seemed to be absent. [233] In addition to this marked fragmentation route, the molecular ions of furanes and thiophenes preferably decompose by C_2H_2 loss, [234] a behavior resembling that of pyrroles and quinones (Chap. 6.9.2). Mass spectra of substituted furanes and thiophenes are discussed in the literature. [232,235,236]

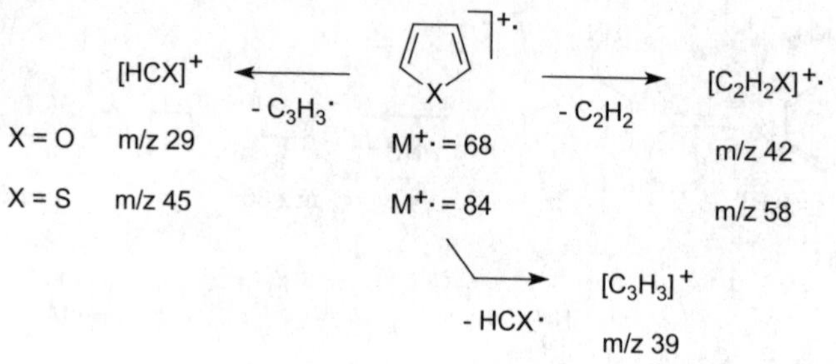

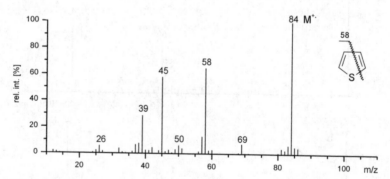

Scheme 6.82.

Fig. 6.57. EI mass spectrum of thiophene. The isotopic pattern of sulfur directly reveals its presence in the ions detected at m/z 45, 58, (69), and 84. Spectrum used by permission of NIST. © NIST 2002.

6.15 Guidelines for the Interpretation of Mass Spectra

6.15.1 Summary of Rules

1. Identify the molecular ion! This is an important inital step, because it is needed to derive the molecular composition (Chap. 6.6.4). If the EI spectrum does not allow for the identification of the molecular ion, soft ionization methods should be employed in addition.

2. The mass differences between the presumed molecular ion and primary fragments must correspond to realistic chemical compositions (Chap. 6.6.4, Table 6.12).

3. The calculated and experimental isotopic patterns have to agree with the molecular formula postulated (Chap. 3.2).

4. The derived molecular formula must obey the nitrogen rule (Chap. 6.2.5). An odd-numbered m/z value of the molecular ion requires 1, 3, 5, ... nitrogens to be contained, whereas an even m/z value belongs to 0, 2, 4, ... nitrogens.

5. Homolytic cleavages cause odd-numbered mass differences between fragment and molecular ion (Chap. 6.2.5). Rearrangement fragmentations cause even-numbered mass differences. This rule toggles if odd numbers of nitrogen are contained in the neutral loss.

6. In general, fragmentations obey the even-electron rule (Chap. 6.1.3). Odd-electron fragments from rearrangement fragmentations behave as if they were molecular ions of the respective smaller molecule.

7. The competition of homolytic cleavages is governed by Stevenson's rule (Chap. 6.2.2). Thermodynamic stability of the pairs of products formed is decisive in selecting the preferred fragmentation channel.

8. Calculate $r + d$ to check formula proposals and to derive some structural characteristics (Chap. 6.4.4).

9. Write down a fragmentation scheme, thereby carefully tracking the origin of primary fragment ions and of characteristic ions used for structure assignment. From the purely analytical point of view this is very useful. However, one should keep in mind that any proposed fragmentation scheme remains a working hypothesis unless experimental confirmation is available.

10. Employ additional techniques, such as measurement of accurate mass (Chap. 3.3), tandem mass spectrometry, or other spectroscopic methods to crosscheck and to refine your assignments.

6.15.2 Systematic Approach to Mass Spectra

1. Collect background information such as origin of the sample, presumed compound class, solubility, thermal stability, or other spectroscopic information.

2. Write *m/z* labels to all relevant peaks and calculate mass differences between prominent peaks. Do you recognize characteristic ion series or mass differences that point to common neutral losses?

3. Check which ionization method was used and examine the general appearance of the mass spectrum. Is the molecular ion peak intensive (as with aromatic, heterocyclic, polycyclic compounds) or weak (as with aliphatic and multifunctional compounds)? Are there typical impurities (solvent, grease, plasticizers) or background signals (residual air, column bleed in GC-MS)?

4. Is accurate mass data available for some of the peaks?

5. Now, follow the above rules to proceed.

6. Derive information on the presence/absence of functional groups.

7. Be careful when using collections of common neutral losses and *m/z*-to-structure relationship tables. They can never be comprehensive. Moreover, one tends to get fixed to the first proposal.

8. Put together the known structural features and try to assign the structure to the unknown sample. Sometimes, only partial structures of the analyte can be derived or isomers cannot be distinguished.

9. Crosscheck proposed molecular structure and mass spectral data. This is also recommended between the single steps of mass spectral interpretation.

10. Are there reference spectra available (at least of similar compounds) either from the literature or from mass spectral data bases (Chap. 5.7)?

11. Never rigidly follow this scheme! Sometimes, a step back or forth may accelerate the process or help to avoid pitfalls.

12. Good luck!

Reference List

1. Biemann, K.; Dibeler, V.H.; Grubb, H.M.; Harrison, A.G.; Hood, A.; Knewstubb, P.F.; Krauss, M.; McLafferty, F.W.; Melton, C.E.; Meyerson, S.; Reed, R.I.; Rosenstock, H.M.; Ryhage, R.; Saunders, R.A.; Stenhagen, E.; Williams, A.E. *Mass Spectrometry of Organic Ions;* 1st ed.; McLafferty, F.W., editor; Academic Press: London, 1963.

2. Budzikiewicz, H.; Djerassi, C.; Williams, D.H. *Mass Spectrometry of Organic Compounds;* 1st ed.; Holden-Day: San Francisco, 1967.

3. Smith, R.M.; Busch, K.L. *Understanding Mass Spectra - A Basic Approach;* 1st ed.; John Wiley & Sons: New York, 1999.

4. Budzikiewicz, H. *Massenspektrometrie - eine Einführung;* 4th ed.; Wiley-VCH: Weinheim, 1998.

5. McLafferty, F.W.; Turecek, F. *Interpretation of Mass Spectra;* 4th ed.; University Science Books: Mill Valley, 1993.

6. Hesse, M.; Meier, H.; Zeeh, B. Massenspektren, in *Spektroskopische Methoden in der Organischen Chemie*, 6th ed.; Georg Thieme Verlag: Stuttgart, 2002; Chapter 4, pp. 243.

7. *Applications of Mass Spectrometry to Organic Stereochemistry;* 1st ed.; Splitter, J.S.; Turecek, F., editors; Verlag Chemie: Weinheim, 1994.

8. Price, P. Standard Definitions of Terms Relating to Mass Spectrometry. A Report From the Committee on Measurements and Standards of the Amercian Society for Mass Spectrometry. *J. Am. Chem. Soc. Mass Spectrom.* **1991,** *2,* 336-348.

9. Todd, J.F.J. Recommendations for Nomenclature and Symbolism for Mass Spectroscopy Including an Appendix of Terms Used in Vacuum Technology. *Int. J. Mass Spectrom. Ion Proc.* **1995,** *142,* 211-240.

10. Svec, H.J.; Junk, G.A. Electron-Impact Studies of Substituted Alkanes. *J. Am. Chem. Soc.* **1967,** *89,* 790-796.

11. Friedman, L.; Long, F.A. Mass Spectra of Six Lactones. *J. Am. Chem. Soc.* **1953,** *75,* 2832-1836.

12. Karni, M.; Mandelbaum, A. The 'Even-Electron Rule'. *Org. Mass Spectrom.* **1980,** *15,* 53-64.

13. Bowen, R.D.; Harrison, A.G. Loss of Methyl Radical From Some Small Immonium Ions: Unusual Violation of the Even-Electron Rule. *Org. Mass Spectrom.* **1981,** *16,* 180-182.

14. Krauss, D.; Mainx, H.G.; Tauscher, B.; Bischof, P. Fragmentation of Trimethylsilyl Derivatives of 2-Alkoxyphenols: a Further Violation of the 'Even-Electron Rule'. *Org. Mass Spectrom.* **1985,** *20,* 614-618.

15. Nizigiyimana, L.; Rajan, P.K.; Haemers, A.; Claeys, M.; Derrick, P.J. Mechanistic Aspects of High-Energy Collision-Induced Dissociation Proximate to the Charge in Saturated Fatty Acid *n*-Butyl Esters Cationized With Lithium. Evidence for Hydrogen Radical Removal. *Rapid Commun. Mass Spectrom.* **1997,** *11,* 1808-1812.

16. Veith, H.J.; Gross, J.H. Alkane Loss From Collisionally Activated Alkylmethyleneimmonium Ions. *Org. Mass Spectrom.* **1991,** *26,* 1061-1064.

17. McLafferty, F.W. Mass Spectrometric Analysis. I. Aliphatic Halogenated Compounds. *Anal. Chem.* **1962,** *34,* 2-15.

18. Stevenson, D.P. Ionization and Dissociation by Electronic Impact. Ionization Potentials and Energies of Formation of Sec-Propyl and Tert-Butyl Radicals. Some Limitations on the Method. *Discuss. Faraday Soc.* **1951,** *10,* 35-45.

19. Audier, H.E. Ionisation et fragmentation en spectrometrie de masse I. sur la répartition de la charge positive entre fragment provenant des mêmes ruptures. *Org. Mass Spectrom.* **1969,** *2,* 283-298.

20. Harrison, A.G.; Finney, C.D.; Sherk, J.A. Factors Determining Relative Ionic Abundances in Competing Fragmentation Reactions. *Org. Mass Spectrom.* **1971,** *5,* 1313-1320.

21. Levsen, K. Reaction Mechanisms, in *Fundamental Aspects of Organic Mass Spectrometry*, 1st ed.; Verlag Chemie: Weinheim, 1978; pp. 152-208.

22. NIST NIST Chemistry Webbook. *http://webbook. nist. gov/* **2002.**

23. Sharkey, A.G., Jr.; Shultz, J.L.; Friedel, R.A. Mass Spectra of Ketones. *Anal. Chem.* **1956,** *28,* 934-940.

24. Gilpin, J.A.; McLafferty, F.W. Mass Spectrometric Analysis: Aliphatic Aldehydes. *Anal. Chem.* **1957,** *29,* 990-994.

25. Liedtke, R.J.; Djerassi, C. Mass Spectrometry in Structural and Stereochemical Problems. CLXXXIII. A Study of the Electron Impact Induced Fragmentation of Aliphatic Aldehydes. *J. Am. Chem. Soc.* **1969,** *91,* 6814-6821.

26. Harrison, A.G. The High-Resolution Mass Spectra of Aliphatic Aldehydes. *Org. Mass Spectrom.* **1970,** *3,* 549-555.

27. Levsen, K.; McLafferty, F.W. Metastable Ion Characteristics. XXVII. Structure and Unimolecular Reactions of $[C_2H_6N]^+$ and $[C_3H_8N]^+$ Ions. *J. Am. Chem. Soc.* **1974,** *96,* 139-144.

28. Bowen, R.D. The Chemistry of $[C_nH_{2n+2}N]^+$ Ions. *Mass Spectrom. Rev.* **1991,** *10,* 225-279.

29. Bowen, R.D. Ion-Neutral Complexes. *Accounts of Chemical Research* **1991,** *24,* 364-371.

30. Djerassi, C.; Fenselau, C. Mass Spectrometry in Structural and Stereochemical Problems. LXXXIV. The Nature of the Cyclic Transition State in Hydrogen Rearrangements of Aliphatic Ethers. *J. Am. Chem. Soc.* **1965,** *87,* 5747-5762.

31. McLafferty, F.W. Mass Spectrometric Analysis of Aliphatic Ethers. *Anal. Chem.* **1957,** *29,* 1782-1789.
32. Friedel, R.A.; Shultz, J.L.; Sharkey, A.G., Jr. Mass Spectra of Alcohols. *Anal. Chem.* **1956,** *28,* 927-934.
33. Gohlke, R.S.; McLafferty, F.W. Mass Spectrometric Analysis of Aliphatic Amines. *Anal. Chem.* **1962,** *34,* 1281-1287.
34. O'Nier, M.J., Jr.; Wier, T.P., Jr. Mass Spectrometry of Heavy Hydrocarbons. *Anal. Chem.* **1951,** 830-843.
35. Audier, H.E.; Milliet, A.; Sozzi, G.; Denhez, J.P. The Isomerization Mechanisms of Alkylamines: Structure of $[C_2H_6N]^+$ and $[C_3H_8N]^+$ Fragment Ions. *Org. Mass Spectrom.* **1984,** *19,* 79-81.
36. Phillips, G.R.; Russell, M.E.; Solka, B.H. Structure of the $[C_2H_5O]^+$ Ion in the Mass Spectrum of Diethyl Ether. *Org. Mass Spectrom.* **1975,** *10,* 819-823.
37. McAdoo, D.J.; Hudson, C.E. Gas Phase Ionic Reactions Are Generally Stepwise Processes. *Int. J. Mass Spectrom. Ion Proc.* **1984,** *62,* 269-276.
38. Williams, D.H.; Budzikiewicz, H.; Pelah, Z.; Djerassi, C. Mass Spectroscopy and Its Application to Structural and Stereochemical Problems. XLIV. Fragmentation Behavior of Monocyclic Ketones. *Monatsh. Chem.* **1964,** *95,* 166-177.
39. Seibl, J.; Gäumann, T. Massenspektren Organischer Verbindungen. 2. Mitteilung: Cyclohexanone. *Helv. Chim. Acta* **1963,** *46,* 2857-2872.
40. Yates, B.F.; Bouma, W.J.; Radom, L. Detection of the Prototype Phosphonium (CH_2PH_3), Sulfonium (CH_2SH_2), and Chloronium (CH_2ClH) Ylides by Neutralization-Reionization Mass Spectrometry: a Theoretical Prediction. *J. Am. Chem. Soc.* **1984,** *106,* 5805-5808.
41. Grützmacher, H.-F. Unimolecular Reaction Mechanisms: the Role of Reactive Intermediates. *Int. J. Mass Spectrom. Ion Proc.* **1992,** *118/119,* 825-855.
42. Hammerum, S. Distonic Radical Cations in the Gaseous and Condensed Phase. *Mass Spectrom. Rev.* **1988,** *7,* 123-202.
43. Stirk, K.M.; Kiminkinen, M.L.K.; Kenttämaa, H.I. Ion-Molecule Reactions of Distonic Radical Cations. *Chem. Rev.* **1992,** *92,* 1649-1665.
44. Yates, B.F.; Bouma, W.J.; Radom, L. Distonic Radical Cations. Guidelines for

the Assessment of Their Stability. *Tetrahedron* **1986,** *42,* 6225-6234.
45. Hammerum, S.; Derrick, P.J. Thermodynamics of Intermediate Ion-Molecule Complexes or Kinetics of Competing Reactions? The Reactions of Low-Energy Isobutylamine and Neopentylamine Molecular Ions. *J. Chem. Soc. ,Perkin Trans. 2* **1986,** 1577-1580.
46. Yates, B.F.; Radom, L. Intramolecular Hydrogen Migration in Ionized Amines: a Theoretical Study of the Gas-Phase Analogues of the Hofmann-Löffler and Related Rearrangements. *J. Am. Chem. Soc.* **1987,** *109,* 2910-2915.
47. Zeller, L.; Farrell, J., Jr.; Vainiotalo, P.; Kenttämaa, H.I. Long-Lived Radical Cations of Simple Organophosphates Isomerize Spontaneously to Distonic Structures in the Gas Phase. *J. Am. Chem. Soc.* **1992,** *114,* 1205-1214.
48. Sack, T.M.; Miller, D.L.; Gross, M.L. The Ring Opening of Gas-Phase Cyclopropane Radical Cations. *J. Am. Chem. Soc.* **1985,** *107,* 6795-6800.
49. Grubb, H.M.; Meyerson, S. Mass Spectra of Alkylbenzenes, in *Mass spectrometry of organic ions,* 1st ed.; McLafferty, F.W., editor; Academic Press: New York, 1963; pp. 453-527.
50. McLafferty, F.W.; Bockhoff, F.M. Collisional Activation and Metastable Ion Characteristics. 67. Formation and Stability of Gaseous Tolyl Ions. *Org. Mass Spectrom.* **1979,** *14,* 181-184.
51. Cone, C.; Dewar, M.J.S.; Landman, D. Gaseous Ions. 1. MINDO/3 Study of the Rearrangement of Benzyl Cation to Tropylium. *J. Am. Chem. Soc.* **1977,** *99,* 372-376.
52. Traeger, J.C.; McLoughlin, R.G. Threshold Photoionization and Dissociation of Toluene and Cycloheptatriene. *J. Am. Chem. Soc.* **1977,** *99,* 7351-7352.
53. Howe, I.; McLafferty, F.W. Unimolecular Decomposition of Toluene and Cycloheptatriene Molecular Ions. Variation of the Degree of Scrambling and Isotope Effect With Internal Energy. *J. Am. Chem. Soc.* **1971,** *93,* 99-105.
54. Kuck, D. Half a Century of Scrambling in Organic Ions: Complete, Incomplete, Progressive and Composite Atom Interchange. *Int. J. Mass Spectrom.* **2002,** *213,* 101-144.
55. Grützmacher, H.-F. Intra- and Intermolecular Reactions of Aromatic Radical Ca-

tions: an Account of Mechanistic Concepts and Methods in Mass Spectrometry. *Org. Mass Spectrom.* **1993**, *28*, 1375-1387.

56. Mormann, M.; Kuck, D. Protonated 1,3,5-Cycloheptatriene and 7-Alkyl-1,3,5-Cycloheptatrienes in the Gas Phase: Ring Contraction to the Isomeric Alkylbenzenium Ions. *J. Mass Spectrom.* **1999**, *34*, 384-394.

57. Rylander, P.N.; Meyerson, S.; Grubb, H.M. Organic Ions in the Gas Phase. II. The Tropylium Ion. *J. Am. Chem. Soc.* **1957**, *79*, 842-846.

58. Cooks, R.G.; Howe, I.; Tam, S.W.; Williams, D.H. Studies in Mass Spectrometry. XXIX. Hydrogen Scrambling in Some Bicyclic Aromatic Systems. Randomization Over Two Rings. *J. Am. Chem. Soc.* **1968**, *90*, 4064-4069.

59. Borchers, F.; Levsen, K. Isomerization of Hydrocarbon Ions. III. $[C_8H_8]^+$, $[C_8H_8]^{2+}$, $[C_6H_6]^+$, and $[C_6H_5]^+$ Ions. *Org. Mass Spectrom.* **1975**, *10*, 584-594.

60. Nishishita, T.; McLafferty, F.W. Metastable Ion Characteristics. XXXXVII. Collisional Activation Mass Spectra of Pentene and Hexene Molecular Ions. *Org. Mass Spectrom.* **1977**, *12*, 75-77.

61. Levsen, K. Isomerization of Hydrocarbon Ions. II. Octenes and Isomeric Cycloalkanes. Collisional Activation Study. *Org. Mass Spectrom.* **1975**, *10*, 55-63.

62. Levsen, K.; Heimbrecht, J. Isomerization of Hydrocarbon Ions. VI. Parameters Determining the Isomerization of Aliphatic Hydrocarbon Ions. *Org. Mass Spectrom.* **1977**, *12*, 131-135.

63. Borchers, F.; Levsen, K.; Schwarz, H.; Wesdemiotis, C.; Winkler, H.U. Isomerization of Linear Octene Cations in the Gas Phase. *J. Am. Chem. Soc.* **1977**, *99*, 6359-6365.

64. Schneider, B.; Budzikiewicz, H. A Facile Method for the Localization of a Double Bond in Aliphatic Compounds. *Rapid Commun. Mass Spectrom.* **1990**, *4*, 550-551.

65. Peake, D.A.; Gross, M.L. Iron(I) Chemical Ionization and Tandem Mass Spectrometry for Locating Double Bonds. *Anal. Chem.* **1985**, *57*, 115-120.

66. Fordham, P.J.; Chamot-Rooke, J.; Guidice, E.; Tortajada, J.; Morizur, J.-P. Analysis of Alkenes by Copper Ion Chemical Ionization Gas Chromatography/Mass Spectrometry and Gas Chro-

matography/Tandem Mass Spectrometry. *J. Mass Spectrom.* **1999**, *34*, 1007-1017.

67. Levsen, K.; Weber, R.; Borchers, F.; Heimbach, H.; Beckey, H.D. Determination of Double Bonds in Alkenes by Field Ionization Mass Spectrometry. *Anal. Chem.* **1978**, *50*, 1655-1658.

68. Buser, H.-R.; Arn, H.; Guerin, P.; Rauscher, S. Determination of Double Bond Position in Mono-Unsaturated Acetates by Mass Spectrometry of Dimethyl Disulfide Adducts. *Anal. Chem.* **1983**, *55*, 818-822.

69. Scribe, P.; Guezennec, J.; Dagaut, J.; Pepe, C.; Saliot, A. Identification of the Position and the Stereochemistry of the Double Bond in Monounsaturated Fatty Acid Methyl Esters by Gas Chromatography/Mass Spectrometry of Dimethyl Disulfide Derivatives. *Anal. Chem.* **1988**, *60*, 928-931.

70. Pepe, C.; Dif, K. The Use of Ethanethiol to Locate the Triple Bond in Alkynes and the Double Bond in Substituted Alkenes by Gas Chromatography/Mass Spectrometry. *Rapid Commun. Mass Spectrom.* **2001**, *15*, 97-103.

71. Pepe, C.; Sayer, H.; Dagaut, J.; Couffignal, R. Determination of Double Bond Positions in Triunsaturated Compounds by Means of Gas Chromatography/Mass Spectrometry of Dimethyl Disulfide Derivatives. *Rapid Commun. Mass Spectrom.* **1997**, *11*, 919-921.

72. Levsen, K.; Heimbach, H.; Shaw, G.J.; Milne, G.W.A. Isomerization of Hydrocarbon Ions. VIII. The Electron Impact Induced Decomposition of *n*-Dodecane. *Org. Mass Spectrom.* **1977**, *12*, 663-670.

73. Lavanchy, A.; Houriet, R.; Gäumann, T. The Mass Spectrometric Fragmentation of *n*-Heptane. *Org. Mass Spectrom.* **1978**, *13*, 410-416.

74. Lavanchy, A.; Houriet, R.; Gäumann, T. The Mass Spectrometric Fragmentation of *n*-Alkane. *Org. Mass Spectrom.* **1979**, *14*, 79-85.

75. Levsen, K. Isomerization of Hydrocarbon Ions. I. Isomeric Octanes. Collisional Activation Study. *Org. Mass Spectrom.* **1975**, *10*, 43-54.

76. Traeger, J.C.; McAdoo, D.J.; Hudson, C.E.; Giam, C.S. Why Are Alkane Eliminations From Ionized Alkanes So Abundant? *J. Am. Chem. Soc. Mass Spectrom.* **1998**, *9*, 21-28.

77. McAdoo, D.J.; Bowen, R.D. Alkane Eliminations From Ions in the Gas Phase. *Eur. Mass Spectrom.* **1999**, *5*, 389-409.

78. Olivella, S.; Solé, A.; McAdoo, D.J.; Griffin, L.L. Unimolecular Reactions of Ionized Alkanes: Theoretical Study of the Potential Energy Surface for CH_3 and CH_4 Losses From Ionized Butane and Isobutane. *J. Am. Chem. Soc.* **1994**, *94*, 11078-11088.

79. Williams, D.H. A Transition State Probe. *Accounts of Chemical Research* **1977**, *10*, 280-286.

80. Ludányi, K.; Dallos, A.; Kühn, Z.; Vékey, D. Mass Spectrometry of Very Large Saturated Hydrocarbons. *J. Mass Spectrom.* **1999**, *34*, 264-267.

81. Biemann, K. Application of Mass Spectrometry in Organic Chemistry, Especially for Structure Determination of Natural Products. *Angew. Chem.* **1962**, *74*, 102-115.

82. Happ, G.P.; Stewart, D.W. Rearrangement Peaks in the Mass Spectra of Certain Aliphatic Acids. *J. Am. Chem. Soc.* **1952**, *74*, 4404-4408.

83. McLafferty, F.W. Mass Spectrometric Analysis. Broad Applicability to Chemical Research. *Anal. Chem.* **1956**, *28*, 306-316.

84. McLafferty, F.W. Mass Spectrometric Analysis: Molecular Rearrangements. *Anal. Chem.* **1965**, *31*, 82-87.

85. Kingston, D.G.I.; Bursey, J.T.; Bursey, M.M. Intramolecular Hydrogen Transfer in Mass Spectra. II. McLafferty Rearrangement and Related Reactions. *Chem. Industry* **1974**, *74*, 215-245.

86. Zollinger, M.; Seibl, J. McLafferty Reactions in Even-Electron Ions? *Org. Mass Spectrom.* **1985**, *11*, 649-661.

87. Djerassi, C.; Tökés, L. Mass Spectrometry in Structural and Stereochemical Problems. XCIII. Further Observations on the Importance of Interatomic Distance in the McLafferty Rearrangement. Synthesis and Fragmentation Behavior of Deuterium-Labeled 12-Oxo Steroids. *J. Am. Chem. Soc.* **1966**, *88*, 536-544.

88. Djerassi, C.; von Mutzenbecher, G.; Fajkos, J.; Williams, D.H.; Budzikiewicz, H. Mass Spectrometry in Structural and Stereochemical Problems. LXV. Synthesis and Fragmentation Behavior of 15-Oxo Steroids. The Importance of Inter-Atomic Distance in the McLafferty Rearrangement. *J. Am. Chem. Soc.* **1965**, *87*, 817-826.

89. Henion, J.D.; Kingston, D.G.I. Mass Spectrometry of Organic Compounds. IX. McLafferty Rearrangements in Some Bicyclic Ketones. *J. Am. Chem. Soc.* **1974**, *96*, 2532-2536.

90. Stringer, M.B.; Underwood, D.J.; Bowie, J.H.; Allison, C.E.; Donchi, K.F.; Derrick, P.J. Is the McLafferty Rearrangement of Ketones Concerted or Stepwise? The Application of Kinetic Isotope Effects. *Org. Mass Spectrom.* **1992**, *27*, 270-276.

91. Dewar, M.J.S. Multibond Reactions Cannot Normally Be Synchronous. *J. Am. Chem. Soc.* **1984**, *106*, 209-219.

92. Holmes, J.L.; Lossing, F.P. Gas-Phase Heats of Formation of Keto and Enol Ions of Carbonyl Compounds. *J. Am. Chem. Soc.* **1980**, *102*, 1591-1595.

93. Hrušák, J. MNDO Calculations on the Neutral and Cationic [CH_3-CO-R] Systems in Relation to Mass Spectrometric Fragmentations. *Z. Phys. Chem.* **1991**, *172*, 217-226.

94. Beynon, J.H.; Saunders, R.A.; Williams, A.E. The High Resolution Mass Spectra of Aliphatic Esters. *Anal. Chem.* **1961**, *33*, 221-225.

95. Harrison, A.G.; Jones, E.G. Rearrangement Reactions Following Electron Impact on Ethyl and Isopropyl Esters. *Can. J. Chem.* **1965**, *43*, 960-968.

96. Wesdemiotis, C.; Feng, R.; McLafferty, F.W. Distonic Radical Ions. Stepwise Elimination of Acetaldehyde From Ionized Benzyl Ethyl Ether. *J. Am. Chem. Soc.* **1985**, *107*, 715-716.

97. Kingston, E.E.; Eichholzer, J.V.; Lyndon, P.; McLeod, J.K.; Summons, R.E. An Unexpected γ-Hydrogen Rearrangement in the Mass Spectra of Di-*Ortho*-Substituted Alkylbenzenes. *Org. Mass Spectrom.* **1988**, *23*, 42-47.

98. Benoit, F.M.; Harrison, A.G. Hydrogen Migrations in Mass Spectrometry. II. Single and Double Hydrogen Migrations in the Electron Impact Fragmentation of Propyl Benzoate. *Org. Mass Spectrom.* **1976**, 1056-1062.

99. Müller, J.; Krebs, G.; Lüdemann, F.; Baumgartner, E. Wasserstoff-Umlagerungen Beim Elektronenstoß-Induzierten Zerfall Von η^6–Benzoesäure-n-Propylester-Tricarbonylchrom. *J. Organomet. Chem.* **1981**, *218*, 61-68.

100. Benoit, F.M.; Harrison, A.G.; Lossing, F.P. Hydrogen Migrations in Mass Spectrometry. III. Energetics of Formation of

[R'CO$_2$H$_2$]$^+$ in the Mass Spectra of R'CO$_2$R. *Org. Mass Spectrom.* **1977**, *12*, 78-82.

101. Tajima, S.; Azami, T.; Shizuka, H.; Tsuchiya, T. An Investigation of the Mechanism of Single and Double Hydrogen Atom Transfer Reactions in Alkyl Benzoates by the Ortho Effect. *Org. Mass Spectrom.* **1979**, *14*, 499-502.

102. Meyerson, S. Formation of [C$_{18}$H$_{36}$]$^{+·}$ and [C$_7$H$_7$O$_2$]$^+$ in the Mass Spectrum of *n*-Octadecyl Benzoate. *Org. Mass Spectrom.* **1989**, *24*, 652-662.

103. Elder, J.F., Jr.; Beynon, J.H.; Cooks, R.G. The Benzoyl Ion. Thermochemistry and Kinetic Energy Release. *Org. Mass Spectrom.* **1976**, *11*, 415-422.

104. McLafferty, F.W.; Gohlke, R.S. Mass Spectrometric Analysis-Aromatic Acids and Esters. *Anal. Chem.* **1959**, *31*, 2076-2082.

105. Yinon, J. Mass Spectral Fragmentation Pathways in Phthalate Esters. A Tandem Mass Spectrometric Collision-Induced Dissociation Study. *Org. Mass Spectrom.* **1988**, *23*, 755-759.

106. Djerassi, C.; Fenselau, C. Mass Spectrometry in Structural and Stereochemical Problems. LXXXVI. The Hydrogen-Transfer Reactions in Butyl Propionate, Benzoate, and Phthalate. *J. Am. Chem. Soc.* **1965**, *87*, 5756-5762.

107. Turecek, F.; Hanus, V. Retro-Diels-Alder Reaction in Mass Spectrometry. *Mass Spectrom. Rev.* **1984**, *3*, 85-152.

108. Turecek, F.; Hanuš, V. Charge Distribution Between Formally Identical Fragments: the Retro-Diels-Alder Cleavage. *Org. Mass Spectrom.* **1980**, *15*, 4-7.

109. Kühne, H.; Hesse, M. The Mass Spectral Retro-Diels-Alder Reaction of 1,2,3,4-Tetrahydronaphthalene, Its Derivatives and Related Heterocyclic Compounds. *Mass Spectrom. Rev.* **1982**, *1*, 15-28.

110. Budzikiewicz, H.; Brauman, J.I.; Djerassi, C. Mass Spectrometry and Its Application to Structural and Stereochemical Problems. LXVII. Retro-Diels-Alder Fragmentation of Organic Molecules Under Electron Impact. *Tetrahedron* **1965**, *21*, 1855-1879.

111. Karpati, A.; Rave, A.; Deutsch, J.; Mandelbaum, A. Stereospecificity of Retro-Diels-Alder Fragmentation Under Electron Impact. *J. Am. Chem. Soc.* **1973**, *90*, 4244.

112. Hammerum, S.; Djerassi, C. Mass Spectrometry in Structural and Stereochemical Problems. CCXXXIII. Stereochemical Dependence of the Retro-Diels-Alder Reaction. *J. Am. Chem. Soc.* **1973**, *95*, 5806-5807.

113. Djerassi, C. Steroids Made It Possible: Organic Mass Spectrometry. *Org. Mass Spectrom.* **1992**, *27*, 1341-1347.

114. Dixon, J.S.; Midgley, I.; Djerassi, C. Mass Spectrometry in Structural and Stereochemical Problems. 248. Stereochemical Effects in Electron Impact Induced Retro-Diels-Alder Fragmentations. *J. Am. Chem. Soc.* **1977**, *99*, 3432-3441.

115. Barnes, C.S.; Occolowitz, J.L. Mass Spectra of Some Naturally Occurring Oxygen Heterocycles and Related Compounds. *Aust. J. Chem.* **1964**, *17*, 975-986.

116. Ardanaz, C.E.; Guidugli, F.H.; Catalán, C.A.N.; Joseph-Nathan, P. Mass Spectral Studies of Methoxynaphthoflavones. *Rapid Commun. Mass Spectrom.* **1999**, *13*, 2071-2079.

117. Ballenweg, S.; Gleiter, R.; Krätschmer, W. Chemistry at Cyclopentene Addends on [60]Fullerene. Matrix-Assisted Laser Desorption-Ionization Time-of-Flight Mass Spectrometry As a Quick and Facile Method for the Characterization of Fullerene Derivatives. *Synth. Met.* **1996**, *77*, 209-212.

118. Ballenweg, S.; Gleiter, R.; Krätschmer, W. Unusual Functionalization of C$_{60}$ Via Hydrozirconation: Reactivity of the C$_{60}$-Zr(IV) Complex Vs. Alkyl-Zr(IV) Complexes. *J. Chem. Soc., Chem. Commun.* **1994**, 2269-2270.

119. Aczel, T.; Lumpkin, H.E. Correlation of Mass Spectra With Structure in Aromatic Oxygenated Compounds. Aromatic Alcohols and Phenols. *Anal. Chem.* **1960**, *32*, 1819-1822.

120. Beynon, J.H. Correlation of Molecular Structure and Mass Spectra, in *Mass Spectrometry and its Applications to Organic Chemistry*, 1st ed.; Elsevier: Amsterdam, 1960; pp. 352.

121. Beynon, J.H.; Lester, G.R.; Williams, A.E. Specific Molecular Rearrangements in the Mass Spectra of Organic Compounds. *J. Phys. Chem.* **1959**, *63*, 1861-1869.

122. Momigny, J. The Mass Spectra of Monosubstituted Benzene Derivatives. Phenol, Monodeuteriophenol, Thiophenol, and Aniline. *Bull. Soc. RoyalSci. Liège* **1953**, *22*, 541-560.

123. Occolowitz, J.L. Mass Spectrometry of Naturally Occurring Alkenyl Phenols and

Their Derivatives. *Anal. Chem.* **1964**, *36*, 2177-2181.

124. Stensen, W.G.; Jensen, E. Structural Determinationof 1,4-Naphthoquinones by Mass Spectrometry/Mass Spectrometry. *J. Mass Spectrom.* **1995**, *30*, 1126-1132.

125. Guigugli, F.H.; Kavka, J.; Garibay, M.E.; Santillan, R.L.; Joseph-Nathan, P. Mass Spectral Studies of Naphthoflavones. *Org. Mass Spectrom.* **1987**, *22*, 479-485.

126. Pelah, Z.; Wilson, J.M.; Ohashi, M.; Budzikiewicz, H.; Djerassi, C. Mass Spectrometry in Structural and Stereochemical Problems. XXXIV. Aromatic Methyl and Ethyl Ethers. *Tetrahedron* **1963**, *19*, 2233-2240.

127. Molenaar-Langeveld, T.A.; Ingemann, S.; Nibbering, N.M.M. Skeletal Rearrangements Preceding Carbon Monoxide Loss From Metastable Phenoxymethylene Ions Derived From Phenoxyacetic Acid and Anisole. *Org. Mass Spectrom.* **1993**, *28*, 1167-1178.

128. Zagorevskii, D.V.; Régimbal, J.-M.; Holmes, J.L. The Heat of Formation of the [C₆H₅O]⁺ Isomeric Ions. *Int. J. Mass Spectrom. Ion Proc.* **1997**, *160*, 211-222.

129. Cooks, R.G.; Bertrand, M.; Beynon, J.H.; Rennekamp, M.E.; Setser, D.W. Energy Partitioning Data As an Ion Structure Probe. Substituted Anisoles. *J. Am. Chem. Soc.* **1973**, *95*, 1732-1739.

130. Alexander, J.J. Mechanism of Photochemical Decarbonylation of Acetyldicarbonyl-η⁵-Cyclopentadienyliron. *J. Am. Chem. Soc.* **1975**, *97*, 1729-1732.

131. Coville, N.J.; Johnston, P. A Mass-Spectral Investigation of Site-Selective Carbon Monoxide Loss From Isotopically Labeled [MnRe(CO)₁₀]⁺. *J. Organomet. Chem.* **1989**, *363*, 343-350.

132. Tobita, S.; Ogino, K.; Ino, S.; Tajima, S. On the Mechanism of Carbon Monoxide Loss From the Metastable Molecular Ion of Dimethyl Malonate. *Int. J. Mass Spectrom. Ion Proc.* **1988**, *85*, 31-42.

133. Moldovan, Z.; Palibroda, N.; Mercea, V.; Mihailescu, G.; Chiriac, M.; Vlasa, M. Mass Spectra of Some β-Keto Esters. A High Resolution Study. *Org. Mass Spectrom.* **1981**, *16*, 195-198.

134. Vairamani, M.; Mirza, U.A. Mass Spectra of Phenoxyacetyl Derivatives. Mechanism of Loss of CO From Phenyl Phenoxyacetates. *Org. Mass Spectrom.* **1987**, *22*, 406-409.

135. Tajima, S.; Siang, L.F.; Fuji-Shige, M.; Nakajima, S.; Sekiguchi, O. Collision-Induced Dissociation Spectra Versus Collision Energy Using a Quadrupole Ion Trap Mass Spectrometer. II. Loss of CO From Ionized *o*-, *m*- and *p*-Anisoyl Fluoride, $CH_3OC_6H_4COF^{+\cdot}$ *J. Mass Spectrom.* **2000**, *35*, 1144-1146.

136. Tajima, S.; Tobita, S.; Mitani, M.; Akuzawa, K.; Sawada, H.; Nakayama, M. Loss of CO From the Molecular Ions of *o*-, *m*- and *p*-Anisoyl Fluorides, $CH_3OC_6H_4COF$, With Fluorine Atom Migration. *Org. Mass Spectrom.* **1991**, *26*, 1023-1026.

137. Tou, J.C. Competitive and Consecutive Eliminations of Molecular Nitrogen and Carbon Monoxide (or Ethene) From Heterocyclics Under Electron Impact. *J. Heterocycl. Chem.* **1974**, *11*, 707-711.

138. McFadden, W.H.; Lounsbury, M.; Wahrhaftig, A.L. The Mass Spectra of Three Deuterated Butanols. *Can. J. Chem.* **1958**, *36*, 990-998.

139. Bukovits, G.J.; Budzikiewicz, H. Mass Spectroscopic Fragmentation Reactions. XXVIII. The Loss of Water From *n*-Alkan-1-ols. *Org. Mass Spectrom.* **1983**, *18*, 219-220.

140. Meyerson, S.; Leitch, L.C. Organic Ions in the Gas Phase. XIV. Loss of Water From Primary Alcohols Under Electron Impact. *J. Am. Chem. Soc.* **1964**, *86*, 2555-2558.

141. Bowen, R.D. The Role of Ion-Neutral Complexes in the Reactions of Onium Ions and Related Species. *Org. Mass Spectrom.* **1993**, *28*, 1577-1595.

142. Bowen, R.D. Potential Energy Profiles for Unimolecular Reactions of Isolated Organic Ions: Some Isomers of [C₄H₁₀N]⁺ and [C₅H₁₂N]⁺. *J. Chem. Soc., Perkin Trans. 2* **1980**, 1219-1227.

143. Bowen, R.D.; Derrick, P.J. Unimolecular Reactions of Isolated Organic Ions: the Chemistry of the Oxonium Ions [CH₃CH₂CH₂CH₂O=CH₂]⁺ and [CH₃CH₂CH₂CH=OCH₃]⁺. *Org. Mass Spectrom.* **1993**, *28*, 1197-1209.

144. Solling, T.I.; Hammerum, S. The Retro-Ene Reaction of Gaseous Immonium Ions Revisited. *J. Chem. Soc., Perkin Trans. 2* **2001**, 2324-2428.

145. Veith, H.J.; Gross, J.H. Alkene Loss From Metastable Methyleneimmonium Ions: Unusual Inverse Secondary Isotope Effect in Ion-Neutral Complex Intermediate Fragmentations. *Org. Mass Spectrom.* **1991**, *26*, 1097-1105.

146. Budzikiewicz, H.; Bold, P. A McLafferty Rearrangement in an Even-Electron System: C_3H_6 Elimination From the α-Cleavage Product of Tributylamine. *Org. Mass Spectrom.* **1991**, *26*, 709-712.

147. Bowen, R.D.; Colburn, A.W.; Derrick, P.J. Unimolecular Reactions of Isolated Organic Ions: Reactions of the Immonium Ions $[CH_2=N(CH_3)CH(CH_3)_2]^+$, $[CH_2=N(CH_3)CH_2CH_2CH_3]^+$ and $[CH_2=N(CH_2CH_2CH_3)_2]^+$. *J. Chem. Soc., Perkin Trans. 2* **1993**, 2363-2372.

148. Gross, J.H.; Veith, H.J. Propene Loss From Phenylpropylmethyleneiminium Ions. *Org. Mass Spectrom.* **1994**, *29*, 153-154.

149. Gross, J.H.; Veith, H.J. Unimolecular Fragmentations of Long-Chain Aliphatic Iminium Ions. *Org. Mass Spectrom.* **1993**, *28*, 867-872.

150. Uccella, N.A.; Howe, I.; Williams, D.H. Structure and Isomerization of Gaseous $[C_3H_8N]^+$ Metastable Ions. *J. Chem. Soc., B* **1971**, 1933-1939.

151. Levsen, K.; Schwarz, H. Influence of Charge Localization on the Isomerization of Organic Ions. *Tetrahedron* **1975**, *31*, 2431-2433.

152. Bowen, R.D. Unimolecular Reactions of Isolated Organic Ions: Olefin Elimination From Immonium Ions $[R^1R^2N=CH_2]^+$. *J. Chem. Soc., Perkin Trans. 2* **1982**, 409-413.

153. Bowen, R.D. Reactions of Isolated Organic Ions. Alkene Loss From the Immonium Ions $[CH_3CH=NHC_2H_5]^+$ and $[CH_3CH=NHC_3H_7]^+$. *J. Chem. Soc., Perkin Trans. 2* **1989**, 913-918.

154. Bowen, R.D.; Colburn, A.W.; Derrick, P.J. Unimolecular Reactions of the Isolated Immonium Ions $[CH_3CH=NHC_4H_9]^+$, $[CH_3CH_2CH=NHC_4H_9]^+$ and $[(CH_3)_2C=NHC_4H_9]^+$. *Org. Mass Spectrom.* **1990**, *25*, 509-516.

155. Bowen, R.D.; Maccoll, A. Unimolecular Reactions of Ionized Ethers. *J. Chem. Soc., Perkin Trans. 2* **1990**, 147-155.

156. Traeger, J.C.; Hudson, C.E.; McAdoo, D.J. Energy Dependence of Ion-Induced Dipole Complex-Mediated Alkane Eliminations From Ionized Ethers. *J. Phys. Chem.* **1990**, *94*, 5714-5717.

157. Bowen, R.D.; Williams, D.H. Non-Concerted Unimolecular Reactions of Ions in the Gas-Phase: the Importance of Ion-Dipole Interactions in Carbonium Ion Isomerizations. *Int. J. Mass Spectrom. Ion Phys.* **1979**, *29*, 47-55.

158. Bowen, R.D.; Williams, D.H. Unimolecular Reactions of Isolated Organic Ions. The Importance of Ion-Dipole Interactions. *J. Am. Chem. Soc.* **1980**, *102*, 2752-2756.

159. Bowen, R.D. Potential Energy Profiles for Unimolecular Reactions of Isolated Organic Ions: $[EtCH=NHMe]^+$ and $[Me_2C=NHMe]^+$. *J. Chem. Soc., Perkin Trans. 2* **1982**, 403-408.

160. Bowen, R.D.; Derrick, P.J. The Mechanism of Ethylene Loss From the Oxonium Ion $[CH_3CH_2O=CHCH_2CH_3]^+$. *J. Chem. Soc., Chem. Commun.* **1990**, 1539-1541.

161. Bowen, R.D.; Colburn, A.W.; Derrick, P.J. Unimolecular Reactions of Isolated Organic Ions: Chemistry of the Unsaturated Oxonium Ion $[CH_2=CHCH=OCH_3]^+$. *Org. Mass Spectrom.* **1992**, *27*, 625-632.

162. Bowen, R.D.; Derrick, P.J. The Mechanism of Ethylene Elimination From the Oxonium Ions $[CH_3CH_2CH=OCH_2CH_3]^+$ and $[(CH_3)_2C=OCH_2CH_3]^+$. *J. Chem. Soc., Perkin Trans. 2* **1992**, 1033-1039.

163. Nguyen, M.T.; Vanquickenborne, L.G.; Bouchoux, G. On the Energy Barrier for 1,2-Elimination of Methane From Dimethyloxonium Cation. *Int. J. Mass Spectrom. Ion Proc.* **1993**, *124*, R11-R14.

164. Lias, S.G.; Liebman, J.F.; Levin, R.D. Evaluated Gas Phase Basicities and Proton Affinities of Molecules; Heats of Formation of Protonated Molecules. *J. Phys. Chem. Ref. Data* **1984**, *13*, 695-808.

165. Lias, S.G.; Bartmess, J.E.; Liebman, J.F.; Holmes, J.L.; Levin, R.D.; Mallard, W.G. Gas-Phase Ion and Neutral Thermochemistry. *J. Phys. Chem. Ref. Data* **1988**, *17, Supplement 1*, 861 pp.

166. Hunter, E.P.L.; Lias, S.G. Evaluated Gas Phase Basicities and Proton Affinities of Molecules: An Update. *J. Phys. Chem. Ref. Data* **1998**, *27*, 413-656.

167. Bowen, R.D.; Stapleton, B.J.; Williams, D.H. Nonconcerted Unimolecular Reactions of Ions in the Gas Phase: Isomerization of Weakly Coordinated Carbonium Ions. *J. Chem. Soc. ,Chem. Commun.* **1978**, 24-26.

168. Morton, T.H. Gas Phase Analogues of Solvolysis Reactions. *Tetrahedron* **1982**, *38*, 3195-3243.

169. Morton, T.H. The Reorientation Criterion and Positive Ion-Neutral Complexes. *Org. Mass Spectrom.* **1992**, *27*, 353-368.

170. Longevialle, P. Ion-Neutral Complexes in the Unimolecular Reactivity of Organic Cations in the Gas Phase. *Mass Spectrom. Rev.* **1992**, *11*, 157-192.

171. McAdoo, D.J. Ion-Neutral Complexes in Unimolecular Decompositions. *Mass Spectrom. Rev.* **1988**, *7*, 363-393.

172. Rylander, P.N.; Meyerson, S. Organic Ions in the Gas Phase. I. The Cationated Cyclopropane Ring. *J. Am. Chem. Soc.* **1956**, *78*, 5799-5802.

173. Meyerson, S. Cationated Cyclopropanes As Reaction Intermediates in Mass Spectra: an Earlier Incarnation of Ion-Neutral Complexes. *Org. Mass Spectrom.* **1989**, *24*, 267-270.

174. Bowen, R.D.; Williams, D.H.; Hvistendahl, G.; Kalman, J.R. Potential Energy Profiles for Unimolecular Reactions of Organic Ions: $[C_3H_8N]^+$ and $[C_3H_7O]^+$. *Org. Mass Spectrom.* **1978**, *13*, 721-728.

175. Longevialle, P.; Botter, R. Electron Impact Mass Spectra of Bifunctional Steroids. The Interaction Between Ionic and Neutral Fragments Derived From the Same Parent Ion. *Org. Mass Spectrom.* **1983**, *18*, 1-8.

176. Longevialle, P.; Botter, R. The Interaction Between Ionic and Neutral Fragments From the Same Parent Ion in the Mass Spectrometer. *Int. J. Mass Spectrom. Ion Phys.* **1983**, *47*, 179-182.

177. Redman, E.W.; Morton, T.H. Product-Determining Steps in Gas-Phase Broensted Acid-Base Reactions. Deprotonation of 1-Methylcyclopentyl Cation by Amine Bases. *J. Am. Chem. Soc.* **1986**, *108*, 5701-5708.

178. Filges, U.; Grützmacher, H.-F. Fragmentations of Protonated Benzaldehydes Via Intermediate Ion/Molecule Complexes. *Org. Mass Spectrom.* **1986**, *21*, 673-680.

179. Hudson, C.E.; McAdoo, D.J. Alkane Eliminations From Radical Cations Through Ion-Radical Complexes. *Int. J. Mass Spectrom. Ion Proc.* **1984**, *59*, 325-332.

180. McAdoo, D.J.; Hudson, C.E. Ion-Neutral Complex-Mediated Hydrogen Exchange in Ionized Butanol: a Mechanism for Nonspecific Hydrogen Migration. *Org. Mass Spectrom.* **1987**, *22*, 615-621.

181. Hammerum, S. Formation and Stabilization of Intermediate Ion-Neutral Complexes in Radical Cation Dissociation Reactions. *J. Chem. Soc., Chem. Commun.* **1988**, 858-859.

182. Hammerum, S.; Audier, H.E. Experimental Verification of the Intermediacy and Interconversion of Ion-Neutral Complexes As Radical Cations Dissociate. *J. Chem. Soc., Chem. Commun.* **1988**, 860-861.

183. Traeger, J.C.; Hudson, C.E.; McAdoo, D.J. Isomeric Ion-Neutral Complexes Generated From Ionized 2-Methylpropanol and n-Butanol: the Effect of the Polarity of the Neutral Partner on Complex-Mediated Reactions. *J. Am. Chem. Soc. Mass Spectrom.* **1991**, *3*, 409-416.

184. Sozzi, G.; Audier, H.E.; Mourgues, P.; Milliet, A. Alkyl Phenyl Ether Radical Cations in the Gas Phase: a Reaction Model. *Org. Mass Spectrom.* **1987**, *22*, 746-747.

185. Morton, T.H. Ion-Molecule Complexes in Unimolecular Fragmentations of Gaseous Cations. Alkyl Phenyl Ether Molecular Ions. *J. Am. Chem. Soc.* **1980**, *102*, 1596-1602.

186. Blanchette, M.C.; Holmes, J.L.; Lossing, F.P. The Fragmentation of Ionized Alkyl Phenyl Ethers. *Org. Mass Spectrom.* **1989**, *24*, 673-678.

187. Harnish, D.; Holmes, J.L. Ion-Radical Complexes in the Gas Phase: Structure and Mechanism in the Fragmentation of Ionized Alkyl Phenyl Ethers. *J. Am. Chem. Soc.* **1991**, *113*, 9729-9734.

188. Zappey, H.W.; Ingemann, S.; Nibbering, N.M.M. Isomerization and Fragmentation of Aliphatic Thioether Radical Cations in the Gas Phase: Ion-Neutral Complexes in the Reactions of Metastable Ethyl Propyl Thioether Ions. *J. Chem. Soc. ,Perkin Trans. 2* **1991**, 1887-1892.

189. Broer, W.J.; Weringa, W.D. Potential Energy Profiles for the Unimolecular Reactions of $[C_3H_7S]^+$ Ions. *Org. Mass Spectrom.* **1980**, 229-234.

190. Schwarz, H. Some Newer Aspects of Mass Spectrometric Ortho Effects. *Top. Curr. Chem.* **1978**, *73*, 231-263.

191. Meyerson, S.; Drews, H.; Field, E.K. Mass Spectra of Ortho-Substituted Diarylmethanes. *J. Am. Chem. Soc.* **1964**, *86*, 4964-4967.

192. Laseter, J.L.; Lawler, G.C.; Griffin, G.W. Influence of Methyl Substitution on Mass Spectra of Diphenylmethanes. Analytical Applications. *Anal. Lett.* **1973**, *6*, 735-744.

193. Martens, J.; Praefcke, K.; Schwarz, H. Spectroscopic Investigations. IX. Analytical Importance of the *Ortho* Effect in Mass Spectrometry. Benzoic and Thiobenzoic

Acid Derivatives. *Z. Naturforsch., B* **1975**, *30*, 259-262.

194. Grützmacher, H.-F. Mechanisms of Mass Spectrometric Fragmentation Reactions. XXXII. The Loss of *Ortho* Halo Substituents From Substituted Thiobenzamide Ions. *Org. Mass Spectrom.* **1981**, *16*, 448-450.

195. Ramana, D.V.; Sundaram, N. Ortho Effects in Organic Molecules on Electron Impact. X. Unusual Ortho Effects of the Methyl Group in *o*-Picolinotoluidide. *Org. Mass Spectrom.* **1982**, *17*, 465-469.

196. Ramana, D.V.; Sundaram, N.; George, N. Ortho Effects in Organic Molecules on Electron Impact. 14. Concerted and Stepwise Ejections of SO_2 and N_2 From *N*-Arylidene-2-nitrobenzenesulfenamides. *Org. Mass Spectrom.* **1987**, *22*, 140-144.

197. Sekiguchi, O.; Noguchi, T.; Ogino, K.; Tajima, S. Fragmentation of Metastable Molecular Ions of Acetylanisoles. *Int. J. Mass Spectrom. Ion Proc.* **1994**, *132*, 172-179.

198. Barkow, A.; Pilotek, S.; Grützmacher, H.-F. Ortho Effects: a Mechanistic Study. *Eur. Mass Spectrom.* **1995**, *1*, 525-537.

199. Danikiewicz, W. *Ortho* Interactions During Fragmentation of *N*-(2-Nitrophenyl)methanesulfonamide and Its *N*-Alkyl Derivatives Upon Electron Ionization. *Eur. Mass Spectrom.* **1997**, *3*, 209-216.

200. Danikiewicz, W. Electron Ionization-Induced Fragmentation of *N*-Alkyl-*o*-Nitroanilines: Observation of New Types of *Ortho* Effects. *Eur. Mass Spectrom.* **1998**, *4*, 167-179.

201. Spiteller, G. The *Ortho* Effect in the Mass Spectra of Aromatic Compounds. *Monatsh. Chem.* **1961**, *92*, 1147-1154.

202. Smith, J.G.; Wilson, G.L.; Miller, J.M. Mass Spectra of Isopropyl Benzene Derivatives. A Study of the *Ortho* Effect. *Org. Mass Spectrom.* **1975**, *10*, 5-17.

203. Schwarz, H.; Köppel, C.; Bohlmann, F. Electron Impact-Induced Fragmentation of Acetylene Compounds. XII. Rearrangement of Bis(Trimethylsilyl) Ethers of Unsaturated α,ω-Diols and Mass Spectrometric Identification of Isomeric Phenols. *Tetrahedron* **1974**, *30*, 689-693.

204. Svendsen, J.S.; Sydnes, L.K.; Whist, J.E. Mass Spectrometric Study of Dimethyl Esters of Trimethylsilyl Ether Derivatives of Some 3-Hydroxy Dicarboxylic Acids. *Org. Mass Spectrom.* **1987**, *22*, 421-429.

205. Svendsen, J.S.; Whist, J.E.; Sydnes, L.K. A Mass Spectrometric Study of the Dimethyl Ester Trimethylsilyl Enol Ether Derivatives of Some 3-Oxodicarboxylic Acids. *Org. Mass Spectrom.* **1987**, *22*, 486-492.

206. Poole, C.F. Recent Advances in the Silylation of Organic Compounds for Gas Chromatography, in *Handbook of derivates for chromatography*, 1st ed.; Blau, G.; King, G.S., editors; Heyden & Son: London, 1977; Chapter 4, pp. 152-200.

207. Halket, H.M.; Zaikin, V.G. Derivatization in Mass Spectrometry -1. Silylation. *Eur. J. Mass Spectrom.* **2003**, *9*, 1-21.

208. Beynon, J.H.; Bertrand, M.; Cooks, R.G. Metastable Loss of Nitrosyl Radical From Aromatic Nitro Compounds. *J. Am. Chem. Soc.* **1973**, *95*, 1739-1745.

209. McLuckey, S.A.; Glish, G.L. The Effect of Charge on Hydroxyl Loss From *Ortho*-Substituted Nitrobenzene Ions. *Org. Mass Spectrom.* **1987**, *22*, 224-228.

210. Beynon, J.H.; Saunders, R.A.; Topham, A.; Williams, A.E. The Dissociation of *o*-Nitrotoluene Under Electron Impact. *J. Chem. Soc.* **1965**, 6403-6405.

211. Herbert, C.G.; Larka, E.A.; Beynon, J.H. The Elimination of Masses 27 and 28 From the [M-OH]$^+$ Ion of 2-Nitrotoluene. *Org. Mass Spectrom.* **1984**, *19*, 306-310.

212. Meyerson, S.; Puskas, I.; Fields, E.K. Organic Ions in the Gas Phase. XVIII. Mass Spectra of Nitroarenes. *J. Am. Chem. Soc.* **1966**, *88*, 4974-4908.

213. Riley, J.S.; Baer, T.; Marbury, G.D. Sequential *Ortho* Effects: Characterization of Novel [M-35]$^+$ Fragment Ions in the Mass Spectra of 2-Alkyl-4,6-Dinitrophenols. *J. Am. Chem. Soc. Mass Spectrom.* **1991**, *2*, 69-75.

214. Yinon, J. Mass Spectral Fragmentation Pathways in 2,4,6-Trinitroaromatic Compounds. A Tandem Mass Spectrometric Collision-Induced Dissociation Study. *Org. Mass Spectrom.* **1987**, *22*, 501-505.

215. Porter, Q.N.; Baldas, J. *Mass Spectrometry of Heterocyclic Compounds;* 1st ed.; Wiley Interscience: New York, 1971.

216. Schwarz, H.; Bohlmann, F. Mass Spectrometric Investigation of Amides. II. Electron-Impact Induced Fragmentation of (Phenylacetyl)Aziridine, -Pyrrolidine, and -Piperidine. *Tetrahedron Lett.* **1973**, *38*, 3703-3706.

217. Nakano, T.; Martin, A. Mass Spectrometric Fragmentation of the Oxetanes of 3,5-

Dimethylisoxazole, 2,4-Dimethylthiazole, and 1-Acetylimidazole. *Org. Mass Spectrom.* **1981**, *16*, 55-61.

218. Grützmacher, H.-F.; Pankoke, D. Rearrangement Reactions of the Molecular Ions of Some Substituted Aliphatic Oxiranes. *Org. Mass Spectrom.* **1989**, *24*, 647-652.

219. Collin, J.E.; Conde-Caprace, G. Ionization and Dissociation of Cyclic Ethers by Electron Impact. *Int. J. Mass Spectrom. Ion Phys.* **1968**, *1*, 213-225.

220. Duffield, A.M.; Budzikiewicz, H.; Williams, D.H.; Djerassi, C. Mass Spectrometry in Structural and Stereochemical Problems. LXIV. A Study of the Fragmentation Processes of Some Cyclic Amines. *J. Am. Chem. Soc.* **1965**, *87*, 810-816.

221. Burgers, P.C.; Holmes, J.L.; Mommers, A.A.; Terlouw, J.K. Neutral Products of Ion Fragmentations: Hydrogen Cyanide and Hydrogen Isocyanide (HNC) Identified by Collisionally Induced Dissociative Ionization. *Chem. Phys. Lett.* **1983**, *102*, 1-3.

222. Hop, C.E.C.A.; Dakubu, M.; Holmes, J.L. Do the Aminopyridine Molecular Ions Display Aniline- or Pyridine-Type Behavior? *Org. Mass Spectrom.* **1988**, *23*, 609-612.

223. Rosenstock, H.M.; Stockbauer, R.; Parr, A.C. Unimolecular Kinetics of Pyridine Ion Fragmentation. *Int. J. Mass Spectrom. Ion Phys.* **1981**, *38*, 323-331.

224. Burgers, P.C.; Holmes, J.L. Kinetic Energy Release in Metastable-Ion Fragmentations. *Rapid Commun. Mass Spectrom.* **1989**, *2*, 279-280.

225. Cardoso, A.M.; Ferrer, A.J. Fragmentation Reactions of Molecular Ions and Dications of Indoleamines. *Eur. Mass Spectrom.* **1999**, *5*, 11-18.

226. Rodríguez, J.G.; Urrutia, A.; Canoira, L. Electron Impact Mass Spectrometry of Indole Derivatives. *Int. J. Mass Spectrom. Ion Proc.* **1996**, *152*, 97-110.

227. Hesse, M. *Indolalkaloide, Teil 1: Text;* 1st ed.; VCH: Weinheim, 1974.

228. Hesse, M. *Indolalkaloide, Teil 2: Spektren;* 1st ed.; VCH: Weinheim, 1974.

229. Biemann, K. Four Decades of Structure Determination by Mass Spectrometry: From Alkaloids to Heparin. *J. Am. Chem. Soc. Mass Spectrom.* **2002**, *13*, 1254-1272.

230. Duffield, A.M.; Beugelmans, R.; Budzikiewicz, H.; Lightner, D.A.; Williams, D.H.; Djerassi, C. Mass Spectrometry in Structural and Stereochemical Problems. LXIII. Hydrogen Rearrangements Induced by Electron Impact on *N-n*-Butyl- and *N-n*-Pentylpyrroles. *J. Am. Chem. Soc.* **1965**, *87*, 805-810.

231. Aubagnac, J.L.; Campion, p. Mass Spectrometry of Nitrogen Heterocycles. X. Contribution to the Behavior of the Aniline Ion and Aminopyridine Ions Prior to Fragmentation by Loss of Hydrogen Cyanide. *Org. Mass Spectrom.* **1979**, *14*, 425-429.

232. Heyns, K.; Stute, R.; Scharmann, H. Mass Spectrometric Investigations. XII. Mass Spectra of Furans. *Tetrahedron* **1966**, *22*, 2223-2235.

233. De Jong, F.; Sinnige, H.J.M.; Janssen, M.J. Carbon Scrambling in Thiophene Under Electron Impact. *Rec. Trav. Chim. Pays-Bas* **1970**, *89*, 225-226.

234. Williams, D.H.; Cooks, R.G.; Ronayne, J.; Tam, S.W. Studies in Mass Spectrometry. XXVII. The Decomposition of Furan, Thiophene, and Deuterated Analogs Under Electron Impact. *Tetrahedron Lett.* **1968**, *14*, 1777-1780.

235. Riepe, W.; Zander, M. The Mass Spectrometric Fragmentation Behavior of Thiophene Benzologs. *Org. Mass Spectrom.* **1979**, *14*, 455-456.

236. Rothwell, A.P.; Wood, K.V.; Gupta, A.K.; Prasad, J.V.N.V. Mass Spectra of Some 2- and 3-Cycloalkenylfurans and -Cycloalkenylthiophenes and Their Oxy Derivatives. *Org. Mass Spectrom.* **1987**, *22*, 790-795.

7 Chemical Ionization

Mass spectrometrists have ever been searching for ionization methods softer than EI, because molecular weight determination is of key importance for structure elucidation. *Chemical ionization* (CI) is the first of the *soft ionization methods* we are going to discuss. Historically, field ionization (FI, Chap. 8) has been applied some years earlier, and thus CI can be regarded as the second soft ionization method introduced to analytical mass spectrometry. Several aspects of CI possess rather close similarity to EI making its discussion next to EI more convenient. CI goes back to experiments of Talrose in the early 1950s [1] and was developed to an analytically useful technique by Munson and Field in the mid-1960s. [2-5] Since then, the basic concept of CI has been extended and applied in numerous different ways, meanwhile providing experimental conditions for a wide diversity of analytical tasks. [5,6] The monograph by Harrison is especially recommended for further reading. [7]

> **Note:** When a positive ion results from CI, the term may be used without qualification; nonetheless *positive-ion chemical ionization* (PICI) is frequently found in the literature. When negative ions are formed, the term *negative-ion chemical ionization* (NICI) should be used. [8]

7.1 Basics of Chemical Ionization

7.1.1 Formation of Ions in Chemical Ionization

In chemical ionization new ionized species are formed when gaseous molecules interact with ions. Chemical ionization may involve the transfer of an electron, proton, or other charged species between the reactants. [8] These reactants are i) the neutral analyte M and ii) ions from a reagent gas.

CI differs from what we have encountered in mass spectrometry so far because *bimolecular processes* are used to generate analyte ions. The occurrence of bimolecular reactions requires a sufficiently large number of ion-molecule collisions during the dwelltime of the reactants in the ion source. This is achieved by significantly increasing the partial pressure of the reagent gas. Assuming reasonable collision cross sections and an ion source residence time of 1 μs, [9] a molecule will undergo 30–70 collisions at an ion source pressure of about 2.5×10^2 Pa. [10] The 10^3–10^4-fold excess of reagent gas also shields the analyte molecules effectively from ionizing primary electrons which is important to suppress competing direct EI of the analyte. There are four general pathways to form ions from a neutral analyte M in CI:

$$M + [BH]^+ \rightarrow [M+H]^+ + B; \qquad \textit{proton transfer} \qquad (7.1)$$

$$[M] + B \rightarrow [M-H]^- + [BH]^+; \qquad \textit{proton transfer} \qquad (7.1a)$$

$$M + X^+ \rightarrow [M+X]^+; \qquad \textit{electrophilic addition} \qquad (7.2)$$

$$M + X^+ \rightarrow [M-A]^+ + AX; \qquad \textit{anion abstraction} \qquad (7.3)$$

$$M + X^{+\bullet} \rightarrow M^{+\bullet} + X; \qquad \textit{charge exchange} \qquad (7.4)$$

Although proton transfer is generally considered to yield protonated analyte molecules, $[M+H]^+$, acidic analytes may also form abundant $[M-H]^-$ ions by protonating some other neutral. Electrophilic addition chiefly occurs by attachment of complete reagent ions to the analyte molecule, e.g., $[M+NH_4]^+$ in case of ammonia reagent gas. Hydride abstractions are abundant representatives of anion abstraction, e.g., aliphatic alcohols rather yield $[M-H]^+$ ions than $[M+H]^+$ ions. [11,12] Whereas reactions 7.1–7.3 result in even electron ions, charge exchange (Eq. 7.4) yields radical ions of low internal energy which behave similar to molecular ions in low-energy electron ionization (Chap. 5.1.5).

> **Note:** It is commonplace to denote $[M+H]^+$ and $[M-H]^+$ ions as *quasimolecular ions* because these ions comprise the otherwise intact analyte molecule and are detected instead of a molecular ion when CI or other soft ionization methods are employed. Usually, the term is also applied to $[M+alkali]^+$ ions created by other soft ionization methods.

7.1.2 Chemical Ionization Ion Sources

CI ion sources exhibit close similarity to EI ion sources (Chap. 5.2.1). In fact, modern EI ion sources can usually be switched to CI operation in seconds, i.e., they are constructed as *EI/CI combination ion sources*. Such a change requires the EI ion source to be modified according to the needs of holding a comparatively high pressure of reagent gas (some 10^2 Pa) without allowing too much leakage into the ion source housing. [13] This is accomplished by axially inserting some inner wall, e.g., a small cylinder, into the ion volume leaving only narrow holes for the entrance and exit of the ionizing primary electrons, the inlets and the exiting ion beam. The ports for the reference inlet, the gas chromatograph (GC) and the direct probe (DIP) need to be tightly connected to the respective inlet system during operation, i.e., an empty DIP is inserted even when another inlet actually provides the sample flow into the ion volume. The reagent gas is introduced directly into the ion volume to ensure maximum pressure inside at minimum losses to the ion source housing (Fig. 7.1). During CI operation, the pressure in the ion source housing typically rises by a factor of 20–50 as compared to the background pressure of the instrument, i.e., to 5×10^{-4}–10^{-3} Pa. Thus, sufficient pumping speed ($\geq 200\,\mathrm{l\,s^{-1}}$) is necessary to maintain stable operation in CI mode. The energy of the primary electrons is preferably adjusted to some 200 eV, because electrons of lower energy experience difficulties in penetrating the reagent gas.

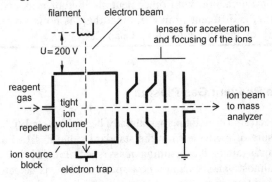

Fig. 7.1. Schematic layout of a chemical ionization ion source. Adapted from Ref. [14] by permission. © Springer-Verlag Heidelberg, 1991.

7.1.3 Sensitivity of Chemical Ionization

Ionization in CI is the result of one or several competing chemical reactions. Therefore, the sensitivity in CI strongly depends on the conditions of the experiment. In addition to primary electron energy and electron current, the reagent gas, the reagent gas pressure, and the ion source temperature have to be stated with the sensitivity data to make a comparison. Modern magnetic sector instruments are specified to have a sensitivity of about 4×10^{-8} C $\mu\mathrm{g}^{-1}$ for the $[M+H]^+$ quasi-molecular ion of methylstearate, m/z 299, at $R = 1000$ in positive-ion CI mode. This is approximately one order of magnitude less than for EI.

7.2 Chemical Ionization by Protonation

7.2.1 Source of Protons

The occurrence of $[M+H]^+$ ions due to bimolecular processes between ions and their neutral molecular counterparts is called *autoprotonation* or *self-CI*. Usually, autoprotonation is an unwanted phenomenon in EI-MS. [M+1] ions from auto-protonation become more probable with increasing pressure and with decreasing temperature in the ion source. Furthermore, the formation of [M+1] ions is promoted if the analyte is of high volatility or contains acidic hydrogens. Thus, self-CI can mislead mass spectral interpretation either by leading to an overestimation of the number of carbon atoms from the ^{13}C isotopic peak (Chap. 3.2.1) or by indicating a by 1 u higher molecular mass (Fig. 7.6 and cf. *nitrogen rule* Chap. 6.2.5). However, in CI-MS with methane or ammonia reagent gas, for example, the process of autoprotonation is employed to generate the reactant ions.

> **Note:** The process $M^* + X \rightarrow MX^{+\bullet} + e^-$, i.e., ionization of internally excited molecules upon interaction with other neutrals is known as *chemi-ionization*. Chemi-ionization is different from CI in that there is no ion-molecule reaction involved [8,15] (cf. Penning ionization, Chap. 2.2.1).

7.2.2 Methane Reagent Gas Plasma

The EI mass spectrum of methane has already been discussed (Chap. 6.1). Rising the partial pressure of methane from the standard value in EI of about 10^{-4} Pa to 10^2 Pa significantly alters the resulting mass spectrum. [1] The molecular ion, $CH_4^{+\bullet}$, m/z 16, almost vanishes and a new species, CH_5^+, is detected at m/z 17 instead. [16] In addition some ions at higher mass occur, the most prominent of which may be assigned as $C_2H_5^+$, m/z 29, [17,18] and $C_3H_5^+$, m/z 41 (Fig. 7.2). The positive ion CI spectrum of methane can be explained as the result of competing and consecutive bimolecular reactions in the ion source: [4,6,10]

$$CH_4 + e^- \rightarrow CH_4^{+\bullet}, CH_3^+, CH_2^{+\bullet}, CH^+, C^{+\bullet}, H_2^{+\bullet}, H^+ \qquad (7.5)$$

$$CH_4^{+\bullet} + CH_4 \rightarrow CH_5^+ + CH_3^\bullet \qquad (7.6)$$

$$CH_3^+ + CH_4 \rightarrow C_2H_7^+ \rightarrow C_2H_5^+ + H_2 \qquad (7.7)$$

$$CH_2^{+\bullet} + CH_4 \rightarrow C_2H_4^{+\bullet} + H_2 \qquad (7.8)$$

$$CH_2^{+\bullet} + CH_4 \rightarrow C_2H_3^+ + H_2 + H^\bullet \qquad (7.9)$$

$$C_2H_3^+ + CH_4 \rightarrow C_3H_5^+ + H_2 \qquad (7.10)$$

$$C_2H_5^+ + CH_4 \rightarrow C_3H_7^+ + H_2 \qquad (7.11)$$

The relative abundances of these product ions change dramatically as the ion source pressure increases from EI conditions to 25 Pa. Above 100 Pa, the relative concentrations stabilize at the levels represented by the CI spectrum of methane reagent gas (Fig. 7.3). [4,19] Fortunately, the ion source pressure of some 10^2 Pa in CI practice is in the plateau region of Fig. 7.3, thereby ensuring reproducible CI conditions. The influence of the ion source temperature is more pronounced than in EI because the high collision rate rapidly effects a thermal equilibrium.

> **Note:** Although the temperature of the ionized reagent gas is by far below that of a plasma, the simultaneous presence of free electrons, protons, numerous ions and radicals lead to its description as a *reagent gas plasma*.

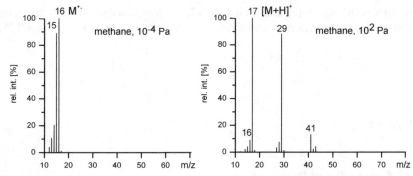

Fig. 7.2. Comparison of the methane spectrum upon electron ionization at different ion source pressures: (**a**) approx. 10^{-4} Pa, (**b**) approx. 10^2 Pa. The latter represents the typical methane reagent gas spectrum in positive-ion CI.

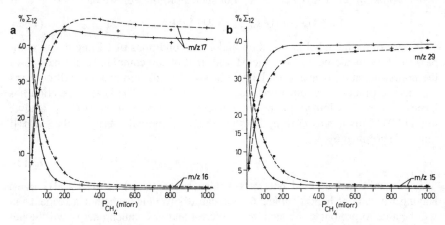

Fig. 7.3. Percentage of total ionization above m/z 12 ($\% \Sigma_{12}$) of (**a**) $CH_4^{+\cdot}$, m/z 16, and CH_5^+, m/z 17, and (**b**) CH_5^+, m/z 15, and $C_2H_5^+$, m/z 29, as a function of CH_4 pressure at 100 eV electron energy and at ion source temperatures 50 °C (——) and 175 °C (-----); 100 mTorr = 13.33 Pa. Adapted from Ref. [19] by permission. © Elsevier Science, 1990.

7.2.2.1 CH_5^+ and Related Ions

The protonated methane ion, CH_5^+, represents a reactive as well as fascinating species in the methane reagent gas plasma. Its structure has been calculated and experimentally verified. [16] The chemical behavior of the CH_5^+ ion appears to be compatible with a stable structure, involving a three-center two-electron bond associating 2 hydrogens and the carbon atom. Rearrangement of this structure due to exchange between one of these hydrogens and one of the three remaining hydrogens appears to be a fast process that is induced by interactions with the chemical ionization gas. In case of the $C_2H_7^+$ intermediate during $C_2H_5^+$ ion formation several isomerizing structures are discussed. [17,18] In protonated fluoromethane, the conditions are quite different, promoting a weak C–F and a strong F–H bond. [20]

Scheme 7.1.

7.2.3 Energetics of Protonation

The tendency of a (basic) molecule B to accept a proton is quantitatively described by its *proton affinity* PA (Chap. 2.11). For such a protonation we have: [3]

$$B_g + H_g^+ \rightarrow [BH]_g^+; \quad -\Delta H_r^0 = PA_{(B)} \tag{7.12}$$

In case of an intended protonation under the conditions of CI one has to compare the PAs of the neutral analyte M with that of the complementary base B of the proton-donating reactant ion $[BH]^+$ (Brønsted acid). Protonation will occur as long as the process is exothermic, i.e., if $PA_{(B)} < PA_{(M)}$. The heat of reaction has basically to be distributed among the degrees of freedom of the $[M+H]^+$ analyte ion. [12,21] This means in turn, that the minimum internal energy of the $[M+H]^+$ ions is determined by:

$$E_{int(M+H)} = -\Delta PA = PA_{(M)} - PA_{(B)} \tag{7.13}$$

Some additional thermal energy will also be contained in the $[M+H]^+$ ions. Having PA data at hand (Table 2.6) one can easily judge whether a reagent ion will be able to protonate the analyte of interest and how much energy will be put into the $[M+H]^+$ ion.

Example: The CH_5^+ reactant ion will protonate C_2H_6 because Eq. 7.13 gives $\Delta PA = PA_{(CH4)} - PA_{(C2H6)} = 552 - 601 = -49$ kJ mol^{-1}. The product, protonated ethane, $C_2H_7^+$, immediately stabilizes by H_2 loss to yield $C_2H_5^+$. [17,18] In case of tetrahydrofurane, protonation is more exothermic: $\Delta PA = PA_{(CH4)} - PA_{(C4H8O)} = 552 - 831 = -279$ kJ mol^{-1}.

7.2.3.1 Impurities of Lower PA than the Reagent Gas

Due to the above energetic considerations, impurities of the reagent gas having a higher PA than the neutral reagent gas are protonated by the reactant ion. [3] Residual water is a frequent source of contamination. Higher concentrations of water in the reagent gas may even alter its properties completely, i.e., H_3O^+ becomes the predominant species in a CH_4/H_2O mixture under CI conditions (Fig. 7.4). [22]

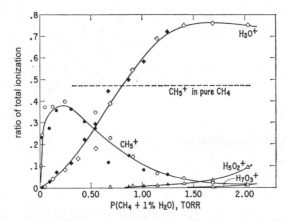

Fig. 7.4. Relative concentrations of CH_5^+ and H_3O^+ ions vs. pressure of a mixture of CH_4 (99 %) and H_2O (1 %). 1 Torr = 133 Pa. Reproduced from Ref. [22] by permission. © American Chemical Society, 1965.

> **Note:** Any analyte of suitable PA may be regarded as basic impurity of the reagent gas, and therefore becomes protonated in excellent yield. Heteroatoms and π-electron systems are the preferred sites of protonation. Nevertheless, the additional proton often moves between several positions of the ion, sometimes accompanied by its exchange with otherwise fixed hydrogens. [23,24]

7.2.4 Methane Reagent Gas PICI Spectra

The $[M+H]^+$ quasimolecular ion in methane reagent gas PICI spectra – generally denoted *methane-CI spectra* – is usually intense and often represents the base peak. [25-27] Although protonation in CI is generally exothermic by 1–4 eV, the degree of fragmentation of $[M+H]^+$ ions is much lower than that observed for the same analytes under 70 eV EI conditions (Fig. 7.5). This is because $[M+H]^+$ ions have i) a narrow internal energy distribution, and ii) fast radical-induced bond cleavages are prohibited, because solely intact molecules are eliminated from these even-electron ions.

Occasionally, hydride abstraction may occur instead of protonation. Electrophilic addition fairly often gives rise to $[M+C_2H_5]^+$ and $[M+C_3H_5]^+$ adduct ions. Thus, [M+29] and [M+41] peaks are sometimes observed in addition to the expected – usually clearly dominating – [M+1] peak.

> **Note:** The occurrence of an [M–1] instead of an [M+1] peak may be recognized from the mass differences to adjacent peaks. In such a case, [M+29] and [M+41] peaks are observed as seemingly [M+31] and [M+43] peaks, respectively. An apparent loss of 16 u might indicate an $[M+H–H_2O]^+$ ion instead of an $[M+H–CH_4]^+$ ion.

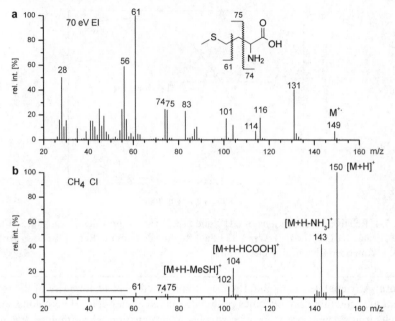

Fig. 7.5. Comparison of (**a**) 70 eV EI spectrum and (**b**) methane reagent gas CI spectrum of the amino acid methionine. Fragmentation is strongly reduced in the CI mass spectrum.

7.2.5 Other Reagent Gases in PICI

As pointed out, the value of ΔPA determines whether a particular analyte can be protonated by a certain reactant ion and how exothermic the protonation will be. Considering other reagent gases than methane therefore allows some tuning of the PICI conditions. The systems employed include molecular hydrogen and hydrogen-containing mixtures, [12,21,28] isobutane, [29-33] ammonia, [30,34-40] dimethylether, [41] diisopropylether, [42] acetone, [11] acetaldehyde, [11], benzene, [43] and iodomethane. [44] Even transition metal ions such as Cu⁺ [45] and Fe⁺ [46] can be employed as reactant ions to locate double bonds. However, nitrous oxide reagent gas better serves that purpose. [39,47,48] The most common reagent gases are summarized in Table 7.1. The EI and CI spectra of ammonia and isobutane are compared in Fig. 7.6.

Isobutane is an especially versatile reagent gas, because i) it provides low-fragmentation PICI spectra of all but the most unpolar analytes, ii) gives almost exclusively one well-defined adduct ($[M+C_4H_9]^+$, [M+57]) if any (Fig. 7.7), and iii) can also be employed for electron capture (Chap. 7.4).

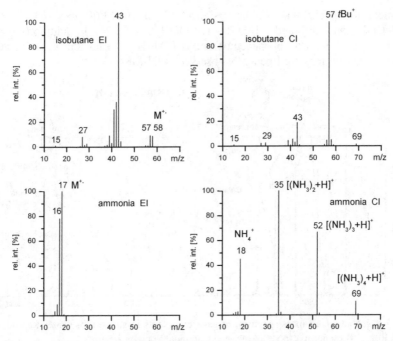

Fig. 7.6. Standard EI versus positive-ion CI spectra of isobutane (upper) and ammonia (lower part). Ammonia forms abundant cluster ions upon CI.

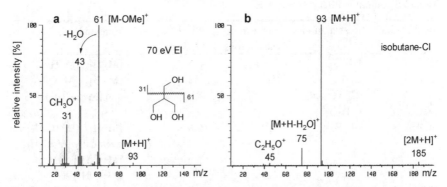

Fig. 7.7. Comparison of (**a**) 70 eV EI and (**b**) isobutane-CI spectrum of glycerol. Instead of a molecular ion, an $[M+H]^+$ quasimolecular ion is observed in EI mode, too. In addition to $[M+H]^+$, the CI spectrum shows few fragment ions and a weak $[2M+H]^+$ cluster ion signal.

Example: In an overdose case where evidence was available for the ingestion of Percodan (a mixture of several common drugs) the isobutane-CI mass spectrum of the gastric extract was obtained (Fig. 7.8). [29] All drugs give rise to an $[M+H]^+$ ion. Due to the low exothermicity of protonation by the *tert*-$C_4H_9^+$ ion, most $[M+H]^+$ ions do not show fragmentation. Solely that of aspirin shows intense fragment ion peaks that can be assigned as $[M+H–H_2O]^+$, m/z 163; $[M+H–H_2C=CO]^+$, m/z 139; and $[M+H–CH_3COOH]^+$, m/z 121. In addition to the $[M+H]^+$ ion at m/z 180, phenacetin forms a $[2M+H]^+$ cluster ion, m/z 163. Such $[2M+H]^+$ cluster ions are frequently observed in CI-MS.

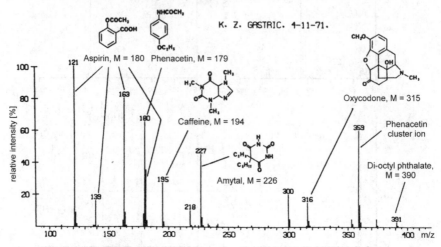

Fig. 7.8. Isobutane CI mass spectrum of gastric contents in an overdose case. Reproduced from Ref. [29] by permission. © American Chemical Society, 1970.

Table 7.1. Common PICI reagent gases

Reagent Gas	Reactant Ions	Neutral from Reactant Ions	PA of Neutral Product	Analyte Ions
H_2	H_3^+	H_2	424	$[M+H]^+$, $[M–H]^+$
CH_4	CH_5^+, ($C_2H_5^+$ and $C_3H_5^+$)	CH_4	552	$[M+H]^+$ ($[M+C_2H_5]^+$ and $[M+C_3H_5]^+$)
i-C_4H_{10}	t-$C_4H_9^+$	i-C_4H_8	820	$[M+H]^+$, ($[M+C_4H_9]^+$, eventually $[M+C_3H_3]^+$, $[M+C_3H_5]^+$ and $[M+C_3H_7]^+$)
NH_3	NH_4^+	NH_3	854	$[M+H]^+$, $[M+NH_4]^+$

Note: Resulting from the large excess of the reagent gas, its spectrum is of much higher intensity than that of the analyte. Therefore, CI spectra are usually acquired starting above the m/z range occupied by reagent ions, e.g., above m/z 50 for methane or above m/z 70 for isobutane.

7.3 Charge Exchange Chemical Ionization

Charge exchange (CE) or *charge transfer ionization* occurs when an ion-neutral reaction takes place in which the ionic charge is transferred to the neutral. [8] In principle, any of the reagent systems discussed so far is capable to effect CE because the respective reagent molecular ions $X^{+\bullet}$ are also present in the plasma:

$$X^{+\bullet} + M \rightarrow M^{+\bullet} + X \qquad (7.14)$$

However, other processes, in particular proton transfer, are prevailing with methane, isobutane, and ammonia, for example. Reagent gases suitable for CE should exhibit abundant molecular ions even under the conditions of CI, whereas potentially protonating species have to be absent or at least of minor abundance.

> **Note:** The acronym CE is also used for *capillary electrophoresis*, a separation method. CE may be coupled to a mass spectrometer via an electrospray interface (Chaps. 11, 12), and thus CE-CI and CE-ESI-MS must not be confused.

7.3.1 Energetics of CE

The energetics of CE are determined by the ionization energy (IE) of the neutral analyte, $IE_{(M)}$, and the *recombination energy* of the reactant ion, $RE_{(X+\bullet)}$. Recombination of an atomic or molecular ion with a free electron is the inverse of its ionization. $RE_{(X+\bullet)}$ is defined as the exothermicity of the gas phase reaction: [7]

$$X^{+\bullet} + e^- \rightarrow X; \qquad -\Delta H_r = RE_{(X+\bullet)} \qquad (7.15)$$

For monoatomic ions, the RE has the same numerical value as the IE of the neutral; for diatomic or polyatomic species differences due to storage of energy in internal modes or electronic excitation may occur. Ionization of the analyte via CE is effected if: [49]

$$RE_{(x+\bullet)} - IE_{(M)} > 0 \qquad (7.16)$$

Now, the heat of reaction, and thus the minimum internal energy of the analyte molecular ion, is given by: [50]

$$E_{int(M+\bullet)} \geq RE_{(x+\bullet)} - IE_{(M)} \qquad (7.17)$$

(The $\geq$ sign indicates the additional contribution of thermal energy.) In summary, no CE is expected if $RE_{(x+\bullet)}$ is less than $IE_{(M)}$; predominantly $M^{+\bullet}$ ions are expected if $RE_{(x+\bullet)}$ is slightly above $IE_{(M)}$; and extensive fragmentation will be observed if $RE_{(x+\bullet)}$ is notably greater than $IE_{(M)}$. [50] Accordingly, the "softness" of CE-CI can be adjusted by choosing a reagent gas of suitable RE. Fortunately, the differences between REs and IEs are small, and unless highest accuracy is required, IE data may be used to estimate the effect of a CE reagent gas (Table 2.1).

7.3.2 Reagent Gases for CE-CI

The gases employed for CE-CI are numerous. Reagent gases such as hydrogen [21] or methane can also affect CE. Typically, pure compounds are employed as CE reagent gases, but occasionally they are diluted with nitrogen to act as an inert or sometimes reactive buffer gas (Table 7.2). Compared to protonating CI conditions, the reagent gas is typically admitted at somewhat lower pressure (15–80 Pa). Primary electron energies are reported in the 100–600 eV range. The major reagent gases include chlorobenzene, [51] benzene, [43,52,53] carbon disulfide, [54,55] xenon, [54,56] carbon oxysulfide, [54-58] carbon monoxide, [50,54,56] nitrogen oxide, [50,59] dinitrogen oxide, [55,59] nitrogen, [54] and argon. [54]

Example: The CE-CI spectra of cyclohexene are compared to the 70 eV EI spectrum. Different CE reagent gases show different degrees of fragmentation (Fig. 7.9). [61] The relative intensity of the molecular ion increases as the RE of the reagent gas decreases. CE-CI mass spectra closely resemble low-energy EI spectra (Chap. 5.1.5) because molecular ions are formed upon CE. As the sensitivity of CE-CI is superior to low-energy EI [49], CE-CI may be preferred over low-energy EI.

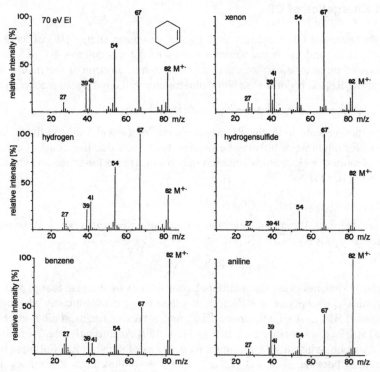

Fig. 7.9. Comparison of 70 eV EI mass spectrum and CE mass spectra of cyclohexene recorded with different reagent gases. Adapted from Ref. [61] by permission. © John Wiley & Sons, 1976.

Table 7.2. Compilation of CE-CI reagent gases [7,50,51,54-56,60]

Reagent Gas	Reactant Ion	RE or RE range [eV]
$C_6H_5NH_2$	$C_6H_5NH_2^{+\bullet}$	7.7
C_6H_5Cl	$C_6H_5Cl^{+\bullet}$	9.0
C_6H_6	$C_6H_6^{+\bullet}$	9.2
$NO^\bullet : N_2 = 1 : 9$	NO^+	9.3
$C_6F_6 : CO_2 = 1 : 9$	$C_6F_6^{+\bullet}$	10.0
$CS_2 : N_2 = 1 : 9$	$CS_2^{+\bullet}$	9.5–10.2
H_2S	$H_2S^{+\bullet}$	10.5
$COS : CO = 1 : 9$	$COS^{+\bullet}$	11.2
Xe	$Xe^{+\bullet}$	12.1, 13.4
N_2O ($: N_2 = 1 : 9$)	$N_2O^{+\bullet}$	12.9
CO_2	$CO_2^{+\bullet}$	13.8
CO	$CO^{+\bullet}$	14.0
Kr	$Kr^{+\bullet}$	14.0, 14.7
N_2	$N_2^{+\bullet}$	15.3
H_2	$H_2^{+\bullet}$	15.4
Ar	$Ar^{+\bullet}$	15.8, 15.9
Ne	$Ne^{+\bullet}$	21.6, 21.7
He	$He^{+\bullet}$	24.6

7.3.4 Compound Class-Selective CE-CI

The energy distribution upon CE largely differs from that obtained upon EI in that the CE process delivers an energetically well-defined ionization of the neutral. Choosing the appropriate reagent gas allows for the selective ionization of a targeted compound class contained in a complicated mixture. [49,51,53,62] The differentiation is possible due to the characteristic range of ionization energies for each compound class. This property of CE-CI can also be applied to construct breakdown graphs from a set of CE-CI mass spectra using reactant ions of stepwise increasing $RE_{(X+\bullet)}$, e.g., $C_6F_6^{+\bullet}$, $CS_2^{+\bullet}$, $COS^{+\bullet}$, $Xe^{+\bullet}$, $N_2O^{+\bullet}$, $CO^{+\bullet}$, $N_2^{+\bullet}$ (Chap. 2.10.5). [56,60]

Example: CE-CI allows for the direct determination of the molecular weight distributions of the major aromatic components in liquid fuels and other petroleum products. [49,51,62] The approach involves selective CE between $C_6H_5Cl^{+\bullet}$ and the substituted benzenes and naphthalenes in the sample. In this application, chlorobenzene also serves as the solvent for the fuel to avoid interferences with residual solvent. Thus, the paraffinic components present in the fuel can be suppressed in the resulting CE-CI mass spectra (Fig. 7.10). [51]

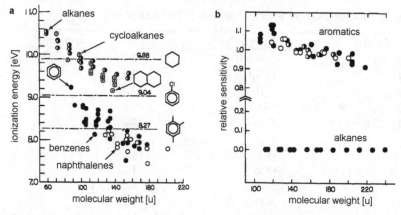

Fig. 7.10. Ionization energies of certain classes of organic molecules (**a**) as a function of their molecular weight, and (**b**) relative sensitivities for (O) alkylbenzenes, (●) polyolefines, and (◑) substituted naphthalenes in chlorobenzene CE-CI. Adapted from Ref. [51] by permission. © American Chemical Society, 1983.

7.3.5 Regio- and Stereoselectivity in CE-CI

Small differences in activation energy (AE) as existing for fragmentations of regioisomers [54,55,57,58] or stereoisomers do not cause significant differences in 70 eV EI spectra. Pathways leading to fragments F_1 and $F_{1'}$ cannot be distinguished, because minor differences in the respective AEs are overridden by the excess energy of the fragmenting ions (Chap. 2.5). The situation changes if an energy-tunable ionization method is applied which in addition offers a narrow energy distribution. If the appearance energies $AE_{(F1)}$ and $AE_{(F1')}$ of the fragments F_1 and $F_{1'}$ from isomeric precursor are just below $RE_{(x+\bullet)}$, a significant alteration of relative intensities of F_1 and $F_{1'}$ would be effected.

   Example: In epimeric 4-methylcyclohexanols the methyl and the hydroxyl group can either both reside in axial position (*cis*) or one is equatorial while the other is axial (*trans*). In the *trans* isomer, stereospecific 1,4–H_2O elimination should proceed easily (Chap. 6.10.3), whereas H_2O loss from the *cis* isomer is more demanding. CE-CI using $C_6F_6^{+\bullet}$ reactant ions clearly distinguishes these stereoisomers by their $M^{+\bullet}/[M–H_2O]^{+\bullet}$ ratio (*trans* : *cis* = 0.09 : 2.0 = 23). [60]

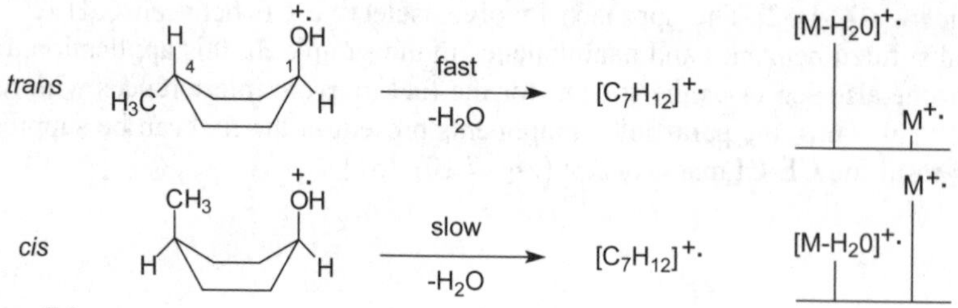

Scheme 7.2.

7.4 Electron Capture

In any CI plasma, both positive and negative ions are formed simultaneously, e.g., $[M+H]^+$ and $[M-H]^-$ ions, and it is just a matter of the polarity of the acceleration voltage which ions are extracted from the ion source. [63] Thus, NICI mass spectra are readily obtained by deprotonation of acidic analytes such as phenols or carboxylic acids or by anion attachment. [64-67]

However, one process of negative-ion formation is of special interest, because it provides superior sensitivity with many toxic and/or environmentally relevant substances: [68-72] this is *electron capture* (EC) or *electron attachment*. [73] EC is a resonance process whereby an external electron is incorporated into an orbital of an atom or molecule. [8] Strictly speaking, EC is not a sub-type of CI because the electrons are not provided by a reacting ion, but are freely moving through the gas at thermal energy. Nonetheless, the ion source conditions to achieve EC are similar to PICI. [74]

7.4.1 Ion Formation by Electron Capture

When a neutral molecule interacts with an electron of high kinetic energy, the positive radical ion is generated by EI. If the electrons have less energy than the IE of the respective neutral, EI is prohibited. As the electrons approach thermal energy, EC can occur instead. Under EC conditions, there are three different mechanisms of ion formation: [65,75-77]

$$M + e^- \rightarrow M^{-\bullet}; \qquad \textit{resonance electron capture} \qquad (7.18)$$

$$M + e^- \rightarrow [M-A]^- + A^\bullet \qquad \textit{dissociative electron capture} \qquad (7.19)$$

$$M + e^- \rightarrow [M-B]^- + B^+ + e^- \qquad \textit{ion-pair formation} \qquad (7.20)$$

Resonance electron capture directly yields the negative molecular ion, $M^{-\bullet}$, whereas even-electron fragment ions are formed by dissociative electron capture and ion-pair formation. Molecular ions are generated by capture of electrons with 0–2 eV kinetic energy, whereas fragment ions are generated by capture of electrons from 0 to 15 eV. Ion-pair formation tends to occur when electron energies exceed 10 eV. [77]

7.4.3 Energetics of EC

The potential energy curves of a neutral molecule AB and the potential ionic products from processes 7.18–7.20 are compared below (Fig. 7.11). These graphs reveal that the formation of negative molecular ions, $AB^{-\bullet}$, is energetically much more favorable than homolytic bond dissociation of AB and that the $AB^{-\bullet}$ ions have internal energies close to the activation energy for dissociation. [65,73,75]

The negative molecular ions from EC are therefore definitely less excited than their positive counterparts from 70 eV EI.

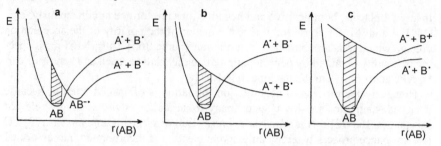

Fig. 7.11. Energetics of (**a**) resonance electron capture, (**b**) dissociative electron capture, and (**c**) ion-pair formation. Adapted from Ref. [75] by permission. © John Wiley & Sons, 1981.

Example: The EC mass spectrum of benzo[a]pyrene, $C_{20}H_{12}$, exclusively shows the negative molecular ion at m/z 252 (Fig. 7.12). The two additional minor signals correspond to impurities of the sample.

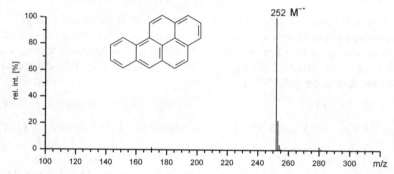

Fig. 7.12. EC spectrum of benzo[a]pyrene; isobutane buffer gas, ion source 200 °C.

The energetics of EC are determined by the *electron affinity* (EA) of the neutral. The EA is the negative of the heat of reaction of the attachment of a zero kinetic energy electron to a neutral molecule or atom:

$$M + e^- \rightarrow M^{-\bullet}, \qquad -\Delta H_r = EA_{(M)} \tag{7.21}$$

As the IE of a molecule is governed by the atom of lowest IE within that neutral (Chap. 2.2.2), the EA of a molecule is basically determined by the atom of highest electronegativity. This is why the presence of halogens, in particular F and Cl, and nitro groups make analytes become attractive candidates for EC (Table 7.3). [78] If EC occurs with a neutral of negative EA, the electron–molecule complex will have a short lifetime (*autodetachment*), but in case of positive EA a negative molecular ion can persist.

Example: Consider the dissociative EC process $CF_2Cl_2 + e^- \rightarrow F^- + CFCl_2^{\bullet}$. Let the potential energy of CF_2Cl_2 be zero. The homolytic bond dissociation energy $D_{(F-CFCl2)}$ has been calculated as 4.93 eV. Now, the potential energy of the products is 4.93 eV less the electron affinity of a fluorine atom ($EA_{(F\bullet)} = 3.45$ eV), i.e., the process is endothermic by 1.48 eV. The experimental AE of the fragments is 1.8 eV. This yields a minimum excess energy of 0.32 eV (Fig. 7.13). [79]

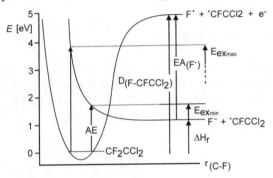

Fig. 7.13. Potential energy diagram of dissociative EC process $CF_2Cl_2 + e^- \rightarrow F^- + CFCl_2^{\bullet}$.

Table 7.3. Selected electron affinities [80]

Compound	EA [eV]	Compound	EA [eV]
carbon dioxide	−0.600	pentachlorobenzene	0.729
naphthalene	−0.200	carbon tetrachloride	0.805
acetone	0.002	biphenylene	0.890
1,2-dichlorobenzene	0.094	nitrobenzene	1.006
benzonitrile	0.256	octafluorocyclobutane	1.049
molecular oxygen	0.451	pentafluorobenzonitrile	1.084
carbon disulfide	0.512	2-nitronaphthalene	1.184
benzo[e]pyrene	0.534	1-bromo-4-nitrobenzene	1.292
tetrachloroethylene	0.640	antimony pentafluoride	1.300

7.4.4 Creating Thermal Electrons

Thermionic emission from a heated metal filament is the standard source of free electrons. However, those electrons usually are significantly above thermal energy and need to be decelerated for EC. Buffer gases such as methane, isobutane, or carbon dioxide serve well for that purpose, but others have also been used. [64,74,81] The gases yield almost no negative ions themselves while moderating the energetic electrons to thermal energy. [69] Despite inverse polarity of the extraction voltage, the same conditions as for PICI can usually be applied (Fig. 7.12). However, EC is comparatively sensitive to ion source conditions. [67,82] The actual ion source temperature, the buffer gas, the amount of sample introduced, and ion source contaminations play important roles each. Regular cleaning of the ion source is important. [70] Lowering the ion source temperature provides lower-energy electrons, e.g., assuming Maxwellian energy distribution the mean electron energy is 0.068 eV at 250 °C and 0.048 eV at 100 °C. [69] Alternatively, electrons of well-defined low energy can be provided by an electron monochromator. [79,82-84]

7.4.5 Appearance of EC Spectra

EC spectra generally exhibit strong molecular ions and some primary fragment ions. As $M^{-\bullet}$ is an odd-electron species, homolytic bond cleavages as well as rearrangement fragmentations may occur (Fig. 7.14) Apart from the changed charge sign, there are close similarities to the fragmentation pathways of positive molecular ions (Chap. 6.). [67,76,77]

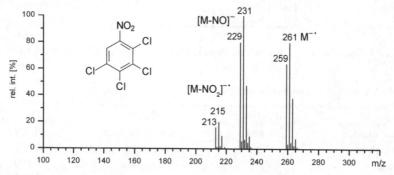

Fig. 7.14. Methane EC spectrum of 2,3,4,5-tetrachloronitrobenzene. Redrawn from Ref. [77] by permission. © John Wiley & Sons, 1988.

7.4.6 Applications of EC

EC, especially when combined with GC-MS, is widespread in monitoring environmental pollutants such as toxaphene, [72] dioxins, [74,79,82,84] pesticides, [79] halogenated metabolites, [71] DNA adducts, [78] explosives, [66,85,86] and others. [68,69,87,88]

7.5 Sample Introduction in CI

In CI, the analyte is introduced into the ion source the same way as described for EI, i.e., via *direct insertion probe* (DIP), *direct exposure probe* (DEP), *gas chromatograph* (GC), or *reservoir inlet* (Chap. 5.3).

7.5.1 Desorption Chemical Ionization

CI in conjunction with a *direct exposure probe* is known as *desorption chemical ionization* (DCI). [30,89,90] In DCI, the analyte is applied from solution or suspension to the outside of a thin resistively heated wire loop or coil. Then, the analyte is directly exposed to the reagent gas plasma while being rapidly heated at rates of several hundred °C s^{-1} and to temperatures up to about 1500 °C (Chap. 5.3.2 and Fig. 5.16). The actual shape of the wire, the method how exactly the sample is applied to it, and the heating rate are of importance for the analytical result. [91,92] The rapid heating of the sample plays an important role in promoting molecular species rather than pyrolysis products. [93] A laser can be used to effect extremely fast evaporation from the probe prior to CI. [94] In case of non-availability of a dedicated DCI probe, a field emitter on a field desorption probe (Chap. 8) might serve as a replacement. [30,95] Different from *desorption electron ionization* (DEI), DCI plays an important role. [92] DCI can be employed to detect arsenic compounds present in the marine and terrestrial environment [96], to determine the sequence distribution of β-hydroxyalkanoate units in bacterial copolyesters [97], to identify additives in polymer extracts [98] and more. [99] Provided appropriate experimental setup, high resolution and accurate mass measurements can also be achieved in DCI mode. [100]

Example: DCI requires fast scanning of the mass analyzer because of the rather sudden evaporation of the analyte. The pyrolysis (Py) DCI mass spectrum of cellulose, $H(C_6H_{10}O_5)_nOH$, acquired using NH_3 reagent gas, 100 eV electron energy, and scanning m/z 140–700 at 2 s per cycle shows a series of distinct signals. The peaks are separated by 162 u, i.e., by $C_6H_{10}O_5$ saccharide units. The main signals are due to ions from anhydro-oligosaccharides, $[(C_6H_{10}O_5)_n+NH_4]^+$, formed upon heating of the cellulose in the ammonia CI plasma (Fig. 7.15). [91]

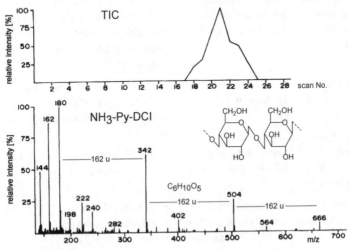

Fig. 7.15. Ammonia-Py-DCI mass spectrum of cellulose and total ion current. Adapted from Ref. [91] by permission. © Research Council of Canada, 1994.

Note: Although *desorption chemical ionization* being the correct term, [92] DCI is sometimes called *direct CI, direct exposure CI, in-beam CI,* or even *surface ionization* in the literature.

7.6 Analytes for CI

Whether an analyte is suitable to be analyzed by CI depends on what particular CI technique is to be applied. Obviously, protonating PICI will be beneficial for other compounds than CE-CI or EC. In general, most analytes accessible to EI (Chap. 5.6) can be analyzed by protonating PICI, and PICI turns out to be especially useful when molecular ion peaks in EI are absent or very weak. CE-CI and EC play a role where selectivity and/or very high sensitivity for a certain compound class is desired (Table 7.4). The typical mass range for CI reaches from 80 to 1200 u. In DCI, molecules up to 2000 u are standard, but up to 6000 u may become feasible. [92]

Table 7.4. CI methods for different groups of analytes

Analyte	Thermodynamic Properties[a]	Example	Suggested CI Method
low polarity, no heteroatoms	low to high IE, low PA, low EA	alkanes, alkenes, aromatic hydrocarbons	CE
low to medium polarity, one or two heteroatoms	low to medium IE, medium to high PA, low EA	alcohols, amines, esters, heterocyclic compounds	PICI, CE
medium to high polarity, some heteroatoms	low to medium IE, high PA and low EA	diols, triols, amino acids, disaccharides, substituted aromatic or heterocyclic compounds	PICI
low to high polarity, halogens (especially F or Cl)	medium IE, low PA, medium to high EA	halogenated compounds, derivatives, e.g., trifluoroacetate, pentafluorobenzyl	EC
high polarity, medium to high molecular mass	low to medium IE, high PA and low EA	mono- to tetrasaccharides, low mass peptides, other polar oligomers	DCI
high polarity, very high molecular mass	decomposition products of low to medium IE, high PA and low EA	polysaccharides, humic compounds, synthetic polymers	Py-DCI

a) IE: ionization energy, PA: proton affinity, EA: electron affinity

7.7 Mass Analyzers for CI

For the choice of a mass analyzer to be operated with a CI ion source the same criteria as for EI apply (Chap. 5.5). As mentioned before, sufficient pumping speed at the ion source housing is a prerequisite.

Reference List

1. Talrose, V.L.; Ljubimova, A.K. Secondary Processes in the Ion Source of a Mass Spectrometer (Reprint From 1952). *J. Mass Spectrom.* **1998**, *33*, 502-504.
2. Munson, M.S.B.; Field, F.H. Reactions of Gaseous Ions. XV. Methane + 1% Ethane and Methane + 1% Propane. *J. Am. Chem. Soc.* **1965**, *87*, 3294-3299.
3. Munson, M.S.B. Proton Affinities and the Methyl Inductive Effect. *J. Am. Chem. Soc.* **1965**, *87*, 2332-2336.
4. Munson, M.S.B.; Field, F.H. Chemical Ionization Mass Spectrometry. I. General Introduction. *J. Am. Chem. Soc.* **1966**, *88*, 2621-2630.
5. Munson, M.S.B. Development of Chemical Ionization Mass Spectrometry. *Int. J. Mass Spectrom.* **2000**, *200*, 243-251.
6. Richter, W.J.; Schwarz, H. Chemical Ionization - a Highly Important Productive Mass Spectrometric Analysis Method. *Angew. Chem.* **1978**, *90*, 449-469.
7. Harrison, A.G. *Chemical Ionization Mass Spectrometry;* 2nd ed.; CRC Press: Boca Raton, 1992.
8. Todd, J.F.J. Recommendations for Nomenclature and Symbolism for Mass Spectroscopy Including an Appendix of Terms Used in Vacuum Technology. *Int. J. Mass Spectrom. Ion. Proc.* **1995**, *142*, 211-240.
9. Griffith, K.S.; Gellene, G.I. A Simple Method for Estimating Effective Ion Source Residence Time. *J. Am. Soc. Mass Spectrom.* **1993**, *4*, 787-791.
10. Field, F.H.; Munson, M.S.B. Reactions of Gaseous Ions. XIV. Mass Spectrometric Studies of Methane at Pressures to 2 Torr. *J. Am. Chem. Soc.* **1965**, *87*, 3289-3294.
11. Hunt, D.F.; Ryan, J.F.I. Chemical Ionization Mass Spectrometry Studies. I. Identification of Alcohols. *Tetrahedron Lett.* **1971**, *47*, 4535-4538.
12. Herman, J.A.; Harrison, A.G. Effect of Reaction Exothermicity on the Proton Transfer Chemical Ionization Mass Spectra of Isomeric C_5 and C_6 Alkanols. *Can. J. Chem.* **1981**, *59*, 2125-2132.
13. Beggs, D.; Vestal, M.L.; Fales, H.M.; Milne, G.W.A. Chemical Ionization Mass Spectrometer Source. *Rev. Sci. Inst.* **1971**, *42*, 1578-1584.
14. Schröder, E. *Massenspektrometrie - Begriffe und Definitionen;* Springer-Verlag: Heidelberg, 1991.
15. Price, P. Standard Definitions of Terms Relating to Mass Spectrometry. A Report From the Committee on Measurements and Standards of the ASMS. *J. Am. Chem. Soc. Mass Spectrom.* **1991**, *2*, 336-348.
16. Heck, A.J.R.; de Koning, L.J.; Nibbering, N.M.M. Structure of Protonated Methane. *J. Am. Soc. Mass Spectrom.* **1991**, *2*, 454-458.
17. Mackay, G.I.; Schiff, H.I.; Bohme, K.D. A Room-Temperature Study of the Kinetics and Energetics for the Protonation of Ethane. *Can. J. Chem.* **1981**, *59*, 1771-1778.
18. Fisher, J.J.; Koyanagi, G.K.; McMahon, T.B. The $C_2H_7^+$ Potential Energy Surface: a Fourier Transform Ion Cyclotron Resonance Investigation of the Reaction of Methyl Cation with Methane. *Int. J. Mass Spectrom.* **2000**, *195/196*, 491-505.
19. Drabner, G.; Poppe, A.; Budzikiewicz, H. The Composition of the Methane Plasma. *Int. J. Mass Spectrom. Ion Proc.* **1990**, *97*, 1-33.
20. Heck, A.J.R.; de Koning, L.J.; Nibbering, N.M.M. On the Structure and Unimolecular Chemistry of Protonated Halomethanes. *Int. J. Mass Spectrom. Ion Proc.* **1991**, *109*, 209-225.
21. Herman, J.A.; Harrison, A.G. Effect of Protonation Exothermicity on the CI Mass Spectra of Some Alkylbenzenes. *Org. Mass Spectrom.* **1981**, *16*, 423-427.
22. Munson, M.S.B.; Field, F.H. Reactions of Gaseous Ions. XVI. Effects of Additives

on Ionic Reactions in Methane. *J. Am. Chem. Soc.* **1965**, *87*, 4242-4247.

23. Kuck, D.; Petersen, A.; Fastabend, U. Mobile Protons in Large Gaseous Alkylbenzenium Ions. The 21-Proton Equilibration in Protonated Tetrabenzylmethane and Related "Proton Dances". *Int. J. Mass Spectrom.* **1998**, *179/180*, 129-146.

24. Kuck, D. Half a Century of Scrambling in Organic Ions: Complete, Incomplete, Progressive and Composite Atom Interchange. *Int. J. Mass Spectrom.* **2002**, *213*, 101-144.

25. Fales, H.M.; Milne, G.W.A.; Axenrod, T. Identification of Barbiturates by CI-MS. *Anal. Chem.* **1970**, *42*, 1432-1435.

26. Milne, G.W.A.; Axenrod, T.; Fales, H.M. CI-MS of Complex Molecules. IV. Amino Acids. *J. Am. Chem. Soc.* **1970**, *92*, 5170-5175.

27. Fales, H.M.; Milne, G.W.A. CI-MS of Complex Molecules. II. Alkaloids. *J. Am. Chem. Soc.* **1970**, *92*, 1590-1597.

28. Herman, J.A.; Harrison, A.G. Energetics and Structural Effects in the Fragmentation of Protonated Esters in the Gas Phase. *Can. J. Chem.* **1981**, *59*, 2133-2145.

29. Milne, G.W.A.; Fales, H.M.; Axenrod, T. Identification of Dangerous Drugs by Isobutane CI-MS. *Anal. Chem.* **1970**, *42*, 1815-1820.

30. Takeda, N.; Harada, K.-I.; Suzuki, M.; Tatematsu, A.; Kubodera, T. Application of Emitter CI-MS to Structural Characterization of Aminoglycoside Antibiotics. *Org. Mass Spectrom.* **1982**, *17*, 247-252.

31. McGuire, J.M.; Munson, B. Comparison of Isopentane and Isobutane as Chemical Ionization Reagent Gases. *Anal. Chem.* **1985**, *57*, 680-683.

32. McCamish, M.; Allan, A.R.; Roboz, J. Poly(Dimethylsiloxane) As Mass Reference for Accurate Mass Determination in Isobutane CI-MS. *Rapid Commun. Mass Spectrom.* **1987**, *1*, 124-125.

33. Maeder, H.; Gunzelmann, K.H. Straight-Chain Alkanes As Reference Compounds for Accurate Mass Determination in Isobutane CI-MS. *Rapid Commun. Mass Spectrom.* **1988**, *2*, 199-200.

34. Hunt, D.F.; McEwen, C.N.; Upham, R.A. CI-MS II. Differentiation of Primary, Secondary, and Tertiary Amines. *Tetrahedron Lett.* **1971**, *47*, 4539-4542.

35. Keough, T.; DeStefano, A.J. Factors Affecting Reactivity in Ammonia CI-MS. *Org. Mass Spectrom.* **1981**, *16*, 527-533.

36. Hancock, R.A.; Hodges, M.G. A Simple Kinetic Method for Determining Ion-Source Pressures for Ammonia CI-MS. *Int. J. Mass Spectrom. Ion Phys.* **1983**, *46*, 329-332.

37. Rudewicz, P.; Munson, B. Effect of Ammonia Partial Pressure on the Sensitivities for Oxygenated Compounds in Ammonia CI-MS. *Anal. Chem.* **1986**, *58*, 2903-2907.

38. Lawrence, D.L. Accurate Mass Measurement of Positive Ions Produced by Ammonia Chemical Ionization. *Rapid Commun. Mass Spectrom.* **1990**, *4*, 546-549.

39. Busker, E.; Budzikiewicz, H. Studies in CI-MS. 2. Isobutane and Nitric Oxide Spectra of Alkynes. *Org. Mass Spectrom.* **1979**, *14*, 222-226.

40. Brinded, K.A.; Tiller, P.R.; Lane, S.J. Triton X-100 As a Reference Compound for Ammonia High-Resolution CI-MS and as a Tuning and Calibration Compound for Thermospray. *Rapid Commun. Mass Spectrom.* **1993**, *7*, 1059-1061.

41. Wu, H.-F.; Lin, Y.-P. Determination of the Sensitivity of an External Source Ion Trap Tandem Mass Spectrometer Using Dimethyl Ether Chemical Ionization. *J. Mass Spectrom.* **1999**, *34*, 1283-1285.

42. Barry, R.; Munson, B. Selective Reagents in CI-MS: Diisopropyl Ether. *Anal. Chem.* **1987**, *59*, 466-471.

43. Allgood, C.; Lin, Y.; Ma, Y.C.; Munson, B. Benzene as a Selective Chemical Ionization Reagent Gas. *Org. Mass Spectrom.* **1990**, *25*, 497-502.

44. Srinivas, R.; Vairamani, M.; Mathews, C.K. Gase-Phase Halo Alkylation of C_{60}-Fullerene by Ion-Molecule Reaction Under Chemical Ionization. *J. Am. Soc. Mass Spectrom.* **1993**, *4*, 894-897.

45. Fordham, P.J.; Chamot-Rooke, J.; Guidice, E.; Tortajada, J.; Morizur, J.-P. Analysis of Alkenes by Copper Ion Chemical Ionization Gas Chromatography/Mass Spectrometry and Gas Chromatography/Tandem Mass Spectrometry. *J. Mass Spectrom.* **1999**, *34*, 1007-1017.

46. Peake, D.A.; Gross, M.L. Iron(I) Chemical Ionization and Tandem Mass Spectrometry for Locating Double Bonds. *Anal. Chem.* **1985**, *57*, 115-120.

47. Budzikiewicz, H.; Blech, S.; Schneider, B. Studies in Chemical Ionization. XXVI. Investigation of Aliphatic Dienes by Chemical Ionization With Nitric Oxide. *Org. Mass Spectrom.* **1991**, *26*, 1057-1060.

48. Schneider, B.; Budzikiewicz, H. A Facile Method for the Localization of a Double Bond in Aliphatic Compounds. *Rapid Commun. Mass Spectrom.* **1990**, *4*, 550-551.

49. Hsu, C.S.; Qian, K. Carbon Disulfide CE-MS as a Low-Energy Ionization Technique for Hydrocarbon Characterization. *Anal. Chem.* **1993**, *65*, 767-771.

50. Einolf, N.; Munson, B. High-Pressure CE-MS. *Int. J. Mass Spectrom. Ion Phys.* **1972**, *9*, 141-160.

51. Sieck, L.W. Determination of Molecular Weight Distribution of Aromatic Components in Petroleum Products by CI-MS with Chlorobenzene As Reagent Gas. *Anal. Chem.* **1983**, *55*, 38-41.

52. Allgood, C.; Ma, Y.C.; Munson, B. Quantitation Using Benzene in Gas Chromatography/CI-MS. *Anal. Chem.* **1991**, *63*, 721-725.

53. Subba Rao, S.C.; Fenselau, C. Evaluation of Benzene As a Charge Exchange Reagent. *Anal. Chem.* **1978**, *50*, 511-515.

54. Li, Y.H.; Herman, J.A.; Harrison, A.G. CE Mass Spectra of Some C_5H_{10} Isomers. *Can. J. Chem.* **1981**, *59*, 1753-1759.

55. Abbatt, J.A.; Harrison, A.G. Low-Energy Mass Spectra of Some Aliphatic Ketones. *Org. Mass Spectrom.* **1986**, *21*, 557-563.

56. Herman, J.A.; Li, Y.-H.; Harrison, A.G. Energy Dependence of the Fragmentation of Some Isomeric $C_6H_{12}^{+\cdot}$ Ions. *Org. Mass Spectrom.* **1982**, *17*, 143-150.

57. Chai, R.; Harrison, A.G. Location of Double Bonds by CI-MS. *Anal. Chem.* **1981**, *53*, 34-37.

58. Keough, T.; Mihelich, E.D.; Eickhoff, D.J. Differentiation of Monoepoxide Isomers of Polyunsaturated Fatty Acids and Fatty Acid Esters by Low-Energy CE-MS. *Anal. Chem.* **1984**, *56*, 1849-1852.

59. Polley, C.W., Jr.; Munson, B. Nitrous Oxide As Reagent Gas for Positive Ion Chemical Ionization Mass Spectrometry. *Anal. Chem.* **1983**, *55*, 754-757.

60. Harrison, A.G.; Lin, M.S. Stereochemical Applications of Mass Spectrometry. 3. Energy Dependence of the Fragmentation of Stereoisomeric Methylcyclohexanols. *Org. Mass Spectrom.* **1984**, *19*, 67-71.

61. Hsu, C.S.; Cooks, R.G. CE-MS at High Energy. *Org. Mass Spectrom.* **1976**, *11*, 975-983.

62. Roussis, S. Exhaustive Determination of Hydrocarbon Compound Type Distributions by High Resolution Mass Spectrometry. *Rapid Commun. Mass Spectrom.* **1999**, *13*, 1031-1051.

63. Hunt, D.F.; Stafford, G.C., Jr.; Crow, F.W. Pulsed Positive- and Negative-Ion CI-MS. *Anal. Chem.* **1976**, *48*, 2098-2104.

64. Dougherty, R.C.; Weisenberger, C.R. Negative Ion Mass Spectra of Benzene, Naphthalene, and Anthracene. A New Technique for Obtaining Relatively Intense and Reproducible Negative Ion Mass Spectra. *J. Am. Chem. Soc.* **1968**, *90*, 6570-6571.

65. Dillard, J.G. Negative Ion Mass Spectrometry. *Chem. Rev.* **1973**, *73*, 589-644.

66. Bouma, W.J.; Jennings, K.R. Negative CI-MS of Explosives. *Org. Mass Spectrom.* **1981**, *16*, 330-335.

67. Budzikiewicz, H. Studies in Negative Ion Mass Spectrometry. XI. Negative Chemical Ionization (NCI) of Organic Compounds. *Mass Spectrom. Rev.* **1986**, *5*, 345-380.

68. Hunt, D.F.; Crow, F.W. Electron Capture Negative-Ion CI-MS. *Anal. Chem.* **1978**, *50*, 1781-1784.

69. Ong, V.S.; Hites, R.A. Electron Capture Mass Spectrometry of Organic Environmental Contaminants. *Mass Spectrom. Rev.* **1994**, *13*, 259-283.

70. Oehme, M. Quantification of fg-pg Amounts by Electron Capture Negative Ion Mass Spectrometry - Parameter Optimization and Practical Advice. *Fresenius J. Anal. Chem.* **1994**, *350*, 544-554.

71. Bartels, M.J. Quantitation of the Tetrachloroethylene Metabolite *N*-Acetyl-*S*-(trichlorovinyl)cysteine in Rat Urine Via Negative Ion Chemical Ionization Gas Chromatography/Tandem Mass Spectrometry. *Biol. Mass Spectrom.* **1994**, *23*, 689-694.

72. Fowler, B. The Determination of Toxaphene in Environmental Samples by Negative Ion Electron Capture (EC) HR-MS. *Chemosphere* **2000**, *41*, 487-492.

73. von Ardenne, M.; Steinfelder, K.; Tümmler, R. *Elektronenanlagerungs-Massenspektrographie Organischer Substanzen;* Springer-Verlag: Heidelberg, 1971.

74. Laramée, J.A.; Arbogast, B.C.; Deinzer, M.L. EC Negative Ion CI-MS of 1,2,3,4-Tetrachlorodibenzo-*p*-dioxin. *Anal. Chem.* **1986**, *58*, 2907-2912.

75. Budzikiewicz, H. Mass Spectrometry of Negative Ions. 3. Mass Spectrometry of

Negative Organic Ions. *Angew. Chem.* **1981**, *93*, 635-649.

76. Bowie, J.H. The Formation and Fragmentation of Negative Ions Derived From Organic Molecules. *Mass Spectrom. Rev.* **1984**, *3*, 161-207.

77. Stemmler, E.A.; Hites, R.A. The Fragmentation of Negative Ions Generated by EC Negative Ion Mass Spectrometry: a Review With New Data. *Biomed. Environ. Mass Spectrom.* **1988**, *17*, 311-328.

78. Giese, R.W. Detection of DNA Adducts by Electron Capture Mass Spectrometry. *Chem. Res. Toxicol.* **1997**, *10*, 255-270.

79. Laramée, J.A.; Mazurkiewicz, P.; Berkout, V.; Deinzer, M.L. Electron Monochromator-Mass Spectrometer for Negative Ion Analysis of Electronegative Compounds. *Mass Spectrom. Rev.* **1996**, *15*, 15-42.

80. NIST, NIST Chemistry Webbook. *http://webbook. nist. gov/* **2002**.

81. Williamson, D.H.; Knighton, W.B.; Grimsrud, E.P. Effect of Buffer Gas Alterations on the Thermal Electron Attachment and Detachment Reactions of Azulene by Pulsed High Pressure Mass Spectrometry. *Int. J. Mass Spectrom.* **2000**, *195/196*, 481-489.

82. Carette, M.; Zerega, Y.; Perrier, P.; Andre, J.; March, R.E. Rydberg EC-MS of 1,2,3,4-Tetrachlorodibenzo-*p*-dioxin. *Eur. Mass Spectrom.* **2000**, *6*, 405-408.

83. Wei, J.; Liu, S.; Fedoreyev, S.A.; Vionov, V.G. A Study of Resonance Electron Capture Ionization on a Quadrupole Tandem Mass Spectrometer. *Rapid Commun. Mass Spectrom.* **2000**, *14*, 1689-1694.

84. Zerega, Y.; Carette, M.; Perrier, P.; Andre, J. Rydberg EC-MS of Organic Pollutants. *Organohal. Comp.* **2002**, *55*, 151-154.

85. Yinon, J. Mass Spectrometry of Explosives: Nitro Compounds, Nitrate Esters, and Nitramines. *Mass Spectrom. Rev.* **1982**, *1*, 257-307.

86. Cappiello, A.; Famiglini, G.; Lombardozzi, A.; Massari, A.; Vadalà, G.G. EC Ionization of Explosives With a Microflow Rate Particle Beam Interface. *J. Am. Soc. Mass Spectrom.* **1996**, *7*, 753-758.

87. Knighton, W.B.; Grimsrud, E.P. High-Pressure EC-MS. *Mass Spectrom. Rev.* **1995**, *14*, 327-343.

88. Aubert, C.; Rontani, J.-F. Perfluoroalkyl Ketones: Novel Derivatization Products for the Sensitive Determination of Fatty Acids by Gas GC-MS in EI and NICI Modes. *Rapid Commun. Mass Spectrom.* **2000**, *14*, 960-966.

89. Cotter, R.J. Mass Spectrometry of Nonvolatile Compounds by Desorption From Extended Probes. *Anal. Chem.* **1980**, *52*, 1589A-1602A.

90. Kurlansik, L.; Williams, T.J.; Strong, J.M.; Anderson, L.W.; Campana, J.E. DCI-MS of Synthetic Porphyrins. *Biomed. Mass Spectrom.* **1984**, *11*, 475-481.

91. Helleur, R.J.; Thibault, P. Optimization of Pyrolysis-Desorption CI-MS and Tandem Mass Spectrometry of Polysaccharides. *Can. J. Chem.* **1994**, *72*, 345-351.

92. Vincenti, M. The Renaissance of DCI-MS: Characterization of Large Involatile Molecules and Nonpolar Polymers. *Int. J. Mass Spectrom.* **2001**, *212*, 505-518.

93. Beuhler, R.J.; Flanigan, E.; Greene, L.J.; Friedman, L. Proton Transfer MS of Peptides. Rapid Heating Technique for Underivatized Peptides Containing Arginine. *J. Am. Chem. Soc.* **1974**, *96*, 3990-3999.

94. Cotter, R.J. Laser Desorption CI-MS. *Anal. Chem.* **1980**, *52*, 1767-1770.

95. Hunt, D.F.; Shabanowitz, J.; Botz, F.K. CI-MS of Salts and Thermally Labile Organics With Field Desorption Emitters as Solids Probes. *Anal. Chem.* **1977**, *49*, 1160-1163.

96. Cullen, W.R.; Eigendorf, G.K.; Pergantis, S.A. DCI-MS of Arsenic Compounds Present in the Marine and Terrestrial Environment. *Rapid Commun. Mass Spectrom.* **1993**, *7*, 33-36.

97. Abate, R.; Garozzo, D.; Rapisardi, R.; Ballistreri, A.; Montaudo, G. Sequence Distribution of β-Hydroxyalkanoate Units in Bacterial Copolyesters Determined by-DCI-MS. *Rapid Commun. Mass Spectrom.* **1992**, *6*, 702-706.

98. Juo, C.G.; Chen, S.W.; Her, G.R. Mass Spectrometric Analysis of Additives in Polymer Extracts by DCI and CID with B/E Linked Scanning. *Anal. Chim. Acta* **1995**, *311*, 153-164.

99. Chen, G.; Cooks, R.G.; Jha, S.K.; Green, M.M. Microstructure of Alkoxy and Alkyl Substituted Isocyanate Copolymers Determined by DCI-MS. *Anal. Chim. Acta* **1997**, *356*, 149-154.

100. Pergantis, S.A.; Emond, C.A.; Madilao, L.L.; Eigendorf, G.K. Accurate Mass Measurements of Positive Ions in the DCI Mode. *Org. Mass Spectrom.* **1994**, *29*, 439-444.

8 Field Ionization and Field Desorption

The first observation of the desorption of positive ions from surfaces by high electrostatic fields was made by means of a field ion microscope. [1,2] The mass spectrometric analysis of some field-ionized gases followed. [2-4] However, it was Beckey putting forth the development of a focusing field ionization ion source. [5] In these early experiments electric field strengths of about 10^8 V cm^{-1} (1 VÅ^{-1}) were generated at sharp tungsten tips. [2,4,5] The method of *field ioniza-tion* (FI) was soon extended to analyze volatile liquids [6-11] and solids intro-duced by evaporation from a sample vial in close proximity to the ionizing tip or wire electrode. [12]. FI, still immature in the mid-1960s, had soon to compete with chemical ionization (CI, Chap. 7). [13] The major breakthrough came from its further development to *field desorption* (FD), because FD circumvents the evapo-ration of the analyte prior to ionization. [14,15] Instead, the processes of ioniza-tion *and* subsequent desorption of the ions formed are united on the surface of the *field emitter*. The specific charm of FI-MS and of FD-MS in particular arises from their extraordinary softness of ionization yielding solely intact molecular ions in most cases, and from the capability of FD to handle neutral as well as ionic ana-lytes. [16-25] FD-MS had its flourishing period from the mid-1970s to the mid-1980s and suffered from the advent of fast-atom bombardment (FAB) and later electrospray ionization (ESI) mass spectrometry. [21] However, it appears that the unique capabilities of FD-MS are currently being re-discovered. [26-29]

8.1 Field Ionization Process

Inghram and Gomer explained the process of *field ionization* for a hydrogen atom. [3,4,30] If a hydrogen atom resides on a metal surface, e.g., tungsten, its proton electron potential is only slightly distorted. However, in the presence of a strong electric field, e.g., 2 VÅ^{-1} with the metal at positive polarity, this distortion be-comes remarkable (Fig. 8.1). As a result, the electron can leave the proton by tun-neling into the bulk metal through a potential barrier that is only a few angstroms wide and some electronvolts high. [18,31] Thereby, the former hydrogen atom is ionized, and the resulting proton is immediately driven away by action of the strong electric field. Interestingly, the situation is quite similar for an isolated hy-drogen atom. Here, the electric field causes analogous distortion of the potential, and with sufficient field strength, the atom is field ionized, too. This means that atoms or molecules can be ionized by action of a strong electric field independent of whether they have been adsorbed to the anodic surface or whether they are moving freely through the space between the electrodes.

Field ionization essentially is a process of the autoionization type, i.e., an internally supra-excited atom or molecular moiety loses an electron spontaneously without further interaction with an energy source. [32] Different from electron ionization, there is no excess energy transferred onto the incipient ion, and thus, dissociation of the ions is reduced to minimum.

$$M \rightarrow M^{+\cdot} + e^-$$ (8.1)

In practice, electric fields sufficient to effect field ionization are only obtained in close proximity to sharp tips, edges, or thin wires. The smaller the radius of the curvature of the anode, the further away (1–10 nm) the field suffices to cause ionization. The importance of sufficient electric field strength is reflected by the half-life calculated for a hydrogen atom: it is in the order of 10^{-1} s at 0.5 VÅ^{-1}, 10^{-10} s at 1.0 VÅ^{-1}, and 10^{-16} s at 2.5 VÅ^{-1}. [18]

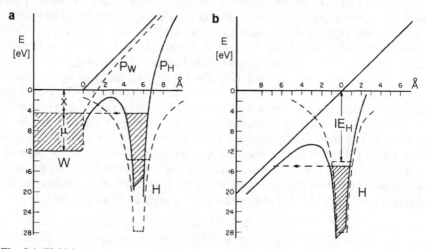

Fig. 8.1. Field ionization of a hydrogen atom (H). (**a**) close to a tungsten surface (W), (**b**) isolated. Conditions and symbols: electric field 2 V Å^{-1}, P$_W$ image potential of W distorted by the field, P$_H$ potential of the hydrogen atom distorted by the field, X work function, μ Fermi level. Broken lines represent potentials in absence of the electric field. Adapted from Ref. [4] by permission. © Verlag der Zeitschrift für Naturforschung, 1955.

Note: The field anode is usually referred to as *field emitter, FI emitter,* or *FD emitter*. The properties of the field emitter are of key importance for FI- and FD-MS. The electrode on the opposite side is called *field cathode* or simply *counter electrode*.

8.2 FI and FD Ion Source

In FI- and FD-MS, a high voltage of 8–12 kV is normally applied between *field emitter* (field anode) and *counter electrode* (field cathode) usually located 2–3 mm in front of the emitter. Consequently, the desorbing ions are accelerated to 8–12 keV kinetic energy. This usually exceeds the tolerance even of the double-focusing magnetic sector mass analyzers which are typically used in conjunction with FI/FD ion sources. Thus, the counter electrode is set to negative potential to establish the high voltage for ion generation while the difference between emitter potential and ground potential at the entrance slit defines the lower acceleration voltage of 5–10 kV (Fig. 8.2). [33]

In FI mode, the analyte is introduced via external inlet systems, i.e., from a direct probe, a reservoir inlet or a gas chromatograph (Chap. 5.3). In FD mode, the

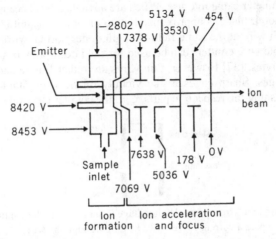

Fig. 8.2. FI/FD ion source with potentials applied to emitter and lenses. Reproduced from Ref. [33] by permission. © American Association for the Advancement of Science, 1979.

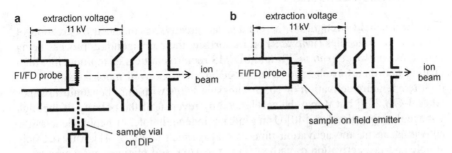

Fig. 8.3. Schematic of an FI/FD ion source (**a**) in FI mode, (**b**) in FD mode. The distance between emitter and counter electrode is shown exaggerated for clarity. Adapted from Ref. [34] by permission. © Springer-Verlag, Heidelberg, 1991.

analyte is supplied directly on the surface of the field emitter. This does not only guarantee a more effective usage of the sample, it also combines the steps of ionization and desorption, thereby minimizing the risk of thermal decomposition prior to ionization (Fig. 8.3). This is especially important in case of highly polar or ionic analytes that cannot be evaporated without thermal degradation.

8.3 Field Emitters

8.3.1 Blank Metal Wires as Emitters

In the first FI experiments, the high electric field strength was obtained at sharp tungsten tips. [2,4,5] Later, edges of sharp blades [35] and wires of a few micrometers in diameter came into use. Wires are advantageous because the emitting surface of a smooth blade is approximately two orders of magnitude smaller than that of a smooth wire at the same field strength under normal working conditions (Fig. 8.4). Simple wire emitters can be used for field desorption of unpolar [36] or electrolytic analytes. [37] However, thin wires are rather fragile and break during electric discharges. Sharp edges in the vicinity of the wire should therefore be avoided by polishing the respective parts, e.g., the counter electrode. [38]

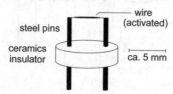

Fig. 8.4. Schematic of a wire emitter. High voltage is supplied via the emitter-holding pins. These also serve to pass a current for resistive heating through the wire.

8.3.2 Activated Emitters

The electric field strength on the emitter can greatly be enhanced by creation of dendritic microneedles (*whiskers*) on its surface, the corresponding process being known as *emitter activation*. FI- and FD-MS require emitters of reproducible high quality, and therefore the activation procedure has received much attention. The high-temperature activation of 10 µm tungsten wires with pure benzonitrile vapor takes 3–7 h, [39] but it may be accelerated by reversal of the polarity of the high voltage during activation. [40] Using indane, indene, indole, or naphthalene as the activating agent, the activation times are also much shorter. [41] An extremely simple and fast activation of carbon fiber, tungsten, and platinum wire emitters is performed in the presence of benzene vapor by heating to 600–900 °C in a high external field for a few seconds. [36,42] The activation procedures with pure benzonitrile and with indene are employed commercially to produce carbon whiskers on emitters with customized properties (Fig. 8.5). Microneedle growth is alternatively achieved by decomposition of hexacarbonyltungsten, $W(CO)_6$, on a cathode producing an electric discharge. [43] A self-controlling mechanism which draws ions preferably to the top of the growing tungsten needles has been suggested. [43] SEM pictures of activated emitter surfaces and single whiskers exhibit an impressing beauty of their own. [40,41,43-46]

Alternatives to activated tungsten wire emitters are also known, but less widespread in use. Cobalt and nickel [44,47] as well as silver [48] can be electrochemically deposited on wires to produce activated FD emitters. Mechanically strong and efficient emitters can be made by growing fine silicon whiskers from silane gas on gold-coated tungsten or tantalum wires of 60 μm diameter. [45] Finally, on the fracture-surface of graphite rods fine microcrystallites are exposed, the sharpness of which provides field strengths sufficient for ionization. [49]

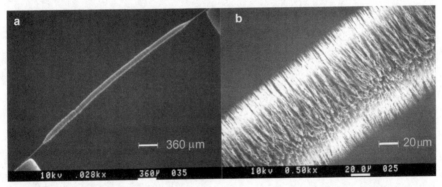

Fig. 8.5. SEM pictures of activated tungsten wire emitters; (**a**) overview showing the thin tungsten wire close to the holders and the whisker-bearing middle section, (**b**) detail of the middle with the whiskers resolved. By courtesy of Carbotec, Bergisch Gladbach, Germany.

8.3.3 Emitter Temperature

Emitters can be heated by passing through an electric current supplied via the emitter holders. It is not trivial to establish a precise calibration of emitter temperature versus *emitter heating current* (EHC). [50,51] The actual temperature not only depends on the emitter material, but also on diameter and length of the emitter, and on length and area density of the whiskers. A useful estimate for tungsten emitters with carbon whiskers is given below (Fig. 8.6). For the purpose of cleaning, baking is usually performed at 50–60 mA in case of activated tungsten emitters of 10 μm diameter and at 80–100 mA for otherwise identical emitters of 13 μm diameter.

In practice, moderate heating of the emitter at constant current serves to reduce adsorption to its surface during FI measurements. Heating at a constant rate (1–8 mA min^{-1}) is frequently employed to enforce desorption of analytes from the emitter in FD-MS. To avoid electric discharges resulting from too massive ion desorption, emission-controlled emitter heating can be advantageous. [52-54] Furthermore, baking of the emitter (2–5 s to 800–1000 °C) is used to clean it for subsequent measurements.

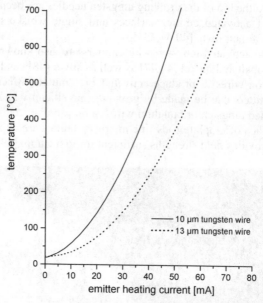

Fig. 8.6. Calibration of EHC vs. temperature for emitters activated with carbon needles on 10 μm and 13 μm tungsten wire. By courtesy of Carbotec, Bergisch Gladbach, Germany.

8.3.4 Handling of Activated Emitters

Activated wire emitters are extremely fragile, because after activation the material rather behaves like ceramics than like a metal wire. The slightest touch with a syringe needle as well as electric discharges during operation cause immediate destruction of the emissive wire. Proper handling of the emitter is therefore a prerequisite; [21] it includes i) manipulation by grasping it at the robust ceramics socket or at both pins simultaneously using tweezers, ii) baking it before first use in order to outgas and clean the emitter, iii) application of analytes by transfer of a drop of solution whereby only the liquid surface may contact the emitter wire (Fig. 8.7), iv) evaporation of the solvent usually before insertion into the vacuum lock (Fig. 8.8), v) switching on the high voltage only after the high vacuum has fully recovered, and finally after the measurement has been completed vi) baking to remove residual sample only after switching off the high voltage. Under conditions of proper handling, an emitter can last for about 20 measurements.

Note: To apply a solution of the analyte, the emitter can either be *dipped into*, or alternatively, a drop of 1–2 µl can be *transferred onto* the emitter by means of a microliter syringe. [46] The latter method exhibits better reproducibility and avoids contamination of the emitter holders. Special micromanipulators are available to handle the syringe, [15] but a skilled operator with some exercise can accomplish it manually.

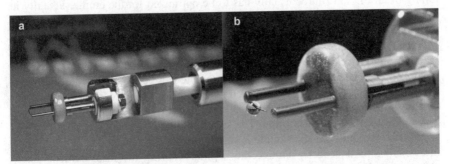

Fig. 8.7. FD probe. (**a**) Emitter holder of a JEOL FD probe tip, (**b**) a drop formed of 1–2 µl analyte solution placed onto the activated emitter by means of a microliter syringe.

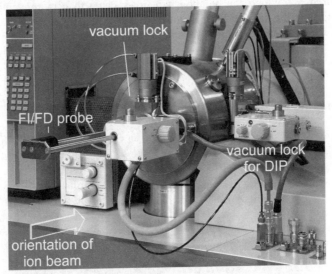

Fig. 8.8. FD probe inserted into the vacuum lock. FD probes are generally inserted in axial position to leave the vacuum lock of the DIP free for FI use. The emitter wire is now oriented vertically to comply with the beam geometry of the magnetic sector analyzer.

8.3.5 Liquid Injection Field Desorption Ionization

Numerous analytes could be good candidates for FD-MS, but undergo immediate decomposition by reacting with ambient air and/or water under the conditions of conventional emitter loading. Inert conditions such as emitter loading in a glove box does not really avoid the problem, because the emitter still needs to be mounted to the probe before insertion into the vacuum lock. Furthermore, the tuning of an FD ion source usually has to be optimized for the emitter actually in use which is not practicable with the sample already loaded onto its surface.

Liquid injection field desorption ionization (LIFDI) – initially introduced as *in-source liquid injection* (ISLI) FD – presents a major breakthrough for FD-MS of reactive analytes. [55] There is no risk of decomposition prior to starting the measurement, because the solution of the analyte can be handled under inert conditions. It is transported through a fused silica capillary merely by the sucking action of the ion source vacuum and then spreads out over the entire emitter driven by capillary forces and adsorption. The small volume of solvent (ca. 40 nl) evaporates within seconds. As the sample is supplied from the "backside" to the emitter, there neither is a need to remove the capillary during the measurement, no to change the adjustment of the emitter in the ion source (Fig. 8.9). Thus, the emitter can be used repeatedly without breaking the vacuum between successive measurements and without repeated focusing of the ion source. LIFDI also speeds up the measurements which can be about ten times faster than in conventional FD-MS. In addition, the use of LIFDI is not restricted to sensitive samples [29] and it simplifies the delicate procedure of emitter loading.

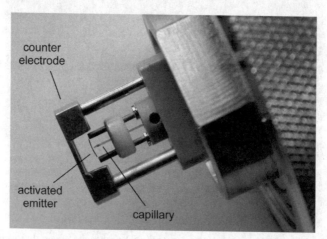

Fig. 8.9. Probe for LIFDI. A fused silica capillary delivers the sample to the "backside" of the activated emitter. Here, the counter electrode is part of the FD probe. By courtesy of Linden CMS, Leeste, Germany.

8.4 FI Spectra

FI mass spectra are normally characterized by intense molecular ion peaks accompanied by no or at least few fragment ions. [7,11,12] Especially in case of unpolar low-mass analytes, FI-MS can serve as molecular ion mass spectrometry (Fig. 8.10). [56] This property made FI-MS become a standard tool for hydrocarbon analysis in the petroleum industry. [6,9,10,13,26,56-58]

Although convenient at first sight, the lack of fragment ion peaks in FI spectra also means a lack of structural information. If more than an estimate of the elemental composition based on the isotopic pattern is desired, collision-induced dissociation (CID, Chap. 2.12.1) can deliver fragment ions for structure elucidation. Fortunately, the fragmentation pathways of $M^{+\bullet}$ ions in CID are the same as in EI-MS (Chap. 6).

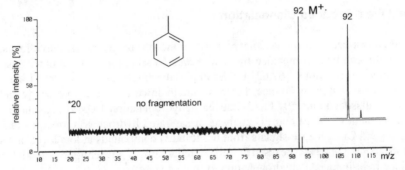

Fig. 8.10. FI spectrum of toluene. Solely the molecular ion and its isotopomer are observed. (The CID spectrum of field-ionized toluene is shown in Chap. 2.12.1, the EI spectrum is discussed in Chap. 6.4.3).

Note: In FI-MS, the ionization efficiency is very low, because of the low probability for a neutral effusing from any inlet system towards the field emitter to come close enough to the whiskers. Consequently, FI-MS produces very low ion currents. The application of FI-MS is therefore restricted to samples that are too volatile for FD-MS or require gas chromatographic separation before.

8.4.1 Origin of [M+H]⁺ Ions in FI-MS

Sometimes, FI mass spectra show signals due to reactions of the analyte with the emitter surface or between molecules adsorbed to that surface. In case of acetone for example, it was demonstrated that $[M+H]^+$ quasimolecular ions are produced mainly by a *field-induced proton-transfer* reaction in the physically adsorbed layer. [59] The mechanism of this field-induced reaction depends on the existence of tautomeric structures of the neutral molecule. Besides the $[M+H]^+$ quasimolecular ions, $[M-H]^{\bullet}$ radicals are formed:

$$M^{+\bullet} + M \rightarrow [M+H]^+ + [M-H]^\bullet \qquad (8.2)$$

Furthermore, the radicals formed upon field-induced hydrogen abstraction can lead to polymerization products on the emitter surface. The mechanism of this "field polymerization" helped to elucidate the phenomenon of activation of field emitters, i.e., the growth of microneedles on the emitter surface under the conditions of field ionization of certain polar organic compounds. [59]

Note: Analytes possessing exchangeable hydrogens also tend to form $[M+H]^+$ ions in FI-MS. Occasionally, the quasimolecular ion occurs in favor of the molecular ion that can be of lower intensity or almost be absent (Fig. 8.13). Criteria to distinguish $M^{+\bullet}$ from $[M+H]^+$ ions have been published. [60]

8.4.2 Field-Induced Dissociation

In certain cases, the advantageous property of the strong electric field to effect soft ionization can be accompanied by *field-induced dissociation*, i.e., another type of field-induced reactions. [61,62] For example, the fragment ions in the FI spectra of low-mass aliphatic amines, ketones and hydrocarbons are formed by field-induced dissociation. [59] Field-induced dehydrogenation [63,64] presents a severe problem to the FI and FD analysis of saturated hydrocarbon mixtures reaching beyond C_{40}, because signals from unsaturated compounds at low levels are superseded by $[M-H_2]^{+\bullet}$ peaks. [65-68] A reduction of the emitter potential can reduce field-induced dehydrogenation to some degree.

8.4.3 Multiply-Charged Ions in FI-MS

Multiply charged ions of minor abundance are frequently observed in FI and FD mass spectra. Their increased abundance as compared to EI spectra can be rationalized by either of the following two-step processes: i) *Post-ionization* of gaseous $M^{+\bullet}$ ions can occur due to the probability for an $M^{+\bullet}$ ion to suffer a second or even third ionization while drifting away from the emitter surface. [69,70] Especially ions generated in locations not in line-of-sight to the counter electrode pass numerous whiskers on their first 10–100 µm of flight:

$$M \rightarrow M^{+\bullet} + e^- \qquad (8.3)$$

$$M^{+\bullet} \rightarrow M^{2+} + e^- \qquad (8.4)$$

ii) Alternatively, a surface-bound ion, $M^{+\bullet}_{(surf)}$, is formed and ionized for a second time before leaving the surface: [71-73]

$$M \rightarrow M^{+\bullet}_{(surf)} + e^- \qquad (8.5)$$

$$M^{+\bullet}_{(surf)} \rightarrow M^{2+} + e^- \qquad (8.6)$$

> **Note:** For the recognition of multiply charged ions, it is important to keep in mind that the m/z scale is compressed by a factor equal to the charge state z (Chap. 3.5 and 4.2.2).

Table 8.1. Ions formed by FI

Analytes	Ions Formed
unpolar	$M^{+\cdot}$, M^{2+}, occasionally $M^{3+\cdot}$, rarely $[M+H]^+$
medium polarity	$M^{+\cdot}$, M^{2+} and/or $[M+H]^+$
polar	$[M+H]^+$
ionic	thermal decomposition

8.5 FD Spectra

Although electrospray ionization and matrix-assisted laser desorption ionization allow to transfer much larger ions into the gas phase, it is FD that can be regarded the softest ionization method in mass spectrometry. [27,74] This is mainly because the ionization process itself puts no excess energy into the incipient ions. Problems normally arise above 3000 u molecular weight when significant heating of the emitter causes thermal decomposition of the sample.

Example: The extraordinary stable trityl ion, Ph_3C^+, m/z 243, tends to dominate mass spectra (Chap. 6.6.2). Thus, neither the EI spectrum of chlorotriphenyl-methane nor that of its impurity triphenylmethanol show molecular ions (Fig. 8.11). An isobutane PICI spectrum also shows the trityl ion almost exclusively, although some hint is obtained from the Ph_2COH^+ ion, m/z 183, that cannot be explained as a fragment of a chlorotriphenylmethane ion. Only FD reveals the presence of the alcohol by its molecular ion at m/z 260 while that of the chloride is detected at m/z 278. Both molecular ions undergo some $OH^\cdot$ or $Cl^\cdot$ loss, respectively, to yield the Ph_3C^+ fragment ion of minor intensity.

8.5.1 Ion Formation in FD-MS

8.5.1.1 Field Ionization

In *field ionization* (as an experimental configuration) *field ionization* (the process) is the major pathway of ion generation. In *field desorption* from activated emitters, the analyte may also undergo *field ionization*. Presuming that the molecules are deposited in layers on the shanks of the whiskers or between them, this requires that i) analytes of low polarity are polarized by action of the electric field, ii) become mobile upon heating, and iii) finally reach the locations of ionizing electric

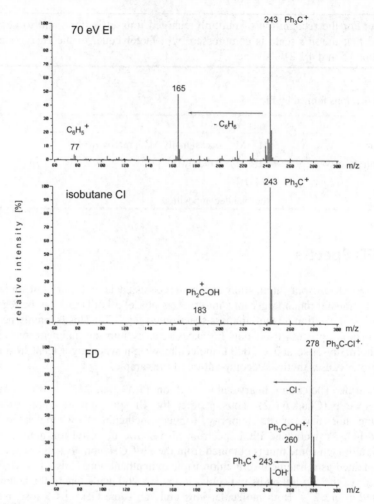

Fig. 8.11. Comparison of EI, PICI, and FD mass spectra of chlorotriphenylmethane containing some triphenylmethanol. By courtesy of C. Limberg, Humboldt University, Berlin.

field strength at the tips of the whiskers (Fig. 8.12). The requirement of mobility of the polarized molecules can either be fulfilled via gas phase transport or surface diffusion. [46] As the level of mobility to travel some ten micrometers along a whisker is much lower than that needed for evaporation from a separate inlet system, thermal decomposition of the field-ionized analyte in FD-MS is much reduced as compared to FI-MS.

The relevance of the FI mechanism of ionization decreases as the polarity of the analyte increases. In certain cases such as sucrose, for example, it is delicate to decide whether gas phase mobility of the neutral and molecular ions jointly formed by FI still play a role [75,76] or not. [77]

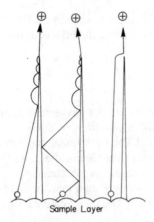

Fig. 8.12. Schematic representation of the transport of neutrals to the tips of the field-enhancing whiskers. Reproduced from Ref. [46] by permission. © Elsevier Science, 1981.

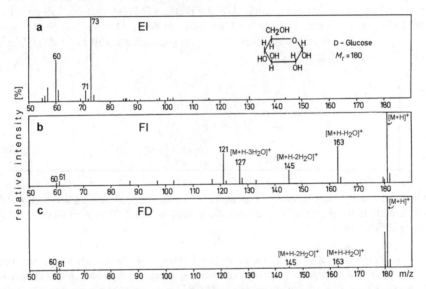

Fig. 8.13. D-Glucose mass spectra: (**a**) EI yields only ions due to decomposition and fragmentation, (**b**) FI still produces several fragments, and (**c**) FD almost exclusively gives ions related to the intact molecule. Adapted from Ref. [14] by permission. © Elsevier Science, 1969.

Example: D-Glucose may be evaporated into the ion source without complete decomposition as demonstrated by its FI spectrum (Fig. 8.13). FD yields a spectrum with a very low degree of fragmentation that is most probably caused by the need for slight heating of the emitter. The occurrence of $M^{+\bullet}$ ions, m/z 180, and $[M+H]^+$ ions, m/z 181, in the FD spectrum suggests that ion formation occurs via field ionization and field-induced proton transfer, respectively. However, thermal energy plus a hard ionization method such as EI effect extreme fragmentation. By comparison of the EI and FI spectra, the effects of thermal energy may roughly be separated from those of EI itself.

8.5.1.2 Desorption of Ions

Analytes of very high polarity are not anymore ionized by field ionization. Here, the prevailing pathways are *protonation* or *cationization*, i.e., the attachment of alkali ions to molecules. [78] The subsequent desorption of the ions from the surface is effected by the action of the electric field. As $[M+Na]^+$ and $[M+K]^+$ quasi-molecular ions are already present in the condensed phase, the field strength required for their desorption is lower than that for field ionization or field-induced $[M+H]^+$ ion formation. [37,79] The desorption of ions is also effective in case of ionic analytes.

Example: FD from untreated wire emitters in the presence of intentionally added alkali metal salts was used to obtain mass spectra of tartaric acid, arginine, pentobarbital and other compounds. [78,80] Besides $[M+H]^+$ quasimolecular ions, m/z 175, the FD mass spectrum of arginine exhibits $[M+Na]^+$, m/z 197, and $[M+K]^+$, m/z 213, ions due to alkali metal cationization as well as $[2M+H]^+$, m/z 349, cluster ions (Fig. 8.14). [37]

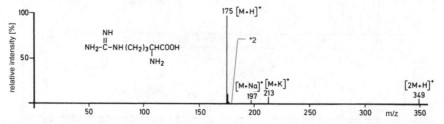

Fig. 8.14. The FD mass spectrum of arginine, $M_r = 174$, desorbed from an untreated metal wire emitter in the presence of alkali ions. Adapted from Ref. [37] by permission. © John Wiley & Sons, 1977.

Two major concepts of ion formation and desorption have been suggested, but it has remained a matter of debate whether the concept of *field-induced desolvation* [81-83] or that of *ion evaporation* [84,85] more appropriately describes the event. Although different in several aspects, the models are in agreement that ions are created in the condensed phase and subsequently desorbed into the gas phase. Both accept the electric field as the driving force to effect extraction of ionic species after charge separation in the surface layer on the emitter. Protuberances are assumed to develop from where ions can escape into the gas phase as a result of the field-enhancing effect of these protrusions. The differences of the models may in part be attributed to the different experimental and theoretical approaches. The model of the Röllgen group is much based on microscopic observation of protuberances from glassy sample layers (Fig. 8.15), [81,82] whereas the Derrick group assumes their size to be a thousand-fold smaller (Fig. 8.16). [84] Thus, the latter

emphasizes the role of mobility on the molecular scale instead of microscopic vis-
cous flow of the surface layer.

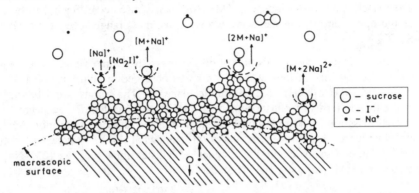

Fig. 8.15. Illustration of the desolvation of ions. (**a**) Charge separation inside a protuberance, (**b**) continuous reconstruction of the surface allows for successive rupture of intermolecular bonds and stepwise desolvation of the ions. Reproduced from Ref. [83] by permission. © Elsevier Science, 1984.

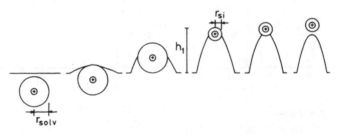

Fig. 8.16. Schematic model for ion evaporation. r_{solv} is the radius of the solvation sphere,
r_{si} the radius of the separating ion, and h_1 the height of the protuberance when the radius of
its parabolic tip equals r_{si}. Reproduced from Ref. [85] by permission. © Elsevier Science,
1987.

8.5.2 Cluster Ion Formation in FD-MS

Analytes of high polarity exhibit a strong tendency towards cationization. Besides
the aforementioned quasimolecular ions, $[nM+H]^+$ and $[nM+alkali]^+$ cluster ions
are frequently found. A priori, there is no reason why these ions should not be interpreted as resulting from additional components of a mixture. However, the sequence of events can serve as a reliable criterion to distinguish components of
higher molecular weight from cluster ions. Cluster ions are preferably formed
when the coverage of the emitter surface is still high, i.e., in the beginning of desorption. As desorption proceeds, the probability for cluster ion formation decreases, because the surface layer is diminished. In addition, a continuously rising
emitter heating current assists thermal decomposition of clusters. Whereas cluster

ions decrease in abundance, real higher-mass components do need higher emitter temperature to become mobile and ionized thereafter.

Doubly-charged cluster ions, e.g., [M+2Na]$^{2+}$, may also occur. Such doubly- or even multiply-charged ions can serve to extend the mass range accessible by FD-MS. [27]

Example: During the acquisition of an FD mass spectrum of a putative clean disaccharide, a series of ions at higher *m/z* was observed in addition to [M+H]$^{+}$, *m/z* 341 and [M+Na]$^{+}$, *m/z* 363, quasimolecular ions. Those ions could be interpreted either in terms of cluster ions or of higher oligosaccharides, respectively. The high-mass ions were only observed soon after the onset of desorption, while their abundance decreased remarkably at somewhat higher emitter current. Thereby, these signals could be assigned to cluster ions such as [2M+Na]$^{+}$, *m/z* 703, and [3M+Na]$^{+}$, *m/z* 1043 (Fig. 8.17).

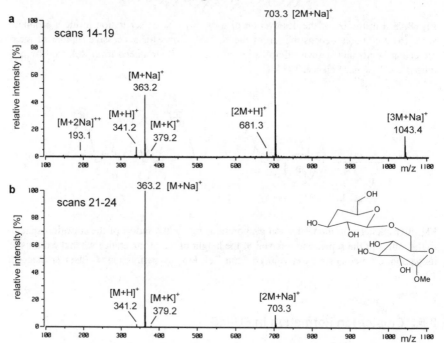

Fig. 8.17. FD mass spectra of a disaccharide (**a**) at the beginning of desorption, (**b**) towards end of desorption. [86] By courtesy of H. Friebolin, University of Heidelberg.

> **Note:** Pure analytes often show a comparatively sharp onset of desorption. Desorption then lasts for several scans until the sample is consumed. Finally, the intensity of the signals rapidly drops to zero again. In case of mixtures, some fractionation by molecular weight of the components is observed (Fig. 8.20).

8.5.3 FD-MS of Ionic Analytes

The intact cation C^+ of an ionic analyte of the general composition $[C^+A^-]$ gives rise to the base peak in positive-ion FD mass spectra. In addition, singly charged cluster ions of the $[C_nA_{n-1}]^+$ type are observed. [87,88] Their abundance and the maximum of n varies depending on the ionic species encountered as well as on the actual experimental parameters such as temperature and sample load of the emitter (Fig. 8.18). The advantage of these cluster ions is that the mass difference between the members of the series corresponds to $[C^+A^-]$. Thus, the counterion A^- can be determined by subtraction of the mass of C^+. Moreover, the isotopic pattern of the anion is reflected in the cluster ion signals, making the identification of chloride and bromide, for example, very easy. Interestingly, ionic species corresponding to some sort of "molecular ion of the salt" can also be formed. Usually, such ions are of much lower intensity than the even-electron cluster ions. Applications of FD-MS to detect cations are numerous and include organic cations [17,19,87-91] as well as inorganic ones [92,93] even down to the trace level, [94] e.g., for trace-metal analysis in physiological fluids of multiple sclerosis patients. [95]

Zwitterions are always a little bit delicate to handle in mass spectrometry. Depending on the acidity of the proton-donating site and on the basicity of the proton-accepting site either the cationic or the anionic species can be formed preferentially. Adding acids to the solution of the analyte prior to loading of the emitter can significantly enhance the signal resulting from the protonated species. [96]

As pointed out, anions can be analyzed indirectly by FD-MS via the formation of cluster ions with their positive counterion. Negative-ion FD-MS for the direct detection of the anion A^- and cluster ions of the general composition $[C_{n-1}A_n]^-$ can be performed, [97,98] but negative-ion FD-MS has remained an exception. This is due to the fact that electrons are easily emitted from activated emitters before negative ions desorb. Then, the strong emission of electrons causes the destruction of the emitter. Low emitter voltage and larger emitter–counter electrode distance help to avoid such problems. [99] Neutral analytes can give rise to $[M-H]^-$ ions or

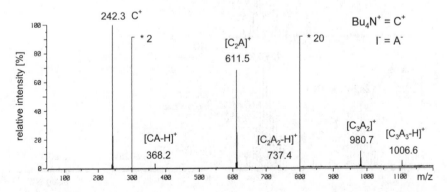

Fig. 8.18. FD mass spectrum of $[(n\text{-}C_4H_9)_4N]^+$ I^-. The intact ammonium ion is detected at m/z 242; additional signals are due to cluster ions. (For the EI spectrum cf. Chap. 6.10.4).

products of nucleophilic addition, e.g., [M+Cl]⁻ ions. [100] Nevertheless, it is recommended to apply other ionization methods such as fast atom bombardment (FAB, Chap. 9), matrix-assisted laser desorption/ionization (MALDI, Chap. 10), or electrospray ionization (ESI, Chap. 11) for the analysis of anions.

8.5.4 Best Anode Temperature and Thermal Decomposition

The onset of desorption of an analyte depends not solely on its intrinsic properties, but also on the extraction voltage and the emitter heating current applied. FD spectra are typically acquired while the *emitter heating current* (EHC) is increased at a constant rate (1–8 mA min⁻¹). Alternatively, the heating can be regulated in a emission-controlled manner. [52-54] The desorption usually begins before the competing thermal decomposition of the analytes becomes significant. Nevertheless, increasing temperature of the emitter causes increasing thermal energy to be loaded onto the desorbing ions, thereby effecting some fragmentation. The optimum temperature of the emitter where a sufficiently intense signal at the lowest level of fragmentation is obtained has been termed *best anode temperature* (BAT). [18,31]

 Example: The FD spectrum of a ruthenium-carbonyl-porphyrin complex shows an isotopic pattern very close to the theoretical distribution (Chap. 3.2.8). The loss of the carbonyl ligand chiefly results from thermal decomposition. A spectrum accumulated close to BAT (scans 19-25, EHC = 25–30 mA) is nearly free from CO loss while a spectrum accumulated of scans 30-36 (35–40 mA)

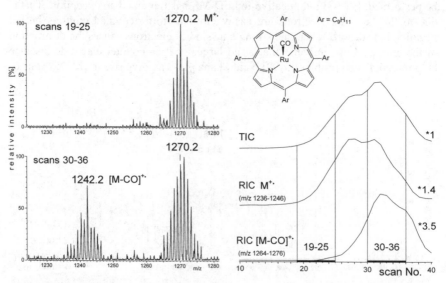

Fig. 8.19. Thermal CO loss during FD measurement of a ruthenium-carbonyl-porphyrin complex. Adapted from Ref. [101] by permission. © IM Publications, 1997.

shows significant CO loss (Fig. 8.19). This is demonstrated by comparison of the total ion chromatogram (TIC) with the reconstructed ion chromatogram (RIC) of $M^{+\bullet}$ and $[M–CO]^{+\bullet}$ (Chap. 5.4). The FD spectrum of a lower mass complex was essentially free from signals belonging to CO loss because lower emitter currents were sufficient to effect desorption. [101]

8.5.5 FD-MS of Polymers

FD-MS is perfectly suited for the analysis of synthetic oligomers and polymers. [27,28,61,65-67,102-104] In advantageous cases, polymer molecules beyond molecular weights of 10,000 u can be measured. [105] Besides the mass analyzer used, limiting factors for the mass range are thermal decomposition of polymer and presence or absence of charge-stabilizing groups. The combination of aromatic rings and low polarity in polystyrene, for example, is almost ideal for FD-MS. Heteroatoms are also useful because of their general capability to serve as proton- or metal ion-accepting sites. The worst case for mass spectrometry is presented by polyethylene (PE). [65-67] The FD mass spectra of PE oligomers can be obtained up to m/z 3500. However, starting at around C_{40} hydrocarbons, field-induced and thermally induced dehydrogenation can no longer be suppressed. Thermal decomposition of the hydrocarbon chain plays a role above 2000 u.

Example: The FD mass spectrum of polyethylene of nominal average molecular weight 1000 u (PE 1000) was obtained by summing of all scans where desorption occurred, i.e., from the 4–45 mA range of emitter heating current (Fig. 8.20). [67] The result represents the molecular weight distribution of the polymer. The experimental average molecular weight can be calculated thereof by means of Eq. 3.1 (Chap. 3.1.4). The fractionating effect of emitter heating and the significant changes in spectral appearance are demonstrated by the spectra in the lower part of Fig. 8.20 showing fractions of the total ion desorption.

8.5.6 Sensitivity of FI-MS and FD-MS

In FD mode, the sensitivity (Chap. 5.2.4) of actual magnetic sector instruments is about 4×10^{-11} C μg^{-1} for the quasimolecular ion of cholesterol, m/z 387, at $R = 1000$. This is 10^4 times less than specified for those instruments in EI mode and 10^3 times less than for CI mode.

In FI mode, the sensitivity of such instruments is about 4×10^{-9} A Pa^{-1} for the molecular ion of acetone, m/z 58, at $R = 1000$. This corresponds to an ion current of 4×10^{-13} A at a realistic ion source pressure of 10^{-4} Pa.

Although the ion currents produced by FI/FD ion sources are by orders of magnitude smaller than those from EI or CI ion sources, the detection limits are quite good. In general, about 0.1 ng of sample yield a sufficient signal-to-noise ratio (S/N ≥ 10) in FD-MS. This is because most of the ion current is collected in a single ionic species (including the isotopomers). Furthermore, the background of FI/FD ion sources is very clean, providing a good signal-to-background ratio.

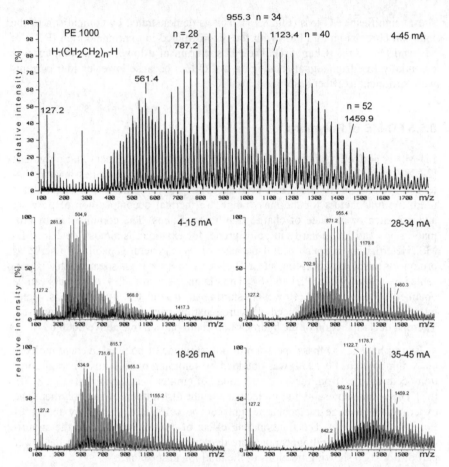

Fig. 8.20. FD spectra of PE 1000. Emitter potential 10 kV, EHC 2–50 mA. Adapted from Ref. [67] by permission. © IM Publications, 2000.

8.5.7 Types of Ions in FD-MS

At first sight, FI and FD produce a disadvantageous variety of ions depending on the polarity of the analyte and on the presence or absence of impurities such as alkali metal ions. However, with some knowledge of the ions formed, the signals can be deconvoluted without difficulty (Table 8.2).

> **Note:** One alkali adduct ion almost never occurs exclusively, i.e., $[M+H]^+$, $[M+Na]^+$ and $[M+K]^+$ (M+1, M+23 and M+39) are observed with varying relative intensities at 22 u and 16 u distance, respectively. This makes the recognition of those peaks straightforward and effectively assists the assignment of the molecular weight.

Table 8.2. Ions formed by FD

Method	Analytes	Ions Formed
FD	unpolar	$M^{+\bullet}$, M^{2+}, occasionally $M^{3+\bullet}$
FD	medium polarity	$M^{+\bullet}$, M^{2+} and/or $[M+H]^+$, $[M+alkali]^+$, occasionally $[2M]^{+\bullet}$ and/or $[2M+H]^+$, $[2M+alkali]^+$, rarely $[M+2H]^{2+}$, $[M+2\,alkali]^{2+}$
FD	polar	$[M+H]^+$, $[M+alkali]^+$, often $[2M+H]^+$, $[2M+alkali]^+$, occasionally $[nM+H]^+$, $[nM+alkali]^+$, rarely $[M+2H]^{2+}$, $[M+2\,alkali]^{2+}$
FD	ionic[a]	C^+, $[C_n+A_{n-1}]^+$, rarely $[CA]^{+\bullet}$

a Comprising of cation C^+ and anion A^-.

8.6 Analytes for FI and FD

Analytes for FI have to be evaporated prior to ionization, and thus any sample suitable for EI (Chap. 5.6) or CI (Chap. 7.6) yields low-fragmentation FI mass spectra. For FD, the analyte should be soluble to at least 0.01 mg ml^{-1} in some solvent. Concentrations of 0.1–2 mg ml^{-1} are ideal, whereas significantly higher concentrations result in overloading the emitter, and in turn cause destruction due to electric discharge. In case of extremely low solubility, repeated application of solution to the emitter is also possible. Fine suspensions or dispersions can be used if necessary. Pure water tends not to wet the emitter surface; a problem that can be circumvented by addition of some methanol before transferring the solution to the emitter. Whatever the solvent, it should be volatile enough to evaporate prior to introduction of the probe into the vacuum lock, while small volumes of dimethylformamide or dimethylsulfoxide, for example, can also be evaporated in the rough vacuum of the vacuum lock.

The analyte may be neutral or ionic. However, anions are usually detected solely in an indirect manner, i.e., from corresponding cluster ions. Solutions containing metal salts, e.g., from buffers or excess of non-complexed metals, are to be avoided, because sudden desorption of the metal ions at higher emitter current often leads to rupture of the emitter. A mass range up to 3000 u is easily covered by FD, examples reaching up to 10,000 u have been presented.

8.7 Mass Analyzers for FI and FD

In FI and FD, ions of 10–12 keV kinetic energy are generated as a continuous ion current. Compared to EI or CI the ion current is weak and exhibits a tendency to fluctuations. The voltage drop across the whiskers (roughly proportional to the whisker length) causes an energy spread of the ions which results in poor resolution with single-focusing magnetic sector instruments. [38] Therefore, double-focusing magnetic sector instruments are the standard in FD-MS. Although linear quadrupoles have been adapted to FI/FD ion sources rather successfully, [106,107] they never became widespread with FI/FD. Mass analyzers that require a pulsed supply of ions (oaTOF, QIT) are also not easily fitted to FI/FD ion sources. The ability to externally accumulate ions before injection can compensate for fluctuation to some degree. This makes FT-ICR analyzers with an external FD ion source very attractive [29] due to the unique resolving power and mass accuracy of FT-ICR instruments.

Reference List

1. Müller, E.W. Feldemission. *Ergebnisse der exakten Naturwissenschaften* **1953**, *27*, 290-360.
2. Gomer, R.; Inghram, M.G. Applications of Field Ionization to Mass Spectrometry. *J. Am. Chem. Soc.* **1955**, *77*, 500.
3. Inghram, M.G.; Gomer, R. Mass-Spectrometric Analysis of Ions From the Field Microscope. *J. Chem. Phys.* **1954**, *22*, 1279-1280.
4. Inghram, M.G.; Gomer, R. Mass-Spectrometric Investigation of the Field Emission of Positive Ions. *Z. Naturforsch.* **1955**, *10A*, 863-872.
5. Beckey, H.D. Mass Spectrographic Investigations, Using a Field Emission Ion Source. *Z. Naturforsch.* **1959**, *14A*, 712-721.
6. Beckey, H.D. Field-Ionization Mass Spectra of Organic Molecules. Normal C_1 to C_9 Paraffins. *Z. Naturforsch.* **1962**, *17A*, 1103-1111.
7. Beckey, H.D. Molecule Dissociation by High Electric Fields. *Z. Naturforsch.* **1963**, *19A*, 71-83.
8. Beckey, H.D. Mass Spectrometric Analysis with a New Source for the Production of Ion Fields at Thin Wires or Metal Edges. *Fresenius Z. Anal. Chem.* **1963**, *197*, 80-90.
9. Beckey, H.D.; Wagner, G. Analytical Use of an Ion Field Mass Spectrometer. *Fresenius Z. Anal. Chem.* **1963**, *197*, 58-80.
10. Beckey, H.D.; Schulze, P. Field Ionization Mass Spectra of Organic Molecules. III. *n*-Paraffins Up to C_{16} and Branched Paraffins. *Z. Naturforsch.* **1965**, *20A*, 1329-1335.
11. Beckey, H.D.; Wagner, G. Field Ionization Mass Spectra of Organic Molecules. II. Amines. *Z. Naturforsch.* **1965**, *20A*, 169-175.
12. Beckey, H.D. Analysis of Solid Organic Natural Products by Field Ionization Mass Spectrometry. *Fresenius Z. Anal. Chem.* **1965**, *207*, 99-104.
13. Beckey, H.D. Comparison of Field Ionization and Chemical Ionization Mass Spectra of Decane Isomers. *J. Am. Chem. Soc.* **1966**, *88*, 5333-5335.
14. Beckey, H.D. Field Desorption Mass Spectrometry: a Technique for the Study of Thermally Unstable Substances of Low Volatility. *Int. J. Mass Spectrom. Ion Phys.* **1969**, *2*, 500-503.
15. Beckey, H.D.; Heindrichs, A.; Winkler, H.U. New Field Desorption Techniques. *Int. J. Mass Spectrom. Ion Phys.* **1970**, *3*, A9-A11.
16. Beckey, H.D. *Field-Ionization Mass Spectrometry;* Pergamon: Elmsford, 1971.

17. Sammons, M.C.; Bursey, M.M.; White, C.K. Field Desorption Mass Spectrometry of Onium Salts. *Anal. Chem.* **1975**, *47*, 1165-1166.

18. Beckey, H.D. *Principles of Field Desorption and Field Ionization Mass Spectrometry;* Pergamon Press: Oxford, 1977.

19. Larsen, E.; Egsgaard, H.; Holmen, H. Field Desorption Mass Spectrometry of 1-Methylpyridinium Salts. *Org. Mass Spectrom.* **1978**, *13*, 417-424.

20. Schulten, H.-R. *Ion Formation From Organic Solids: Analytical Applications of Field Desorption Mass Spectrometry*; Springer Series in Chemical Physics [25], Benninghoven, A., editor; Springer-Verlag: Heidelberg, 1983; pp. 14-29.

21. Lattimer, R.P.; Schulten, H.-R. Field Ionization and Field Desorption Mass Spectrometry: Past, Present, and Future. *Anal. Chem.* **1989**, *61*, 1201A-1215A.

22. Beckey, H.D.; Schulten, H.-R. Field Desorption Mass Spectrometry. *Angew. Chem.* **1975**, *87*, 425-438.

23. Lehmann, W.D.; Schulten, H.-R. Physikalische Methoden in der Chemie: Massenspektrometrie II. CI-, FI- und FD-MS. *Chem. Unserer Zeit.* **1976**, *10*, 163-174.

24. Wood, G.W. Field Desorption Mass Spectrometry: Applications. *Mass Spectrom. Rev.* **1982**, *1*, 63-102.

25. Olson, K.L.; Rinehart, K.L. Field Desorption, Field Ionization, and Chemical Ionization Mass Spectrometry. *Methods Carbohyd. Chem.* **1993**, *9*, 143-164.

26. Del Rio, J.C.; Philp, R.P. Field Ionization Mass Spectrometric Study of High Molecular Weight Hydrocarbons in a Crude Oil and a Solid Bitumen. *Org. Geochem.* **1999**, *30*, 279-286.

27. Guo, X.; Fokkens, R.H.; Peeters, H.J.W.; Nibbering, N.M.M.; de Koster, C.G. Multiple Cationization of Polyethylene Glycols in Field Desorption Mass Spectrometry: a New Approach to Extend the Mass Scale on Sector Mass Spectrometers. *Rapid Commun. Mass Spectrom.* **1999**, *13*, 2223-2226.

28. Lattimer, R.P. Field Ionization (FI-MS) and Field Desorption (FD-MS), in *Mass Spectrometry of Polymers*, Montaudo, G.; Lattimer, R.P., editors; CRC Press: Boca Raton, 2001; pp. 237-268.

29. Schaub, T.M.; Hendrickson, C.L.; Qian, K.; Quinn, J.P.; Marshall, A.G. High-Resolution Field Desorption/Ionization Fourier Transform Ion Cyclotron Resonance Mass Analysis of Nonpolar Molecules. *Anal. Chem.* **2003**, *75*, 2172-2176.

30. Gomer, R. Field Emission, Field Ionization, and Field Desorption. *Surf. Sci.* **1994**, *299/300*, 129-152.

31. Prókai, L. *Field Desorption Mass Spectrometry;* Marcel Dekker: New York, 1990.

32. Todd, J.F.J. Recommendations for Nomenclature and Symbolism for Mass Spectroscopy Including an Appendix of Terms Used in Vacuum Technology. *Int. J. Mass Spectrom. Ion. Proc.* **1995**, *142*, 211-240.

33. Ligon, W.V., Jr. Molecular Analysis by Mass Spectrometry. *Science* **1979**, *205*, 151-159.

34. Schröder, E. *Massenspektrometrie - Begriffe und Definitionen;* Springer-Verlag: Heidelberg, 1991.

35. Derrick, P.J.; Robertson, A.J.B. Field Ionization Mass Spectrometry With Conditioned Razor Blades. *Int. J. Mass Spectrom. Ion Phys.* **1973**, *10*, 315-321.

36. Giessmann, U.; Heinen, H.J.; Röllgen, F.W. Field Desorption of Nonelectrolytes Using Simply Activated Wire Emitters. *Org. Mass Spectrom.* **1979**, *14*, 177-179.

37. Heinen, H.J.; Giessmann, U.; Röllgen, F.W. Field Desorption of Electrolytic Solutions Using Untreated Wire Emitters. *Org. Mass Spectrom.* **1977**, *12*, 710-715.

38. Beckey, H.D.; Krone, H.; Röllgen, F.W. Comparison of Tips, Thin Wires, and Sharp Metal Edges As Emitters for Field Ionization Mass Spectrometry. *J. Sci. Instrum.* **1968**, *1*, 118-120.

39. Beckey, H.D.; Hilt, E.; Schulten, H.-R. High Temperature Activation of Emitters for Field Ionization and Field Desorption Spectrometry. *J. Phys. E: Sci. Instrum.* **1973**, *6*, 1043-1044.

40. Linden, H.B.; Hilt, E.; Beckey, H.D. High-Rate Growth of Dendrites on Thin Wire Anodes for Field Desorption Mass Spectrometry. *J. Phys. E: Sci. Instrum.* **1978**, *11*, 1033-1036.

41. Rabrenovic, M.; Ast, T.; Kramer, V. Alternative Organic Substances for Generation of Carbon Emitters for Field Desorption Mass Spectrometry. *Int. J. Mass Spectrom. Ion Phys.* **1981**, *37*, 297-307.

42. Okuyama, F.; Hilt, E.; Röllgen, F.W.; Beckey, H.D. Enhancement of Field Ionization Efficiency of 10-μm Tungsten Filaments Caused by the Nucleation of

Carbide Particles. *J. Vac. Sci. Technol.* **1977**, *14*, 1033-1035.

43. Linden, H.B.; Beckey, H.D.; Okuyama, F. On the Mechanism of Cathodic Growth of Tungsten Needles by Decomposition of Hexacarbonyltungsten Under High-Field Conditions. *Appl. Phys.* **1980**, *22*, 83-87.

44. Bursey, M.M.; Rechsteiner, C.E.; Sammons, M.C.; Hinton, D.M.; Colpitts, T.; Tvaronas, K.M. Electrochemically Deposited Cobalt Emitters for Field Ionization and Field Desorption Mass Spectrometry. *J. Phys. E: Sci. Instrum.* **1976**, *9*, 145-147.

45. Matsuo, T.; Matsuda, H.; Katakuse, I. Silicon Emitter for Field Desorption Mass Spectrometry. *Anal. Chem.* **1979**, *51*, 69-72.

46. Okuyama, F.; Shen, G.-H. Ion Formation Mechanisms in Field-Desorption Mass Spectrometry of Semirefractory Metal Elements. *Int. J. Mass Spectrom. Ion Phys.* **1981**, *39*, 327-337.

47. Rechsteiner, C.E., Jr.; Mathis, D.E.; Bursey, M.M.; Buck, R.P. A Novel Inexpensive Device for the Electrochemical Generation of Metallic Emitters for Field Desorption. *Biomed. Mass Spectrom.* **1977**, *4*, 52-54.

48. Goldenfeld, I.V.; Veith, H.J. A New Emitter for Field-Desorption Mass Spectrometry with Silver-Electroplated Microneedles. *Int. J. Mass Spectrom. Ion Phys.* **1981**, *40*, 361-363.

49. Kosevich, M.V.; Shelkovsky, V.S. A New Type of Graphite Emitter for Field Ionization/Field Desorption Mass Spectrometry. *Rapid Commun. Mass Spectrom.* **1993**, *7*, 805-811.

50. Kümmler, D.; Schulten, H.-R. Correlation Between Emitter Heating Current and Emitter Temperature in Field Desorption Mass Spectrometry. *Org. Mass Spectrom.* **1975**, *10*, 813-816.

51. Winkler, H.U.; Linden, B. On the Determination of Desorption Temperatures in Field Desorption Mass Spectrometry. *Org. Mass Spectrom.* **1976**, *11*, 327-329.

52. Schulten, H.-R.; Lehmann, W.D. High-Resolution Field Desorption Mass Spectrometry. Part VII. Explosives and Explosive Mixtures. *Anal. Chim. Acta* **1977**, *93*, 19-31.

53. Schulten, H.-R.; Nibbering, N.M.M. An Emission-Controlled Field Desorption and Electron Impact Spectrometry Study of Some *N*-Substituted Propane and Butane

Sultams. *Biomed. Mass Spectrom.* **1977**, *4*, 55-61.

54. Schulten, H.-R. Recent Advances in Field Desorption Mass Spectrometry. *Adv. Mass Spectrom.* **1978**, *7A*, 83-97.

55. Linden, H. B. Germany DE 99-19963317, **2001**, 4 pp.

56. Severin, D. Molecular Ion Mass Spectrometry for the Analysis of High- and Nonboiling Hydrocarbon Mixtures. *Erdöl und Kohle - Erdgas - Petrochemie vereinigt mit Brennstoff-Chemie* **1976**, *29*, 13-18.

57. Mead, L. Field Ionization Mass Spectrometry of Heavy Petroleum Fractions. Waxes. *Anal. Chem.* **1968**, *40*, 743-747.

58. Scheppele, S.E.; Hsu, C.S.; Marriott, T.D.; Benson, P.A.; Detwiler, K.N.; Perreira, N.B. Field-Ionization Relative Sensitivities for the Analysis of Saturated Hydrocarbons From Fossil-Energy-Related Materials. *Int. J. Mass Spectrom. Ion Phys.* **1978**, *28*, 335-346.

59. Röllgen, F.W.; Beckey, H.D. Surface Reactions Induced by Field Ionization of Organic Molecules. *Surf. Sci.* **1970**, *23*, 69-87.

60. Schulten, H.-R.; Beckey, H.D. Criteria for Distinguishing Between $M^{+\cdot}$ and $[M+H]^+$ Ions in Field Desorption Mass Spectra. *Org. Mass Spectrom.* **1974**, *9*, 1154-1155.

61. Neumann, G.M.; Cullis, P.G.; Derrick, P.J. Mass Spectrometry of Polymers: Polypropylene Glycol. *Z. Naturforsch.* **1980**, *35A*, 1090-1097.

62. McCrae, C.E.; Derrick, P.J. The Role of the Field in Field Desorption Fragmentation of Polyethylene Glycol. *Org. Mass Spectrom.* **1983**, *18*, 321-323.

63. Heine, C.E.; Geddes, M.M. Field-Dependent $[M-2H]^{+\cdot}$ Formation in the Field Desorption Mass Spectrometric Analysis of Hydrocarbon Samples. *Org. Mass Spectrom.* **1994**, *29*, 277-284.

64. Klesper, G.; Röllgen, F.W. Field-Induced Ion Chemistry Leading to the Formation of $[M-2_nH]^{+\cdot}$ and $[2M-2_mH]^{+\cdot}$ Ions in Field Desorption Mass Spectrometry of Saturated Hydrocarbons. *J. Mass Spectrom.* **1996**, *31*, 383-388.

65. Evans, W.J.; DeCoster, D.M.; Greaves, J. Field Desorption Mass Spectrometry Studies of the Samarium-Catalyzed Polymerization of Ethylene Under Hydrogen. *Macromolecules* **1995**, *28*, 7929-7936.

66. Evans, W.J.; DeCoster, D.M.; Greaves, J. Evaluation of Field Desorption Mass

Spectrometry for the Analysis of Polyethylene. *J. Am. Soc. Mass Spectrom.* **1996**, *7*, 1070-1074.

67. Gross, J.H.; Weidner, S.M. Influence of Electric Field Strength and Emitter Temperature on Dehydrogenation and C-C Cleavage in Field Desorption Mass Spectrometry of Polyethylene Oligomers. *Eur. J. Mass Spectrom.* **2000**, *6*, 11-17.

68. Gross, J.H.; Vékey, K.; Dallos, A. Field Desorption Mass Spectrometry of Large Multiply Branched Saturated Hydrocarbons. *J. Mass Spectrom.* **2001**, *36*, 522-528.

69. Goldenfeld, I.V.; Korostyshevsky, I.Z.; Nazarenko, V.A. Multiple Field Ionization. *Int. J. Mass Spectrom. Ion Phys.* **1973**, *11*, 9-16.

70. Helal, A.I.; Zahran, N.F. Field Ionization of Indene on Tungsten. *Int. J. Mass Spectrom. Ion Proc.* **1988**, *85*, 187-193.

71. Röllgen, F.W.; Heinen, H.J. Formation of Multiply Charged Ions in a Field Ionization Mass Spectrometer. *Int. J. Mass Spectrom. Ion Phys.* **1975**, *17*, 92-95.

72. Röllgen, F.W.; Heinen, H.J. Energetics of Formation of Doubly Charged Benzene Ions by Field Ionization. *Z. Naturforsch.* **1975**, *30A*, 918-920.

73. Röllgen, F.W.; Heinen, H.J.; Levsen, K. Doubly-Charged Fragment Ions in the Field Ionization Mass Spectra of Alkylbenzenes. *Org. Mass Spectrom.* **1976**, *11*, 780-782.

74. Kane-Maguire, L.A.P.; Kanitz, R.; Sheil, M.M. Comparison of Electrospray Mass Spectrometry With Other Soft Ionization Techniques for the Characterization of Cationic π-Hydrocarbon Organometallic Complexes. *J. Organomet. Chem.* **1995**, *486*, 243-248.

75. Rogers, D.E.; Derrick, P.J. Mechanisms of Ion Formation in Field Desorption of Oligosaccharides. *Org. Mass Spectrom.* **1984**, *19*, 490-495.

76. Derrick, P.J.; Nguyen, T.-T.; Rogers, D.E.C. Concerning the Mechanism of Ion Formation in Field Desorption. *Org. Mass Spectrom.* **1985**, *11*, 690.

77. Veith, H.J.; Röllgen, F.W. On the Ionization of Oligosaccharides. *Org. Mass Spectrom.* **1985**, *11*, 689-690.

78. Schulten, H.-R.; Beckey, H.D. Field Desorption Mass Spectrometry With High-Temperature Activated Emitters. *Org. Mass Spectrom.* **1972**, *6*, 885-895.

79. Davis, S.C.; Neumann, G.M.; Derrick, P.J. Field Desorption Mass Spectrometry With Suppression of the High Field. *Anal. Chem.* **1987**, *59*, 1360-1362.

80. Veith, H.J. Alkali Ion Addition in FD Mass Spectrometry. Cationization and Protonation-Ionization Methods in the Application of Nonactivated Emitters. *Tetrahedron* **1977**, *33*, 2825-2828.

81. Giessmann, U.; Röllgen, F.W. Electrodynamic Effects in Field Desorption Mass Spectrometry. *Int. J. Mass Spectrom. Ion Phys.* **1981**, *38*, 267-279.

82. Röllgen, F. W. *Ion Formation From Organic Solids: Principles of Field Desorption Mass Spectrometry*; Springer Series in Chemical Physics [25]; Benninghoven, A., editor; Springer-Verlag: Heidelberg, 1983; pp. 2-13.

83. Wong, S.S.; Giessmann, U.; Karas, M.; Röllgen, F.W. Field Desorption of Sucrose Studied by Combined Optical Microscopy and Mass Spectrometry. *Int. J. Mass Spectrom. Ion Proc.* **1984**, *56*, 139-150.

84. Derrick, P.J. Mass Spectroscopy at High Mass. *Fresenius Z. Anal. Chem.* **1986**, *324*, 486-491.

85. Davis, S.C.; Natoli, V.; Neumann, G.M.; Derrick, P.J. A Model of Ion Evaporation Tested Through Field Desorption Experiments on Glucose Mixed With Alkali Metal Salts. *Int. J. Mass Spectrom. Ion Proc.* **1987**, *78*, 17-35.

86. Reichert, H. Monodesoxygenierte Glucosid-Synthese und Kinetikstudien der enzymatischen Spaltung. *Dissertation*, Organisch-Chemisches Institut, Universität Heidelberg, 1997.

87. Veith, H.J. Field Desorption Mass Spectrometry of Quaternary Ammonium Salts: Cluster Ion Formation. *Org. Mass Spectrom.* **1976**, *11*, 629-633.

88. Veith, H.J. Mass Spectrometry of Ammonium and Iminium Salts. *Mass Spectrom. Rev.* **1983**, *2*, 419-446.

89. Veith, H.J. Collision-Induced Fragmentations of Field-Desorbed Cations. 4. Collision-Induced Fragmentations of Alkylideneammonium Ions. *Angew. Chem.* **1980**, *92*, 548-550.

90. Fischer, M.; Veith, H.J. Collision-Induced Fragmentations of Field-Desorbed Cations. 5. Reactions of 2- and 3-Phenyl Substituted Alkylalkylidene Iminium Ions in the Gas Phase. *Helv. Chim. Acta* **1981**, *64*, 1083-1091.

91. Miermans, C.J.H.; Fokkens, R.H.; Nibbering, N.M.M. A Study of the Applicability of Various Ionization Methods and Tandem Mass Spectrometry in the Analyses of Triphenyltin Compounds. *Anal. Chim. Acta* **1997**, *340*, 5-20.

92. Röllgen, F.W.; Giessmann, U.; Heinen, H.J. Ion Formation in Field Desorption of Salts. *Z. Naturforsch.* **1976**, *31A*, 1729-1730.

93. Röllgen, F.W.; Ott, K.H. On the Formation of Cluster Ions and Molecular Ions in Field Desorption of Salts. *Int. J. Mass Spectrom. Ion Phys.* **1980**, *32*, 363-367.

94. Lehmann, W.D.; Schulten, H.-R. Determination of Alkali Elements by Field Desorption Mass Spectrometry. *Anal. Chem.* **1977**, *49*, 1744-1746.

95. Schulten, H.-R.; Bohl, B.; Bahr, U.; Mueller, R.; Palavinskas, R. Qualitative and Quantitative Trace-Metal Analysis in Physiological Fluids of Multiple Sclerosis Patients by Field Desorption Mass Spectrometer. *Int. J. Mass Spectrom. Ion Phys.* **1981**, *38*, 281-295.

96. Keough, T.; DeStefano, A.J. Acid-Enhanced Field Desorption Mass Spectrometry of Zwitterions. *Anal. Chem.* **1981**, *53*, 25-29.

97. Ott, K.H.; Röllgen, F.W.; Zwinselmann, J.J.; Fokkens, R.H.; Nibbering, N.M.M. Negative Ion Field Desorption Mass Spectra of Some Inorganic and Organic Compounds. *Org. Mass Spectrom.* **1980**, *15*, 419-422.

98. Daehling, P.; Röllgen, F.W.; Zwinselmann, J.J.; Fokkens, R.H.; Nibbering, N.M.M. Negative Ion Field Desorption Mass Spectrometry of Anionic Surfactants. *Fresenius Z. Anal. Chem.* **1982**, *312*, 335-337.

99. Mes, G.F.; Van der Greef, J.; Nibbering, N.M.M.; Ott, K.H.; Röllgen, F.W. The Formation of Negative Ions by Field Ionization. *Int. J. Mass Spectrom. Ion Phys.* **1980**, *34*, 295-301.

100. Dähling, P.; Ott, K.H.; Röllgen, F.W.; Zwinselmann, J.J.; Fokkens, R.H.; Nibbering, N.M.M. Ionization by Proton Abstraction in Negative Ion Field Desorption Mass Spectrometry. *Int. J. Mass Spectrom. Ion Phys.* **1983**, *46*, 301-304.

101. Frauenkron, M.; Berkessel, A.; Gross, J.H. Analysis of Ruthenium Carbonyl-Porphyrin Complexes: a Comparison of Matrix-Assisted Laser Desorption/Ionization Time-of-Flight, Fast-Atom Bombardment and Field Desorption Mass Spectrometry. *Eur. Mass. Spectrom.* **1997**, *3*, 427-438.

102. Lattimer, R.P.; Harmon, D.J.; Welch, K.R. Characterization of Low Molecular Weight Polymers by Liquid Chromatography and Field Desorption Mass Spectroscopy. *Anal. Chem.* **1979**, *51*, 293-296.

103. Craig, A.C.; Cullis, P.G.; Derrick, P.J. Field Desorption of Polymers: Polybutadiene. *Int. J. Mass Spectrom. Ion Phys.* **1981**, *38*, 297-304.

104. Lattimer, R.P.; Schulten, H.-R. Field Desorption of Hydrocarbon Polymers. *Int. J. Mass Spectrom. Ion Phys.* **1983**, *52*, 105-116.

105. Matsuo, T.; Matsuda, H.; Katakuse, I. Use of Field Desorption Mass Spectra of Polystyrene and Polypropylene Glycol As Mass References Up to Mass 10000. *Anal. Chem.* **1979**, *51*, 1329-1331.

106. Heinen, H.J.; Hötzel, C.; Beckey, H.D. Combination of a Field Desorption Ion Source With a Quadrupole Mass Analyzer. *Int. J. Mass Spectrom. Ion Phys.* **1974**, *13*, 55-62.

107. Gierlich, H.H.; Heinen, H.J.; Beckey, H.D. The Application of Quadrupole Mass Filters in Field Desorption Mass Spectrometry. *Biomed. Mass Spectrom.* **1975**, *2*, 31-35.

9 Fast Atom Bombardment

When particles of kiloelectronvolt kinetic energy are impinging on a surface, they cause the ejection of neutrals and secondary ions from that surface. *Secondary ion mass spectrometry* (SIMS) employing the sputtering effects of a beam of impacting ions on bulk, inorganic materials, [1-3] ^{252}Cf *plasma desorption* (PD) time-of-flight (TOF) mass spectrometry effecting desorption/ionization of biomolecules by impact of single megaelectronvolt nuclear fission fragments, [4-6] and *molecular beam solid analysis* (MBSA) using energetic neutrals [7,8] had already been known when SIMS was applied to organic solids for the first time. [9,10] However, the organic surfaces tended to cause electrostatic charging upon ion impact, thereby disturbing the ion source potentials. Employing a beam of energetic neutral atoms in analogy to the MBSA technique circumvented such problems and gave an impetus to the further development of this promising method. [11,12] The term *fast atom bombardment* (FAB) was coined [11-13] and prevailed. [7] It turned out that intact molecular or quasimolecular ions could be generated even in case of highly polar compounds that were definitely not candidates for electron ionization (EI, Chaps. 5, 6) or chemical ionization (CI, Chap. 7). Those FAB spectra still suffered from rapid radiolytic decomposition of the samples upon irradiation and from the comparatively harsh conditions of desorption/ionization. The use of a *liquid matrix* where the analyte was dissolved for analysis brought the awaited improvement. [14,15] Thus, "matrix-assisted fast atom bombardment" – as one now would probably term it – started its career in organic mass spectrometry [16,17] and soon became a powerful competitor of field desorption (Chap. 8). It turned out that the properties of the liquid matrix are of key importance for the resulting FAB spectra. [18-20] Due to some electric conductivity of the matrix, primary ions could now be successfully employed again. [21-24] If primary ions instead of neutrals are used to provide the energy for secondary ion ejection from the liquid matrix, the technique is termed *liquid secondary ion mass spectrometry* (LSIMS). Parallel to FAB and LSIMS, inorganic SIMS has tremendously developed to become a highly regarded method for surface analysis. [25-27] Although being soft, extremely versatile and still relevant ionization methods, FAB and LSIMS have widely been replaced by matrix-assisted laser desorption/ionization (MALDI, Chap. 10) and electrospray ionization (ESI, Chap. 11).

Note: Besides their momentum, the nature of the primary particles is of minor relevance for the spectral appearance [28] because little difference is observed between FAB and LSIMS spectra. Otherwise not explicitly distinguished from LSIMS, here the usage of the term FAB will also implicate LSIMS.

9.1 Ion Sources for FAB and LSIMS

9.1.1 FAB Ion Sources

A FAB ion source basically is an EI ion source (Chap. 5.2.1) modified to give free access to the fast atom beam. The electron-emitting filament and the ion source heaters are switched off during FAB operation (Fig. 9.1). The FAB gas is introduced via a needle valve into the lower part of the *FAB gun* mounted above the ion source. From there, it effuses into the ionization chamber of the FAB gun and into the ion source housing. The *saddle field gun* [29] is the most common type of FAB gun, delivering a primary particle flux of some 10^{10} s^{-1} mm^{-2}. [30,31] The gas is ionized and the ions are accelerated by a high voltage (4–8 kV) and focused onto the sample. [11,13,17,32] Neutralization of the energetic noble gas ions is effected by charge exchange with incoming neutrals (Chaps. 2.12.3, 7.3). The kinetic energy of the atoms is mostly conserved during charge exchange, and thus the neutrals hit the exposed surface with high kinetic energy. Of course, neutralization is not quantitative, but this is not an issue as long electrostatic charging of the sample is avoided. [28] Ion guns for the generation of energetic noble gas ions can therefore be employed without disadvantage. [33] Xenon is preferred over argon and neon as FAB gas, [34,35] because it transfers a higher momentum when impacting onto the surface at equal kinetic energy (Fig. 9.2). Consequently, even mercury has been tried as FAB gas. [36]

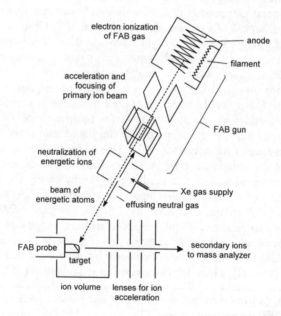

Fig. 9.1. Schematic of a FAB ion source and a FAB gun.

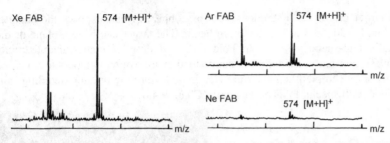

Fig. 9.2. Comparison of the efficiency of xenon, argon and neon FAB gas for the FAB mass spectrum of Met-enkephalin, a small peptide. The intensities are in scale. Reproduced from Ref. [34] by permission. © Elsevier Science, 1983.

> **Note:** Effusing FAB gas and evaporating matrix are an additional load to the high vacuum pump of the ion source housing, affording sufficient pumping speed (300–500 l s^{-1}) for stable operation. Different from EI and CI ion sources, FAB ion sources are operated without heating to reduce evaporation of the matrix and thermal stress of the analyte. Accordingly, the ion source is contaminated with matrix. Often, ion sources are constructed as EI/CI/FAB combination ion sources. After FAB measurements, it is therefore recommended to heat and pump the ion source over night prior to EI or CI operation.

9.1.2 LSIMS Ion Sources

As mentioned before, primary ions can also be employed to provide the energy for secondary ion emission when organic compounds are admixed to a liquid matrix. [21-23] Cs$^+$ ions are preferably used in organic *liquid secondary ion mass spectrometry* (LSIMS). The Cs$^+$ ions are produced by evaporation from a surface coated with cesium alumina silicate or other cesium salts. [21] Temperatures of about 1000 °C are necessary to generate a sufficient flow of primary ions, and thus precautions must be taken to shield the LSIMS ion source from that heat. The Cs$^+$ ions are extracted, accelerated and focused onto the target as usual by electrostatic lenses. [37-39] It is an advantage of Cs$^+$ ion guns that the beam energy can be more widely varied, e.g., in the 5–25 keV range, in order to adjust for optimized secondary ion emission. [40] Especially with high-mass analytes, Cs$^+$ ion guns generally yield superior ion emission as compared to Xe FAB. [28] In order to further increase the momentum of the primary ions, gold negative atomic ions, [41] as well as molecular and massive cluster ions (Chap. 9.5) have been used as primary ions.

9.1.3 FAB Probes

The analyte, either solid or admixed to some liquid matrix, is introduced into the FAB ion source by means of a probe bearing a sample holder or *FAB target*. The

FAB target usually is a stainless steel or copper tip that exposes the analyte at some angle (30–60°) to the fast atom beam. The target can have a plane or more specially cup-shaped surface to hold a 1–3 µl drop of matrix/analyte mixture (Fig. 9.3). Normally, the target is maintained at ion source temperature, i.e., only slightly above ambient temperature. Heating or – more important – cooling can be provided with special FAB probes only (Chap. 9.4.6).

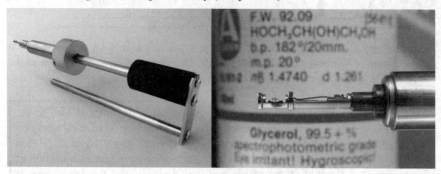

Fig. 9.3. FAB probe of a JEOL JMS-700 magnetic sector instrument (left). The probe tip with a drop of glycerol placed onto the exchangeable stainless steel FAB target (right).

9.2 Ion Formation in FAB and LSIMS

9.2.1 Ion Formation from Inorganic Samples

The energy provided by the impacting primary particle causes a collision cascade in the upper atomic or molecular layers of the sample. Within 30–60 ps, a cylindrical expansion is effected in the sample along the path of penetration. [22] Not all of this energy is dissipated and absorbed in deeper sample layers. A portion is directed towards the surface, where it effects ejection of material into the vacuum (Fig. 9.4). [24] In case of a bulk inorganic salt such as cesium iodide, Cs^+ and I^- ions are heading away from the surface. [42] Those ions having the suitable polarity are attracted by the extraction/acceleration voltage, those of opposite charge sign are pushed back onto the surface. Held together by strong inter-ionic forces, ionic clusters may desorb as such or dissociate in the selvedge region due to their internal energy content.

Example: Bombardment of inorganic solids such as cesium iodide or gold can produce series of cluster ions which are very useful for mass calibration of the instrument over a wide mass range. CsI works equally well in positive- and negative-ion mode to yield $[(CsI)_nCs]^+$ and $[(CsI)_nI]^-$ cluster ions, respectively (Fig. 9.5). Starting from $n = 0$, $[(CsI)_nCs]^+$ cluster ions have been observed up to m/z 90,000. [42] Larger $[(CsI)_nCs]^+$ cluster ions dissociate to yield smaller ones: $[(CsI)_nCs]^+ \rightarrow [(CsI)_{n-x}Cs]^+ + (CsI)_x$. [43] Gold produces a negative $Au_n^{-\bullet}$ cluster ion series up to about m/z 10,000. [44]

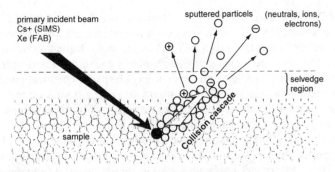

Fig. 9.4. Simple illustration of an instantaneous collision cascade generated as a result of primary particle impact in desorption/ionization mass spectrometry. Adapted from Ref. [24] by permission. © John Wiley & Sons, 1995.

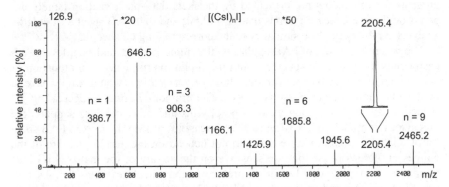

Fig. 9.5. Negative-ion FAB spectrum of solid CsI. The monoisotopic $[(CsI)_nI]^-$ cluster ion series (cf. expanded view of m/z 2205.4) is well suited for calibrating a wide mass range.

> **Note:** CsI and Au both bear the advantage of being monoisotopic. This insures the peak top to exactly represent the theoretical isotopic mass of the respective cluster ion, independent of its m/z ratio or actual resolution (Chap. 3.3.5, 3.4). CsI, KI, and other alkali salts providing more narrow-spaced cluster ion series can alternatively be employed as saturated solutions in glycerol. [45-47]

9.2.2 Ion Formation from Organic Samples

According to a paper by Todd "it is a common feature of FAB and LSIMS that they defy any generally acceptable mechanistic description." [48] Unlike EI or CI, where gaseous ions are generated from gaseous molecules, desorption/ionization techniques combine a state transition from liquid or solid to the gas phase and ionization of neutral molecules. Nonetheless, reviews dealing with the processes of desorption and ion formation under FAB and LSIMS conditions offer some insight. [22-24] Basically, there are two major concepts, the *chemical ionization model* on one side, [48-52] and the *precursor model* on the other. [52-56]

The *chemical ionization model* of FAB assumes the formation of analyte ions to occur in the selvedge region some micrometers above the liquid matrix. In this space, a plasma state similar to the reagent gas plasma in chemical ionization can exist powered by the quasi-continuous supply of energy from a stream of impacting primary particles. The reactive species present may undergo numerous bimolecular reactions, the most interesting of them being the protonation of analyte molecules to yield $[M+H]^+$ quasimolecular ions. Plasma conditions could also explain the observation of $M^{+\bullet}$ and $M^{-\bullet}$ radical ions as formed in case of low-polarity analytes. Here, the primary ion beam serves to sputter material from the liquid surface and to subsequently ionize neutrals by particle impact (cf. EI, Chap 5.1) in the gas phase. As matrix molecules are preferably ionized for statistical reasons, these may then act as reagent ions to effect CI of the gaseous analyte. Striking arguments for this model are presented by the facts that ion formation largely depends on the presence of gaseous matrix, [49,50] and that FAB spectra of volatile analytes exhibit very close similarity to the corresponding CI spectra. [52]

The *precursor model* of FAB applies well to ionic analytes and samples that are easily converted to ionic species within the liquid matrix, e.g., by protonation or deprotonation or due to cationization. Those *preformed ions* would simply have to be desorbed into the gas phase (Fig. 9.6). The promoting effect of decreasing pH (added acid) on $[M+H]^+$ ion yield of porphyrins and other analytes supports the precursor ion model. [55,56] The relative intensities of $[M+H]^+$ ions in FAB spectra of aliphatic amine mixtures also do not depend on the partial pressure of the amines in the gas phase, but are sensitive on the acidity of the matrix. [57] Furthermore, incomplete desolvation of preformed ions nicely explains the observation of matrix (Ma) adducts such as $[M+Ma+H]^+$ ions. The precursor model bears some similarities to ion evaporation in field desorption (Chap. 8.5.1).

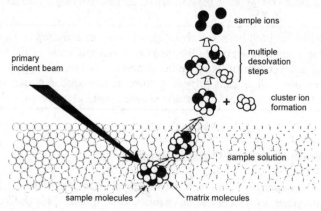

Fig. 9.6. In LSIMS and FAB, sample–matrix cluster ion formation and desolvation processes occur on a longer time scale. Adapted from Ref. [24] by permission. © John Wiley & Sons, 1995.

It has been estimated that a single impact causes the eruption of about 10^3 secondary neutrals, but yields only 0.02–1.5 ions. [46,48,57,58] The ions are then heading away from the surface in a *supersonic expansion* at speeds of about 1000 m s^{-1}. [22,46]

9.3 FAB Matrices

9.3.1 The Role of the Liquid Matrix

Soon after the first use of a glycerol matrix, the importance of a liquid matrix for FAB was recognized. [17] Other organic solvents of low volatility were explored in order to obtain better spectra. The tasks of the matrix are numerous: [22,23,35,59,60] i) It has to absorb the primary energy. ii) By solvation it helps to overcome intermolecular forces between analyte molecules or ions. iii) The liquid matrix provides a continuously refreshing and long-lasting supply of analyte. iv) Finally, it assists analyte ion formation, e.g., by yielding proton donating/accepting or electron donating/accepting species upon bombardment. Nowadays, numerous matrices are in use, with 3-nitrobenzyl alcohol (NBA), [61-63] 2-nitrophenyl octylether (NPOE), [18,61] and thioglycerol [16] being the most widespread. In particular NBA is a good choice for a first trial with an unknown sample. Others, such as di-, tri- and tetraethyleneglycols, [16,18] and triethanolamine (TEA), [18] followed. Liquid paraffin, [64-66] sulfolane, [67,68] concentrated sulfuric acid, [69] and others are also in use. [18,70,71] Several reviews on FAB matrices have appeared. [19,20,59,60,72] In summary, a good FAB matrix has to fulfill the following criteria: [22,23,35,59,60] i) The analyte must be soluble in the matrix. Otherwise, addition of co-solvents, e.g., dimethylformamide (DMF), dimethylsulfoxide (DMSO) or other additives, [73,74] can become necessary.

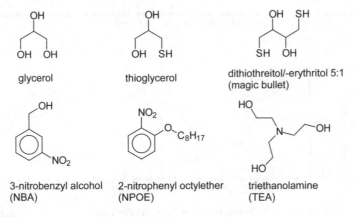

glycerol   thioglycerol   dithiothreitol/-erythritol 5:1
                                          (magic bullet)

3-nitrobenzyl alcohol   2-nitrophenyl octylether   triethanolamine
(NBA)   (NPOE)   (TEA)

Scheme 9.1.

ii) Only low-vapor pressure solvents can be easily used as a matrix in FAB. In principle, volatile solvents could be employed, provided a stable surface could be obtained on the time scale of recording a mass spectrum. iii) The viscosity of the solvent must be low enough to ensure the diffusion of the solutes to the surface. [75] iv) Ions from the matrix itself should be as unobtrusive as possible in the resulting FAB spectrum. v) The matrix itself has to be chemically inert. However, specific ion formation reactions promoting the secondary ion yield are beneficial.

> **Note:** A great advantage of FAB is that the matrix can be perfectly adapted to the analyte's requirements. On the other hand, using the wrong matrix can result in complete suppression of analytically useful signals.

9.3.2 Characteristics of FAB Matrix Spectra

FAB matrix spectra are generally characterized by a series of matrix (Ma) cluster ions accompanied by some more abundant fragment ions in the lower m/z range. In positive-ion FAB, $[\text{Ma}_n+\text{H}]^+$ cluster ions predominate, while $[\text{Ma}_n-\text{H}]^-$ cluster ions are preferably formed in negative-ion FAB (Fig. 9.7). The principal ion series

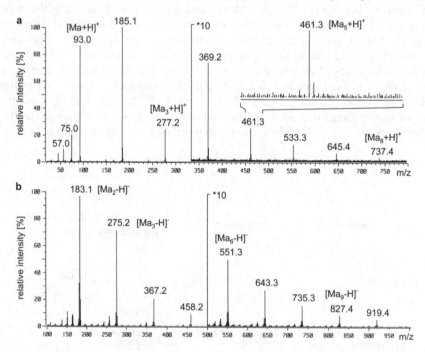

Fig. 9.7. FAB spectra of neat glycerol, $M_r = 92$ u. (**a**) Positive ions (for the positive-ion CI spectrum of glycerol cf. Chap. 7.2.5); (**b**) negative ions. The expanded view in (**a**) shows the "peak at every m/z-character" of FAB spectra.

may be accompanied by $[Ma_n+alkali]^+$ ions and some fragments of minor intensity, e.g., $[Ma_n+H-H_2O]^+$. The fragment ions detected below the $[Ma+H]^+$ ion, which normally also gives rise to the base peak, are almost the same as observed in the positive-ion CI mass spectrum of the respective matrix compound. [46]

In addition to the prominent cluster ions, radiolytic decomposition of the matrix generates an enormous number of different ions, radicals, and cluster ions resulting thereof. [76,77] Despite being of minor intensity, they contribute to the "peak at every m/z-character" of FAB spectra, i.e., there is significant *chemical noise* (Chap. 5.2.4). [78,79] During elongated measurements, the changes of the matrix spectrum due to increasing radiolytic decay are clearly visible. [76] High kinetic energy of impacting primary particles combined with reduced particle flux seem to diminish destructive effects of irradiation. [80]

9.3.3 Unwanted Reactions in FAB-MS

The conditions of the FAB process also promote unwanted reactions between analyte and matrix. Even though such processes are not relevant in the majority of FAB measurements, one should be aware of them. Besides addition or condensation reactions with matrix fragment ions, [81,82] reduction [83-86] and dehalogenation [87,88] of the analyte represent the more prominent side-reactions in FAB. Electron transfer to cause the reduction of otherwise doubly charged ions have also been observed. [47]

9.4 Applications of FAB-MS

9.4.1 FAB-MS of Analytes of Low to Medium Polarity

The FAB plasma provides conditions that allow to ionize molecules by either loss or addition of an electron to form positive molecular ions, $M^{+\bullet}$, [52,89] or negative molecular ions, $M^{-\bullet}$, respectively. Alternatively, protonation or deprotonation may result in $[M+H]^+$ or $[M-H]^-$ quasimolecular ions. Their occurrence is determined by the respective basicity or acidity of analyte and matrix. Cationization, preferably with alkali metal ions, is also frequently observed. Often, $[M+H]^+$ ions are accompanied by $[M+Na]^+$ and $[M+K]^+$ ions as already noted with FD-MS (Chap. 8.5.7). Furthermore, it is not unusual to observe $M^{+\bullet}$ and $[M+H]^+$ ions in the same FAB spectrum. [52] In case of simple aromatic amines, for example, the peak intensity ratio $M^{+\bullet}/[M+H]^+$ increases as the ionization energy of the substrate decreases, whereas 4-substituted benzophenones show preferential formation of $[M+H]^+$ ions, regardless of the nature of the substituents. [90] It can be assumed that protonation is initiated when the benzophenone carbonyl groups form hydrogen bonds with the matrix.

Exchangeable hydrogens can be replaced by alkali ions without affecting the charge state of a molecule. Thus, $[M-H_n+alkali_{n+1}]^+$ and $[M-H_n+alkali_{n-1}]^-$ ions can also be observed if one or more acidic hydrogens are easily exchanged. [10,21,47,74,91,92] The addition of cation exchange resins or crown ethers may help to reduce alkali ion contaminations. [93] Double protonation to yield $[M+2H]^{2+}$ ions or double cationization [47] are only observed with high-mass analytes [94,95] and otherwise remain an exception in FAB. Which of the above processes will most effectively contribute to the total ion yield strongly depends on the actual analyte–matrix pair.

Example: The positive-ion FAB spectrum of tetramesitylporphyrin, $C_{56}H_{54}N_4$, in NBA matrix exhibits $M^{+\cdot}$ and $[M+H]^+$ ions (Fig. 9.8). [96] The presence of both species can be recognized by comparison of experimental and calculated M+1 intensity of the isotopic pattern. This difference can only be explained by assuming about 20 % $[M+H]^+$ ion formation. The diffuse groups of signals around m/z 900 reveal the formation of some adduct ions with the matrix, e.g., $[M+Ma+H-H_2O]^+$ at m/z 918.

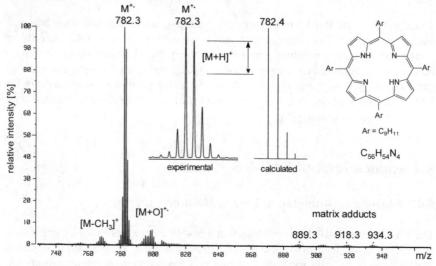

Fig. 9.8. Partial positive-ion FAB spectrum of a tetramesitylporphyrin in NBA matrix. Comparison of the experimental and calculated isotopic patterns reveals the presence of $M^{+\cdot}$ and $[M+H]^+$ ions. Adapted from Ref. [96] by permission. © IM Publications, 1997.

Example: Polyethyleneglycols (PEGs) of average molecular weights up to about 2000 u are well soluble in NBA. Resulting from their flexible polyether chain, $H(OCH_2CH_2)_nOH$, PEGs are easily cationized by loose complexation with Na^+ or K^+ ions. Traces of alkali salts are sufficient to prefer $[M+alkali]^+$ over $[M+H]^+$ ions. The positive-ion FAB spectrum of PEG 600 in NBA nicely shows the molecular weight distribution of the oligomer (Fig. 9.9). The peaks belonging to the same series are displayed at 44 u distance.

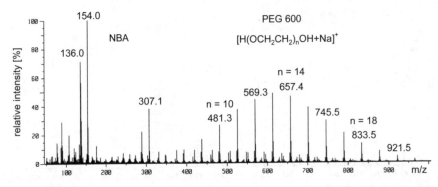

Fig. 9.9. Positive-ion FAB spectrum of polyethyleneglycol of average molecular weight 600 u (PEG 600) in NBA matrix.

Note: PEGs ranging from PEG 300 to PEG 2000 are often used for mass calibration. They are particularly useful as internal reference (Chap. 3.3.5) for accurate mass measurements in positive-ion FAB-MS.

9.4.2 FAB-MS of Ionic Analytes

FAB is well suited for the analysis of ionic analytes. In positive-ion mode, the spectrum is usually dominated by the cationic species, C^+, which is accompanied by cluster ions of the general composition $[C_n+A_{n-1}]^+$. Thus, the distance between these signals corresponds to the complete salt CA, i.e., yields its "molecular" weight. This behavior is perfectly analogous to FD (Chap. 8.5.3). In negative-ion FAB, the anion A^- will cause the base peak of the spectrum, and accordingly, cluster ions of the type $[C_{n-1}+A_n]^-$ are formed in addition. Consequently, both cation and anion are usually identified from the same FAB spectrum, irrespective of the chosen polarity. Nonetheless, it is common practice to select the polarity of the more interesting ion for the measurement.

Provided the salt is sufficiently soluble in the matrix, the signals normally exhibit high intensity as compared to those of the matrix. This result is consistent with the model of preformed ions in solution that only need to be desorbed into the gas phase.

Example: The positive-ion FAB spectrum of an immonium perchlorate, $[C_{14}H_{16}N]^+ ClO_4^-$, [97] dissolved in NBA is dominated by the immonium ion (C^+) at m/z 198 (Fig. 9.10). The perchlorate counterion can well be identified from the cluster ions $[2C+A]^+$, m/z 495, and $[3C+2A]^+$, m/z 792. The signal at m/z 495 is shown expanded to demonstrate that chlorine is readily recognized from its isotopic pattern (Chap. 3.2.3).

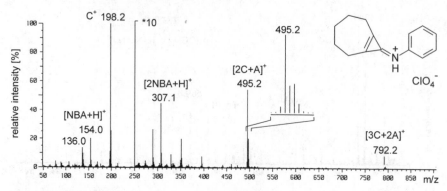

Fig. 9.10. Positive-ion FAB spectrum of an immonium salt. [97] The perchlorate counter-ion can well be identified from the first and second cluster ion. By courtesy of H. Irn-gartinger, University of Heidelberg.

Example: The Bunte salt $[CH_3(CH_2)_{15}S\text{-}SO_3]^- Na^+$ yields a very useful negative-ion FAB spectrum from NBA matrix (Fig. 9.11). NBA forms $[Ma–H]^-$ and $Ma^{-\cdot}$ ions. The salt anion contributes the base peak at m/z 337.3. $[C+2A]^-$, m/z 697.5, and $[2C+3A]^-$, m/z 1057.6, cluster ions are observed in addition, their isotopic patterns being in good agreement with theoretical expectation. It is noteworthy that the matrix adduct at m/z 513.4 is a negative radical ion.

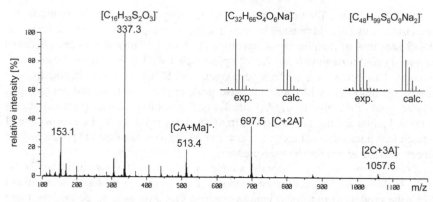

Fig. 9.11. Negative-ion FAB mass spectra of a Bunte salt. The insets compare experimental and calculated isotopic patterns of the $[C+2A]^-$ and $[2C+3A]^-$ cluster ions. By courtesy of M. Grunze, University of Heidelberg.

9.4.3 High-Mass Analytes in FAB-MS

FAB is chiefly applied to analytes up to about m/z 3000, but significantly heavier ions are sometimes accessible. The upper limit surely has been demonstrated by the detection of $[(CsI)_nCs]^+$ cluster ions up to m/z 90,000. [42] In case of organic molecules, the detection of $[M+H]^+$, $[M+2H]^{2+}$ and $[M+3H]^{3+}$ ions of porcine trypsin, a peptide of M_r = 23,463 u, presents the record. [95] FAB spectra of peptides in the m/z 3000–6000 range are more commonly reported, [16,94,98] and the FAB spectrum of a dendrimer of 7000 u has also been published. [99]

Example: Inclusion complexes of [60]fullerene, C_{60}, [100] of its oxides $C_{60}O$, $C_{60}O_2$ and $C_{60}O_3$ [101] and of several cycloaddition products of the fullerene [102] in γ-cyclodextrin (γ-CD) can be analyzed by FAB. Best results are obtained in the negative-ion mode using "magic bullet" matrix (eutectic 5:1 mixture of dithiothreitol/dithioerythritol). [100] As one fullerene molecule is enclosed between two γ-CD units, the [M–H]$^-$ ions of these host-guest complexes [103] are detected starting from [$C_{156}H_{159}O_{80}$]$^-$, monoisotopic peak at m/z 3311.8, and above (Fig. 9.12).

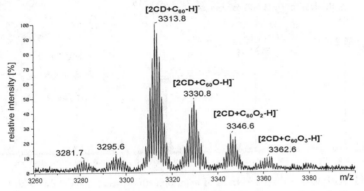

Fig. 9.12. Partial negative-ion FAB spectrum of γ–CD fullerene complexes in "magic bullet" matrix. Reproduced from Ref. [101] by permission. © IM Publications, 1998.

Note: The choice of the matrix is determined by the outer sphere of the analyte molecules, i.e., by γ-CD in case of the above inclusion complexes. [60]Fullerene and its oxides yield poor spectra in "magic bullet", but work well with less polar matrices such as NBA and NPOE.

9.4.4 Accurate Mass Measurements in FAB

9.4.4.1 Admixture of Mass Calibration Standards

FAB produces long-lasting signals of sufficient intensity, thereby allowing to set magnetic sector instruments to 5,000–10,000 resolution as needed for accurate mass measurements. In the range up to about m/z 600, internal calibration can sometimes be obtained by using the matrix peaks as mass reference. Nevertheless, the admixture of some other mass calibrant to the matrix–analyte solution is normally preferred. PEGs are frequently employed for calibration purposes. Then, mass reference peaks are evenly spaced over the m/z range of interest (44 u distant in case of PEG) and their intensity can be adjusted to approximate those of the analyte. However, the mass calibrant can only be admixed if unwanted reactions with the analyte do not occur. In particular, in case of badly soluble analytes, complete suppression of the analyte signals by the added calibrant often presents a problem.

Example: The positive-ion high-resolution (HR) FAB spectrum of a cationic fluorescent marker dye shows the signal of the analyte ion enclosed by a set of mass reference peaks due to the admixture of PEG 600 (Fig. 9.13). The elemental composition of the analyte can be assigned with good accuracy: exp. *m/z* 663.3182, calc. *m/z* 663.3158 for $[C_{37}H_{42}O_4N_4F_3]^+$; exp. *m/z* 664.3189, calc. *m/z* 664.3186 for $[^{13}C^{12}C_{36}H_{42}O_4N_4F_3]^+$.

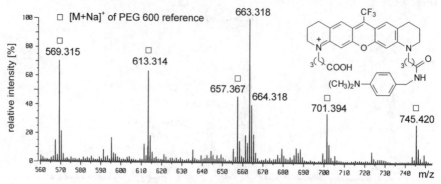

Fig. 9.13. Positive-ion FAB spectrum of a cationic fluorescent marker dye with PEG 600 admixed for internal mass calibration. By courtesy of K. H. Drexhage, University of Siegen and J. Wolfrum, University of Heidelberg.

9.4.3.2 Pseudo-Internal Calibration

Provided a mass spectrometer offers very good scan-to-scan reproducibility of mass calibration, there is no absolute necessity to admix the mass calibration compound to the analyte. Instead, alternating scans on analyte and mass reference can yield almost the same level of accuracy. The simplest technical realization of such a procedure of *pseudo-internal calibration* is presented by changing the target without interruption of the measurement. [104] The use of a *dual-target* FAB probe (DTP), i.e., of a split or double-sided target that offers two separated positions, is much better. [105] Switching between these positions is then achieved by rotating the probe axially between successive scans. The resulting *total ion chromatogram* (TIC, Chap. 5.4) typically has a saw-tooth appearance (Fig. 9.14). Calibration of the analyte spectra is then performed by transferring the internal calibration of the mass reference scan(s) to the successive scan(s) on the analyte side. The advantages of a DTP are: i) independent of resolution, the reference peaks do not interfere with those from the analyte, ii) mutual suppression of reference compound and analyte are excluded, iii) there is no need to adjust the relative intensities of analyte and calibrant peaks very closely, and iv) otherwise reactive mass calibration compounds can be employed. [106]

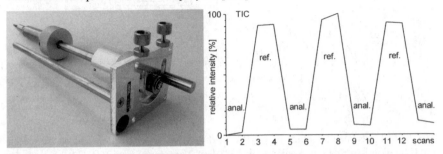

Fig. 9.14. FAB dual-target probe with handle for 180° axial turns (left) and TIC of a HR-FAB measurement using this probe (right).

> **Note:** Scanning of a magnet is affected by hysteresis. This causes the reproducibility of mass calibration to improve after several scan cycles have passed. For best results with dual-target probes, it is therefore recommended to skip the first few scans.

9.4.5 Continuous-Flow FAB

In standard FAB, the surface of the matrix solution is depleted of analyte and suffers from radiational damage during elongated measurements. Refreshment of the surface proceeds by diffusion (limited by the viscosity of the matrix) or evaporation. *Continuous-flow fast atom bombardment* (CF-FAB) continuously refreshes the surface exposed to the atom beam. [107,108] The same effect is obtained in slightly different way by the *frit-fast atom bombardment* (frit-FAB) technique. [109,110] In addition, both CF-FAB and frit-FAB can be used for online-coupling of *liquid chromatography* (LC, Chap. 12) [111] or *capillary electrophoresis* (CE) to a FAB ion source. [112]

In CF-FAB, a carrier solution which is typically composed of 95 % water and 5 % glycerol is passed through the CF-FAB interface via a fused silica capillary of 75 µm inner diameter at flow rates of 5–20 µl min^{-1}. Into this flow, a volume of 0.5–5 µl of liquid sample is injected. As the liquid reaches the end of the capillary inside the FAB ion source, it becomes exposed to the atom beam. The water evaporates while excess glycerol is typically absorbed by a small wick attached to the probe tip. The signal persists as long as it takes to completely elute the analyte from the interface (Fig. 9.15). [107,113] Due to lower flow rates of 1–2 µl min^{-1}, the frit-FAB design does not need an absorbing pad (Fig 9.15). Generally, the open capillary yields less tailing after injections and can accommodate higher flow rates, whereas the frit-FAB technique operates at lower flow rates and is easier to stabilize during operation. [113-115]

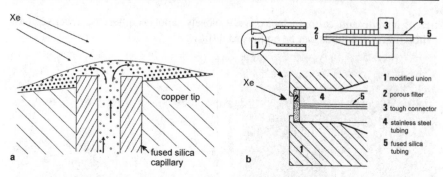

Fig. 9.15. Experimental setups of CF-FAB (**a**) and frit-FAB (**b**). Reproduced from Ref. [108] by permission (**a**). © Academic Press, 1986. Reproduced from Ref. [110] by permission (**b**). © Elsevier Science, 1988.

Example: The significant increase in sensitivity and the improvements of signal-to-background ratio (Chap. 5.2.4) in CF-FAB spectra are demonstrated by comparison of the standard FAB and CF-FAB mass spectra of peptides. [108,116] In case of 10 pmol of the peptide des(gln)-substance P (Fig. 9.16) solely CF-FAB reveals the low-intensity signals due to some loss of side chain groups and several a-type fragments.

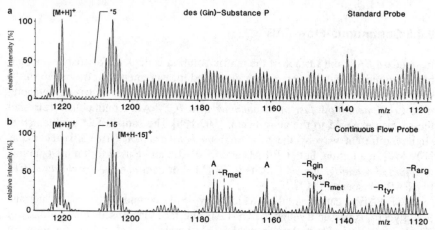

Fig. 9.16. Comparison of standard FAB (**a**) and CF-FAB spectra (**b**) of 10 pmol of the peptide des(gln)-substance P. Reproduced from Ref. [108] by permission. © Academic Press, 1986.

9.4.6 Low-Temperature FAB

Cooled FAB probes have been designed to prolong the acquisition time for FAB measurements with more volatile matrices. [117] Research on sputtering processes from solid gases has contributed to FAB at cryogenic temperatures. [118,119]

More recently, studies concerning cluster ion formation from solid or deeply cooled liquid alcohols [120-122] have gained new interest. [123,124] *Low-temperature fast atom bombardment* (LT-FAB) of frozen aqueous solutions of metal salts provides a source of abundant hydrated metal ions. [125-127] Organic molecules can also be detected from their frozen solutions. [128] Such LT-FAB applications are particularly interesting when enabling the detection of species that would otherwise not be accessible by mass spectrometry, because they are either extremely air- and/or water-sensitive [129,130] as the phosphaoxetane intermediate of the Wittig reaction [131] or insoluble in standard FAB matrices. [106,132]

LT-FAB mass spectra are obtained during thawing of the frozen solution in the ion source of the mass spectrometer, thereby allowing to employ almost any solvent as matrix in LT-FAB-MS. Consequently, neither volatility nor unwanted chemical reactions with the matrix restrict the choice of a matrix. Instead, the solvent matrix may be tailored to the analyte's requirements.

LT-FAB needs somewhat higher amounts of sample than FAB at ambient temperature where more effective matrices than standard solvents are used. Thus, the analytes should be dissolved in the respective solvent to yield 0.5–3.0 μg μl^{-1} solutions. About 3–4 μl of solution are deposited on the FAB probe tip and frozen. Freezing can be achieved in two ways: i) by cooling the target with cold nitrogen gas inside a specially designed vacuum lock before application of the sample [129,131] or ii) by simply immersing the target with the drop of solution into liquid nitrogen for about 30 s prior to transfer into the vacuum lock. [106,123,124,130,133]

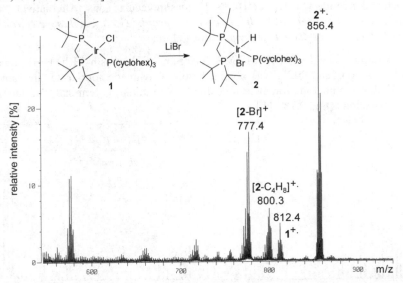

Fig. 9.17. Partial LT-FAB mass spectrum of the reaction mixture containing the iridium complexes **1** and **2** in toluene. In addition to the changes in mass, the isotopic pattern changes upon exchange of Cl by Br. By courtesy of P. Hofmann, University of Heidelberg.

Example: Selective activation of C–H bonds is rarely observed in saturated alkyl groups, but the iridium complex **1** does react by C–H insertion of the metal into a ligand bond upon treatment with LiBr in solution. The reaction can be tracked by LT-FAB-MS (Fig. 9.17). A decreasing intensity of the molecular ion of **1**, *m/z* 812.4, and increasing M$^{+\bullet}$ of **2**, *m/z* 856.4, indicate the progress of this reaction. Furthermore, the halogen exchange is indicated by the isotopic pattern.

9.4.7 FAB-MS and Peptide Sequencing

The possibility to obtain peptide sequence information from FAB mass spectra was soon recognized. [15,134] Initially, the sequence was derived from fragment ions observed in the full scan FAB spectra. [15,98] Another approach to sequence information is to subject the peptide to enzymatic hydrolysis by a mixture of several carboxypeptidases to produce a series of truncated molecules. The FAB spectrum of the mixture then reveals the *C*-terminal sequence. [135,136] In the MALDI community, this approach became known as *peptide ladder sequencing*. [137] Nowadays, sequencing of peptides and other biopolymers by mass spectrometry makes use of tandem MS methods and represents the major field of work for many mass spectrometrists. [138-141] As protein ions are too big to effect fragmentation by collision-induced dissociation (CID, Chap. 2.12.1), they are enzymatically degraded to peptides prior to their mass spectrometric examination, e.g., by tryptic digestion. [142] The digest may be used directly to obtain MS/MS spectra of peptide [M+H]$^+$ or [M+Na]$^+$ quasimolecular ions. Alternatively, the peptides may be separated by *liquid chromatography* (LC), *capillary electrophoresis* (CE) [143], or *2D gel electrophoresis* before.

Example: The FAB-CID-MS/MS spectrum of thymosin-T1 [M+H]$^+$ quasimolecular ions, *m/z* 1427.7, as obtained from a magnetic four-sector instrument [144] shows numerous fragment ions due to *N*-terminal, *C*-terminal, and internal fragmentations (Fig. 9.18). [145]

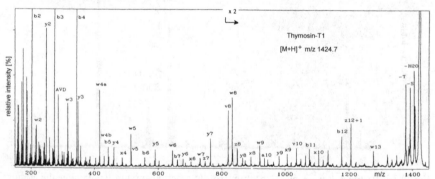

Fig. 9.18. FAB-CID-MS/MS spectrum of thymosin-T1 [M+H]$^+$ ions, *m/z* 1427.7. Reproduced from Ref. [145] by permission. © American Chemical Society, 1993.

Note: One of the advantages of MS/MS techniques is that they do not require the full isolation of all compounds of interest, because the precursor ion selection of MS1 excludes accompanying ions from contributing to the CID spectrum of the actually selected precursor ion as acquired by MS2 (Chaps. 4.2.7, 4.3.6, 4.4.5 and 12.3).

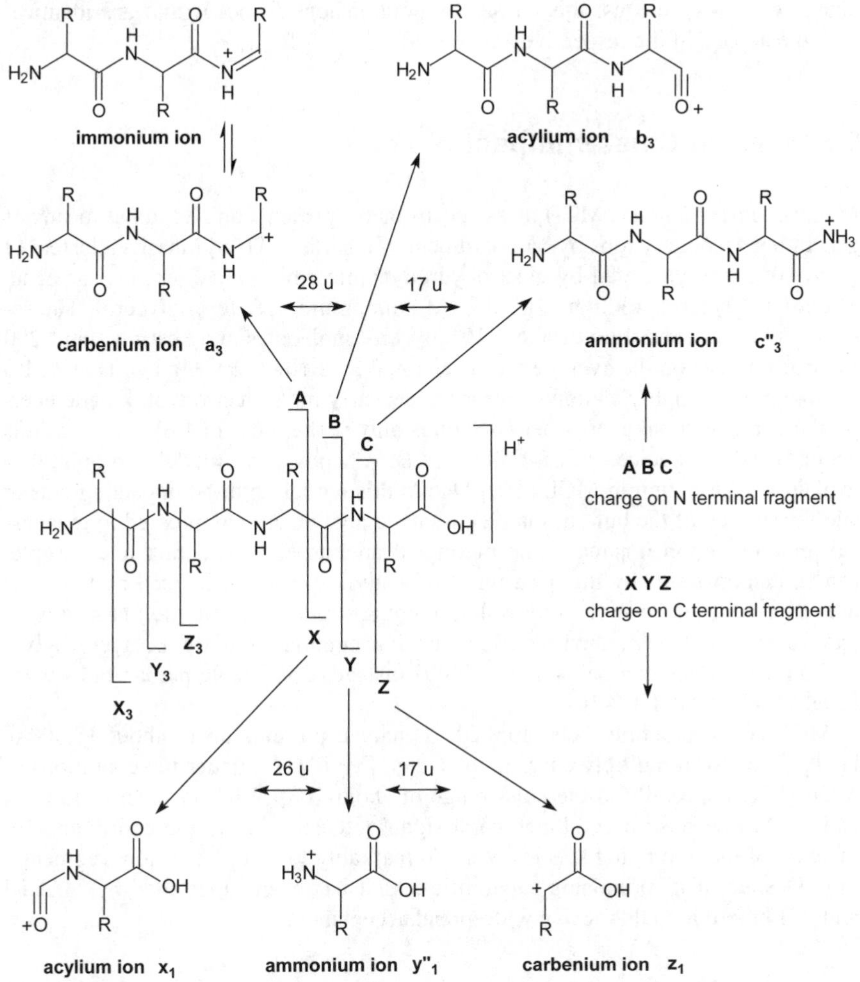

Scheme 9.2.

It would be complicating if not impossible having to obtain sequence information from such a spectrum without rules to follow. [138-141,146,147] The most abundant ions obtained from the fragmenting peptide ion usually belong to six series named **a**, **b**, and **c** if the proton (charge) is kept in the *N*-terminus or **x**, **y**, and **z**, respectively, where the proton is located in the *C*-terminal part. Within each series the mass difference should be just that of one amino acid. Ideally, one can count down amino acid by amino acid. Sometimes, a fragment ion may contain more hydrogens than expected from the "clear cut". Those additional hydrogens are indicated as y''_1 for example, where the index $_1$ denotes the presence of 1 amino acid residue. However, reality is not so simple. Besides these ions, internal fragments of the chain may be formed by cleavage of two or more bonds. Among others, e.g., w type ions, this causes immonium ions to occur that are identical with the a_1 ions of the respective amino acid.

9.5 Massive Cluster Impact

Massive cluster impact (MCI) mass spectrometry presents an additional means of generating secondary ions by bombardment of a surface. [148] Massive clusters of up to 10^8 u are generated by electrohydrodynamic emission (Chap. 11.1.3) of an electrolyte/glycerol solution, e.g., 0.75 M ammonium acetate in glycerol. The resulting clusters consisting of about 10^6 glycerol molecules and bearing about 200 electron charges on the average are accelerated by a 10–20 kV high voltage. [149] Although those highly charged microdroplets carry megaelectronvolt kinetic energies, the translational energy per nucleon is only in the order of 1 eV, whereas it is about 50 eV per nucleon in case of Xe FAB. A shock wave model is proposed to explain ion formation in MCI. [150] Due to this model both the impacting cluster and the surface of the bulk or matrix-dissolved analyte are compressed to gigapascal pressures upon impact. Some mixing of analyte and impacting microdroplet can be demonstrated by the occurrence of analyte species cationized by the same ions as used as electrolyte in the solution employed to generate the massive clusters from. Nonetheless, a matrix effect due to accumulation of a thin layer of glycerol on the surface can be excluded. [149] Instead, dry sample preparations work equally well in MCI. [149]

MCI has successfully been applied to analyze proteins up to about 17,000 u. [151] These form multiply charged ions, e.g., $[M+6Na]^{6+}$, under the conditions of MCI. [151] Especially in the mass range of about 10^4 u, MCI is superior to FAB and LSIMS because it combines good signal intensity due to the enormous momentum of the impinging species with a remarkably low degree of ion fragmentation. Despite of its promising capabilities, MCI has been superseded by MALDI and ESI before it could receive widespread acceptance.

9.6 ^{252}Californium Plasma Desorption

^{252}*Californium plasma desorption* (^{252}Cf-PD) dates back to 1973 [4-6,22,154-156] and was the first method to yield quasimolecular ions of bovine insulin. [157] Practically, ^{252}Cf-PD served for protein characterization, a field of application which is now almost fully transferred to MALDI or ESI (Chaps. 10, 11). [158]

In ^{252}Cf-PD-MS, particles of megaelectronvolt translational energy are created from radioactive decay of ^{252}Cf nuclides, the nuclear process being the source of their kinetic energy. Each event yields two nuclides of varying identity, the sum of their masses being the mass of the former nucleus. The fission fragments of similar mass are travelling in opposite direction. Thus, only one of each pair can be employed to effect desorption of ions from a thin film of analyte on a support foil. The ionization process in ^{252}Cf-PD is different from FAB with some closer relations to dry SIMS. The incident particles are normally travelling from the backside through a thin sample layer on a support foil. [24,159] The initial interaction of the fission fragment with a solid produces approximately 300 electron–hole pairs per angstrom along its track. Recombination of these pairs releases a large amount of energy to the surrounding medium. Within an organic layer, the resulting sudden heat is dissipated by lattice vibrations (phonons) that finally effect desorption of ions from that layer. [6,160,161] In addition to spontaneous ion desorption, ionization processes can occur on the nanosecond timescale in the gas phase. [162]

Deposition of the analyte on nitrocellulose films instead of metal foils allows the removal of alkali ion contaminations by washing of the sample layer which results in better PD spectra. [163] Further improvements can be achieved by adsorption of the analyte molecules on top of an organic low-molecular weight matrix layer. [164,165]

Obviously, ^{252}Cf-PD creates ions in a pulsed manner – one burst of ions per fission event – analogous to laser desorption, a fact that restricted the adaptation of ^{252}Cf-PD to time-of-flight (TOF) analyzers (Chap. 4.2). The second fission fragment is not useless, however, because it serves to trigger the time measurement of the TOF analyzer if the fission fragment source is placed between sample and fission fragment detector (Fig. 9.19). [24,160]

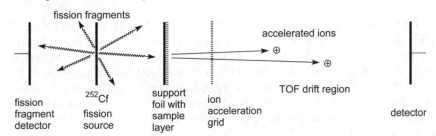

Fig. 9.19. Schematic of a ^{252}Cf-plasma desorption TOF instrument. [160]

Example: The ^{252}Cf-PD mass spectrum of bovine insulin exhibits the [M+H]$^+$ quasimolecular ion as well as the doubly charged [M+2H]$^{2+}$ and triply charged [M+3H]$^{3+}$ ion (Fig. 9.20). [156] Fragment ions corresponding to the A and B chain as well as some **a**-type peptide fragments ions are observed in addition.

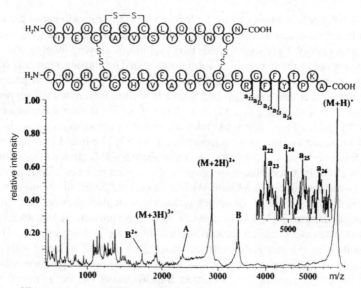

Fig. 9.20. ^{252}Cf-PD mass spectrum of oxidized insulin. Reproduced from Ref. [156] by permission. © John Wiley and Sons, 1994.

9.7 General Characteristics of FAB and LSIMS

9.7.1 Sensitivity of FAB-MS

In FAB mode, the sensitivity (Chap. 5.2.4) is more difficult to specify than for other ionization methods, because the intensity of a signal also strongly depends on the actual preparation on the target. Magnetic sector instruments yield ion currents of about 10^{-11}–10^{-10} A on matrix ions at $R = 1000$. Significantly lower figures (10^{-15}–10^{-14} A) are obtained for the quasimolecular ion of bovine insulin, m/z 5734.6, at $R = 6000$. Accordingly, the detection limits vary depending on the solubility of the analyte and the ease to achieve some sort of ionization if not already ionic at all.

9.7.2 Types of Ions in FAB-MS

FAB produces a variety of ions depending on the polarity and on the ionization energy of the analyte as well as on the presence or absence of impurities such as alkali metal ions. [138] However, with some knowledge of the types of ions formed, reasonable compositions can be assigned to the signals (Table 9.1).

Table 9.1. Ions formed by FAB/LSIMS

Analytes	Positive Ions	Negative Ions
non-polar	$M^{+\bullet}$	$M^{-\bullet}$
medium polarity	$M^{+\bullet}$ and/or $[M+H]^+$, $[M+alkali]^+$, *clusters* $[2M]^{+\bullet}$ and/or $[2M+H]^+$, $[2M+alkali]^+$, *adducts* $[M+Ma+H]^+$, $[M+Ma+alkali]^+$	$M^{-\bullet}$ and/or $[M-H]^-$, *clusters* $[2M]^{-\bullet}$ and/or $[2M-H]^-$ *adducts* $[M+Ma]^{-\bullet}$, $[M+Ma-H]^-$
polar	$[M+H]^+$, $[M+alkali]^+$, *clusters* $[nM+H]^+$, $[nM+alkali]^+$, *adducts* $[M+Ma+H]^+$, $[M+Ma+alkali]^+$ *exchange* $[M-H_n+alkali_{n+1}]^+$ *high-mass anal.* $[M+2H]^{2+}$, $[M+2alkali]^{2+}$	$[M-H]^-$, *clusters* $[nM-H]^-$ *adducts* $[M+Ma-H]^-$ *exchange* $[M-H_n+alkali_{n-1}]^-$
ionic[a]	C^+, $[C_n+A_{n-1}]^+$, rarely $[CA]^{+\bullet}$	A^-, $[C_{n-1}+A_n]^-$, rarely $[CA]^{-\bullet}$

[a] Comprising cation C^+ and anion A^-.

9.7.3 Analytes for FAB-MS

For FAB/LSIMS, the analyte should be soluble to at least 0.1 mg ml^{-1} in some solvent or even better directly in the matrix; concentrations of 0.1–3 µg µl^{-1} in the matrix are ideal. In case of extremely low solubility, additives such as other solvents, acids or surfactants can help. [74]

The analyte may be neutral or ionic. Solutions containing metal salts, e.g., from buffers or excess of noncomplexed metals, may cause a confusingly large number of signals due to multiple proton/metal exchange and adduct ion formation. [91] The mass range up to 3000 u is easily covered by FAB, samples reaching up to about twice that mass still may work if sufficient solubility and some ease of ionization are combined.

9.7.4 Mass Analyzers for FAB-MS

Double-focusing magnetic sector instruments represent the standard in FAB-MS, because they combine a suitable mass range with the ability to perform high-resolution and accurate mass measurements. Until the advent of ESI and MALDI, FAB-MS/MS on magnetic four-sector instruments was the method of choice for biomolecule sequencing (Chaps. 4.3.6, 9.4.7). [144,145,152] Linear quadrupoles have also been adapted to FAB ion sources. [153] Triple quadrupole instruments in particular were used for peptide sequencing in the mid 1980s. However, this type of FAB application for the most part has been replaced by matrix-assisted laser desorption/ionization (MALDI) and electrospray ionization (ESI). Other types of mass analyzers are rare exceptions with FAB or LSIMS ion sources.

Reference List

1. Benninghoven, A.; Kirchner, F. The Energy Distribution of Atomized Neutral and Charged Products. *Z. Naturforsch.* **1963**, *18A*, 1008-1010.

2. Honig, R.E. The Development of Secondary Ion-MS (SIMS): a Retrospective. *Int. J. Mass Spectrom. Ion Proc.* **1985**, *66*, 31-54.

3. Honig, R.E. Stone-Age Mass Spectrometry: the Beginnings of "SIMS" at RCA Laboratories, Princeton. *Int. J. Mass Spectrom. Ion Proc.* **1995**, *143*, 1-10.

4. Macfarlane, R.D.; Torgerson, D.F. Californium-252-Plasma Desorption Time-of-Flight-MS. *Int. J. Mass Spectrom. Ion Phys.* **1976**, *21*, 81-92.

5. Macfarlane, R.D.; Torgerson, D.F. Californium-252 Plasma Desorption Mass Spectroscopy. *Science* **1976**, *191*, 920-925.

6. Macfarlane, R. D. Ion Formation from Organic Solids: High Energy Heavy-ion Induced Desorption. Benninghoven, A., ed., Springer Series in Chemical Physics 25, 32-46, 1983, Springer-Verlag Heidelberg.

7. Devienne, F.M.; Roustan, J.-C. "Fast Atom Bombardment" - a Rediscovered Method for Mass Spectrometry. *Org. Mass Spectrom.* **1982**, *17*, 173-181.

8. Devienne, F.M. Different Uses of High Energy Molecular Beams. *Entropie* **1967**, *18*, 61-67.

9. Barber, M.; Vickerman, J.C.; Wolstenholme, J. Secondary Ion Mass Spectra of Some Simple Organic Molecules. *J. Chem. Soc., Faraday Trans. 1* **1980**, *76*, 549-559.

10. Benninghoven, A. Ion Formation from Organic Solids: Secondary ion-MS of organic compounds. Benninghoven, A., ed., Springer Series in Chemical Physics 25, 64-89, 1983, Springer-Verlag Heidelberg.

11. Barber, M.; Bordoli, R.S.; Sedgwick, R.D.; Tyler, A.N. FAB of Solids as an Ion Source in Mass Spectrometry. *Nature* **1981**, *293*, 270-275.

12. Barber, M.; Bordoli, R.S.; Sedgwick, R.D.; Tyler, A.N. FAB-MS of Cobalamines. *Biomed. Mass Spectrom.* **1981**, *8*, 492-495.

13. Surman, D.J.; Vickerman, J.C. FAB Quadrupole-MS. *J. Chem. Soc. , Chem. Commun.* **1981**, 324-325.

14. Barber, M.; Bordoli, R.S.; Sedgwick, R.D.; Tyler, A.N.; Bycroft, B.W. FAB-MS of Bleomycin A2 and B2 and Their Metal Complexes. *Biochem. Biophys. Res. Commun.* **1981**, *101*, 632-638.

15. Morris, H.R.; Panico, M.; Barber, M.; Bordoli, R.S.; Sedgwick, R.D.; Tyler, A.N. FAB: a New Mass Spectrometric Method for Peptide Sequence Analysis. *Biochem. Biophys. Res. Commun.* **1981**, *101*, 623-631.

16. Barber, M.; Bordoli, R.S.; Elliott, G.J.; Sedgwick, R.D.; Tyler, A.N.; Green, B.N. FAB-MS of Bovine Insulin and Other Large Peptides. *J. Chem. Soc. , Chem. Commun.* **1982**, 936-938.

17. Barber, M.; Bordoli, R.S.; Elliott, G.J.; Sedgwick, R.D.; Tyler, A.N. FAB-MS. *Anal. Chem.* **1982**, *54*, 645A-657A.

18. Meili, J.; Seibl, J. Matrix Effects in FAB-MS. *Int. J. Mass Spectrom. Ion Phys.* **1983**, *46*, 367-370.

19. Gower, L.J. Matrix Compounds for FAB-MS. *Biomed. Mass Spectrom.* **1985**, *12*, 191-196.

20. De Pauw, E.; Agnello, A.; Derwa, F. Liquid Matrices for Liquid Secondary Ion-MS-FAB [LSIMS-FAB]: an Update. *Mass Spectrom. Rev.* **1991**, *10*, 283-301.

21. Aberth, W.; Straub, K.M.; Burlingame, A.L. Secondary Ion-MS With Cesium Ion Primary Beam and Liquid Target Matrix for Analysis of Bioorganic Compounds. *Anal. Chem.* **1982**, *54*, 2029-2034.

22. Sundqvist, B.U.R. Desorption Methods in-MS. *Int. J. Mass Spectrom. Ion Proc.* **1992**, *118/119*, 265-287.

23. Sunner, J. Ionization in Liquid Secondary Ion-MS (LSIMS). *Org. Mass Spectrom.* **1993**, *28*, 805-823.

24. Busch, K.L. Desorption Ionization-MS. *J. Mass Spectrom.* **1995**, *30*, 233-240.

25. *Secondary Ion Mass Spectrometry: Basic Concepts, Instrumental Aspects, Applications;* 1st ed.; Benninghoven, A.; Werner, H.W.; Rudenauer, F.G., editors; Wiley Interscience: New York, 1986.

26. Wilson, R.G.; Stevie, F.A.; Magee, C.W. *Secondary Ion Mass Spectrometry: A Practical Handbook for Depth Profiling and Bulk Impurity Analysis;* John Wiley & Sons: Chichester, 1989.

27. Becker, S.; Dietze, H.-J. State-of-the-Art in Inorganic Mass Spectrometry for

Analysis of High-Purity Materials. *Int. J. Mass Spectrom.* **2003,** *228*, 127-150.

28. Miller, J.M. FAB-MS of Organometallic, Coordination, and Related Compounds. *Mass Spectrom. Rev.* **1989,** *9*, 319-347.

29. Franks, J.; Ghander, A.M. Saddle Field Ion Source of Spherical Configuration for Etching and Thinning Applications. *Vacuum* **1974,** *24*, 489-491.

30. Alexander, A.J.; Hogg, A.M. Characterization of a Saddle-Field Discharge Gun for FABMS Using Different Discharge Vapors. *Int. J. Mass Spectrom. Ion Proc.* **1986,** *69*, 297-311.

31. Boggess, B.; Cook, K.D. Determination of Flux From a Saddle Field FAB Gun. *J. Am. Soc. Mass Spectrom.* **1994,** *5*, 100-105.

32. Barber, M.; Bordoli, R.S.; Sedgwick, R.D.; Tyler, A.N. FAB of Solids: a New Ion Source for Mass Spectrometry. *J. Chem. Soc. , Chem. Commun.* **1981,** 325-327.

33. McDowell, R.A.; Morris, H.R. FAB-MS: Biological Analysis Using an Ion Gun. *Int. J. Mass Spectrom. Ion Phys.* **1983,** *46*, 443-446.

34. Morris, H.R.; Panico, M.; Haskins, N.J. Comparison of Ionization Gases in FAB Mass Spectra. *Int. J. Mass Spectrom. Ion Phys.* **1983,** *46*, 363-366.

35. Fenselau, C. Ion Formation from Organic Solids: FAB. Benninghoven, A., ed., Springer Series in Chemical Physics **25**, 90-100, 1983, Springer-Verlag Heidelberg.

36. Stoll, R.; Schade, U.; Röllgen, F.W.; Giessmann, U.; Barofsky, D.F. Fast Atom and Ion Bombardment of Organic Samples Using Mercury. *Int. J. Mass Spectrom. Ion Phys.* **1982,** *43*, 227-229.

37. Burlingame, A. L.; Aberth, W. Ion Formation from Organic Solids: Use of a Cesium Primary Beam for Liquid SIMS Analysis of Bio-Organic Compounds. Benninghoven, A., ed., Springer Series in Chemical Physics **25**, 167-171, 1983, Springer-Verlag Heidelberg.

38. Burlingame, A.L. Comparison of Three Geometries for a Cesium Primary Beam Liquid Secondary Ion-MS Source. *Anal. Chem.* **1984,** *56*, 2915-2918.

39. Aberth, W.; Burlingame, A.L. Comparison of Three Geometries for a Cesium Primary Beam Liquid Secondary Ion-MS Source. *Anal. Chem.* **1984,** *56*, 2915-2918.

40. Aberth, W.H.; Burlingame, A.L. Effect of Primary Beam Energy on the Secondary-Ion Sputtering Efficiency of Liquid Secondary-Ionization-MS in the 5-30-KeV Range. *Anal. Chem.* **1988,** *60*, 1426-1428.

41. McEwen, C.N.; Hass, J.R. Negative Gold Ion Gun for Liquid Secondary Ion-MS. *Anal. Chem.* **1985,** *57*, 890-892.

42. Katakuse, I.; Nakabushi, H.; Ichihara, T.; Sakurai, T.; Matsuo, T.; Matsuda, H. Generation and Detection of Cluster Ions $[(CsI)_NCs]^+$ Ranging Up to $m/z = 90,000$. *Int. J. Mass Spectrom. Ion Proc.* **1984,** *57*, 239-243.

43. Katakuse, I.; Nakabushi, H.; Ichihara, T.; Sakurai, T.; Matsuo, T.; Matsuda, H. Metastable Decay of Cesium Iodide Cluster Ions. *Int. J. Mass Spectrom. Ion Proc.* **1984,** *62*, 17-23.

44. Sim, P.G.; Boyd, R.K. Calibration and Mass Measurement in Negative-Ion FAB-MS. *Rapid Commun. Mass Spectrom.* **1991,** *5*, 538-542.

45. Rapp, U.; Kaufmann, H.; Höhn, M.; Pesch, R. Exact Mass Determinations Under FAB Conditions. *Int. J. Mass Spectrom. Ion Phys.* **1983,** *46*, 371-374.

46. Sunner, J. Role of Ion-Ion Recombination for Alkali Chloride Cluster Formation in Liquid Secondary Ion-MS. *J. Am. Soc. Mass Spectrom.* **1993,** *4*, 410-418.

47. Cao, Y.; Haseltine, J.N.; Busch, K.L. Double Cationization With Alkali Ions in Liquid Secondary Ion-MS. *Spectroscopy Lett.* **1996,** *29*, 583-589.

48. Todd, P.J. Secondary Ion Emission From Glycerol Under Continuous and Pulsed Primary Ion Current. *Org. Mass Spectrom.* **1988,** *23*, 419-424.

49. Schröder, E.; Münster, H.; Budzikiewicz, H. Ionization by FAB - a Chemical Ionization (Matrix) Process Ion the Gas Phase? *Org. Mass Spectrom.* **1986,** *21*, 707-715.

50. Münster, H.; Theobald, F.; Budzikiewicz, H. The Formation of a Matrix Plasma in the Gas Phase Under FAB Conditions. *Int. J. Mass Spectrom. Ion Proc.* **1987,** *79*, 73-79.

51. Rosen, R.T.; Hartmann, T.G.; Rosen, J.D.; Ho, C.-T. Fast-Atom-Bombardment Mass Spectra of Low-Molecular-Weight Alcohols and Other Compounds. Evidence for a Chemical-Ionization Process in the Gas Phase. *Rapid Commun. Mass Spectrom.* **1988,** *2*, 21-23.

52. Miller, J.M.; Balasanmugam, K. FAB-MS of Some Nonpolar Compounds. *Anal. Chem.* **1989**, *61*, 1293-1295.

53. Benninghoven, A. Organic Secondary Ion-MS (SIMS) and Its Relation to FAB (FAB). *Int. J. Mass Spectrom. Ion Phys.* **1983**, *46*, 459-462.

54. van Breemen, R.B.; Snow, M.; Cotter, R.J. Time-Resolved Laser Desorption-MS. I. Desorption of Preformed Ions. *Int. J. Mass Spectrom. Ion Phys.* **1983**, *49*, 35-50.

55. Musselman, B.; Watson, J.T.; Chang, C.K. Direct Evidence for Preformed Ions of Porphyrins in the Solvent Matrix for FAB-MS. *Org. Mass Spectrom.* **1986**, *21*, 215-219.

56. Shiea, J.; Sunner, J. The Acid Effect in FAB. *Org. Mass Spectrom.* **1991**, *26*, 38-44.

57. Todd, P.J. Solution Chemistry and Secondary Ion Emission From Amine-Glycerol Solutions. *J. Am. Soc. Mass Spectrom.* **1991**, *2*, 33-44.

58. Wong, S.S.; Röllgen, F.W. Sputtering of Large Molecular Ions by Low Energy Particle Impact. *Nulc. Instrum. Methods Phys. Res. ,B* **1986**, *B14*, 436-447.

59. Baczynskyi, L. New Matrices for FAB-MS. *Adv. Mass Spectrom.* **1985**, *10*, 1611-1612.

60. De Pauw, E.; Agnello, A.; Derwa, F. Liquid Matrices for Liquid Secondary Ion-MS. *Mass Spectrom. Rev.* **1986**, *5*, 191-212.

61. Meili, J.; Seibl, J. A New Versatile Matrix for FAB Analysis. *Org. Mass Spectrom.* **1984**, *19*, 581-582.

62. Barber, M.; Bell, D.; Eckersley, M.; Morris, M.; Tetler, L. The Use of *M*-Nitrobenzyl Alcohol As a Matrix in FAB-MS. *Rapid Commun. Mass Spectrom.* **1988**, *2*, 18-21.

63. Aubagnac, J.-L. Use of *M*-Nitrobenzyl Alcohol As a Matrix in Fast-Atom-Bombardment Negative-Ion Mass Spectrometry of Polar Compounds. *Rapid Commun. Mass Spectrom.* **1990**, *4*, 114-116.

64. Dube, G. The Behavior of Aromatic Hydrocarbons Under FAB. *Org. Mass Spectrom.* **1984**, *19*, 242-243.

65. Abdul-Sada, A.K.; Greenway, A.M.; Seddon, K.R. The Extent of Aggregation of Air-Sensitive Alkyllithium Compounds As Determined by Fast-Atom-Bombardment-MS. *J. Organomet. Chem.* **1989**, *375*, C17-C19.

66. Abdul-Sada, A.K.; Greenway, A.M.; Seddon, K.R. The Application of Liquid Paraffin and 3,4-Dimethoxybenzyl Alcohol As Matrix Compounds to FAB-MS. *Eur. Mass Spectrom.* **1996**, *2*, 77-78.

67. Bandini, A.L.; Banditelli, G.; Minghetti, G.; Pelli, B.; Traldi, P. FAB Induced Decomposition Pattern of the Gold(III) Bis(Carbene) Complex [[(*p*-MeC$_6$H$_4$NH)(EtO)C]$_2$AuI$_2$]ClO$_4$, a Retrosynthetic Process? *Organometallics* **1989**, *8*, 590-593.

68. Dobson, J.C.; Taube, H. Coordination Chemistry and Redox Properties of Polypyridyl Complexes of Vanadium(II). *Inorg. Chem.* **1989**, *28*, 1310-1315.

69. Leibman, C.P.; Todd, P.J.; Mamantov, G. Enhanced Positive Secondary Ion Emission From Substituted Polynuclear Aromatic Hydrocarbon/Sulfuric Acid Solutions. *Org. Mass Spectrom.* **1988**, *23*, 634-642.

70. Staempfli, A.A.; Schlunegger, U.P. A New Matrix for FAB Analysis of Corrins. *Rapid Commun. Mass Spectrom.* **1991**, *5*, 30-31.

71. Visentini, J.; Nguyen, P.M.; Bertrand, M.J. The Use of 4-Hydroxybenzenesulfonic Acid As a Reduction-Inhibiting Matrix in Liquid Secondary-Ion-MS. *Rapid Commun. Mass Spectrom.* **1991**, *5*, 586-590.

72. Gower, L.J. Matrix Compounds for FAB: a Further Review. *Adv. Mass Spectrom.* **1985**, *10*, 1537-1538.

73. Rozenski, J.; Herdewijn, P. The Effect of Addition of Carbon Powder to Samples in Liquid Secondary Ion-MS: Improved Ionization of Apolar Compounds. *Rapid Commun. Mass Spectrom.* **1995**, *9*, 1499-1501.

74. Huang, Z.-H.; Shyong, B.-J.; Gage, D.A.; Noon, K.R.; Allison, J. *N*-Alkylnicotinium Halides: a Class of Cationic Matrix Additives for Enhancing the Sensitivity in Negative Ion FAB-MS of Polyanionic Analytes. *J. Am. Soc. Mass Spectrom.* **1994**, *5*, 935-948.

75. Shiea, J.T.; Sunner, J. Effects of Matrix Viscosity on FAB Spectra. *Int. J. Mass Spectrom. Ion Proc.* **1990**, *96*, 243-265.

76. Field, F.H. FAB Study of Glycerol: Mass Spectra and Radiation Chemistry. *J. Phys. Chem.* **1982**, *86*, 5115-5123.

77. Caldwell, K.A.; Gross, M.L. Origins and Structures of Background Ions Produced by FAB of Glycerol. *J. Am. Soc. Mass Spectrom.* **1994**, *5*, 72-91.

78. Busch, K.L. Chemical Noise in-MS. Part I. *Spectroscopy* **2002**, *17*, 32-37.

79. Busch, K.L. Chemical Noise in Mass Spectrometry. Part II - Effects of Choices in Ionization Methods on Chemical Noise. *Spectroscopy* **2003**, *18*, 56-62.

80. Reynolds, J.D.; Cook, K.D. Improving FAB Mass Spectra: the Influence of Some Controllable Parameters on Spectral Quality. *J. Am. Soc. Mass Spectrom.* **1990**, *1*, 149-157.

81. Barber, M.; Bell, D.J.; Morris, M.; Tetler, L.W.; Woods, M.D.; Monaghan, J.J.; Morden, W.E. The Interaction of *M*-Nitrobenzyl Alcohol With Compounds Under Fast-Atom-Bombardment Conditions. *Rapid Commun. Mass Spectrom.* **1988**, *2*, 181-183.

82. Tuinman, A.A.; Cook, K.D. FAB-Induced Condensation of Glycerol With Ammonium Surfactants. I. Regioselectivity of the Adduct Formation. *J. Am. Soc. Mass Spectrom.* **1992**, *3*, 318-325.

83. Aubagnac, J.-L.; Claramunt, R.-M.; Sanz, D. Reduction Phenomenon on the FAB Mass Spectra of *N*-Aminoazoles With a Glycerol Matrix. *Org. Mass Spectrom.* **1990**, *25*, 293-295.

84. Murthy, V.S.; Miller, J.M. Suppression Effects on a Reduction Process in FAB-MS. *Rapid Commun. Mass Spectrom.* **1993**, *7*, 874-881.

85. Aubagnac, J.-L.; Gilles, I.; Lazaro, R.; Claramunt, R.-M.; Gosselin, G.; Martinez, J. Reduction Phenomenon in Frit FAB-MS. *Rapid Commun. Mass Spectrom.* **1995**, *9*, 509-511.

86. Aubagnac, J.-L.; Gilles, I.; Claramunt, R.M.; Escolastico, C.; Sanz, D.; Elguero, J. Reduction of Aromatic Fluorine Compounds in FAB-MS. *Rapid Commun. Mass Spectrom.* **1995**, *9*, 156-159.

87. Théberge, R.; Bertrand, M.J. Beam-Induced Dehalogenation in LSIMS: Effect of Halogen Type and Matrix Chemistry. *J. Mass Spectrom.* **1995**, *30*, 163-171.

88. Théberge, R.; Bertrand, M.J. An Investigation of the Relationship Between Analyte Surface Concentration and the Extent of Beam-Induced Dehalogenation in Liquid Secondary Ion-MS. *Rapid Commun. Mass Spectrom.* **1998**, *12*, 2004-2010.

89. Vetter, W.; Meister, W. FAB Mass Spectrum of β-Carotene. *Org. Mass Spectrom.* **1985**, *20*, 266-267.

90. Nakata, H.; Tanaka, K. Structural and Substituent Effects on $M^{+\cdot}$ Vs. $[M+H]^+$ Formation in FAB Mass Spectra of Simple Organic Compounds. *Org. Mass Spectrom.* **1994**, *29*, 283-288.

91. Moon, D.-C.; Kelly, J.A. A Simple Desalting Procedure for FAB-MS. *Biomed. Environ. Mass Spectrom.* **1988**, *17*, 229-237.

92. Prókai, L.; Hsu, B.-H.; Farag, H.; Bodor, N. Desorption Chemical Ionization, Thermospray, and FAB-MS of Dihydropyridine - Pyridinium Salt-Type Redox Systems. *Anal. Chem.* **1989**, *61*, 1723-1728.

93. Kolli, V.S.K.; York, W.S.; Orlando, R. FAB-MS of Carbohydrates Contaminated With Inorganic Salts Using a Crown Ether. *J. Mass Spectrom.* **1998**, *33*, 680-682.

94. Desiderio, D.M.; Katakuse, I. FAB-MS of Insulin, Insulin A-Chain, Insulin B-Chain, and Glucagon. *Biomed. Mass Spectrom.* **1984**, *11*, 55-59.

95. Barber, M.; Green, B.N. The Analysis of Small Proteins in the Molecular Weight Range 10-24 kDa by Magnetic Sector Mass Spectrometry. *Rapid Commun. Mass Spectrom.* **1987**, *1*, 80-83.

96. Frauenkron, M.; Berkessel, A.; Gross, J.H. Analysis of Ruthenium Carbonyl-Porphyrin Complexes: a Comparison of Matrix-Assisted Laser Desorption/Ionization Time-of-Flight, FAB and Field Desorption-MS. *Eur. Mass Spectrom.* **1997**, *3*, 427-438.

97. Irngartinger, H.; Altreuther, A.; Sommerfeld, T.; Stojanik, T. Pyramidalization in Derivatives of Bicyclo[5.1.0]Oct-1(7)-Enes and 2,2,5,5-Tetramethylbicyclo-[4.1.0]Hept-1(6)-Enes. *Eur. J. Org. Chem.* **2000**, 4059-4070.

98. Barber, M.; Bordoli, R.S.; Elliott, G.J.; Tyler, A.N.; Bill, J.C.; Green, B.N. FAB-MS: a Mass Spectral Investigation of Some of the Insulins. *Biomed. Mass Spectrom.* **1984**, *11*, 182-186.

99. Rajca, A.; Wongsriratanakul, J.; Rajca, S.; Cerny, R. A Dendritic Macrocyclic Organic Polyradical With a Very High Spin of $S = 10$. *Angew. Chem., Int. Ed.* **1998**, *37*, 1229-1232.

100. Andersson, T.; Westman, G.; Stenhagen, G.; Sundahl, M.; Wennerström, O. A Gas Phase Container for C_{60}; a γ-Cyclodextrin Dimer. *Tetrahedron Lett.* **1995**, *36*, 597-600.

101. Giesa, S.; Gross, J.H.; Krätschmer, W.; Gleiter, R. Experiments Towards an Analytical Application of Host-Guest Complexes of [60] Fullerene and Its Deriva-

tives. *Eur. Mass. Spectrom.* **1998**, *4*, 189-196.

102. Juo, C.-G.; Shiu, L.-L.; Shen, C.K.F.; Luh, T.-J.; Her, G.-R. Analysis of C_{60} Derivatives by FAB-MS As γ-Cyclodextrin Inclusion Complexes. *Rapid Commun. Mass Spectrom.* **1995**, *9*, 604-608.

103. Vincenti, M. Host-Guest Chemistry in the Mass Spectrometer. *J. Mass Spectrom.* **1995**, *30*, 925-939.

104. Morgan, R.P.; Reed, M.L. FAB Accurate Mass Measurement. *Org. Mass Spectrom.* **1982**, *17*, 537.

105. Münster, H.; Budzikiewicz, H.; Schröder, E. A Modified Target for FAB Measurements. *Org. Mass Spectrom.* **1987**, *22*, 384-385.

106. Gross, J.H.; Giesa, S.; Krätschmer, W. Negative-Ion Low-Temperature FAB-MS of Monomeric and Dimeric [60]Fullerene Compounds. *Rapid Commun. Mass Spectrom.* **1999**, *13*, 815-820.

107. Caprioli, R.M.; Fan, T.; Cottrell, J.S. A Continuous-Flow Sample Probe for FAB-MS. *Anal. Chem.* **1986**, *58*, 2949-2954.

108. Caprioli, R.M.; Fan, T. High Sensitivity Mass Spectrometric Determination of Peptides: Direct Analysis of Aqueous Solutions. *Biochem. Biophys. Res. Commun.* **1986**, *141*, 1058-1065.

109. Ito, Y.; Takeuchi, T.; Ishii, D.; Goto, M. Direct Coupling of Micro High-Performance Liquid Chromatography With FAB-MS. *J. Chromatogr.* **1985**, *346*, 161-166.

110. Takeuchi, T.; Watanabe, S.; Kondo, N.; Ishii, D.; Goto, M. Improvement of the Interface for Coupling of FAB-MS and Micro High-Performance Liquid Chromatography. *J. Chromatogr.* **1988**, *435*, 482-488.

111. Siethoff, C.; Nigge, W.; Linscheid, M.W. The Determination of Ifosfamide in Human Blood Serum Using LC/MS. *Fresenius J. Anal. Chem.* **1995**, *352*, 801-805.

112. Suter, M.J.F.; Caprioli, R.M. An Integral Probe for Capillary Zone Electrophoresis/Continuous-Flow FAB-MS. *J. Am. Soc. Mass Spectrom.* **1992**, *3*, 198-206.

113. Caprioli, R.M. Continuous-Flow FAB-MS. *Anal. Chem.* **1990**, *62*, 477A-485A.

114. *Continuous-Flow FAB Mass Spectrometry;* Caprioli, R.M., editor; John Wiley & Sons: Chichester, 1990.

115. Caprioli, R.M.; Suter, M.J.F. Continuous-Flow FAB: Recent Advances and Appli-

cations. *Int. J. Mass Spectrom. Ion Proc.* **1992**, *118/119*, 449-476.

116. Caprioli, R.M.; Moore, W.T.; Fan, T. Improved Detection of Suppressed Peptides in Enzymic Digests Analyzed by FAB-MS. *Rapid Commun. Mass Spectrom.* **1987**, *1*, 15-18.

117. Falick, A.M.; Walls, F.C.; Laine, R.A. Cooled Sample Introduction Probe for Liquid Secondary Ionization-MS. *Anal. Biochem.* **1986**, *159*, 132-137.

118. Jonkman, H.T.; Michl, J. Secondary Ion-MS of Small-Molecule Solids at Cryogenic Temperatures. 1. Nitrogen and Carbon Monoxide. *J. Am. Chem. Soc.* **1981**, *103*, 733-737.

119. Orth, R.G.; Jonkman, H.T.; Michl, J. Secondary Ion-MS of Small-Molecule Solids at Cryogenic Temperatures. 2. Rare Gas Solids. *J. Am. Chem. Soc.* **1981**, *103*, 6026-6030.

120. Katz, R.N.; Chaudhary, T.; Field, F.H. Particle Bombardment (FAB) Mass Spectra of Methanol at Sub-Ambient Temperatures. *J. Am. Chem. Soc.* **1986**, *107*, 3897-3903.

121. Katz, R.N.; Chaudhary, T.; Field, F.H. Particle Bombardment (keV) Mass Spectra of Ethylene Glycol, Glycerol, and Water at Sub-Ambient Temperatures. *Int. J. Mass Spectrom. Ion Proc.* **1987**, *78*, 85-97.

122. Johnstone, R.A.W.; Wilby, A.H. FAB at Low Temperatures. Part 2. Polymerization in the Matrix. *Int. J. Mass Spectrom. Ion Proc.* **1989**, *89*, 249-264.

123. Kosevich, M.V.; Czira, G.; Boryak, O.A.; Shelkovsky, V.S.; Vékey, K. Comparison of Positive and Negative Ion Clusters of Methanol and Ethanol Observed by Low Temperature Secondary Ion-MS. *Rapid Commun. Mass Spectrom.* **1997**, *11*, 1411-1416.

124. Kosevich, M.V.; Czira, G.; Boryak, O.A.; Shelkovsky, V.S.; Vékey, K. Temperature Dependences of Ion Currents of Alcohol Clusters Under Low-Temperature Secondary Ion Mass Spectrometric Conditions. *J. Mass Spectrom.* **1998**, *33*, 843-849.

125. Magnera, T.F.; David, D.E.; Stulik, D.; Orth, R.G.; Jonkman, H.T.; Michl, J. Production of Hydrated Metal Ions by Fast Ion or Atom Beam Sputtering. Collision-Induced Dissociation and Successive Hydration Energies of Gaseous Cu^+ With 1-4 Water Molecules. *J. Am. Chem. Soc.* **1989**, *111*, 5036-5043.

126. Boryak, O.A.; Stepanov, I.O.; Kosevich, M.V.; Shelkovsky, V.S.; Orlov, V.V.; Blagoy, Y.P. Origin of Clusters. I. Correlation of Low Temperature FAB Mass Spectra With the Phase Diagram of NaCl-Water Solutions. *Eur. Mass Spectrom.* **1996,** *2,* 329-339.

127. Kosevich, M.V.; Boryak, O.A.; Stepanov, I.O.; Shelkovsky, V.S. Origin of Clusters. II. Distinction of Two Different Processes of Formation of Mixed Metal/Water Clusters Under Low-Temperature FAB. *Eur. Mass Spectrom.* **1997,** *3,* 11-17.

128. Boryak, O.A.; Kosevich, M.V.; Shelkovsky, V.S.; Blagoy, Y.P. Study of Frozen Solutions of Nucleic Acid Nitrogen Bases by Means of Low Temperature Fast-Atom-Bombardment-MS. *Rapid Commun. Mass Spectrom.* **1996,** *10,* 197-199.

129. Huang, M.-W.; Chei, H.-L.; Huang, J.P.; Shiea, J. Application of Organic Solvents As Matrixes to Detect Air-Sensitive and Less Polar Compounds Using Low-Temperature Secondary Ion-MS. *Anal. Chem.* **1999,** *71,* 2901-2907.

130. Hofmann, P.; Volland, M.A.O.; Hansen, S.M.; Eisenträger, F.; Gross, J.H.; Stengel, K. Isolation and Characterization of a Monomeric, Solvent Coordinated Ruthenium(II) Carbene Cation Relevant to Olefin Metathesis. *J. Organomet. Chem.* **2000,** *606,* 88-92.

131. Wang, C.H.; Huang, M.-W.; Lee, C.-Y.; Chei, H.-L.; Huang, J.P.; Shiea, J. Detection of a Thermally Unstable Intermediate in the Wittig Reaction Using Low-Temperature Liquid Secondary Ion and Atmospheric Pressure Ionization-MS. *J. Am. Soc. Mass Spectrom.* **1998,** *9,* 1168-1174.

132. Giesa, S.; Gross, J.H.; Hull, W.E.; Lebedkin, S.; Gromov, A.; Krätschmer, W.; Gleiter, R. C$_{120}$OS: the First Sulfur-Containing Dimeric [60]Fullerene Derivative. *Chem. Commun.* **1999,** 465-466.

133. Gross, J.H. Use of Protic and Aprotic Solvents of High Volatility As Matrixes in Analytical Low-Temperature FAB-MS. *Rapid Commun. Mass Spectrom.* **1998,** *12,* 1833-1838.

134. König, W.A.; Aydin, M.; Schulze, U.; Rapp, U.; Höhn, M.; Pesch, R.; Kalikhevitch, V.N. Fast-Atom-Bombardment for Peptide Sequencing - a Comparison With Conventional Ionization Techniques. *Int. J. Mass Spectrom. Ion Phys.* **1983,** *46,* 403-406.

135. Caprioli, R.M. Enzymes and Mass Spectrometry: a Dynamic Combination. *Mass Spectrom. Rev.* **1987,** *6,* 237-287.

136. Caprioli, R.M. Analysis of Biochemical Reactions With Molecular Specificity Using FAB-MS. *Biochemistry* **1988,** *27,* 513-521.

137. Chait, B.T.; Wang, R.; Beavis, R.C.; Kent, S.B. Protein Ladder Sequencing. *Science* **1993,** *262,* 89-92.

138. Lehmann, W.D. *Massenspektrometrie in der Biochemie;* Spektrum Akademischer Verlag: Heidelberg, 1996.

139. *Mass Spectrometry of Proteins and Peptides;* Chapman, J.R., editor; Humana Press: Totowa, 2000.

140. Snyder, A.P. *Interpreting Protein Mass Spectra;* 1st ed.; Oxford University Press: New York, 2000.

141. Kinter, M.; Sherman, N.E. *Protein Sequencing and Identification Using Tandem Mass Spectrometry;* John Wiley & Sons: Chichester, 2000.

142. Naylor, S.; Findeis, A.F.; Gibson, B.W.; Williams, D.H. An Approach Towards the Complete FAB Analysis of Enzymic Digests of Peptides and Proteins. *J. Am. Chem. Soc.* **1985,** *108,* 6359-6363.

143. Deterding, L.J.; Tomer, K.B.; Wellemans, J.M.Y.; Cerny, R.L.; Gross, M.L. Capillary Electrophoresis/Tandem Mass Spectrometry With Array Detection. *Eur. Mass Spectrom.* **1999,** *5,* 33-40.

144. Bordaz-Nagy, J.; Despeyroux, D.; Jennings, K.R.; Gaskell, S.J. Experimental Aspects of the Collision-Induced Decomposition of Ions in a Four-Sector Tandem Mass Spectrometer. *Org. Mass Spectrom.* **1992,** *27,* 406-415.

145. Stults, J.T.; Lai, J.; McCune, S.; Wetzel, R. Simplification of High-Energy Collision Spectra of Peptides by Amino-Terminal Derivatization. *Anal. Chem.* **1993,** *65,* 1703-1708.

146. Biemann, K.; Papyannopoulos, I.A. Amino Acid Sequencing of Proteins. *Acc. Chem. Res.* **1994,** *27,* 370-378.

147. Papyannopoulos, I.A. The Interpretation of Collision-Induced Dissociation Tandem Mass Spectra of Peptides. *Mass Spectrom. Rev.* **1995,** *14,* 49-73.

148. Mahoney, J.F.; Perel, J.; Ruatta, S.A.; Martino, P.A.; Husain, S.; Lee, T.D. Massive Cluster Impact Mass Spectrometry: a New Desorption Method for the Analysis of Large Biomolecules. *Rapid Commun. Mass Spectrom.* **1991,** *5,* 441-445.

149. Cornett, D.S.; Lee, T.D.; Mahoney, J.F. Matrix-Free Desorption of Biomolecules Using Massive Cluster Impact. *Rapid Commun. Mass Spectrom.* **1994**, *8*, 996-1000.

150. Mahoney, J.F.; Perel, J.; Lee, T.D.; Martino, P.A.; Williams, P. Shock Wave Model for Sputtering Biomolecules Using Massive Cluster Impact. *J. Am. Soc. Mass Spectrom.* **1992**, *3*, 311-317.

151. Mahoney, J.F.; Cornett, D.S.; Lee, T.D. Formation of Multiply Charged Ions From Large Molecules Using Massive-Cluster Impact. *Rapid Commun. Mass Spectrom.* **1994**, *8*, 403-406.

152. Karlsson, N.G.; Karlsson, H.; Hansson, G.C. Sulfated Mucin Oligosaccharides From Porcine Small Intestine Analyzed by Four-Sector Tandem Mass Spectrometry. *J. Mass Spectrom.* **1996**, *31*, 560-572.

153. Caprioli, R.M.; Beckner, C.F.; Smith, L.A. Performance of a FAB Source on a Quadrupole Mass Spectrometer. *Biomed. Mass Spectrom.* **1983**, *10*, 94-97.

154. Sundqvist, B.; Macfarlane, R.D. Californium-252-Plasma Desorption-MS. *Mass Spectrom. Rev.* **1985**, *4*, 421-460.

155. Macfarlane, R.D. Californium-252-Plasma Desorption-MS I - a Historical Perspective. *Biol. Mass Spectrom.* **1993**, *22*, 677-680.

156. Macfarlane, R.D.; Hu, Z.-H.; Song, S.; Pittenauer, E.; Schmid, E.R.; Allmaier, G.; Metzger, J.O.; Tuszynski, W. ²⁵²Cf-Plasma Desorption-MS II - a Perspective of New Directions. *Biol. Mass Spectrom.* **1994**, *23*, 117-130.

157. Håkansson, P.; Kamensky, I.; Sundqvist, B.; Fohlman, J.; Peterson, P.; McNeal, C.J.; Macfarlane, R.D. Iodine-127-Plasma Desorption-MS of Insulin. *J. Am. Chem. Soc.* **1982**, *104*, 2948-2949.

158. Macfarlane, R.D. Mass Spectrometry of Biomolecules: From PDMS to MALDI. *Brazilian J. Phys.* **1999**, *29*, 415-421.

159. Sundqvist, B.; Håkansson, P.; Kamensky, I.; Kjellberg, J. Ion Formation from Organic Solids: Fast heavy ion induced desorption of molecular ions from small proteins. Benninghoven, A., ed., Springer Series in Chemical Physics **25**, 52-57, 1983, Springer-Verlag Heidelberg.

160. Macfarlane, R.D. ²⁵²Californium Plasma Desorption-MS. *Biomed. Mass Spectrom.* **1981**, *8*, 449-453.

161. Wien, K.; Becker, O. Ion Formation from Organic Solids: Secondary ion emission from metals under fission fragment bombardment. Benninghoven, A., ed., Springer Series in Chemical Physics **25**, 47-51, 1983, Springer-Verlag Heidelberg.

162. Zubarev, R.A.; Abeywarna, U.K.; Demirev, P.; Eriksson, J.; Papaléo, R.; Håkansson, P.; Sundqvist, B.U.R. Delayed, Gas-Phase Ion Formation in Plasma Desorption-MS. *Rapid Commun. Mass Spectrom.* **1997**, *11*, 963-972.

163. Jonsson, G.P.; Hedin, A.B.; Håkansson, P.; Sundqvist, B.U.R.; Säve, B.G.S.; Nielsen, P.; Roepstorff, P.; Johansson, K.-E.; Kamensky, I.; Lindberg, M.S.L. Plasma Desorption-MS of Peptides and Proteins Adsorbed on Nitrocellulose. *Anal. Chem.* **1986**, *58*, 1084-1087.

164. Wolf, B.; Macfarlane, R.D. Small Molecules As Substrates for Adsorption/Desorption in Californium-252 Plasma Desorption-MS. *J. Am. Soc. Mass Spectrom.* **1990**, *2*, 29-32.

165. Song, S.; Macfarlane, R.D. PDMS-Chemistry of Angiotensin II and Insulin in Glucose Glass Thin Films. *Anal. Bioanal. Chem.* **2002**, *373*, 647-655.

10 Matrix-Assisted Laser Desorption/Ionization

The technique of *laser desorption/ionization* (LDI) was introduced in the late 1960s, [1-3] i.e., before the advent of field desorption (FD, Chap. 8), *californium plasma desorption* (^{252}Cf-PD, Chap. 9.6) or *fast atom bombardment* (FAB, Chap. 9). While low-mass organic salts and laser light-absorbing organic molecules are easily accessible by LDI, [2,3] it takes some effort to obtain useful spectra of biomolecules, [4] in particular when the mass of the analytes approaches some 10^3 u. [5,6] Until the ending 1980s, FAB and ^{252}Cf-PD were by far more effective than LDI in generating good mass spectra from biological samples. [7]

The introduction of light-absorbing compounds that are admixed to the sample preparations for laser desorption mass spectrometry effected a change. Two approaches were developed: i) the admixture of ultrafine cobalt power (particle size about 30 nm) to analyte solutions in glycerol, [8,9] and ii) the co-crystallization of the analyte with an organic matrix. [10-13] When combined with a *time-of-flight* (TOF, Chap. 4.2) mass analyzer, both methods are capable of producing mass spectra of proteins of about 100,000 u molecular weight. Nonetheless, the application of the *"ultrafine-metal-plus-liquid-matrix"* method remained an exception because the versatility of an organic matrix and the sensitivity of *matrix-assisted laser desorption/ionization* (MALDI) [10,11,14] made it by far superior to the admixture of cobalt powder (Fig. 10.1). [15-18] In its present status, MALDI represents a major analytical tool in modern biochemistry [19-21] and polymer science. [22,23]

10.1 Ion Sources for LDI and MALDI

Both LDI and MALDI make use of the absorption of laser light by a solid sample layer. The energy uptake upon laser irradiation then causes evaporation and ionization of the sample. Wavelengths ranging from ultraviolet (UV) to infrared (IR) have been employed, e.g., nitrogen lasers (337 nm), excimer lasers (193, 248, 308 and 351 nm), Q-switched, frequency-tripled and quadrupled Nd:Yag lasers (355 and 266 nm, respectively), [24] Er:Yag lasers (2.94 µm) [24,25] and TEA-CO$_2$ lasers (10.6 µm). [16,26]

The general setup of LDI/MALDI ion sources is comparatively simple (Fig. 10.2). [27] The pulse of laser light is focused onto a small spot which is typically 0.05–0.2 mm in diameter. [28] As laser irradiance is a critical parameter in MALDI, a variable beam attenuator in the laser optical path is employed to adjust

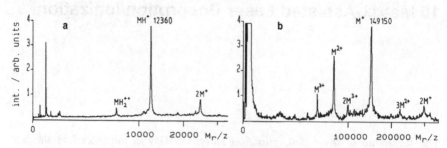

Fig. 10.1. MALDI-TOF mass spectra of (**a**) porcine cytochrome C from 2,5-dihydroxy-benzoic acid matrix at 337 nm and (**b**) a monoclonal antibody from nicotinic acid matrix at 266 nm. Reproduced from Ref. [15] by permission. © John Wiley & Sons, 1991.

the irradiance, e.g., by means of a rotating UV filter of variable transmittance from close to 100 % down to about 1 %. Then, the laser attenuation is individually optimized for each measurement. LDI/MALDI ion sources are generally operated at ambient temperature.

UV lasers are emitting pulses of 3–10 ns duration, while those of IR lasers are in the range of 6–200 ns. Short pulses are needed to effect sudden ablation of the sample layer. In addition, an extremely short time interval of ion generation basically avoids thermal degradation of the analyte. On the other hand, longer irradiation would simply cause heating of the bulk material. In case of IR-MALDI, a slight decrease in threshold fluence has been observed for shorter laser pulses. [29] Furthermore, a short time interval of ion generation means a better definition of the starting pulse for the TOF measurement. However, since the introduction of *delayed extraction*, i.e., of a time delay between laser pulse and the onset of ion acceleration (Chap. 4.2.5), the latter disadvantage of IR lasers has been largely diminished. [29]

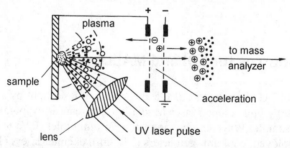

Fig. 10.2. Schematic of a laser desorption ion source for non-resonance light absorption by a solid. Reproduced from Ref. [27] by permission. © Elsevier Science, 1994.

> **Note:** The vast majority of MALDI instruments use UV nitrogen lasers (337 nm, 3 ns). IR-MALDI has been restricted to applications where its deeper penetration offers advantages, e.g., for the direct desorption of analytes from sodium dodecyl sulfate (SDS) gels or thin layer chromatographic (TLC) plates.

10.2 Ion Formation

The mechanisms of ion formation in MALDI are a subject of continuing research. [30-34] The major concerns are the relationship between ion yield and laser fluence, [28,35] the temporal evolution of the desorption process and its implications upon ion formation, [36] the initial velocity of the desorbing ions, [29,37,38] and the question whether preformed ions or ions generated in the gas phase provide the major source of the ionic species detected in MALDI. [39,40]

10.2.1 Ion Yield and Laser Fluence

Below a threshold of *laser irradiance* at about 10^6 W cm^{-2} no ion production is observed. At threshold, a sharp onset of desorption/ionization occurs and ion abundances rise to a high power (5th to 9th) of laser irradiance. [16,35,41] The threshold *laser fluence* for the detection of matrix and analyte ions not only depends on the actual matrix, but also on the molar matrix-to-analyte ratio. A minimum of the threshold fluence for cytochrome C was found at a ratio of 4000 : 1. Significantly higher or lower ratios require almost double laser fluence (Fig. 10.3). [42] The increase at low analyte concentrations can be attributed to a decreasing detection efficiency because a larger volume of material has to be ablated in order to generate a sufficient number of analyte ions for detection. At high analyte concentrations, the energy absorption per volume is reduced as the matrix becomes diluted with analyte molecules causing a higher threshold fluence. The total particle yield from laser desorption as a function of *laser fluence* has been determined by collecting the desorbed neutrals on a quartz crystal microbalance. [43] The study by Quist *et al.* indicates that the desorption of neutrals occurs by thermal evaporation [28] starting at laser fluences of about 11 mJ cm^{-2}. However, the ion-to-neutral ratio of the MALDI process was determined to be less than 10^{-5}. [43]

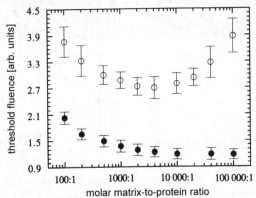

Fig. 10.3. Threshold fluence for positive ions of (O) cytochrome c and (●) sinapinic acid as function of the molar matrix-to-protein ratio. Reproduced from Ref. [42] by permission. © John Wiley & Sons, 1994.

> **Note:** The *fluence* is defined as energy per unit area; in MALDI typical fluences are in the range of 10–100 mJ cm^{-2}. The *irradiance* is fluence divided by the laser pulse duration; in MALDI the irradiances are in the range of 10^6–10^7 W cm^{-2}. [33]

10.2.2 Effect of Laser Irradiation on the Surface

Best MALDI spectra in terms of resolution and low to absent ion fragmentation are obtained slightly above threshold for analyte ion formation. [16,41] An evenly distributed shallow ablation of material from the upper layers of the sample is achieved if a comparatively homogeneous laser fluence is irradiated onto the target. [28,35,44] Laser spot sizes of 100–200 µm as realized by 100–200 mm focal length of the lenses commonly employed are ideal. [45] Numerous single-shot spectra are then obtained from one spot. Such a laser spot size is also advantageous because a number of micrometer-sized crystals are illuminated simultaneously thereby averaging out the effects of mutual orientation of crystal surfaces and laser beam axis. [44,46,47] On the other hand, an extremely sharp laser spot causes the eruption of material from a small area upon formation of a deep crater (Fig. 10.4). The MALDI spectra of cytochrome C ($M_r = 12,360$ u) demonstrate the superior quality of spectra obtained using an optimized spot size. [48]

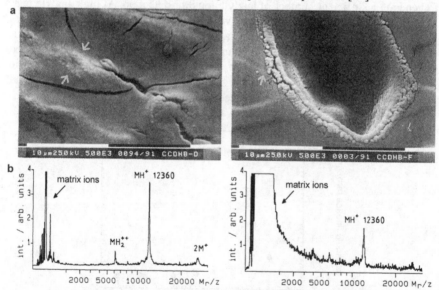

Fig. 10.4. Effect of focused and defocused laser beam. (**a**) SEM micrographs of DHB single crystals after exposure to 10 laser shots (337 nm) under focused (right column) and defocused (left column) irradiation with corresponding sum spectra of horse heart cytochrome C ($M_r = 12,360$ u); black and white bars correspond to 10 µm. (**b**) Resulting MALDI spectra. Reproduced from Ref. [48] by permission. © Elsevier Science, 1991.

> **Note:** MALDI spectra are acquired just above the threshold laser fluence for ion formation. Thus, single-shot spectra normally show a low signal-to-noise ratio (Chap. 5.2.3) due to poor ion statistics. Therefore, 50–200 single-shot spectra are usually accumulated to produce the final spectrum. [47]

10.2.3 Temporal Evolution of a Laser Desorption Plume

The desorption of ions and neutrals into the vacuum upon irradiation of a laser pulse onto a surface proceeds as a jet-like *supersonic expansion*: [38] a small, but initially hot and very rapidly expanding plume is generated. [49] As the expansion is adiabatic, the process is accompanied by fast cooling of the plume. [38]

Although the initial velocity of the desorbed ions is difficult to measure, reported values generally are in the range of 400–1200 m s^{-1}. The initial velocity is almost independent of the ionic mass but dependent on the matrix. [33,36-38,46,50,51] On the other hand, the initial ion velocity is not independent of the compound class, i.e., peptides show a behavior different from oligosaccharides. [51]

The essential independence of mean ion velocities on the molecular weight of the analyte leads to an approximate linear increase of the mean initial kinetic energies of the analyte ions with mass. High-mass ions therefore carry tens of electronvolts of translational energy *before* ion acceleration. [33,41,50] The initial velocity of the ions is superimposed onto that obtained from ion acceleration, thereby causing considerable losses in resolution with continuous extraction TOF analyzers, in particular when operated in the linear mode.

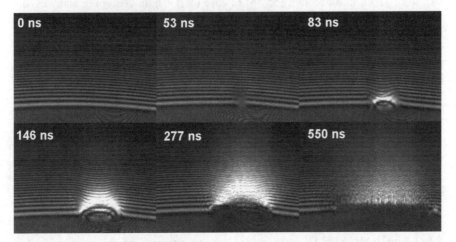

Fig. 10.5. Temporal evolution of a laser desorption plume generated by a 100 ns Er:Yag (2.94 μm) laser pulse from neat glycerol. [53] By courtesy of F. Hillenkamp and A. Leisner, University of Münster.

Example: Laser flash light photographs of the temporal evolution of a laser desorption plume are highly illustrative. [52,53] The plume shown in Fig. 10.5 was generated from neat glycerol by an Er:Yag (2.94 μm) laser pulse of 100 ns pulse width. Pulses of a frequency-doubled ND:YAG laser (532 nm, 15 ns duration) served as the flash light source for obtaining the photographs by the dark field illumination technique.

10.2.4 Ion Formation in MALDI

The locations and processes of ion formation in MALDI seem to be numerous, i.e., no single process applies. [40] The promoting effect of decreasing pH of matrix-analyte solutions upon peptide ion yield indicates that the desorption of pre-formed $[M+H]^+$ ions plays a role. Similar observations are made for quasi-molecular ions generated by cation attachment, e.g., $[M+alkali]^+$ ions in case of oxygen-rich analytes. However, gas phase processes cannot be excluded, because those species may also be generated in the initially formed plasma plume a few hundred micrometers above the sample surface. [54,55] A recent study reveals a gradual increase of the initial ion velocities with increasing mass for neutral oligosaccharides and synthetic polymers to the high level characteristic for peptides and proteins which is also obtained for a small oligosaccharide by introduction of a charged functional group via derivatization. This indicates that typical MALDI analytes need incorporation into the matrix crystal to be detected, whereas gas phase cationization is viable for small neutral analytes. [38] When non-carboxylic acid matrices are being used to protonate slightly basic analyte molecules such as peptides, proton transfer reactions of excited states of the matrix molecules have also to be taken into account. [55,56] In case of UV light-absorbing analytes, direct photoionization can also occur. The frequently observed positive and negative radical ions, $M^{+\bullet}$ [57,58] and $M^{-\bullet}$, [59-61] can only be generated by removal or capture of an electron. Thus, $M^{+\bullet}$ and $M^{-\bullet}$ ions point towards the occurrence of photoionization, [57] charge exchange, and electron capture in the gas phase. [57,62] Which of the above processes contributes most to ion formation depends on the actual combination of matrix, analyte, and eventually present additives or contaminations.

10.3 MALDI Matrices

10.3.1 Role of the Solid Matrix

The role of the matrix in MALDI is analogous to that in FAB (Chap. 9.3.1). Different from FAB, MALDI matrices are generally crystalline solids of low vapor pressure in order not to be volatalized in the ion source vacuum. While basically any liquid can serve as a FAB matrix, the matrix in MALDI has to absorb light of the wavelength which is intended to be used for the experiment. [63] In UV-

MALDI, the molecules must possess a suitable chromophore because energy absorption is based on the strong absorption, and thus the resulting electronic excitation of the matrices. Therefore, the structure of UV-MALDI matrices is based on some aromatic core suitably functionalized to achieve the desired properties.

In case of IR-MALDI, fewer restrictions apply because wavelengths around 3 μm are effectively absorbed by O–H and N–H stretch vibrations, while wavelengths around 10 μm cause excitation of C–O stretch and O–H bending vibrations. [29,32] Therefore, malonic acid, succinic acid, malic acid, urea, and glycerol serve well as matrices in IR-MALDI. [25,26] A matrix can serve as protonating or deprotonating agent or as electron-donating or -accepting agent.

10.3.2 Matrices in UV-MALDI

Nicotinic acid (NA) was the first organic compound that was successfully employed as a matrix in UV-MALDI of peptides and proteins. [12-14] Ever since, better matrices have been sought, the following now being widespread in use: picolinic acid (PA), [64] 3-hydroxypicolinic acid (HPA), [65] and 3-aminopicolinic acid (3-APA) [66] for oligonucleotides and DNA; [67] 2,5-dihydroxybenzoic acid (DHB, gentisic acid) [16,48] and combined matrices having DHB as the main component [68-71], e.g., "super-DHB", [72] for oligosaccharides; α-cyano-4-hydroxycinnamic acid (α-CHC, α-CHCA, 4-HCCA, CCA,) for peptides, smaller proteins, [73], triacylglycerols, [74] and numerous other compounds; [75,76] 3,5-dimethoxy-4-hydroxycinnamic acid (SA, sinapinic acid) for proteins; [77] 2-(4-hydroxyphenylazo)benzoic acid (HABA) for peptides, proteins, glycoproteins, [63] and polystyrene; [78] 2-mercaptobenzothiazole (MBT) and 5-chloro-2-mercaptobenzothiazole (CMBT) for peptides, proteins, and synthetic polymers; [79] 2,6-dihydroxyacetophenone (DHAP) for glycopeptides, phosphopeptides, and proteins; [80,81] 2,4,6-trihydroxyacetophenone (THAP) for solid-supported oligonucleotides; [82] 1,8,9-anthracenetriol (dithranol) for synthetic polymers; [83,84] and 9-nitroanthracene (9-NA), [61,85,86] benzo[a]pyrene, [60] and 2-[(2 E)-3-(4-*tert*-butylphenyl)-2-methylprop-2-enylidene]malonitrile (DCTB) [87] for fullerenes and their derivatives. Even [60]fullerene [88] and porphyrins [89] have been used as matrix. Liquid matrices have been used either due to the specific advantages of a certain matrix or to employ the self-refreshing surface for achieving long-lasting signals in MALDI-magnetic sector experiments. [90,91]

For a first approach to a new analytical problem, it is recommended to try a matrix from this collection; those highlighted by a frame represent the most frequently used matrices. In general, highly polar analytes work better with highly polar matrices, and nonpolar analytes are preferably combined with nonpolar matrices. In unfortunate cases, only one specific analyte-matrix combination might yield useful MALDI spectra.

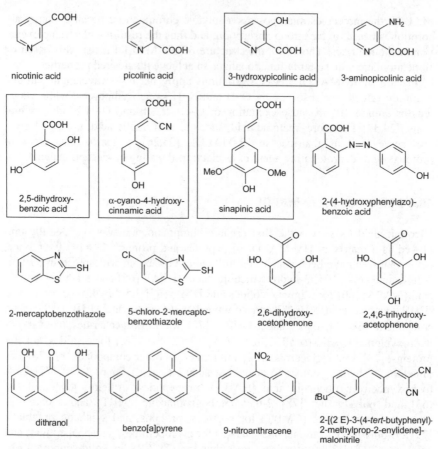

Scheme 10.1.

Note: It is commonplace to use acronyms rather than compound names for matrices. However, these are not always consistently used, e.g., α-CHC, 4-HCCA, CHCA, and CCA all refer to α-cyano-4-hydroxycinnamic acid. Others may be easily confused, e.g., nicotinic acid (NA) and 9-nitroanthracene (9-NA).

10.3.2 Characteristics of MALDI Matrix Spectra

MALDI matrix spectra are characterized by strong molecular and/or quasi-molecular ion signals accompanied by series of matrix (Ma) cluster ions and some more abundant fragment ions. [32] In positive-ion MALDI, $[Ma_n+H]^+$ cluster ions predominate, while $[Ma_n–H]^-$ cluster ions are preferably formed in negative-ion MALDI. The principal ion series may be accompanied by $[Ma_n+alkali]^+$ ions and some fragments of minor intensity, e.g., $[Ma_n+H–H_2O]^+$. In particular with aprotic matrices, radical ions may predominate. In addition, a "continuous" background is formed by clustering of radiolytic decomposition products of the matrix. In general, the spectrum of the neat matrix, i.e., its LDI spectrum, strongly depends on the actual laser fluence and on the presence of impurities. Thus, the "correct" LDI spectrum of the matrix compound largely differs from what is obtained under conditions applied to form analyte ions from that matrix.

10.4 Sample Preparation

10.4.1 Standard Sample Preparation

The standard method of sample preparation in LDI and MALDI involves deposition and subsequent evaporation of 0.5–2 µl of solution on the surface of a *stainless-steel sample holder* or *MALDI target*, as it is often termed. Therefore, the analyte should be soluble to at least about 0.1 mg ml^{-1} in some solvent. If a matrix is used, the matrix is dissolved to yield either a saturated solution or a concentration of about 10 mg ml^{-1}. The solution of the analyte is then admixed to that of the matrix. For optimized MALDI spectra, the molar matrix-to-analyte ratio is normally adjusted not higher than 500 : 1 and above 5000 : 1, i.e., 1 µl of analyte solution are added to 5–50 µl of matrix solution. [15,16,48] In this range, a good signal-to-noise ratio and a low degree of ion fragmentation are preserved. At very high sample concentrations the "matrix effect" is dimished and the spectra start resembling LDI spectra. Too low sample concentrations require additional laser irradiance for sufficient analyte ion production. [42] However, given a proper preparation, even a molar matrix-to-analyte ratio of 10^8 : 1 will produce useful results – conditions that lead to a sample load of 1 fmol. A sufficient miscibility of analyte and matrix is also required. [83]

The crystallization process is a critical parameter in LDI and MALDI sample preparation. [48,92,93] Slow evaporation, e.g., from aqueous solutions, yields comparatively large crystals, which in turn are detrimental for good shot-to-shot reproducibility and mass accuracy. Evenly distributed thin layers of microcrystallites are therefore preferred. [92,94] The formation of such layers can be achieved by i) using volatile solvent(s) such as acetone, ii) by eventually enforcing evaporation by gentle heating of the target or by a softly blowing hair dryer, and finally iii) by using polished targets. Thus, the so-called *thin layer technique* almost revolutionized MALDI sample preparation. [95,96]

Example: The choice of a matrix and optimized conditions of sample preparation have substantial influence on the appearance of MALDI spectra. Even when employing standard matrices such as CHCA or DHB, significant improvements can be achieved, e.g., by appropriate mixing of the two substances (Fig. 10.6). [71]

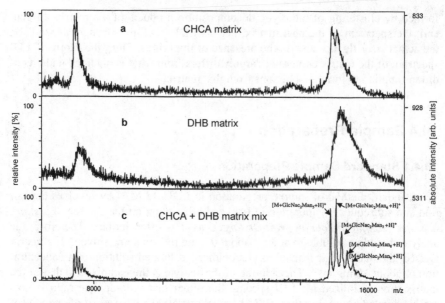

Fig. 10.6. Linear mode positive-ion MALDI-TOF spectra of ribonuclease B in 80 mM urea. (**a**) 300 fmol in CHCA, (**b**) 600 fmol in DHB, and (**c**) 300 fmol in CHCA/DHB matrix mix. Reproduced from Ref. [71] by permission. © Elsevier Science, 2003.

> **Note:** The conventional co-crystallization is usually termed *dried droplet preparation.* The original *thin layer technique* involves preparation of a thin HCHA layer from solution in acetone on top of which the analyte is placed in a second step without re-dissolving the matrix. [95,97]

10.4.2 Cationization and Cation Removal

Metal ions, in particular singly charged ions such as Na^+, K^+, Cs^+, and Ag^+ are sometimes added to the matrix-analyte solution to effect cationization of the neutral analyte. [98] This is advantageous when the analyte has a high affinity to a certain metal ion, e.g., towards alkali ions in case of oligosaccharides. [6] Addition of a certain cation can also help to concentrate the ions in one species, e.g., to promote $[M+K]^+$ ions in favor of all other alkali ion adducts upon addition of a potassium salt.

Silver ions (as silver trifluoroacetate or trifluoromethanesulfonate), Cu^+, and other transition metal ions in their 1+ oxidation state [99,100] are frequently employed to obtain $[M+metal]^+$ ions from non-functionalized or at least nonpolar hydrocarbons, [101] polyethylene, [102,103] or polystyrene (for an example see Chap. 10.5.1). [99,100,104-106]

If an analyte molecule possesses several acidic hydrogens, these can exchange with alkali ions without generating a charged species, e.g., [M–H+K] or

[M–2H+2Na]. As a result, a single analyte will form numerous ionic species thereby significantly decreasing the abundance of each species involved, e.g., [M–2H+Na]⁻, [M–2H+K]⁻, [M–3H+Na+K]⁻ etc. Thus, hydrogen-metal exchange may even result in complete suppression of a signal. In such a case, cation exchange resins should be added to substitute alkali ions for ammonium (Fig. 10.7). [24] There is an analogous necessity to remove sodium dodecyl sulfate contaminations before subjecting samples to MALDI-MS. [107]

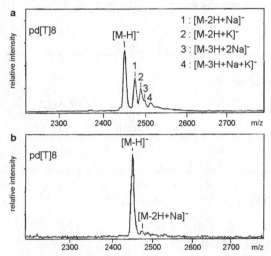

Fig. 10.7. Negative-ion MALDI spectra of the oligonucleotide pd[T]₈; (**a**) 5 pmol, (**b**) same as (**a**) after addition of 5–10 cation exchange beads to the sample preparation. Adapted from Ref. [24] by permission. © John Wiley & Sons, 1992.

10.4.3 Solvent-Free Sample Preparation

If an analyte is definitely insoluble or only soluble in solvents that are not acceptable for the standard MALDI sample preparation technique, it can alternatively be ground together with the solid matrix, preferably in a vibrating ball mill. The resulting fine powder is then spread onto the target. To avoid contamination, nonadherent material should be gently blown away from the target before insertion into the ion source. [103,108,109]

Example: The organic dye pigment red 144, has been subjected to mass analysis by LDI, solvent-based MALDI, and solvent-free MALDI. [109] Its monoisotopic molecular ion, $[C_{40}H_{23}Cl_5O_4N_6]^{+\bullet}$, is expected at m/z 826.0. Due to the strong light absorption of the pigment, the uptake of energy in LDI causes quantitative fragmentation to yield solely $[M–OH]^+$ ions. Here, solvent-based MALDI results in a poor sample preparation because of the unfavorable solvents needed. In this case, solvent-free sample preparation yields the best spectrum exhibiting mainly $M^{+\bullet}$ and $[M+Na]^+$ ions of the pigment (Fig. 10.8).

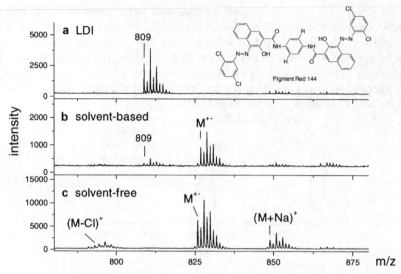

Fig. 10.8. Comparison of spectra of the organic dye pigment red 144 as obtained by (**a**) LDI, (**b**) solvent-based and (**c**) solvent-free MALDI sample preparation. Adapted from Ref. [109] by permission. © John Wiley & Sons, 2001.

10.4.4 Sample Introduction

Sample introduction has undergone a dramatic change in MALDI. In the first experiments, single samples were supplied on MALDI probes designed similar to FAB probes (Chap. 9.1.3). Soon, multi-sample probes came into use. Early commercial products provided approximately twenty spots on one target which was rotated, shifted or preferably moved freely in *x*- and *y*-directions to bring any spot on its surface at the point of laser focus. Driven by the needs of combinatorial chemistry, 96-spot targets were developed to allow for the transfer of samples from a complete standard well plate (Fig. 10.9). More recently, 384-spot and even 1536-spot targets have become available. To take full advantage of such targets it is necessary to combine robotic sample preparation with automated measurement of the MALDI spectra.

The spot size of MALDI preparations and thus the amount of sample necessary to yield a useful layer can be further reduced by so-called *anchor targets* (Bruker Daltonik). Anchor targets exhibit small hydrophilic spots on a hydrophobic surface. As a result, the evaporating drop of matrix-analyte solution is "anchored" to such a point where it shrinks until the onset of crystallization exactly within this hydrophilic area. [110] The resulting preparation covers an about 100fold smaller surface than obtained from a freely spreading drop. In addition to improved detection limits, this technique simplifies automated spot finding due to their precisely defined location on the target.

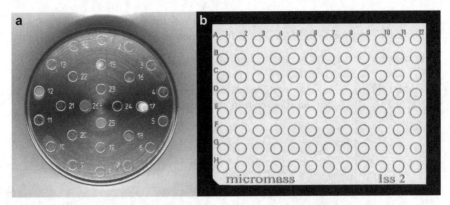

Fig. 10.9. MALDI targets: (**a**) Bruker Scout26 target with some sample preparations using different matrices and (**b**) Micromass target suitable for 96 samples as delivered from standard well plates. (**b**) By courtesy of Waters Corporation, MS Technologies, Manchester.

10.4.5 Additional Methods of Sample Supply

Surface-adsorbed analytes can be examined by laser desorption techniques if they are supplied on a metal foil, a TLC plate, [111] or at least on semiconducting material. Even the foil itself can be accessible to LDI. This requires the foil to be fixed on top of a sample target, e.g., by means of (conducting) double-sided adhesive tape or some general-purpose adhesive. Care has to be taken not to produce sharp edges protruding from the surface because these might cause discharges in the ion source when the accelerating voltage is switched on. Furthermore, the mass calibration can be affected by such an unusually thick "sample layer".

> **Note:** Normally, commercial MALDI instruments tolerate such a sample supply; however, great care is recommended when performing this sort of experiments in order not to loose the sample inside the ion source or to damage the instrument's electronics from electric discharges.

10.4 Applications of LDI

Although examples of LDI of peptides [5] and oligosaccharides [4,6] are known, LDI is much better suited for the analysis of organic and inorganic salts, [112-114] molecules with large conjugated π-electron systems, [115-117] organic dyes as contained in ball point inks, [118], porphyrins [119] or UV light-absorbing synthetic polymers. [5,49] As interferences with matrix ions are excluded, LDI presents a useful alternative to MALDI in the low-mass range. In addition, solvent-free sample preparation can be employed with insoluble analytes. However, LDI is a "harder" method than MALDI and fragmentation has to be taken into account.

Example: Polycyclic aromatic hydrocarbons (PAHs) are easily detected by LDI. The positive-ion LDI-TOF mass spectrum of 1,2,3,4,5,6-hexahydro-phenanthro[1,10,9,8-*opqra*]perylene exclusively exhibits the molecular ion at m/z 356 (Fig. 10.10; for the EI spectrum cf. Chap. 5.1). [117]

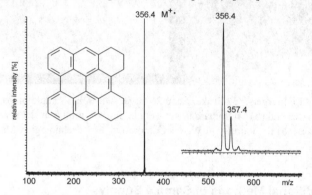

Fig. 10.10. Positive-ion LDI-TOF mass spectrum of 1,2,3,4,5,6-hexahydrophenanthro-[1,10,9,8-*opqra*]perylene. The inset shows an expanded view of the molecular ion signal. Adapted from Ref. [117] with permission. © Elsevier Science, 2002.

Example: Fullerene soots as obtained by the Huffman-Krätschmer synthesis of fullerenes can be characterized by positive- as well as negative-ion LDI. [115] The LDI-TOF spectrum of such a sample exhibits fullerene molecular ion signals well beyond m/z 3000; among these, $C_{60}^{+\bullet}$ and $C_{70}^{+\bullet}$ are clearly preferred (Fig. 10.11). Furthermore, such samples provide experimental carbon-only isotopic patterns over a wide mass range (Chap. 3.2.1).

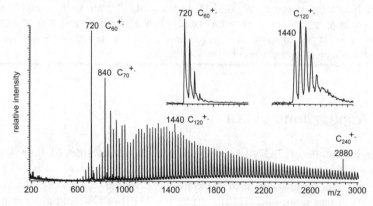

Fig. 10.11. Positive-ion LDI-TOF spectrum of a fullerene soot. The insets show expanded views of the isotopic patterns of $C_{60}^{+\bullet}$ and $C_{120}^{+\bullet}$. By courtesy of W. Krätschmer, Max Planck Institute for Nuclear Physics, Heidelberg.

10.5 Applications of MALDI

10.5.1 MALDI-MS of Synthetic Polymers

MALDI is the method of choice for the analysis of synthetic polymers because it usually provides solely intact and singly charged [62] quasimolecular ions over an essentially unlimited mass range.   [22,23] While polar polymers such as poly(methylmethacrylate) (PMMA), [83,120] polyethylene glycol (PEG), [120,121] and others [79,122,123] readily form [M+H]$^+$ or [M+alkali]$^+$ ions, non-polar polymers like polystyrene (PS) [99,100,105,106] or non-functionalized polymers like polyethylene (PE) [102,103] can only be cationized by transition metal ions in their 1+ oxidation state. [99,100] The formation of evenly spaced oligomer ion series can also be employed to establish an internal mass calibration of a spectrum. [122]

The most important parameters that can be determined by MALDI are *number-average molecular weight* (M_n), *weight-average molecular weight* (M_w), and the molecular weight distribution expressed as *polydispersity* (*PD*): [106,124]

$$M_n = \frac{\sum M_i I_i}{\sum I_i} \tag{10.1}$$

$$M_w = \frac{\sum M_i^2 I_i}{\sum M_i I_i} \tag{10.2}$$

$$PD = \frac{M_w}{M_n} \tag{10.3}$$

where M_i and I_i represent the molecular weights of the oligomeric components and their signal intensities (assuming a linear relationship between number of ions and signal intensity) of the detected species. The formula for M_n is identical to that used for the calculation of the molecular weight from isotopic masses and their abundances as represented by an isotopic pattern (Eq. 3.1 in Chap. 3.1.4).

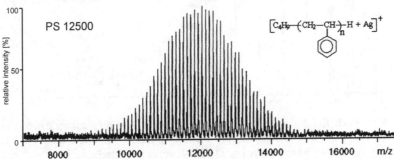

Fig. 10.12. MALDI-TOF spectra of polystyrene 12500 doped with Ag$^+$ ions. Adapted from Ref. [106] by permission. © John Wiley & Sons, 2001.

Example: Polystyrenes ranging from PS 2200 to PS 12500 form [M+metal]$^+$ ions with Ag$^+$ and Cu$^+$ ions when silver or copper(I) salts are admixed to the sample preparation. In case of PS 12500, both metal ions were found to effect cationization equally well, i.e., without causing differences in average molecular weight or ionic abundances (Fig. 10.12). [106]

In addition, the determination of a polymer's endgroup(s) [125,126] and the analysis of random and block-copolymers [127,128] can be achieved by MALDI. However, care has to be taken when judging the MALDI spectra because of the mass-dependent desorption and detection characteristics of the experiment. In case of higher polydispersity (*PD* > 1.1) high-mass ions are underestimated from MALDI spectra. [93,124] The current practice to deal with such samples is to fractionate them by *gel permeation chromatography* (GPC) [123] or *size-exclusion chromatography* (SEC) prior to MALDI analysis. [124,129]

Example: Extracted and synthesized oligo(ethylene terephthalate)s were compared by MALDI-MS. [126] Using the symbols G for ethylene glycol units, GG for diethylene glycol units, and T for terephthalic acid units, the detected oligomers were i) cyclic oligomers [GT]$_n$, ii) linear chains H-[GT]$_n$-G, and iii) some other distributions such as linear H-[GH]$_n$-OH and H-[GGT]$_1$-[GT]$_{n-1}$-G oligomers and cyclic H-[GGT]$_1$-[GT]$_{n-1}$ oligomers. Type i) was mainly contained in technical yarns and tiles, whereas types ii) and iii) were constituents of the model oligomers (Fig. 10.13).

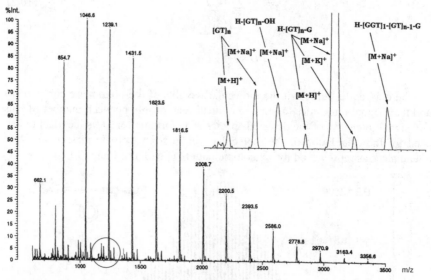

Fig. 10.13. MALDI-TOF spectrum of the model oligo(ethylene terephthalate diol)s. The inset shows an expanded view of the low-intensity peaks in the circle (*m/z* 940–1120). Adapted from Ref. [126] by permission. © John Wiley & Sons, 1995.

10.5.2 Fingerprints by MALDI-MS

The proteins or carbohydrates contained in material of biological origin can supply a characteristic fingerprint of a species. Protein and carbohydrate fingerprints can be readily obtained by means of MALDI-MS, the degree of preceding purification steps depending on the actual type of sample. The composition of the proteins or a certain fraction of these as isolated by some well-defined precipitation procedure can be used to identify bacteria, [130] to reveal whether a Mozzarella cheese was obtained from water buffalo or bovine milk, [131] or whether bovine milk has been fraudulently added in the production of marketed ewe cheese. [132] MALDI-MS of carbohydrates from fungal spores [133] allows for the characterization of the corresponding fungus.

Example: MALDI spectra of protein extracts from *Bacillus* species can be used to distinguish pathogenic and non-pathogenic bacteria. The protein fingerprint obtained from chemically lysed *B. anthracis* (Sterne), *B. thuringiensis* (4A1), *B. cereus* (6E1), and *B. subtilis* (3A1) are clearly different (Fig. 10.14). [130] Even the strains (mentioned in parenthesis) can be assigned to some degree due to the presence of specific biomarker proteins.

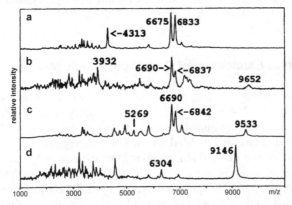

Fig. 10.14. MALDI spectra of protein extracts from *Bacillus* species (matrix α-CHCA). (**a**) *B. anthracis* (Sterne), (**b**) *B. thuringiensis* (4A1), (**c**) *B. cereus* (6E1), (**d**) *B. subtilis* (3A1). Reproduced from Ref. [130] by permission. © John Wiley & Sons, 1996.

10.5.3 Carbohydrates by MALDI-MS

Starting from simple mono- and disaccharides to oligo- and polysaccharides, carbohydrates play an important role in organisms and nutrition. MALDI-MS, typically using DHB or some DHB-containing matrix, [16,68-70,72] presents a powerful tool for their characterization. [134] Applications include the characterization of maltose chains in "gummy bears", [68] fructans in onions, [135] high-molecular-weight oligosaccharides in human milk, [69,70] and others. [98,136]

Example: The maltose chains in the confectionery "gummy bears" extend up to about 30 maltose units. [68] The signals of the ion series originating from $[M+Na]^+$ ions of the saccharide are spaced 162 u distant, i.e., by $C_6H_{10}O_5$ units apart from each other.

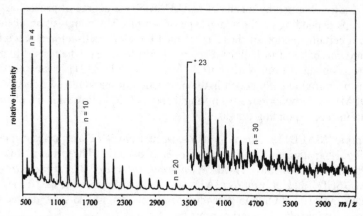

Fig. 10.15. Positive-ion MALDI spectrum of the maltose chains in "gummy bears". Reproduced from Ref. [68] by permission. © John Wiley & Sons, 1995.

10.5.4 Structure Elucidation of Carbohydrates by MALDI

Analogous to peptide sequencing (Chap. 9.4.7), tandem mass spectrometric methods (Chaps. 2.12, 4.2.7, 4.3.6, and 4.4.5) can be employed to elucidate the structure of linear as well as complex branched oligosaccharides. [134,137] If the carbohydrate ions are generated by MALDI, sufficient energy for their fragmentation can be provided in two ways: i) Higher laser irradiance can effect *in-source decay* (ISD) or *metastable dissociation* (2.7.1); [105] the latter being termed *post-source decay* (PSD) in the MALDI-TOF community. ii) *Collision-induced dissociation* (CID) of selected precursor ions can be employed alternatively. (Of course, other compound classes such as peptides, oligonucleotides or synthetic polymers can be treated analogously.) The general scheme of carbohydrate fragmentation is as follows (from Ref. [134] by permission, © Elsevier Science, 2003):

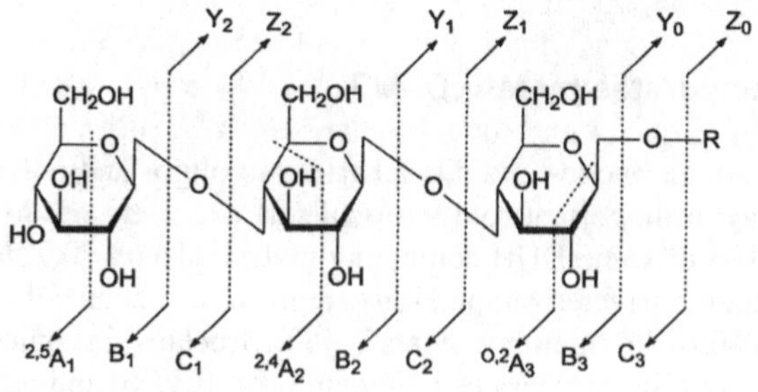

Scheme 10.2.

Example: The PSD-MALDI-TOF spectrum of the $[M+Na]^+$ ion, m/z 1418.9, of high-mannose N-linked glycan $(Man)_6(GlcNAc)_2$ from chicken ovalbumin recorded from DHB shows distinct cleavages of the branched carbohydrate skeleton (Fig. 10.16). [134]

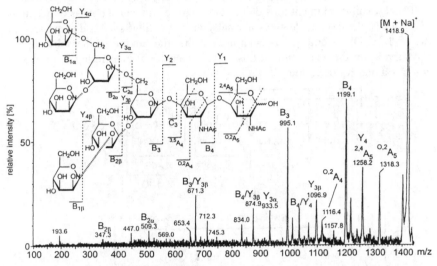

Fig. 10.16. PSD-MALDI-TOF spectrum of N-linked glycan $(Man)_6(GlcNAc)_2$ from chicken ovalbumin. Reproduced from Ref. [134] by permission. © Elsevier Science, 2003.

Note: Carbohydrates possess a high affinity towards alkali metal ions, and thus in MALDI spectra $[M+Na]^+$ and/or $[M+K]^+$ are normally observed instead of or in addition to $[M+H]^+$ ions of very low abundance. Radical ions are not observed. It basically depends on the relative amount of alkali ion impurities or dopant which quasimolecular ion will be dominant.

10.5.5 Oligonucleotides in MALDI

Oligonucleotides and DNA represent the highest polarity class of biopolymers. Therefore, it is of special importance that isolation in an organic matrix allows to overcome their strong intermolecular interaction. MALDI analysis of oligonucleotides is further complicated by the numerous acidic hydrogens present in a single molecule. In particular the phosphate groups easily exchange protons with the ubiqitous alkali ions (Chap. 10.4.2). [24] Thus, MALDI of this compound class requires to follow proven experimental protocols to obtain clean spectra of intact quasimolecular ions. The acidity of the phosphates makes oligonucleotides and DNA easier accessible as $[M–H]^-$ ions in the negative ion mode. [24,67,82,138,139]

Due to their numerous acidic hydrogens, oligonucleotides require desalting prior to MALDI analysis, e.g. by using cation exchange resins (Chap. 10.4.2). [24]

Similar procedures are necessary when other ionization methods are applied to this compound class.

Example: The negative- and positive-ion mode MALDI-TOF spectra of the solid-supported 5-mer oligodeoxynucleotide *po*-CNE 5'-GACTT-3' are compared. [82] Both exhibit fragment ions due to cleavages of the phosphotriester backbone (Fig. 10.17). As oligonucleotides normally do not exhibit such a distinct level of ISD in MALDI spectra, it has been argued that the linking to the solid support plays a role for the generation of this mass ladder of peaks differing from one another by a nucleotide residue.

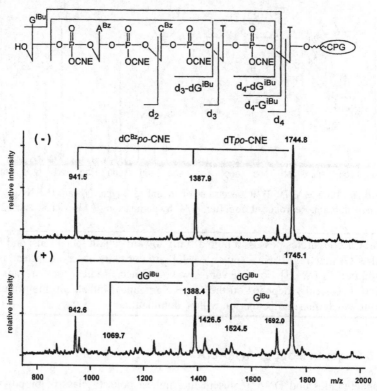

Fig. 10.17. Comparison of the negative- and positive-ion mode MALDI-TOF spectra of the 5-mer *po*-CNE 5'-GACTT-3' oligodeoxynucleotide. Both show fragment ions by ISD. Adapted from Ref. [82] by permission. © John Wiley & Sons, 2000.

10.6 Desorption/Ionization on Silicon

In *desorption/ionization on silicon* (DIOS), the analyte is absorbed by a micrometers-thick porous surface layer on a silicon chip, i.e., the porous silicon is used as substitute of an organic matrix. [140] Porous silicon surfaces possessing high absorptivity in the UV can be generated with varying properties from flat crystalline silicon by using a galvanostatic etching procedure. [141] Stabilization of the freshly prepared surfaces is achieved by hydrosilylation. Porous silicon surfaces can be reused repeatedly after washing. Arrays of 100–1000 sample positions can be realized on a 3×3 cm silicon chip. [141] The DIOS technique offers picomole detection limits for peptides, simple sample preparation, and the absence of matrix peaks in the spectra. [142]

10.7 Atmospheric Pressure MALDI

In *atmospheric pressure MALDI* (AP-MALDI) the MALDI process takes place under atmospheric pressure in dry nitrogen gas. The desorbed ions are then transferred into the vacuum of the mass analyzer by means of an *atmospheric pressure ionization* (API) interface which is typically provided by an electrospray ionization (ESI, Chap. 11) source. AP-MALDI has first been presented in combination with an orthogonal acceleration TOF (oaTOF, Chap. 4.2.6) analyzer where the original ESI ion source was modified to accommodate a simple MALDI target plus laser instead of the ESI spray capillary. [143] AP-MALDI has also been adapted to a quadrupole ion trap (QIT, Chap. 4.5) [144] where an improved design was realized by extending the heated transfer capillary of a Finnigan LCQ ion trap instrument. Thus, a multi-sample target on an *xy*-movable target holder and obser-

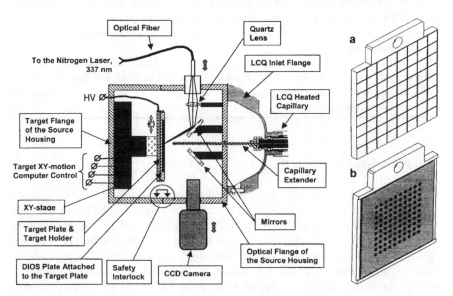

Fig. 10.18. AP-MALDI ion source with extended transfer capillary. Insets: (**a**) the target holder can be equipped with a 64-spot MALDI target or (**b**) a 10×10-spot DIOS chip. Adapted from Ref. [142] by permission. © John Wiley & Sons, 2002.

vation optics could be incorporated on the atmospheric pressure side (Fig. 10.18). [145] The entrance of the capillary extender is held at 1.5–3 kV to attract the ions from the target surface located about 2 mm distant. By modifying the target, AP-DIOS can also be realized on such an AP-MALDI source. [142]

Compared to vacuum MALDI, AP-MALDI has a larger tolerance to laser fluence variations and exhibits reduced fragmentation due to the collisional cooling of the expanding plume. As a result of this cooling process, clustering between matrix and analyte ions is more pronounced. Declustering can be achieved by employing higher laser fluences or adapting the parameters of the atmospheric pressure interface. [146]

10.8 General Characteristics of MALDI

10.8.1 Sample Consumption and Detection Limit

In MALDI, the minimum sample load or the detection limit is usually specified instead of sensitivity (Chap. 5.2.4). Sample loads of 50–500 fmol of a protein for one sample preparation have been communicated earlier [13,14] and 1 fmol can normally be achieved. As thousands of single-shot spectra can be obtained from one preparation spread over some squaremillimeters, the sample consumption has been estimated to approximate 10^{-17} mol per laser shot, i.e., normally more than 99 % of the sample could theoretically be recovered from the target. In between, improved ion extraction and detection as well as miniaturized sample preparation, e.g., by means of the anchor target technology, can provide attomole detection limits for peptides.

> Note: In MALDI-MS, the combination of the actual analyte and the procedure of sample preparation represent the true limiting factors for sample consumption and detection limit.

10.8.2 Analytes for MALDI

For standard MALDI sample preparation, the analyte should be soluble to about 0.1 mg ml^{-1} in some solvent. If an analyte is completely insoluble, solvent-free sample preparation may alternatively be applied (Chap. 10.4.3). The analyte may be neutral or ionic. Solutions containing metal salts, e.g., from buffers or excess of non-complexated metals, may cause a confusingly large number of signals due to multiple proton/metal exchange and adduct ion formation; even complete suppression of the analyte can occur. The mass range of MALDI is theoretically almost unlimited; in practice, limits can be as low as 3000 u, e.g., with polyethylene, or as high as 300,000 u in case of antibodies.

10.8.3 Types of Ions in LDI and MALDI-MS

Very similar to FAB/LSIMS, LDI and MALDI produce a variety of ions depending on the polarity of the analyte, its ionization energy, the characteristics of the matrix (if any) and on the presence or absence of impurities such as alkali metal ions. [19,32,33] The tendency to form radical ions is somewhat lower than in case of FAB/LSIMS (Table 10.1).

Table 10.1. Ions formed by LDI and MALDI

Analytes	Positive Ions	Negative Ions
non-polar	$M^{+\bullet}$	$M^{-\bullet}$
medium polarity	$M^{+\bullet}$ and/or $[M+H]^+$, $[M+alkali]^+$, {*clusters* $[2M]^{+\bullet}$ and/or $[2M+H]^+$, $[2M+alkali]^+$, *adducts* $[M+Ma+H]^+$, $[M+Ma+alkali]^+$}[b]	$M^{-\bullet}$ and/or $[M–H]^-$, {*clusters* $[2M]^{-\bullet}$ and/or $[2M–H]^-$ *adducts* $[M+Ma]^{-\bullet}$, $[M+Ma–H]^-$}
polar	$[M+H]^+$, $[M+alkali]^+$, *exchange* $[M–H_n+alkali_{n+1}]^+$ *high-mass anal.* $[M+2H]^{2+}$, $[M+2alkali]^{2+}$ {*clusters* $[nM+H]^+$, $[nM+alkali]^+$, *adducts* $[M+Ma+H]^+$, $[M+Ma+alkali]^+$}	$[M–H]^-$, *exchange* $[M–H_n+alkali_{n-1}]^-$ {*clusters* $[nM–H]^-$ *adducts* $[M+Ma–H]^-$}
ionic[a]	C^+, $[C_n+A_{n-1}]^+$, {$[CA]^+$}	A^-, $[C_{n-1}+A_n]^-$, {$[CA]^-$}

[a] Comprising of cation C^+ and anion A^-.
[b] Enclosure in parentheses denotes rarely observed species.

10.8.4 Mass Analyzers for MALDI-MS

Laser desorption intrinsically is a pulsed ionization process, which is therefore ideally combined with time-of-flight (TOF) analyzers (Chap. 4.2). [16,49] Ever since the first MALDI experiments, MALDI and TOF have been forming a unit, and the majority of MALDI applications are MALDI-TOF measurements. Vice versa, it was the success of MALDI that pushed forth the tremendous delevopment of TOF mass analyzers. More recently, MALDI has also been adapted to orthogonal acceleration TOF analyzers. [147]

The limited resolution and mass accuracy of the early MALDI-TOF instruments made the combination of MALDI with magnetic sector instruments (Chap. 4.3) desirable, [148,149] but this set-up suffered from low shot-to-shot reproducibility and poor sensitivity; getting a full scan spectrum required thousands of laser shots while scanning the magnet. Even though eutectic matrix mixtures were introduced to circumvent such problems, [90,91] the MALDI-magnetic sector combination never became established.

Fourier transform ion cyclotron resonance (FT-ICR, Chap. 4.6) is much better suited for laser desorption. [5,6] Best results are obtained when the ions are generated in an external ion source and subsequently transferred into the ICR cell for mass analysis. [150] MALDI-FT-ICR has become a mature combination. [151,152] Modern MALDI-FT-ICR instruments make use of collisional cooling of the plasma plume before transferring the ions into the ICR cell. [153,154]

Reference List

1. Fenner, N.C.; Daly, N.R. Laser Used for Mass Analysis. *Rev. Sci. Instrum.* **1966,** *37*, 1068-1070.
2. Vastola, F.J.; Pirone, A.J. Ionization of Organic Solids by Laser Irradiation. *Adv. Mass Spectrom.* **1968,** *4*, 107-111.
3. Vastola, F.J.; Mumma, R.O.; Pirone, A.J. Analysis of Organic Salts by Laser Ionization. *Org. Mass Spectrom.* **1970,** *3*, 101-104.
4. Posthumus, M.A.; Kistemaker, P.G.; Meuzelaar, H.L.C.; Ten Noever de Brauw, M.C. Laser Desorption-Mass Spectrometry of Polar Nonvolatile Bio-Organic Molecules. *Anal. Chem.* **1978,** *50*, 985-991.
5. Wilkins, C.L.; Weil, D.A.; Yang, C.L.C.; Ijames, C.F. High Mass Analysis by LD-FTMS. *Anal. Chem.* **1985,** *57*, 520-524.
6. Coates, M.L.; Wilkins, C.L. LD-FT-MS of Malto-Oligosaccharides. *Biomed. Mass Spectrom.* **1985,** *12*, 424-428.
7. Macfarlane, R.D. Mass Spectrometry of Biomolecules: From PDMS to MALDI. *Brazilian J. Phys.* **1999,** *29*, 415-421.
8. Tanaka, K.; Waki, H.; Ido, Y.; Akita, S.; Yoshida, Y.; Yhoshida, T. Protein and Polymer Analyses Up to *M/z* 100,000 by Laser Ionization-TOF-MS. *Rapid Commun. Mass Spectrom.* **1988,** *2*, 151-153.
9. Tanaka, K. The Origin of Macromolecule Ionization by Laser Irradiation (Nobel Lecture). *Angew. Chem., Int. Ed.* **2003,** *42*, 3861-3870.
10. Karas, M.; Bachmann, D.; Hillenkamp, F. Influence of the Wavelength in High-Irradiance Ultraviolet Laser Desorption-MS of Organic Molecules. *Anal. Chem.* **1985,** *57*, 2935-2939.
11. Karas, M.; Bachmann, D.; Bahr, U.; Hillenkamp, F. Matrix-Assisted Ultraviolet Laser Desorption of Non-Volatile Compounds. *Int. J. Mass Spectrom. Ion Proc.* **1987,** *78*, 53-68.
12. Karas, M.; Hillenkamp, F. Laser Desorption Ionization of Proteins With Molecular Masses Exceeding 10,000 Daltons. *Anal. Chem.* **1988,** *60*, 2299-2301.
13. Karas, M.; Bahr, U.; Ingendoh, A.; Hillenkamp, F. Laser-Desorption-MS of 100,000-250,000-Dalton Proteins. *Angew. Chem.* **1989,** *101*, 805-806.
14. Karas, M.; Ingendoh, A.; Bahr, U.; Hillenkamp, F. Ultraviolet-Laser Desorption/Ionization-MS of Femtomolar Amounts of Large Proteins. *Biomed. Environ. Mass Spectrom.* **1989,** *18*, 841-843.
15. Karas, M.; Bahr, U.; Gießmann, U. MALDI-MS. *Mass Spectrom. Rev.* **1991,** *10*, 335-357.
16. Hillenkamp, F.; Karas, M.; Beavis, R.C.; Chait, B.T. MALDI-MS of Biopolymers. *Anal. Chem.* **1991,** *63*, 1193A-1203A.
17. Beavis, R.C. Matrix-Assisted Ultraviolet Laser Desorption: Evolution and Principles. *Org. Mass Spectrom.* **1992,** *27*, 653-659.
18. Hillenkamp, F.; Karas, M. Matrix-Assisted Laser Desorption/Ionisation, an Experience. *Int. J. Mass Spectrom.* **2000,** *200*, 71-77.
19. Lehmann, W.D. *Massenspektrometrie in der Biochemie;* Spektrum Akademischer Verlag: Heidelberg, 1996.
20. Cotter, R.J. *Time-of-Flight Mass Spectrometry: Instrumentation and Applications in Biological Research;* American Chemical Society: Washington, DC, 1997.
21. Siuzdak, G. *The Expanding Role of Mass Spectrometry in Biotechnology;* MCC Press: San Diego, 2003.
22. *Mass Spectrometry of Polymers;* 1st ed.; Montaudo, G.; Lattimer, R.P., editors; CRC Press: Boca Raton, 2001.
23. Murgasova, R.; Hercules, D.M. MALDI of Synthetic Polymers - an Update. *Int. J. Mass Spectrom.* **2003,** *226*, 151-162.
24. Nordhoff, E.; Ingendoh, A.; Cramer, R.; Overberg, A.; Stahl, B.; Karas, M.; Hillenkamp, F.; Crain, P.F. MALDI-MS of Nucleic Acids With Wavelengths in the Ultraviolet and Infrared. *Rapid Commun. Mass Spectrom.* **1992,** *6*, 771-776.

25. Overberg, A.; Karas, M.; Bahr, U.; Kaufmann, R.; Hillenkamp, F. Matrix-Assisted Infrared-Laser (2.94 Mm) Desorption/Ionization-MS of Large Biomolecules. *Rapid Commun. Mass Spectrom.* **1990,** *4,* 293-296.

26. Overberg, A.; Karas, M.; Hillenkamp, F. Matrix-Assisted Laser Desorption of Large Biomolecules With a TEA-CO$_2$-Laser. *Rapid Commun. Mass Spectrom.* **1991,** *5,* 128-131.

27. Mamyrin, B.A. Laser Assisted Reflectron-TOF-MS. *Int. J. Mass Spectrom. Ion Proc.* **1994,** *131,* 1-19.

28. Dreiwerd, K.; Schürenberg, M.; Karas, M.; Hillenkamp, F. Influence of the Laser Intensity and Spot Size on the Desorption of Molecules and Ions in MALDI With a Uniform Beam Profile. *Int. J. Mass Spectrom. Ion Proc.* **1995,** *141,* 127-148.

29. Berkenkamp, S.; Menzel, C.; Hillenkamp, F.; Dreiwerd, K. Measurements of Mean Initial Velocities of Analyte and Matrix Ions in Infrared MALDI-MS. *J. Am. Soc. Mass Spectrom.* **2002,** *13,* 209-220.

30. Zenobi, R.; Knochenmuss, R. Ion Formation in MALDI-MS. *Mass Spectrom. Rev.* **1999,** *17,* 337-366.

31. Menzel, C.; Dreiwerd, K.; Berkenkamp, S.; Hillenkamp, F. Mechanisms of Energy Deposition in Infrared MALDI-MS. *Int. J. Mass Spectrom.* **2001,** *207,* 73-96.

32. Dreiwerd, K.; Berkenkamp, S.; Leisner, A.; Rohlfing, A.; Menzel, C. Fundamentals of MALDI-MS With Pulsed Infrared Lasers. *Int. J. Mass Spectrom.* **2003,** *226,* 189-209.

33. Dreiwerd, K. The Desorption Process in MALDI. *Chem. Rev.* **2003,** *103,* 395-425.

34. Karas, M.; Krüger, R. Ion Formation in MALDI: The Cluster Ionization Mechanism. *Chem. Rev.* **2003,** *103,* 427-439.

35. Westmacott, G.; Ens, W.; Hillenkamp, F.; Dreiwerd, K.; Schürenberg, M. The Influence of Laser Fluence on Ion Yield in MALDI-MS. *Int. J. Mass Spectrom.* **2002,** *221,* 67-81.

36. Menzel, C.; Dreiwerd, K.; Berkenkamp, S.; Hillenkamp, F. The Role of the Laser Pulse Duration in Infrared MALDI-MS. *J. Am. Soc. Mass Spectrom.* **2002,** *13,* 975-984.

37. Juhasz, P.; Vestal, M.L.; Martin, S.A. On the Initial Velocity of Ions Generated by MALDI and Its Effect on the Calibration of Delayed Extraction-TOF Mass Spectra.

J. Am. Soc. Mass Spectrom. **1997,** *8,* 209-217.

38. Karas, M.; Bahr, U.; Fournier, I.; Glückmann, M.; Pfenninger, A. The Initial Ion Velocity As a Marker for Different Desorption-Ionization Mechanisms in MALDI. *Int. J. Mass Spectrom.* **2003,** *226,* 239-248.

39. Horneffer, V.; Dreiwerd, K.; Ludemann, H.-C.; Hillenkamp, F.; Lage, M.; Strupat, K. Is the Incorporation of Analytes into Matrix Crystals a Prerequisite for MALDI-MS? A Study of Five Positional Isomers of Dihydroxybenzoic Acid. *Int. J. Mass Spectrom.* **1999,** *185/186/187,* 859-870.

40. Glückmann, M.; Pfenninger, A.; Krüger, R.; Thierolf, M.; Karas, M.; Horneffer, V.; Hillenkamp, F.; Strupat, K. Mechanisms in MALDI Analysis: Surface Interaction or Incorporation of Analytes? *Int. J. Mass Spectrom.* **2001,** *210/211,* 121-132.

41. Ens, W.; Mao, Y.; Mayer, F.; Standing, K.G. Properties of Matrix-Assisted Laser Desorption. Measurements With a Time-to-Digital Converter. *Rapid Commun. Mass Spectrom.* **1991,** *5,* 117-123.

42. Medina, N.; Huth-Fehre, T.; Westman, A.; Sundqvist, B.U.R. Matrix-Assisted Laser Desorption: Dependence of the Threshold Fluence on Analyte Concentration. *Org. Mass Spectrom.* **1994,** *29,* 207-209.

43. Quist, A.P.; Huth-Fehre, T.; Sundqvist, B.U.R. Total Yield Measurements in Matrix-Assisted Laser Desorption Using a Quartz Crystal Microbalance. *Rapid Commun. Mass Spectrom.* **1994,** *8,* 149-154.

44. Beavis, R.C.; Chait, B.T. Factors Affecting the Ultraviolet Laser Desorption of Proteins. *Rapid Commun. Mass Spectrom.* **1989,** *3,* 233-237.

45. Ingendoh, A.; Karas, M.; Hillenkamp, F.; Giessmann, U. Factors Affecting the Resolution in Matrix-Assisted Laser Desorption-Ionization-MS. *Int. J. Mass Spectrom. Ion Proc.* **1994,** *131,* 345-354.

46. Aksouh, F.; Chaurand, P.; Deprun, C.; Della-Negra, S.; Hoyes, J.; Le Beyec, Y.; Pinho, R.R. Influence of the Laser Beam Direction on the Molecular Ion Ejection Angle in MALDI. *Rapid Commun. Mass Spectrom.* **1995,** *8,* 515-518.

47. Liao, P.-C.; Allison, J. Dissecting MALDI Mass Spectra. *J. Mass Spectrom.* **1995,** *30,* 763-766.

48. Strupat, K.; Karas, M.; Hillenkamp, F. 2,5-Dihydroxybenzoic Acid: a New Matrix for

Laser Desorption-Ionization-MS. *Int. J. Mass Spectrom. Ion Proc.* **1991**, *111*, 89-102.

49. Cotter, R.J. Laser-MS: an Overview of Techniques, Instruments and Applications. *Anal. Chim. Acta* **1987**, *195*, 45-59.

50. Pan, Y.; Cotter, R.J. Measurement of Initial Translational Energies of Peptide Ions in Laser Desorption/Ionization-MS. *Org. Mass Spectrom.* **1992**, *27*, 3-8.

51. Glückmann, M.; Karas, M. The Initial Ion Velocity and Its Dependence on Matrix, Analyte and Preparation Method in Ultraviolet MALDI. *J. Mass Spectrom.* **1999**, *34*, 467-477.

52. Puretzky, A.A.; Geohegan, D.B.; Hurst, G.B.; Buchanan, M.V.; Luk'yanchuk, B.S. Imaging of Vapor Plumes Produced by Matrix Assisted Laser Desorption: A Plume Sharpening Effect. *Phys. Rev. Lett.* **1999**, *83*, 444-447.

53. Leisner, A.; Rohlfing, A.; Berkenkamp, S.; Röhling, U.; Dreisewerd, K.; Hillenkamp, F. IR-MALDI With the Matrix Glycerol: Examination of the Plume Expansion Dynamics for Lasers of Different Pulse Duration. *36. DGMS Jahrestagung* **2003**, Poster.

54. Wang, B.H.; Dreisewerd, K.; Bahr, U.; Karas, M.; Hillenkamp, F. Gas-Phase Cationization and Protonation of Neutrals Generated by Matrix-Assisted Laser Desorption. *J. Am. Soc. Mass Spectrom.* **1993**, *4*, 393-398.

55. Liao, P.-C.; Allison, J. Ionization Processes in MALDI-MS: Matrix-Dependent Formation of $[M+H]^+$ Vs. $[M+Na]^+$ Ions of Small Peptides and Some Mechanistic Comments. *J. Mass Spectrom.* **1995**, *30*, 408-423.

56. Gimon, M.E.; Preston, L.M.; Solouki, T.; White, M.A.; Russel, D.H. Are Proton Transfer Reactions of Excited States Involved in UV Laser Desorption Ionization? *Org. Mass Spectrom.* **1992**, *27*, 827-830.

57. Juhasz, P.; Costello, C.E. Generation of Large Radical Ions From Oligometallocenes by MALDI. *Rapid Commun. Mass Spectrom.* **1993**, *7*, 343-351.

58. Lidgard, R.O.; McConnell, B.D.; Black, D.S.C.; Kumar, N.; Duncan, M.W. Fragmentation Observed in Continuous Extraction Linear MALDI: a Cautionary Note. *J. Mass Spectrom.* **1996**, *31*, 1443-1445.

59. Irngartinger, H.; Weber, A. Twofold Cycloaddition of [60]Fullerene to a Bifunctional Nitrile Oxide. *Tertrahedron Lett.* **1996**, *37*, 4137-4140.

60. Gromov, A.; Ballenweg, S.; Giesa, S.; Lebedkin, S.; Hull, W.E.; Krätschmer, W. Preparation and Characterization of C_{119}. *Chem. Phys. Lett.* **1997**, *267*, 460-466.

61. Giesa, S.; Gross, J.H.; Hull, W.E.; Lebedkin, S.; Gromov, A.; Krätschmer, W.; Gleiter, R. $C_{120}OS$: the First Sulfur-Containing Dimeric [60]Fullerene Derivative. *Chem. Commun.* **1999**, 465-466.

62. Karas, M.; Glückmann, M.; Schäfer, J. Ionization in MALDI: Singly Charged Molecular Ions Are the Lucky Survivors. *J. Mass Spectrom.* **2000**, *35*, 1-12.

63. Juhasz, P.; Costello, C.E.; Biemann, K. MALDI-MS With 2-(4-Hydroxyphenylazo)benzoic Acid Matrix. *J. Am. Soc. Mass Spectrom.* **1993**, *4*, 399-409.

64. Tang, K.; Taranenko, N.I.; Allman, S.L.; Chen, C.H.; Chang, L.Y.; Jacobson, K.B. Picolinic Acid As a Matrix for Laser-MS of Nucleic Acids and Proteins. *Rapid Commun. Mass Spectrom.* **1994**, *8*, 673-677.

65. Wu, K.J.; Steding, A.; Becker, C.H. Matrix-Assisted Laser Desorption-TOF-MS of Oligonucleotides Using 3-Hydroxypicolinic Acid As an Ultraviolet-Sensitive Matrix. *Rapid Commun. Mass Spectrom.* **1993**, *7*, 142-146.

66. Taranenko, N.I.; Tang, K.; Allman, S.L.; Ch'ang, L.Y.; Chen, C.H. 3-Aminopicolinic Acid As a Matrix for Laser Desorption-MS of Biopolymers. *Rapid Commun. Mass Spectrom.* **1994**, *8*, 1001-1006.

67. Taranenko, N.I.; Potter, N.T.; Allman, S.L.; Golovlev, V.V.; Chen, C.H. Gender Identification by MALDI-TOF-MS. *Anal. Chem.* **1999**, *71*, 3974-3976.

68. Mohr, M.D.; Börnsen, K.O.; Widmer, H.M. MALDI-MS: Improved Matrix for Oligosaccharides. *Rapid Commun. Mass Spectrom.* **1995**, *9*, 809-814.

69. Finke, B.; Stahl, B.; Pfenninger, A.; Karas, M.; Daniel, H.; Sawatzki, G. Analysis of High-Molecular-Weight Oligosaccharides From Human Milk by Liquid Chromatography and MALDI-MS. *Anal. Chem.* **1999**, *71*, 3755-3762.

70. Pfenninger, A.; Karas, M.; Finke, B.; Stahl, B.; Sawatzki, G. Matrix Optimization for MALDI-MS of Oligosaccharides

From Human Milk. *J. Mass Spectrom.* **1999**, *34*, 98-104.

71. Laugesen, S.; Roepstorff, P. Combination of Two Matrices Results in Improved Performance of Maldi MS for Peptide Mass Mapping and Protein Analysis. *J. Am. Soc. Mass Spectrom.* **2003**, *14*, 992-1002.

72. Karas, M.; Ehring, H.; Nordhoff, E.; Stahl, B.; Strupat, K.; Hillenkamp, F.; Grehl, M.; Krebs, B. MALDI-MS With Additives to 2,5-Dihydroxybenzoic Acid. *Org. Mass Spectrom.* **1993**, *28*, 1476-1481.

73. Beavis, R.C.; Chaudhary, T.; Chait, B.T. α-Cyano-4-Hydroxycinnamic Acid As a Matrix for Matrix-Assisted Laser Desorption-MS. *Org. Mass Spectrom.* **1992**, *27*, 156-158.

74. Ayorinde, F.O.; Elhilo, E.; Hlongwane, C. MALDI-TOF-MS of Canola, Castor and Olive Oils. *Rapid Commun. Mass Spectrom.* **1999**, *13*, 737-739.

75. George, M.; Wellemans, J.M.Y.; Cerny, R.L.; Gross, M.L.; Li, K.; Cavalieri, E.L. Matrix Design for MALDI: Sensitive Determination of PAH-DNA Adducts. *J. Am. Soc. Mass Spectrom.* **1994**, *5*, 1021-1025.

76. Lidgard, R.O.; Duncan, M.W. Utility of MALDI-TOF-MS for the Analysis of Low Molecular Weight Compounds. *Rapid Commun. Mass Spectrom.* **1995**, *9*, 128-132.

77. Beavis, R.C.; Chait, B.T. Cinnamic Acid Derivatives As Matrices for Ultraviolet Laser Desorption-MS of Proteins. *Rapid Commun. Mass Spectrom.* **1989**, *3*, 432-435.

78. Montaudo, G.; Montaudo, M.S.; Puglisi, C.; Samperi, F. 2-(4-Hydroxyphenylazo)-Benzoic Acid: a Solid Matrix for MALDI of Polystyrene. *Rapid Commun. Mass Spectrom.* **1994**, *8*, 1011-1015.

79. Xu, N.; Huang, Z.-H.; Watson, J.T.; Gage, D.A. Mercaptobenzothiazoles: a New Class of Matrixes for Laser Desorption Ionization-MS. *J. Am. Soc. Mass Spectrom.* **1997**, *8*, 116-124.

80. Pitt, J.J.; Gorman, J.J. MALDI-TOF-MS of Sialylated Glycopeptides and Proteins Using 2,6-Dihydroxyacetophenone As a Matrix. *Rapid Commun. Mass Spectrom.* **1996**, *10*, 1786-1788.

81. Gorman, J.J.; Ferguson, B.L.; Nguyen, T.B. Use of 2,6-Dihydroxyacetophenone for Analysis of Fragile Peptides, Disulfide Bonding and Small Proteins by MALDI. *Rapid Commun. Mass Spectrom.* **1996**, *10*, 529-536.

82. Meyer, A.; Spinelli, N.; Imbach, J.-L.; Vasseur, J.-J. Analysis of Solid-Supported Oligonucleotides by MALDI-TOF-MS. *Rapid Commun. Mass Spectrom.* **2000**, *14*, 234-242.

83. Kassis, C.M.; DeSimone, J.M.; Linton, R.W.; Lange, G.W.; Friedman, R.M. An Investigation into the Importance of Polymer-Matrix Miscibility Using Surfactant Modified MALDI-MS. *Rapid Commun. Mass Spectrom.* **1997**, *11*, 1462-1466.

84. Carr, R.H.; Jackson, A.T. Preliminary MALDI-TOF and Field Desorption Mass Spectrometric Analyses of Polymeric Methylene Diphenylene Diisocyanate, Its Amine Precursor and a Model Polyether Prepolymer. *Rapid Commun. Mass Spectrom.* **1998**, *12*, 2047-2050.

85. Lebedkin, S.; Ballenweg, S.; Gross, J.H.; Taylor, R.; Krätschmer, W. Synthesis of $C_{120}O$: a New Dimeric [60]Fullerene Derivative. *Tetrahedron Lett.* **1995**, *36*, 4971-4974.

86. Ballenweg, S.; Gleiter, R.; Krätschmer, W. Chemistry at Cyclopentene Addends on [60]Fullerene. Matrix-Assisted Laser Desorption-Ionization-TOF-MS (MALDI-TOF MS) As a Quick and Facile Method for the Characterization of Fullerene Derivatives. *Synth. Met.* **1996**, *77*, 209-212.

87. Brown, T.; Clipston, N.L.; Simjee, N.; Luftmann, H.; Hungebühler, H.; Drewello, T. MALDI of Amphiphilic Fullerene Derivatives. *Int. J. Mass Spectrom.* **2001**, *210/211*, 249-263.

88. Hopwood, F.G.; Michalak, L.; Alderdice, D.S.; Fisher, K.J.; Willett, G.D. C_{60}-Assisted Laser Desorption/Ionization-MS in the Analysis of Phosphotungstic Acid. *Rapid Commun. Mass Spectrom.* **1994**, *8*, 881-885.

89. Jones, R.M.; Lamb, J.H.; Lim, C.K. 5,10,15,20-*Meso*-Tetra(Hydroxyphenyl)-Chlorin As a Matrix for the Analysis of Low Molecular Weight Compounds by MALDI-TOF-MS. *Rapid Commun. Mass Spectrom.* **1995**, *9*, 968-969.

90. Kumar, N.; Kolli, V.S.; Orlando, R. A New Matrix for MALDI on Magnetic Sector Instruments With Point Detectors. *Rapid Commun. Mass Spectrom.* **1996**, *10*, 923-926.

91. Harvey, D.J.; Hunter, A.P. Use of a Conventional Point Detector to Record MALDI Spectra From a Magnetic Sector Instrument. *Rapid Commun. Mass Spectrom.* **1998**, *12*, 1721-1726.

92. Westman, A.; Huth-Fehre, T.; Demirev, P.A.; Sundqvist, B.U.R. Sample Morphology Effects in MALDI-MS of Proteins. *J. Mass Spectrom.* **1995**, *30*, 206-211.

93. Arakawa, R.; Watanabe, S.; Fukuo, T. Effects of Sample Preparation on MALDI-TOF Mass Spectra for Sodium Polystyrene Sulfonate. *Rapid Commun. Mass Spectrom.* **1999**, *13*, 1059-1062.

94. Chan, P.K.; Chan, T.-W.D. Effect of Sample Preparation Methods on the Analysis of Dispersed Polysaccharides by MALDI-TOF-MS. *Rapid Commun. Mass Spectrom.* **2000**, *14*, 1841-1847.

95. Vorm, O.; Mann, M. Improved Mass Accuracy in MALDI-TOF-MS of Peptides. *J. Am. Soc. Mass Spectrom.* **1994**, *5*, 955-958.

96. Vorm, O.; Roepstorff, P.; Mann, M. Improved Resolution and Very High Sensitivity in MALDI-TOF of Matrix Surfaces Made by Fast Evaporation. *Anal. Chem.* **1994**, *66*, 3281-3287.

97. Ayorinde, F.O.; Keith, Q.L., Jr.; Wan, L.W. MALDI-TOF-MS of Cod Liver Oil and the Effect of Analyte/Matrix Concentration on Signal Intensities. *Rapid Commun. Mass Spectrom.* **1999**, *13*, 1762-1769.

98. Bashir, S.; Derrick, P.J.; Critchley, P.; Gates, P.J.; Staunton, J. MALDI-TOF-MS of Dextran and Dextrin Derivatives. *Eur. J. Mass Spectrom.* **2003**, *9*, 61-70.

99. Rashidezadeh, H.; Baochuan, G. Investigation of Metal Attachment to Polystyrenes in MALDI. *J. Am. Soc. Mass Spectrom.* **1998**, *9*, 724-730.

100. Rashidezadeh, H.; Hung, K.; Baochuan, G. Probing Polystyrene Cationization in Matrix-Assisted Laser/Desorption Ionization. *Eur. Mass Spectrom.* **1998**, *4*, 429-433.

101. Kühn, G.; Weidner, S.; Decker, R.; Holländer, A. Derivatization of Double Bonds Investigated by MALDI-MS. *Rapid Commun. Mass Spectrom.* **1997**, *11*, 914-918.

102. Kühn, G.; Weidner, S.; Just, U.; Hohner, S. Characterization of Technical Waxes. Comparison of Chromatographic Techniques and Matrix-Assisted Laser-Desorption/Ionization-MS. *J. Chromatogr. A* **1996**, *732*, 111-117.

103. Pruns, J.K.; Vietzke, J.-P.; Strassner, M.; Rapp, C.; Hintze, U.; König, W.A. Characterization of Low Molecular Weight Hydrocarbon Oligomers by Laser Desorption/Ionization-TOF-MS Using a Solvent-Free Sample Preparation Method. *Rapid Commun. Mass Spectrom.* **2002**, *16*, 208-211.

104. Mowat, I.A.; Donovan, R.J. Metal-Ion Attachment to Non-Polar Polymers During Laser Desorption/Ionization at 337 Nm. *Rapid Commun. Mass Spectrom.* **1995**, *9*, 82-90.

105. Goldschmitt, R.J.; Wetzel, S.J.; Blair, W.R.; Guttman, C.M. Post-Source Decay in the Analysis of Polystyrene by MALDI-TOF-MS. *J. Am. Soc. Mass Spectrom.* **2000**, *11*, 1095-1106.

106. Kéki, S.; Deák, G.; Zsuga, M. Copper(I) Chloride: a Simple Salt for Enhancement of Polystyrene Cationization in MALDI-MS. *Rapid Commun. Mass Spectrom.* **2001**, *15*, 675-678.

107. Puchades, M.; Westman, A.; Blennow, K.; Davidsson, P. Removal of Sodium Dodecyl Sulfate From Protein Samples Prior to MALDI-MS. *Rapid Commun. Mass Spectrom.* **1999**, *13*, 344-349.

108. Trimpin, S.; Grimsdale, A.C.; Räder, H.J.; Müllen, K. Characterization of an Insoluble Poly(9,9-Diphenyl-2,7-Fluorene) by Solvent-Free Sample Preparation for MALDI-TOF-MS. *Anal. Chem.* **2002**, *74*, 3777-3782.

109. Trimpin, S.; Rouhanipour, A.; Az, R.; Räder, H.J.; Müllen, K. New Aspects in MALDI-TOF-MS: a Universal Solvent-Free Sample Preparation. *Rapid Commun. Mass Spectrom.* **2001**, *15*, 1364-1373.

110. Nordhoff, E.; Schürenberg, M.; Thiele, G.; Lübbert, C.; Kloeppel, K.-D.; Theiss, D.; Lehrach, H.; Gobom, J. Sample Preparation Protocols for MALDI-MS of Peptides and Oligonucleotides Using Prestructured Sample Supports. *Int. J. Mass Spectrom.* **2003**, *226*, 163-180.

111. Guittard, J.; Hronowski, X.L.; Costello, C.E. Direct MALDI Mass Spectrometric Analysis of Glycosphingolipids on Thin Layer Chromatographic Plates and Transfer Membranes. *Rapid Commun. Mass Spectrom.* **1999**, *13*, 1838-1849.

112. McCrery, D.A.; Ledford, E.B., Jr.; Gross, M.L. Laser Desorption Fourier Transform-MS. *Anal. Chem.* **1982**, *54*, 1435-1437.

113. Claereboudt, J.; Claeys, M.; Geise, H.; Gijbels, R.; Vertes, A. Laser Microprobe-MS of Quaternary Phosphonium Salts: Direct Versus Matrix-Assisted Laser Desorption. *J. Am. Soc. Mass Spectrom.* **1993**, *4*, 798-812.

114. Gromer, S.; Gross, J.H. Methylseleninate Is a Substrate Rather Than an Inhibitor of

Mammalian Thioredoxin Reductase: Implications for the Antitumor Effects of Selenium. *J. Biol. Chem.* **2002**, *277*, 9701-9706.

115. Wood, T.D.; Van Cleef, G.W.; Mearini, M.A.; Coe, J.V.; Marshall, A.G. Formation of Giant Fullerene Gas-Phase Ions (C_{2n}^+, n = 60-500): Laser Desorption/Electron Ionization Fourier-Transform Ion Cyclotron Resonance Mass Spectrometric Evidence. *Rapid Commun. Mass Spectrom.* **1993**, *7*, 304-311.

116. Beck, R.D.; Weis, P.; Hirsch, A.; Lamparth, I. Laser Desorption-MS of Fullerene Derivatives: Laser-Induced Fragmentation and Coalescence Reactions. *J. Phys. Chem.* **1994**, *98*, 9683-9687.

117. Wolkenstein, K.; Gross, J.H.; Oeser, T.; Schöler, H.F. Spectroscopic Characterization and Crystal Structure of the 1,2,3,4,5,6-Hexahydrophenanthro-[1,10,9,8-*opqra*]perylene. *Tetrahedron Lett.* **2002**, *43*, 1653-1655.

118. Grim, D.M.; Siegel, J.; Allison, J. Does Ink Age Inside of a Pen Cartridge? *J. Forensic Sci.* **2002**, *47*, 1294-1297.

119. Jones, R.M.; Lamb, J.H.; Lim, C.K. Urinary Porphyrin Profiles by Laser Desorption/Ionization-TOF-MS Without the Use of Classical Matrixes. *Rapid Commun. Mass Spectrom.* **1995**, *9*, 921-923.

120. Jackson, A.T.; Yates, H.T.; Scrivens, J.H.; Critchley, G.; Brown, J.; Green, M.R.; Bateman, R.H. The Application of MALDI Combined With Collision-Induced Dissociation to the Analysis of Synthetic Polymers. *Rapid Commun. Mass Spectrom.* **1996**, *10*, 1668-1674.

121. Tang, X.; Dreifuss, P.A.; Vertes, A. New Matrixes and Accelerating Voltage Effects in MALDI of Synthetic Polymers. *Rapid Commun. Mass Spectrom.* **1995**, *9*, 1141-1147.

122. Montaudo, G.; Montaudo, M.S.; Puglisi, C.; Samperi, F. Self-Calibrating Property of MALDI-TOF Spectra of Polymeric Materials. *Rapid Commun. Mass Spectrom.* **1994**, *8*, 981-984.

123. Williams, J.B.; Chapman, T.M.; Hercules, D.M. MALDI-MS of Discrete Mass Poly(Butylene Glutarate) Oligomers. *Anal. Chem.* **2003**, *75*, 3092-3100.

124. Nielen, M.F.W.; Malucha, S. Characterization of Polydisperse Synthetic Polymers by Size-Exclusion Chromatography/MALDI-TOF-MS. *Rapid Commun. Mass Spectrom.* **1997**, *11*, 1194-1204.

125. de Koster, C.G.; Duursma, M.C.; van Rooij, G.J.; Heeren, R.M.A.; Boon, J.J. Endgroup Analysis of Polyethylene Glycol Polymers by MALDI Fourier-Transform Ion Cyclotron Resonance-MS. *Rapid Commun. Mass Spectrom.* **1995**, *9*, 957-962.

126. Weidner, S.; Kühn, G.; Just, U. Characterization of Oligomers in Poly(Ethylene Terephthalate) by MALDI-MS. *Rapid Commun. Mass Spectrom.* **1995**, *9*, 697-702.

127. Montaudo, M.S. Sequence Constraints in a Glycine-Lactic Acid Copolymer Determined by MALDI-MS. *Rapid Commun. Mass Spectrom.* **1999**, *13*, 639-644.

128. Montaudo, M.S. Mass Spectra of Copolymers. *Mass Spectrom. Rev.* **2002**, *21*, 108-144.

129. Murgasova, R.; Hercules, D.M. Quantitative Characterization of a Polystyrene/Poly(α-methylstyrene) Blend by MALDI-MS and Size-Exclusion Chromatography. *Anal. Chem.* **2003**, *75*, 3744-3750.

130. Krishnamurthy, T.; Ross, P.L.; Rajamani, U. Detection of Pathogenic and Non-Pathogenic Bacteria by MALDI-TOF-MS. *Rapid Commun. Mass Spectrom.* **1996**, *10*, 883-888.

131. Angeletti, R.; Gioacchini, A.M.; Seraglia, R.; Piro, R.; Traldi, P. The Potential of MALDI-MS in the Quality Control of Water Buffalo Mozzarella Cheese. *J. Mass Spectrom.* **1998**, *33*, 525-531.

132. Fanton, C.; Delogu, G.; Maccioni, E.; Podda, G.; Seraglia, R.; Traldi, P. MALDI-MS in the Dairy Industry 2. The Protein Fingerprint of Ewe Cheese and Its Application to Detection of Adulteration by Bovine Milk. *Rapid Commun. Mass Spectrom.* **1998**, *12*, 1569-1573.

133. Welham, K.J.; Domin, M.A.; Johnson, K.; Jones, L.; Ashton, D.S. Characterization of Fungal Spores by Laser Desorption/Ionization-TOF-MS. *Rapid Commun. Mass Spectrom.* **2000**, *14*, 307-310.

134. Harvey, D.J. MALDI-MS of Carbohydrates and Glycoconjugates. *Int. J. Mass Spectrom.* **2003**, *226*, 1-35.

135. Stahl, B.; Linos, A.; Karas, M.; Hillenkamp, F.; Steup, M. Analysis of Fructans From Higher Plants by MALDI-MS. *Anal. Biochem.* **1997**, *246*, 195-204.

136. Garrozzo, D.; Impallomeni, G.; Spina, E.; Sturiale, L.; Zanetti, F. MALDI-MS of

Polysaccharides. *Rapid Commun. Mass Spectrom.* **1995**, *9*, 937-941.

137. Harvey, D.J.; Naven, T.J.P.; Küster, B.; Bateman, R.; Green, M.R.; Critchley, G. Comparison of Fragmentation Modes for the Structural Determination of Complex Oligosaccharides Ionized by MALDI-MS. *Rapid Commun. Mass Spectrom.* **1995**, *9*, 1556-1561.

138. Lin, H.; Hunter, J.M.; Becker, C.L. Laser Desorption of DNA Oligomers Larger Than One Kilobase From Cooled 4-Nitrophenol. *Rapid Commun. Mass Spectrom.* **1999**, *13*, 2335-2340.

139. Bartolini, W.P.; Johnston, M.V. Characterizing DNA Photo-Oxidation Reactions by High-Resolution Mass Measurements With MALDI-TOF-MS. *J. Mass Spectrom.* **2000**, *35*, 408-416.

140. Wei, J.; Buriak, J.M.; Siuzdak, G. Desorption-Ionization-MS on Porous Silicon. *Nature* **1999**, *399*, 243-246.

141. Shen, Z.; Thomas, J.J.; Averbuj, C.; Broo, K.M.; Engelhard, M.; Crowell, J.E.; Finn, M.G.; Siuzdak, G. Porous Silicon As a Versatile Platform for Laser Desorption/Ionization-MS. *Anal. Chem.* **2001**, *73*, 612-619.

142. Laiko, V.V.; Taranenko, N.I.; Berkout, V.D.; Musselman, B.D.; Doroshenko, V.M. Atmospheric Pressure Laser Desorption/Ionization on Porous Silicon. *Rapid Commun. Mass Spectrom.* **2002**, *16*, 1737-1742.

143. Laiko, V.V.; Baldwin, M.A.; Burlingame, A.L. Atmospheric Pressure MALDI-MS. *Anal. Chem.* **2000**, *72*, 652-657.

144. Laiko, V.V.; Moyer, S.C.; Cotter, R.J. Atmospheric Pressure MALDI/Ion Trap-MS. *Anal. Chem.* **2000**, *72*, 5239-5243.

145. Moyer, S.C.; Marzilli, L.A.; Woods, A.S.; Laiko, V.V.; Doroshenko, V.M.; Cotter, R.J. Atmospheric Pressure MALDI (AP MALDI) on a Quadrupole Ion Trap Mass Spectrometer. *Int. J. Mass Spectrom.* **2003**, *226*, 133-150.

146. Doroshenko, V.M.; Laiko, V.V.; Taranenko, N.I.; Berkout, V.D.; Lee, H.S. Recent Developments in Atmospheric Pressure MALDI-MS. *Int. J. Mass Spectrom.* **2002**, *221*, 39-58.

147. Krutchinsky, A.N.; Loboda, A.V.; Spicer, V.L.; Dworschak, R.; Ens, W.; Standing, K.G. Orthogonal Injection of MALDI Ions into a TOF Spectrometer Through a Collisional Damping Interface. *Rapid Commun. Mass Spectrom.* **1998**, *12*, 508-518.

148. Hill, J.A.; Annan, R.S.; Biemann, K. MALDI With a Magnetic Mass Spectrometer. *Rapid Commun. Mass Spectrom.* **1991**, *5*, 395-399.

149. Annan, R.S.; Köchling, H.J.; Hill, J.A.; Biemann, K. Matrix-Assisted Laser Desorption Using a Fast-Atom Bombardment Ion Source and a Magnetic Mass Spectrometer. *Rapid Commun. Mass Spectrom.* **1992**, *6*, 298-302.

150. McIver, R.T., Jr.; Li, Y.; Hunter, R.L. MALDI With an External Ion Source Fourier-Transform Mass Spectrometer. *Rapid Commun. Mass Spectrom.* **1994**, *8*, 237-241.

151. Li, Y.; McIver, R.T., Jr.; Hunter, R.L. High-Accuracy Molecular Mass Determination for Peptides and Proteins by Fourier Transform-MS. *Anal. Chem.* **1994**, *66*, 2077-2083.

152. Li, Y.; Tang, K.; Little, D.P.; Koester, H.; McIver, R.T., Jr. High-Resolution MALDI Fourier Transform-MS of Oligonucleotides. *Anal. Chem.* **1996**, *68*, 2090-2096.

153. Baykut, G.; Jertz, R.; Witt, M. MALDI FT-ICR-MS With Pulsed in-Source Collision Gas and in-Source Ion Accumulation. *Rapid Commun. Mass Spectrom.* **2000**, *14*, 1238-1247.

154. O'Connor, P.B.; Costello, C.E. A High Pressure MALDI Fourier Transform-MS Ion Source for Thermal Stabilization of Labile Biomolecules. *Rapid Commun. Mass Spectrom.* **2001**, *15*, 1862-1868.

11 Electrospray Ionization

Electrospray ionization (ESI) "is a soft ionization technique that accomplishes the transfer of ions from solution to the gas phase. The technique is extremely useful for the analysis of large, non-volatile, chargeable molecules such as proteins and nucleic acid polymers." [1] Different from fast atom bombardment (FAB, Chap. 9) the solution is composed of a volatile solvent and the ionic analyte at very low concentration, typically 10^{-6}–10^{-3} M. In addition, the transfer of ions from the condensed phase into the state of an isolated gas phase ion starts at atmospheric pressure and leads continuously into the high vacuum of the mass analyzer. [2-5] This results in a marked softness of ionization and makes electrospray the "wings for molecular elephants". [6] Another reason for the extraordinary high-mass capability of ESI [7,8] is found in the characteristic formation of multiply charged ions in case of high-mass analytes. [4,9] Multiple charging folds up the *m/z* scale by the number of charges and thus compresses the ions into the *m/z* range of standard mass analyzers (Fig. 11.1 and Chap. 3.5)

Nowadays, ESI is the leading member of the group of *atmospheric pressure ionization* (API) methods and the method of choice for *liquid chromatography-mass spectrometry coupling* (LC-MS, Chap. 12). [10-13] Currently, ESI and MALDI (Chap. 10) are the most commonly employed ionization methods and they opened doors to the widespread biological and biomedical application of mass spectrometry. [5,10,11,13-17] Moreover, ESI serves well for the analysis of ionic metal complexes [18,19] and other inorganic analytes. [20-22]

> **Note:** Although the range up to *m/z* 3000 is normally employed for the detection of ions generated by ESI, ions of much higher *m/z* can be formed. [23,24] Even ions at *m/z* 85,000 have been observed. [25]

11.1 Development of ESI and Related Methods

11.1.1 Atmospheric Pressure Ionization

Atmospheric pressure ionization (API) was the first technique to directly interface solution phase with a mass analyzer. [26] In API, a solution of the analyte is injected into a stream of hot nitrogen to rapidly evaporate the solvent. The vapor passes through a ^{63}Ni source where electrons emitted from the radioactive ^{63}Ni isotope initiate a complex series of ionizing processes. Beginning with the ionization of N_2, consecutive ion molecule reactions finally lead to the formation of $[(H_2O)_n+H]^+$ cluster ions that are able to protonate the analyte via ion molecule reactions. [26-28] (Also see APCI in Chap. 11.8).

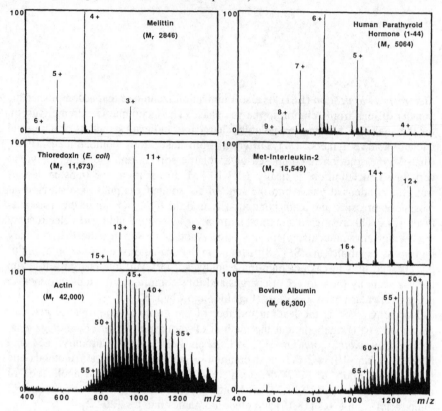

Fig. 11.1. Relationship between the mass of peptides and proteins and the number of ionic charges under ESI conditions. Adapted from Ref. [4] by permission. © American Chemical Society, 1990.

11.1.2 Thermospray

Thermospray (TSP) [29-31] unites three modes of operation. In pure TSP, a solution of the analyte and a volatile buffer, usually 0.1 M ammonium acetate, is evaporated from a heated capillary at a flow rate of $1-2$ ml min^{-1} into a heated chamber, hence the term thermospray. As the solvent evaporates, the analyte is forming adducts with ions from the buffer salt. While most of the neutrals are removed by a vacuum pump, the ions are extracted orthogonally from their main axis of motion by use of an electrostatic potential. The ions are transferred into a quadrupole mass analyzer through a pinhole of about 25 µm in diameter (Fig. 11.2). The quadrupole was employed according to its tolerance to poor vacuum conditions. As the pure TSP mode only works from high-polarity solvents in the presence of a buffer salt, modified modes of operation were developed to expand the use of TSP to lower-polarity systems: ii) use of an electrical discharge in the vapor phase [32] or iii) an electron-emitting filament as used in EI (Chap. 5) or CI (Chap. 7). TSP meant a breakthrough for LC-MS. [33]

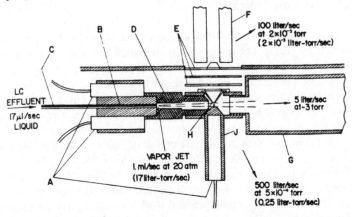

Fig. 11.2. Schematic of a thermospray interface. A: cartridge heater; B: copper block brazed to stainless steel capillary; C: capillary; D: copper tube; E: ion lenses; F: quadrupole mass analyzer; G: line to rotary vane pump; H: ion exit aperture; J: source heater. Reproduced from Ref. [30] by permission. © American Chemical Society, 1983.

Note: Thermospray is not a true API method because the liquid is sprayed into a vacuum of several hundred Pa instead of spraying at atmospheric pressure.

11.1.3 Electrohydrodynamic Ionization

An electrolytic solution of sufficiently low volatility can be transferred into the vacuum and sprayed from a fine capillary by the action of a strong electrostatic field. This is known as *electrohydrodynamic ionization* (EHI). [34,35] EHI results from the interaction of the field with the liquid meniscus at the end of the capillary tube. [36,37] A mist of micrometer-sized electrically charged droplets expands into the vacuum at supersonic speed. The droplets shrink upon evaporation of solvent. Shrinking causes the charge density on their surface to exceed the *Rayleigh limit* of stability, [38] i.e., the surface tension is overcome by electrostatic repulsion. The electric forces then tear the droplets apart. The sequence of droplet shrinking and subsequent disintegration into smaller sub-units occurs repeatedly and in the end leads to the formation of isolated ions in the gas phase. Although optimized EHI conditions [39] and convenient ion sources are at hand, e.g., EHI is compatible to FD ion sources upon replacement of the field emitter by a capillary tube, (Chap. 8.2), [40] EHI never became widespread in organic mass spectrometry. This is most probably due to its limitation to low-volatility solvents. Nevertheless, EHI can be applied to analyze polymers [41] and is used to generate primary ions in massive cluster impact (MCI) mass spectrometry (Chap. 9.5.1).

11.1.4 Electrospray Ionization

The development of electrospray ionization by Dole [42,43] indeed preceded API, TSP, and EHI by years. The underlying principle of ESI which it has in common with EHI even goes back to work by Zeleny [36] and Taylor. [37] Again, a mist of micrometer-sized electrically charged droplets is generated and the repetitive shrinking disintegration is also observed under ESI conditions. In contrast to EHI, the electrostatic sprayer is operated at atmospheric pressure. Thus, sufficient energy to vaporize volatile solvents without freezing the aerosol droplets is supplied by the surrounding gas. Presumably, the ability of ESI to work with standard solvents is the key to the tremendous success of ESI. [6]

The limitation of Dole's experiments was that the ions of the electrosprayed high-molecular weight polystyrene used in his group could not be detected by mass spectrometry. [42-44] It took years of work and lasted until the ending 1980s for the Fenn group to fully realize that analytes of 100–2000 u molecular weight can be readily analyzed with a quadrupole analyzer attached to a properly constructed ESI ion source. [2,6,45]

11.2 Ion Sources for ESI

11.2.1 Basic Design Considerations

An ESI source of sufficiently advanced design for mass spectrometry was designed by the Fenn group in the mid 1980s (Fig. 11.3). [2,45-47] The dilute sample solution is forced by a syringe pump through a hypodermic needle at a rate of 5–20 μl min^{-1}. The needle is held at a potential of 3–4 kV relative to a surrounding cylindrical electrode. Then, the electrosprayed aerosol expands into a countercurrent stream of dry nitrogen gas that serves as a heat supply for vaporization of the solvent. A small portion of the sprayed material enters the aperture of a short capillary (0.2 mm inner diameter, 60 mm length) that interfaces the atmospheric pressure spray to the first pumping stage (ca. 10^2 Pa). Most of the gas expanding from the desolvating aerosol is pumped off by a rotary vane pump as it exits from the capillary. A minor portion passes through the orifice of a skimmer into the high vacuum behind (ca. 10^{-3}–10^{-4} Pa). At this stage, desolvation of the ions is completed, while the ions are focused into a mass analyzer. Suitable potentials applied to capillary, skimmer, and lenses behind provide an effective transfer of ions through the interface, whereas neutral gas is not affected.

Modern ESI sources – also termed *ESI interfaces* – are constructed in many variations of this basic design. [48,49] They may have a heated transfer capillary or a countercurrent stream of heated nitrogen to enforce solvent evaporation. [50] These differences can influence a system's robustness and the degree of cluster ion formation with a particular ESI interface. [51,52] Whatever the details, they are all derived from a *nozzle-skimmer system* proposed by Kantrowitz and Grey [53] that delivers an intense cool molecular jet into the high vacuum environment. [42,45,54] The adiabatic expansion of the gas on the first pumping stage reduces random motion of the particles due to extensive cooling. Furthermore, a portion of the thermal motion is converted into directed flow by the nozzle-skimmer arrangement. In summary, this causes the heavier, analyte ion-containing solvent clusters to travel close to the center of the flight path through the interface, whereas light solvent molecules escape from the jet. [43]

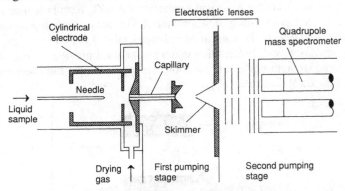

Fig. 11.3. Schematic of an early electrospray ion source. Reproduced from Ref. [2] by permission. © American Association for the Advancement of Science, 1989.

11.2.2 ESI with Modified Sprayers

The pure electrospray process of dispersing a liquid into an aerosol works best at flow rates of 1–20 µl min^{-1}. Conventional unassisted ESI has also limitations as a LC-MS interface due to the solvent properties in terms of volatility and polarity which can be electrosprayed without some type of assistance. Therefore, a number of sprayer modifications including a heated sprayer [55] have been developed to expand the range of ESI applications (Fig. 11.4).

The design of a *pneumatically assisted* ESI interface differs from the pure electrospray interface in that it provides a pneumatic assistance for the spray process. This is achieved by admitting a concentric flow of an inert gas such as nitrogen around the electrospray plume. [56-58] Pneumatic assistance allows for higher flow rates and for a reduced influence of the surface tension of the solvent used. [59] Pneumatically assisted ESI can accommodate flow rates of 10–200 µl min^{-1}.

In addition to high LC flow rates, solutions of high conductivity, and/or high surface tension are unsuitable for use with conventional ESI. An *ultrasonic nebulizer* can reduce such problems because it mechanically creates the spray. Unfortunately, the ultrasonically created droplets are comparatively large and this hinders ion formation, i.e., the ion yield is reduced. Nonetheless, the ultrasonic nebulizer handles liquid flow rates of 50–1000 µl min^{-1} and reduces the restrictions of the range of mobile-phase compositions amenable to ESI. For example, a mobile-phase gradient beginning at water : methanol = 95 : 5 falls outside the solvent range that can be used with conventional ESI, while an ultrasonic nebulizer will work. [60,61] With both pneumatically assisted ESI and ultrasonic nebulizer ESI the role of the high voltage almost reduces to the mere supply of electric charging of the droplets. If in the absence of high flow rates, high polarity alone is an issue, nanoESI provides the better tool to deal with (next section).

For *capillary zone electrophoresis* (CZE) mass spectrometry coupling, another modification of an ESI interface has been developed. This interface uses a *sheath flow* of liquid to make the electrical contact at the CZE terminus, thus defining both the CZE and electrospray field gradients. This way, the composition of the electrosprayed liquid can be controlled independently of the CZE buffer, thereby providing operation with buffers that could not be used previously, e.g., aqueous and high ionic strength buffers. In addition, the interface operation becomes independent of the CZE flow rate. [62]

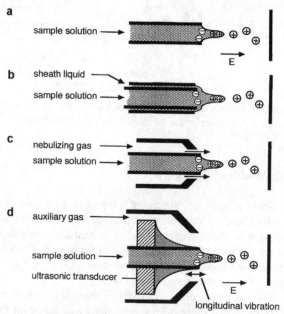

Fig. 11.4. Different sprayers for ESI. (**a**) Pure electrospray, (**b**) ESI with sheath liquid, (**c**) pneumatically assisted ESI, and (**d**) ultrasonic nebulizer. Adapted from Ref. [5] (p. 109) by permission. © John Wiley & Sons, Inc. 1997.

Note: *Pneumatically assisted electrospray* is also termed *ion spray* (ISP). However, the term ISP is not recommended instead of pneumatically assisted ESI because ISP i) represents a mere modification of the ESI setup and ii) is a company-specific term. [63]

11.2.3 Nano-Electrospray

Miniaturization of the electrospray is even more attractive than enlarging the flow rates. It has been theoretically described and experimentally demonstrated by Wilm and Mann that a more narrow spray capillary results in smaller droplets, and moreover in much reduced flow rates. [64] Such a downscaling can be achieved by replacing the spray needle by a borosilicate glass capillary of some microliters volume to which a fine tip is pulled with a micropipette puller. The tip has a narrow bore exit of 1–4 μm diameter making flow rates of 20–50 nl min^{-1} sufficient to provide a stable electrospray. [65] Derived from the nanoliter flow rates, the term *nano-electrospray* (nanoESI) has become established for that technique. While conventional ESI produces initial droplets of 1–2 μm in diameter, the droplet size from nanoESI is less than 200 nm, i.e., their volume is about 100–1000 times smaller. NanoESI allows for high-polarity solvents such as pure water in both positive- and negative-ion mode, has extremely low sample consumption [66], and tolerates even higher loads with buffer salts than conventional ESI. [50,67]

For the measurement, the nanoESI capillary is adjusted at about 1 mm distance to the entrance of the counter electrode by means of a micromanipulator. Thus, precise optical control is needed during positioning to prevent crashing of the tip or electric discharges during operation. Commercial nanoESI sources are therefore equipped with a built-in microscope or camera (Fig. 11.5). The spray voltage of 0.7–1.1 kV is normally applied via an electrically conducting coating on the outer surface of the spray capillaries, usually a sputtered gold film. Occasionally, wider capillaries with a fine metal filament inside are used. With the high voltage switched on, the liquid sample flow is solely driven by capillary forces refilling the aperture as droplets are leaving the tip. Sometimes, liquid flow is slightly supported by a gentle backing pressure on the capillary.

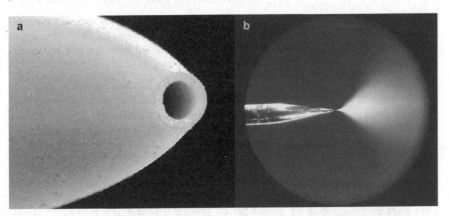

Fig. 11.5. Nano-electrospray; (**a**) SEM micrograph of the open end of a glass nanoESI capillary having a 2-μm aperture, (**b**) microscopic view of the spray from a nanoESI capillary as provided by observations optics. By courtesy of New Objective, Woburn, MA.

Example: The reduced sample consumption of nanoESI allows for the sequencing of 800 fmol peptides (Chap. 9.4.6) obtained by tryptic digestion of the protein bovine serum albumin (BSA, Fig. 11.6). [66] The experiment depicted below requires each of the BSA-derived peptide ions in the full scan spectrum to be subjected to fragment ion analysis by means of CID-MS/MS on a triple quadrupole instrument (Chaps. 2.12 and 4.4.5).

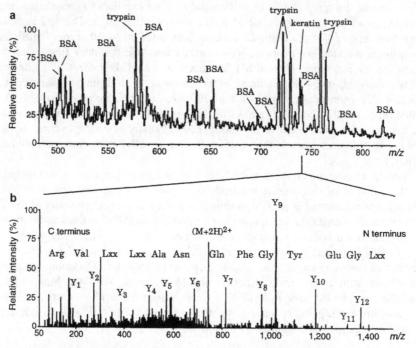

Fig. 11.6. Peptide sequencing by nanoESI-CID-MS/MS from a tryptic digest of bovine serum albumin (BSA); 800 fmol of BSA were used. (**a**) Full scan spectrum, (**b**) fragmentation of the selected doubly charged peptide ion at m/z 740.5. Adapted from Ref. [66] by permission. © Nature Publishing Group, 1996.

> **Note:** Besides its low sample consumption, nanoESI is free of *memory effects* because each sample is supplied in a fresh capillary by means of disposable micropipettes. Furthermore, the narrow exits of nanoESI capillaries prevent air-sensitive samples from rapid decomposition.

11.2.3.1 Nano-Electrospray from a Chip

The sample throughput of nanoESI is limited by the comparatively time-consuming procedure of manual capillary loading. A chip-based nanoESI sprayer on an etched silicon wafer allows for the automated loading of the sprayer array by a pipetting robot (Fig. 11.7). The chip provides a 10×10 array of nanoESI spray nozzles of 10 μm inner diameter. Volumes up to 10 μl are supplied directly from a pipette contacting the chip from the backside. An electrically conducting coating of the pipette tip is used to connect the sprayer to high voltage. Pipetting robot and automated chip handling are united in a common housing that replaces the conventional (nano)ESI spray unit.

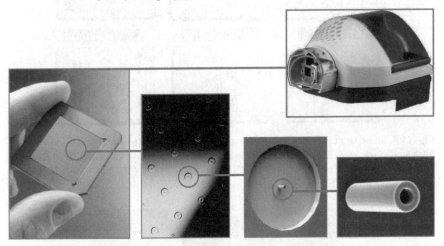

Fig. 11.7. Illustration of the chip-based Advion nanoESI system. The pictures stepwise zoom in from the pipetting unit to the spray capillary on the silicon chip. By courtesy of G. Schultz, Advion BioSciences, Ithaca, NY.

11.2.4 ESI with Modified Spray Geometries

Clogging of capillaries and skimmers is one of the prominent problems with the otherwise easy-to-use ESI interfaces. In particular when analytes are accompanied by involatile impurities such as buffer salts used to improve liquid chromatography or organic material as present in blood or urine samples, for example, the deposition of material can cause the rapid breakdown of the ESI interface. Numerous modifications of spray geometry and skimmer design were developed to circumvent this problem. The principle of such designs is to achieve spatial separation of the deposition site of non-volatized material and the location of the ion entrance into the mass spectrometer. Such designs include i) spraying *off-axis*, ii) spraying at some angle up to *orthogonal*, and iii) guiding the desolvating microdroplets through *bent paths* (Fig. 11.8). [63] Probably the most successful design of modern commercial ESI interfaces is the Micromass *z-spray* interface (Fig. 11.9). Another advantage of orthogonal spraying is that it selectively collects small and highly charged droplets which are the preferred source of analyte ions. Larger and less-charged droplets are not sufficiently attracted by the extraction field at 90° angle and therefore pass by.

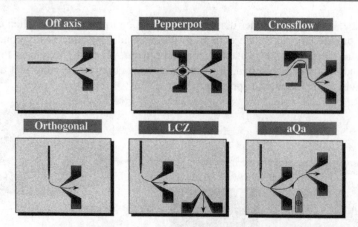

Fig. 11.8. Some strategies used in commercial API ion sources to increase solvent compatibility and system robustness. Some of these designs are exclusive of particular commercial brands: Pepperpot, Crossflow, and LCZ (Micromass); AQA (Thermo Finnigan). Reproduced from Ref. [63] by permission. © John Wiley & Sons, 1999.

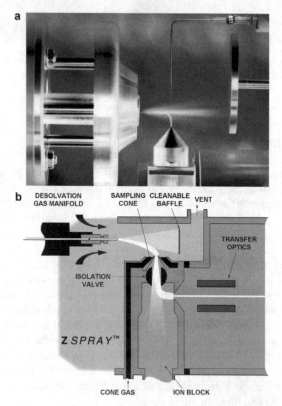

Fig. 11.9. Micromass z-spray interface. (**a**) Photograph of the actual spray, (**b**) schematic drawing. By courtesy of Waters Corporation, MS Technologies, Manchester, UK.

11.2.5 Skimmer CID

The intermediate pressure region between nozzle and skimmer of an electrospray interface does not only provide space for effective desolvation, [68,69] but can also be used to achieve fragmentation of the ions by CID (Chap. 2.12). While a comparatively high pressure in this region can effect collisional cooling rather than dissociative collisions, [70] a higher voltage difference between nozzle and skimmer enhances ionic fragmentation by CID. [68,70,71]

Skimmer CID or *nozzle/skimmer CID* can i) strip off residual solvent molecules, ii) achieve fragmentation of electrosprayed ions resulting in spectra similar to CI mode, [72] and iii) generate first generation fragment ions for further tandem MS experiments. The latter method provides a *pseudo MS^3* operation on triple quadrupole mass spectrometers [73,74] (real MS^3 requires mass-selection prior to the first CID stage in addition). A programmable skimmer CID routine delivers ESI mass spectra with a variable degree of fragmentation from a single run. [71,73-75]

Example: Even a moderate voltage drop between nozzle and skimmer can cause the elimination of weakly bonded substituents such as CO_2 in case of carbon dioxide-protected deprotonated *N*-heterocycles. In particular SnMe$_3$-substituted anions such as 2-(trismethylstannyl)pyrrole-*N*-carbamate exhibit variations in the [A–CO$_2$]$^-$/A$^-$ ratio of up to a factor of 30 (Fig. 11.10). [76]

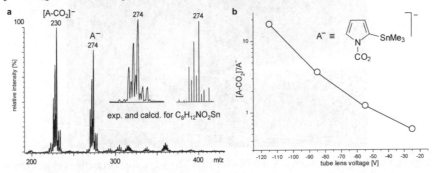

Fig. 11.10. (a) Partial negative-ion ESI spectrum of 2-(trismethylstannyl)pyrrole-*N*-carbamate (A$^-$) from tetrahydrofurane at low nozzle-skimmer voltage drop and (b) dependence of the [A–CO$_2$]$^-$/A$^-$ ratio variation of this voltage.

11.3 Ion Formation

11.3.1 Formation of an Electrospray

Ion formation in ESI can be regarded as divided into three steps: i) creation of an electrically charged spray, ii) dramatic reduction of the droplets' size, and finally iii) liberation of fully desolvated ions. To understand the formation of a continuous spray, consider the surface of an ion-containing liquid at the end of an electri-

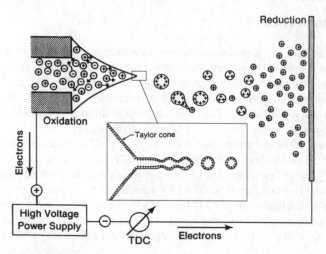

Fig. 11.11. Schematic of Taylor cone formation, ejection of a jet, and its disintegration into a fine spray. The electrochemical processes of ESI [77,78] are also assigned. Adapted from Ref. [49] by permission of the authors.

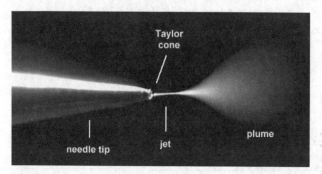

Fig. 11.12. Electrospray from a nanoESI capillary. The jet emitted from the Taylor cone is clearly visible and separate from the region of rapid expansion into a plume of microdroplets. By courtesy of New Objective, Woburn, MA.

cally conducting capillary of about 75 µm inner diameter which is held at an electric potential of 3–4 kV with reference to a counter electrode at 1–2 cm distance. At the open end of the capillary this liquid is exposed to an electric field of about 10^6 V m^{-1}. The electric field causes charge separation in the liquid and finally deformation of the meniscus into a cone, the formation of which has been discovered by Zeleny [36] and theoretically first described by Taylor. [37]

In detail, the surface starts forming an oval under the influence of increasing field strength; and in turn, a sharper curvature of the oval increases the field strength. When a certain field strength is reached, the equilibrium of surface tension and electrostatic forces becomes independent of the curvature's radius, and mathematically, the radius could become zero. However, in a real system infinite field strength is impossible. Instead, at the moment the critical electric field strength is reached, the *Taylor cone* instantaneously forms and immediately starts ejecting a fine jet of liquid from its apex towards the counter electrode. [64] This mode of operation is termed *cone-jet mode*. [5] The jet carries a large excess of ions of one particular charge sign, because it emerges from the point of highest charge density, i.e., from the cone's tip. However, such a jet cannot remain stable for an elongated period, but breaks up into small droplets. Due to their charge, these droplets are driven away from each other by Coulombic repulsion. Overall, this process causes the generation of a fine spray, and thus gave rise to the term *electrospray* (Figs. 11.11 and 11.12).

11.3.2 Disintegration of Charged Droplets

When a micrometer-sized droplet carrying a large excess of ions of one particular charge sign – some 10^4 charges are a realistic value – evaporates some solvent, the charge density on its surface is continuously increased. As soon as electrostatic repulsion exceeds the conservative force of surface tension, disintegration of the droplet into smaller sub-units will occur. The point at which this occurs is known as *Rayleigh limit*. [38] Originally, it has been assumed that the droplets would then suffer a *Coulomb fission* (or *Coulomb explosion*). This process should occur repeatedly to generate increasingly smaller microdoplets. While the model of a cascading reduction in size holds valid, more recent work has demonstrated that the microdroplets do not explode, but eject a series of much smaller microdroplets from an elongated end (Fig. 11.13). [49,79,80] The ejection from an elongated end can be explained by deformation of the flying microdroplets, i.e., they have no

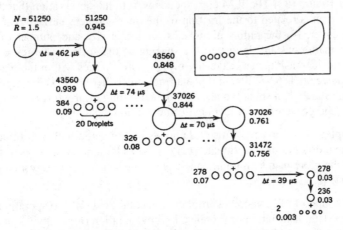

Fig. 11.13. Illustration of droplet jet fission. The average number of charges on a droplet, the radii of the droplets [μm], and the timescale of events are assigned. The inset shows a drawing of droplet jet fission based on an actual flash microphotograph. Reproduced from Ref. [49] by permission of the authors.

perfect spherical shape. Thus, the charge density on their surface is not homogeneous, but significantly increased in the region of sharper curvature. The smaller offspring droplets carry off only about 1–2 % of the mass, but 10–18 % of the charge of the parent droplet. [79] This process resembles the initial ejection of a jet from the Taylor cone. The concept of this so-called *droplet jet fission* is not only based on theoretical considerations but can be proven by flash microphotographs. [80,81] The total series of events from the initially sprayed droplet to the isolated ion takes less than one millisecond.

11.3.3 Formation of Ions from Charged Droplets

The elder model of ion formation, the *charged-residue model* (CRM), assumes the complete desolvation of ions by successive loss of all solvent molecules from droplets that are sufficiently small to contain just one analyte molecule in the end of a cascade of Coulomb fissions. [9,42,84] The charges (protons) of this ultimate droplet are then transferred onto the molecule. This would allow that even large protein molecules can form singly charged ions, and indeed, CRM is supported by this fact. [23]

A later theory, the *ion evaporation model* (IEM), [82,83] describes the formation of desolvated ions as direct evaporation from the surface of highly charged microdroplets. [85] (Ion evaporation is also discussed in FD-MS, Chap. 8.5.1). Ionic solvation energies are in the range of 3–6 eV, but thermal energy can only contribute about 0.03 eV at 300 K to their escape from solution. Thus, the electric force has to provide the energy needed. It has been calculated that a field of 10^9 V m^{-1} is required for ion evaporation which corresponds to a final droplet diameter of 10 nm. [83] The IEM corresponds well to the observation that the number of charges is related to the fraction of the microdroplet's surface that a molecule can cover. As the radius diminishes, molecule size and number of droplet charges remain constant; however, the spacing of the surface charges decreases, and thus the increasing charge density brings more charges within the reach of an analyte molecule. [86,87] Flat and planar molecules therefore exhibit higher average charge states than spheric ones, e.g., the unfolding of proteins is accompanied by higher charge states under identical ESI conditions. [88,89]

Example: The cleavage of disulfide bonds by reduction with 1,4-dithiothreitol causes the unfolding of the protein. This exposes additional basic sites to protonation, and therefore results in higher average charge states in the corresponding ESI spectrum (Fig. 11.14). [88]

Further support of IEM comes from the effect of the droplet evaporation rate on the charge state distribution of proteins. Fast evaporation (more drying gas, higher temperature) favors higher average charge states, while slower evaporation results in fewer charges. This is in accordance with the reduced time available for ion evaporation from the shrinking droplet leading to a relative enrichment of charge on the droplet and thus on the leaving ions. [86]

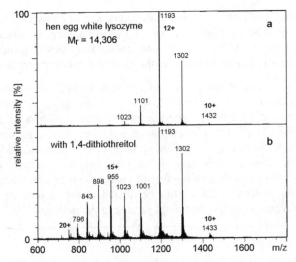

Fig. 11.14. Positive-ion ESI spectra of (**a**) hen egg white lysozyme and (**b**) the protein after addition of 1,4-dithiothreitol. Reproduced from Ref. [88] by permission. © American Chemical Society, 1990.

On the contrary, it may be argued that the electric field strength locally necessary to evaporate ions from a droplet cannot be attained because of the prior fission of the droplet due to crossing the Rayleigh limit. [23,90]

More recent work revealed the importance of gas phase proton transfer reactions. [91-94] This implies that multiply charged peptide ions do not exist as preformed ions in solution, but are generated by gas phase ion-ion reactions (Chap. 11.4.4). The proton exchange is driven by the difference in proton affinities (PA, Chap. 2.11) of the species encountered, e.g., a protonated solvent molecule of low PA will protonate a peptide ion with some basic sites left. Under equilibrium conditions, the process would continue until the peptide ion is "saturated" with protons, a state that also marks its maximum number of charges.

> **Note:** There is a continuing debate about ion formation in ESI. [79,87,95] In summary, it may be assumed that CRM holds valid for large molecules [9] while the formation of smaller ions is better described by IEM. [79,95]

11.4 Charge Deconvolution

11.4.1 Problem of Multiple Charging

The discussion of ion formation in ESI has revealed that the appearance of an ESI spectrum can largely be influenced by the actual experimental conditions as defined by the pH of the sprayed solution, the flow of nebulizing or drying gas, and the temperature of these or of a desolvation capillary. In particular, the degree of multiple charging is affected by those parameters, and of course, by the compound class under study (also cf. Figs. 11.1 and 11.14). In addition, insufficient spraying conditions (like the use of a nebulizer) can lead to a net lack of protons which reduces the achievable charge state due to the limited number of protons available.

Example: The average charge state of a protein depends on whether it is denatured or not and on the actual solvent; lower pH causes more protons to be attached to the protein than neutral conditions. The degree of denaturization in turn depends on the pH of the electrosprayed solution. Resulting conformational changes of the protein, e.g., unfolding upon protonation, make additional basic sites accessible, thereby effecting an increase of the average charge state (Fig. 11.15.) [89] The maximum number of charges that can be placed upon peptide and protein molecules can clearly be related to the number of basic amino acid residues (arginine, lysine, histidine) present in the molecule (Fig. 11.16). [88]

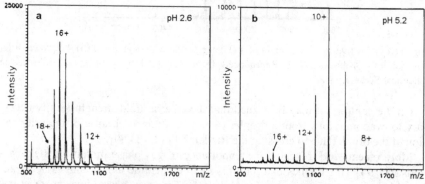

Fig. 11.15. Positive-ion ESI mass spectra of cytochrome c at different pH of the sprayed solution: (**a**) at pH 2.6, (**b**) at pH 5.2. Adapted from Ref. [89] by permission. © American Chemical Society, 1990.

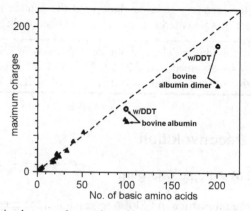

Fig. 11.16. Correlation between the number of basic amino acid residues and the maximum number of charges observed for a set of peptides and proteins under ESI conditions. Reproduced from Ref. [88] by permission. © American Chemical Society, 1990.

Example: In case of synthetic polymers, multiple charging causes the simultaneous occurrence of several interfering ion series each of them representing the molecular weight distribution of the polymer. [87] Whereas low-mass polyethylene glycol exhibits only singly charged ions, PEGs of higher mass form doubly, triply, and multiply charged ions under ESI conditions (Fig. 11.17). [2] Furthermore, those charge distributions depend on the concentration of the sample, e.g., PEG 1450 yields triply and few doubly charged ions at 0.005 mg ml^{-1} in MeOH : H$_2$O = 1 : 1, triply and doubly charged ions of equal abundance plus few singly charged ions at 0.05 mg ml^{-1}, but mainly doubly charged ions accompanied by few singly and triply charged ions at 0.5 mg ml^{-1}. [87] This demonstrates that a comparatively fixed number of charges in a droplet is distributed among few or many analytes molecules contained and thus supports IEM. PEGs and related compounds form [M+alkali$_n$]$^{n+}$ ions. [2,96,97]

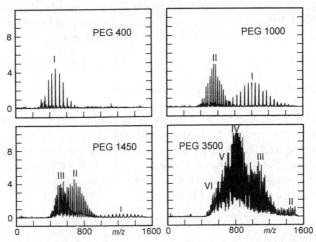

Fig. 11.17. Positive-ion ESI spectra of different PEGs at 0.05 mg ml^{-1}. (**a**) PEG 400 yields singly charged ions, (**b**) PEG 1000 forms singly and doubly charged ions, (**c**) with PEG 1450 the charge states 2+ and 3+ predominate over 1+, and (**d**) with PEG 3500 the spectrum is a complex superimposition of charge states 6+ to 2+. Reproduced from Ref. [2] by permission. © American Association for the Advancement of Science, 1989.

In general, the number of charges on a molecule in ESI depends on its molecular weight [9,98] and on the number of sites available for charge localization, e.g., sites that can be protonated, [1,9,88,89] cationized, [97] or deprotonated. [99,100] On one hand, this behavior is advantageous as it folds up the *m/z* scale and makes even extremely large molecules accessible to standard mass analyzers (Fig. 11.18). On the other hand, this produces a confusingly large number of peaks and requires tools to deal with in order to enable reliable mass assignments of unknown samples.

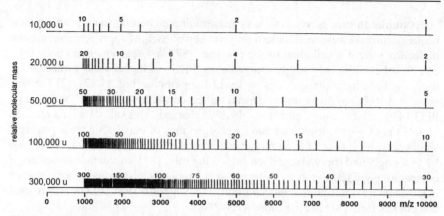

Fig. 11.18. Calculated m/z values for the different charge states of molecules of different molecular weight. Representative peaks are labeled with their corresponding charge state. Adapted from Ref. [98] by permission. © John Wiley & Sons, 1992.

11.4.2 Mathematical Charge Deconvolution

The calculation of the number of charges corresponding to individual peaks and an unknown molecular weight M_r is straightforward. [58,98,101] The procedure is based on the fact that adjacent peaks in an ESI mass spectrum of a single compound have charge states differing by one, i.e., there are no gaps or jumps in the pattern. Thus, for a pair of peaks at m/z_1 (higher value) and m/z_2 (lower value) we have for the charge states of neighboring peaks:

$$n_2 = n_1 + 1 \tag{11.1}$$

Using m_H for the mass of a proton, m/z_1 is determined by

$$m/z_1 = \frac{M_r + n_1 m_H}{n_1} \tag{11.2}$$

and m/z_2 of the peak at lower mass is given by

$$m/z_2 = \frac{M_r + n_2 m_H}{n_2} = \frac{M_r + (n_1 + 1)\, m_H}{n_1 + 1} \tag{11.2a}$$

where n_2 can be expressed by inserting Eq. 11.1. The charge state n_1 can then be obtained from

$$n_1 = \frac{m/z_2 - m_H}{m/z_1 - m/z_2} \tag{11.3}$$

Having calculated n_1, M_r is given by

$$M_r = n_1 \left(m/z_1 - m_H \right) \tag{11.4}$$

In case of cationization instead of protonation, m_H has to be replaced by the corresponding mass of the cationizing agent, e.g., NH_4^+, Na^+, or K^+. (The recognition of cationized species is addressed in Chap. 8.5.7).

Example: For the ease of calculation we use nominal mass (Chap. 3.1.4) and assume the charges from protonation. Consider the first peak at m/z 1001, the second at m/z 501. Now n_1 is obtained according to Eq. 11.3 from $n_1 = (501 - 1) / (1001 - 501) = 500 / 500 = 1$. Therefore, M_r is calculated from Eq. 11.4 to be $M_r = 1 \times (1001 - 1) = 1000$. (The doubly protonated ion is detected at m/z 501 because $(1000 + 2) / 2 = 501$.)

The above algorithm works well for pure compounds and simple mixtures, but it becomes increasingly difficult to assign all peaks properly when complex mixtures are to be addressed. Additional problems arise from the simultaneous presence of peaks due to protonation and alkali ion attachment etc. Therefore, numerous refined procedures have been developed to cope with these requirements. [102] Modern ESI instrumentation is normally equipped with elaborate software for *charge deconvolution*.

Example: The ESI mass spectrum and the charge-deconvoluted molecular weights (inset) of bovine serum albumine (BSA) as obtained from a quadrupole ion trap instrument are compared below (Fig. 11.19). Ion series A belongs to the noncovalent BSA dimer, series B results from the monomer. [24]

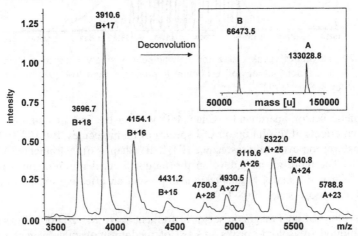

Fig. 11.19. Partial ESI mass spectrum of BSA and molecular weights after charge deconvolution (inset). Charge states are assigned to both series of peaks. Reproduced from Ref. [24] by permission. © John Wiley & Sons, 2000.

Note: As each pair of signals delivers an independent mass value for the hypothetical singly charged ion, mass accuracy can greatly be enhanced in ESI by multiple determination of this value and subsequent calculation of the average.

11.4.3 Hardware Charge Deconvolution

The most effective technique to deal with complex spectra due to multiply charged ions is to achieve the full separation between signals corresponding to different charge states and to resolve their isotopic patterns. Beyond a molecular weight of about 2000 u this requires high-resolving mass analyzers.

Example: At molecular weights of some 10^3 u the isotopic distribution of organic ions becomes several masses wide (Fig. 11.20). The minimum resolution for its full separation is always equal to the ion's mass number (Chaps. 3.3.4 and 3.4.3). Lower resolution can only provide an envelope over the distribution. At insufficient resolution, the resulting peak may even be wider than the envelope. [98]

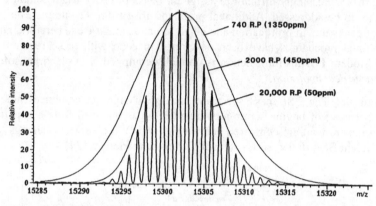

Fig. 11.20. Theoretical peak shape for a hypothetical singly charged protein ion of $M_r = 15,300$ at different settings of resolution. Reproduced from Ref. [98] by permission. © John Wiley & Sons, 1992.

Magnetic sector instruments (Chap. 4.3) were used first to demonstrate the beneficial effects of resolution on ESI spectra of biomolecules. [96,103,104] Fourier transform ion cyclotron resonance (FT-ICR, Chap. 4.6) instruments followed. [105-107] The more recently developed orthogonal acceleration of time-of-flight (oaTOF, Chap. 4.2 and 4.7) analyzers also present an effective means to resolve all or at least most peaks. [108-110]

Example: ESI on a magnetic sector instrument set to $R = 20,000$ allows for the full resolution of isotopic peaks in case of medium-molecular weight proteins (Fig. 11.21). This enables the direct determination of the charge state of the ions from the spacing of the isotopic peaks, i.e., 11+ for the lysozyme ion due to the average spaces of $\Delta m = 0.091$ u and 13+ for the myoglobin ion due to $\Delta m = 0.077$ u. In this particular case, the lysozyme $[M+11H]^{11+}$ ion serves as a mass reference for the accurate mass measurement of the "unknown" $[M+13H]^{13+}$ ion. [103]

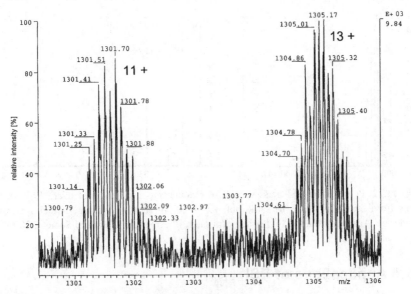

Fig. 11.21. Partial high-resolution ESI spectrum ($R = 20,000$) of a mixture of lysozyme and myoglobin. Reproduced from Ref. [103] by permission. © John Wiley & Sons, 1993.

11.4.4 Controlled Charge Reduction in ESI

The complexity of ESI spectra of mixtures where all components form series of multiply charged ions is apparent. An alternative approach to high resolution is presented by the controlled reduction of the charge state of the ions. *Charge reduction electrospray* (CRE) results in a significantly reduced number of peaks per component at the cost of their detection being required at substantially higher *m/z*. [111] The aforementioned oaTOF analyzers provide a sufficient mass range for such an experimental approach.

Charge reduction can be accomplished by neutralizing ion-molecule reactions during the desolvation step of ESI. The reducing ions needed for such neutralizations can either be generated by irradiating the gas with a ^{210}Po α-particle source [111,112] or more conveniently by a corona discharge. [113,114] Besides being nonradiative, the corona discharge offers the advantage of being tunable to achieve varying degrees of neutralization (Fig. 11.22). [113,114]

Ion-ion chemistry of oppositely charged ions presents another approach to charge reduction and charge state determination. [91] Such studies were done i) by employing the region preceding the skimmer of an ESI interface as a flow reactor and ii) inside a quadrupole ion trap. In the first case, both the substrate ion and the oppositely charged reactant ion are created by means of two separate ESI sprayers attached to a common interface. [115] The approach in a quadrupole ion trap makes use of the in-trap generation of proton-transferring reactant ions, while substrate ions are admitted via the ESI interface. [116]

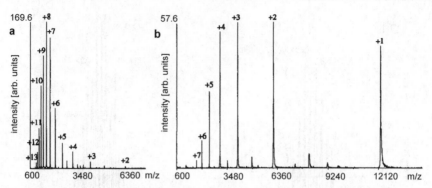

Fig. 11.22. Positive-ion ESI mass spectra of cytochrome c (**a**) under standard ESI conditions from acidic solution, (**b**) same but with medium setting of charge-reducing corona discharge. Adapted from Ref. [113] by permission. © American Chemical Society, 2000.

11.5 Applications of ESI

ESI is not only a versatile tool for any aspects of peptide and protein characterization including their complete sequencing, it also offers numerous other fields of application [5,17] some of which are highlighted below.

11.5.1 ESI of Small Molecules

Polar analytes in the m/z 100–1500 range are often involved in pharmaceutical analytics including metabolism studies. The types of ions formed are various and depend on the ion polarity, the pH of the solution, the presence of salts, and the concentration of the sprayed solution. Multiply charged ions are rarely observed.

Example: The compound below (Fig. 11.23) represents the functional part of an effective drug (BM 50.0341) inhibiting HIV-1 infection by suppression of the unfolding of the gp120 glycoprotein. Its positive-ion nanoESI spectrum from ethanol in the presence of ammonium chloride exhibits signals due to the formation of $[M+H]^+$, $[M+NH_4]^+$, and $[M+Na]^+$ ions. In addition, cluster ions of the type $[2M+H]^+$, $[2M+NH_4]^+$, and $[2M+Na]^+$ are observed. In negative-ion mode, the $[M-H]^-$ ion is accompanied by $[M+Cl]^-$ [117] and $[M+EtO]^-$ adduct ions; the corresponding cluster ion series is also observed.

11.5.2 ESI of Metal Complexes

In general, ESI can be well applied to ionic metal complexes and related compounds if these are soluble to at least 10^{-6} M in solvents suitable for the method. [18,19] Whether conventional or nanoESI should be employed basically depends

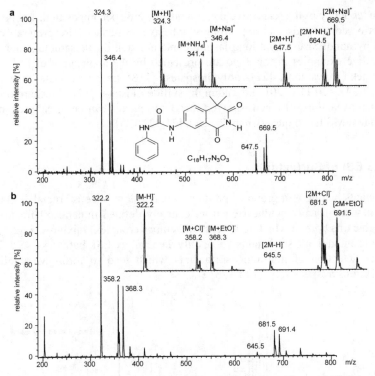

Fig. 11.23. Positive- (**a**) and negative-ion (**b**) nanoESI spectra of an anti-HIV drug from ethanol in the presence of ammonium chloride. By courtesy of H.-C. Kliem and M. Wiessler, German Cancer Research Center, Heidelberg.

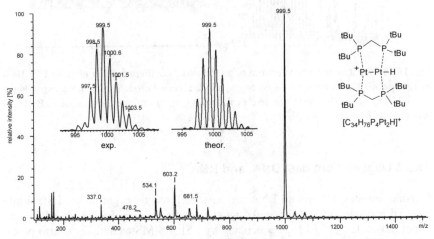

Fig. 11.24. Positive-ion ESI spectrum of a cationic dinuclear platinum hydride complex from dichloromethane solution. The insets compare experimental and theoretical isotopic patterns. By courtesy of P. Hofmann, University of Heidelberg.

on the tendency of the respective compounds towards decomposition. Labile complexes or compounds that are strongly adhesive to surfaces are preferably analyzed by nanoESI to avoid long-lasting contamination of the sample supply line. Illustrative examples in this field are presented by the application of ESI to isopoly metal oxyanions, [22], polyphospates, [118] transition metal complexes (Fig. 11.24), [20,119,120] and cadmium sulfide clusters. [121] Furthermore, the gas phase reactions of electrosprayed metal complexes can be directly examined by tandem MS techniques (Chap. 2.12.3). [21,122,123]

11.5.3 ESI of Surfactants

Surfactants belong to a group of products where a low price is crucial, and therefore they are usually synthesized from coarsely defined mineral oil fractions or vegetable oils both of which represent (sometimes complex) mixtures (Fig. 11.25). Cationic and anionic surfactants are readily detected by ESI, but it also serves well for the detection of non-ionic surfactants which tend to form $[M+alkali]^+$ or $[M-H]^-$ ions, respectively. [124-128]

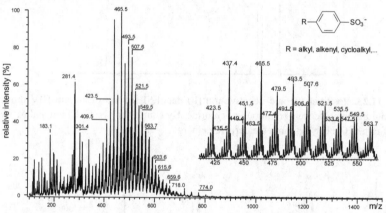

Fig. 11.25. Negative-ion ESI spectrum of an industrial cooling lubricant dissolved 1 : 1000 in 1-propanol. The dominant ions belong to alkylbenzene sulfonates. The inset expands the *m/z* 420–555 range, the most intensive peaks belonging to saturated alkyl chains. By courtesy of OMTEC GmbH, Eberbach.

11.5.4 Oligonucleotides, DNA, and RNA

Oligonucleotides, DNA, and RNA are best analyzed by negative-ion ESI. While MALDI becomes difficult beyond oligonucleotide 20-mers, ESI can handle much larger molecules. [129-131] Sequencing by ESI-MS/MS techniques is also possible. [129,131] However, the problems associated with the multiple exchange of protons versus alkali ions remain (Chap. 10.5.5). Instead of ion exchange beads, nitrogen bases have proven very helpful in removing alkali ions from solutions.

[132] In particular the addition of 25 mM imidazole and piperidine yields very clean spectra (Fig. 11.26). [133]

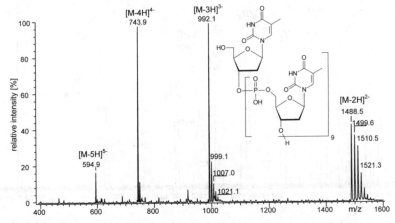

Fig. 11.26. Negative-ion ESI spectrum of the oligonucleotide dT_{10} in the presence of imidazole and piperidine. Here, H^+/Na^+ exchange is not completely suppressed. By courtesy of T. Krüger, University of Heidelberg.

11.5.5 ESI of Oligosaccharides

Oligosaccharides [50,70,134,135] as well as closely related compounds such as glycoproteins, [69,136] gangliosides, [137] liposaccharides etc. are similar to oligonucleotides in that they require polar solvents and very soft ionization, in particular when the molecules are branched. As demonstrated by a large number of applications, ESI permits molecular weight determination and structure elucidation in these cases (Fig. 11.27; for a general fragmentation scheme cf. Chap. 10.5.4). [5,13,14,17]

11.6 Atmospheric Pressure Chemical Ionization

In *atmospheric pressure chemical ionization* (APCI) ion-molecule reactions occurring at atmospheric pressure are employed to generate the ions, i.e., it represents a high-pressure version of conventional chemical ionization (CI, Chap. 7). The CI plasma is maintained by a corona discharge between a needle and the spray chamber serving as the counter electrode. The ions are transferred into the mass analyzer by use of the same type of vacuum interface as employed in ESI. Therefore, ESI ion sources can easily be switched to APCI: instead of an ESI sprayer, a unit comprising a heated pneumatic nebulizer and the spray chamber with the needle electrode are put in front of the orifice, while the atmospheric pressure-to-vacuum interface remains unchanged. [48,138]

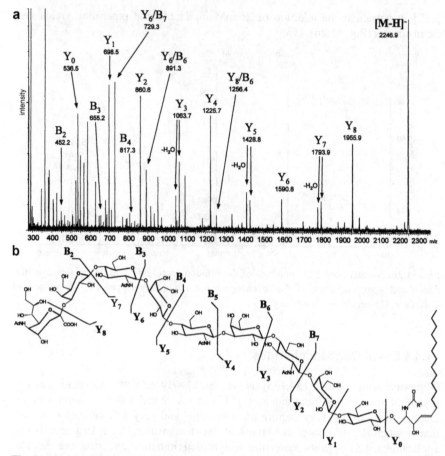

Fig. 11.27. (a) NanoESI-CID-MS/MS spectrum of the [M–H]⁻ ion, m/z 2246.9, of a modified nonasaccharide obtained in a Q-TOF hybrid instrument and **(b)** proposed structure with fragments indicated. Reproduced from Ref. [137] by permission. © Elservier Science, 2001.

The significant enhancement of ion formation by a corona discharge as compared to a ^{63}Ni source has already been implemented in early API sources. [139,140] The nature of the APCI plasma varies widely as both solvent and nebulizing gas contribute to the composition of the CI plasma, i.e., APCI spectra can resemble PICI, CECI, NICI, or EC spectra (Chap. 7.2–7.4) depending on the actual conditions and ion polarity. This explains why APCI conditions suffer from comparatively low reproducibility as compared to other ionization methods.

It is the great advantage of APCI that it – different from ESI – actively generates ions from neutrals. Thus, APCI makes low- to medium-polarity analytes eluting from a liquid chromatograph accessible for mass spectrometry. In contrast to its development as an ionization method, the application of APCI has a backlog behind ESI. The use of APCI rapidly grew in the mid-1990, perhaps because elaborate vacuum interfaces were then available from ESI technology. Nowadays, APCI is used were LC separation is required, but ESI is not applicable to the compound class of interest. [141-143]

11.7 Atmospheric Pressure Photoionization

Recently, *atmospheric pressure photoionization* (APPI) has been introduced. [144] APPI can serve as a complement or alternative to APCI. [142,145] In APPI, a UV light source replaces the corona discharge-powered plasma, while the pneumatic heated sprayer remains almost unaffected (Fig. 11.28). [146-148]

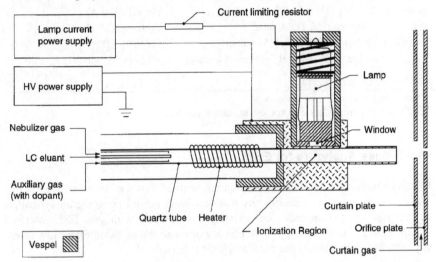

Fig. 11.28. Schematic of an APPI source, including the heated nebulizer probe, photoionization UV lamp and mounting bracket. Reproduced from Ref. [144] by permission. © American Chemical Society, 2000.

11.8 General Characteristics of ESI

11.8.1 Sample Consumption

Sample consumption is chiefly determined by the concentration of the analyte solution and the liquid flow. For example, during a 2.5-min measurement conventional ESI consumes 10 pmol when a 10^{-6} M solution at a flow rate of 4 µl min^{-1} is employed. For nanoESI this reduces to 100 fmol for the same solution at 40 nl min^{-1}.

11.8.2 Types of Ions in ESI

Very similar to the previously introduced desorption methods ESI produces a variety of ions depending on the polarity of the analyte, the characteristics of the solvent, and on the presence or absence of impurities such as alkali metal ions. Radical ions are normally not observed. (Table 11.1).

Table 11.1. Ions formed by ESI

Analytes	Positive Ions	Negative Ions
non-polar	$[M+H]^+$, $[M+alkali]^+$ if any	$[M-H]^-$, $[M+A]^-$ if any
medium to high polarity	$[M+H]^+$, $[M+alkali]^+$ *exchange* $[M-H_n+alkali_{n+1}]^+$ {*clusters* $[2M+H]^+$, $[2M+alkali]^+$, *adducts* $[M+solv+H]^+$, $[M+solv+alkali]^+$}[b]	$[M-H]^-$, $[M+A]^-$ *exchange* $[M-H_n+alkali_{n-1}]^-$ {*clusters* $[2M-H]^-$ *adducts* $[M+solv-H]^-$}
ionic[a]	C^+, $[C_n+A_{n-1}]^+$	A^-, $[C_{n-1}+A_n]^-$

[a] Comprising cation C^+ and anion A^-.
[b] Enclosure in parentheses denotes less abundant species.

11.8.3 Mass Analyzers for ESI

ESI ion sources are offered in combination with all types of mass analyzers, i.e., one is essentially free to choose any mass analyzer according to the analytical requirements or to upgrade an existing instrument with an ESI interface (Chap. 11.4.3). As fragment ions in ESI are often absent or exhibit very low intensity, MS/MS capability of the mass analyzer is beneficial.

Reference List

1. Amad, M.H.; Cech, N.B.; Jackson, G.S.; Enke, C.G. Importance of Gas-Phase Proton Affinities in Determining the ESI Response for Analytes and Solvents. *J. Mass Spectrom.* **2000**, *35*, 784-789.
2. Fenn, J.B.; Mann, M.; Meng, C.K.; Wong, S.F.; Whithouse, C.M. ESI for MS of Large Biomolecules. *Science* **1989**, *246*, 64-71.
3. Fenn, J.B.; Mann, M.; Meng, C.K.; Wong, S.F. Electrospray Ionization - Principles and Practice. *Mass Spectrom. Rev.* **1990**, *9*, 37-70.
4. Smith, R.D.; Loo, J.A.; Edmonds, C.G.; Barinaga, C.J.; Udseth, H.R. New Developments in Biochemical MS: ESI. *Anal. Chem.* **1990**, *62*, 882-899.
5. *Electrospray Ionization Mass Spectrometry - Fundamentals, Instrumentation and Applications;* 1st ed.; Dole, R.B., editor; John Wiley & Sons: Chichester, 1997.
6. Fenn, J.B. Electrospray: Wings for Molecular Elephants (Nobel Lecture). *Angew. Chem., Int. Ed.* **2003**, *42*, 3871-3894.
7. Fuerstenau, S.D.; Benner, W.H. Molecular Weight Determination of Megadalton DNA Electrospray Ions Using Charge Detection Time-of-Flight-MS. *Rapid Commun. Mass Spectrom.* **1995**, *9*, 1528-1538.

8. Fuerstenau, S.D.; Benner, W.H.; Thomas, J.J.; Brugidou, C.; Bothner, B.; Suizdak, G. Mass Spectrometry of an Intact Virus. *Angew. Chem., Int. Ed.* **2001**, *40*, 541-544.

9. Felitsyn, N.; Peschke, M.; Kebarle, P. Origin and Number of Charges Observed on Multiply-Protonated Native Proteins Produced by ESI. *Int. J. Mass Spectrom.* **2002**, *219*, 39-62.

10. *Mass Spectrometry in Drug Discovery;* Rossi, D.T.; Sinz, M.W., editors; Marcel Dekker: New York, 2002.

11. *Modern Mass Spectrometry;* Schalley, C.A., editor; Springer: New York, 2003.

12. Ardrey, R.E. *Liquid Chromatography-Mass Spectrometry - An Introduction;* John Wiley & Sons: Chichester, 2003.

13. Siuzdak, G. *The Expanding Role of Mass Spectrometry in Biotechnology;* MCC Press: San Diego, 2003.

14. Lehmann, W.D. *Massenspektrometrie in der Biochemie;* Spektrum Akademischer Verlag: Heidelberg, 1996.

15. Roboz, J. *Mass Spectrometry in Cancer Research;* 1st ed.; CRC Press: Boca Raton, 1999.

16. *Mass Spectrometry of Proteins and Peptides;* Chapman, J.R., editor; Humana Press: Totowa, 2000.

17. *Applied Electrospray Mass Spectrometry;* Pramanik, B.N.; Ganguly, A.K.; Gross, M.L., editors; Marcel Dekker: New York, 2002.

18. Colton, R.; D'Agostino, A.; Traeger, J.C. ESI-MS Applied to Inorganic and Organometallic Chemistry. *Mass Spectrom. Rev.* **1995**, *14*, 79-106.

19. Traeger, J.C. ESI-MS of Organometallic Compounds. *Int. J. Mass Spectrom.* **2000**, *200*, 387-401.

20. Poon, G.K.; Bisset, G.M.F.; Mistry, P. ESI-MS for Analysis of Low-Molecular-Weight Anticancer Drugs and Their Analogs. *J. Am. Soc. Mass Spectrom.* **1993**, *4*, 588-595.

21. Hinderling, C.; Feichtinger, D.; Plattner, D.A.; Chen, P. A Combined Gas-Phase, Solution-Phase, and Computational Study of C-H Activation by Cationic Iridium(III) Complexes. *J. Am. Chem. Soc.* **1997**, *119*, 10793-10804.

22. Truebenbach, C.S.; Houalla, M.; Hercules, D.M. Characterization of Isopoly Metal Oxyanions Using ESI Time-of-Flight-MS. *J. Mass Spectrom.* **2000**, *35*, 1121-1127.

23. Winger, B.E.; Light-Wahl, K.J.; Ogorzalek Loo, R.R.; Udseth, H.R.; Smith, R.D. Observation and Implications of High Mass-to-Charge Ratio Ions From ESI-MS. *J. Am. Soc. Mass Spectrom.* **1993**, *4*, 536-545.

24. Wang, Y.; Schubert, M.; Ingendoh, A.; Franzen, J. Analysis of Non-Covalent Protein Complexes Up to 290 KDa Using ESI and Ion Trap-MS. *Rapid Commun. Mass Spectrom.* **2000**, *14*, 12-17.

25. Sobott, F.; Hernández, H.; McCammon, M.G.; Tito, M.A.; Robinson, C.V. A Tandem Mass Spectrometer for Improved Transmission and Analysis of Large Macromolecular Assemblies. *Anal. Chem.* **2002**, *74*, 1402-1407.

26. Horning, E.C.; Horning, M.G.; Carroll, D.I.; Dzidic, I.; Stillwell, R.N. New Picogram Detection System Based on a Mass Spectrometer With an External Ionization Source at Atmospheric Pressure. *Anal. Chem.* **1973**, *45*, 936-943.

27. Horning, E.C.; Carroll, D.I.; Dzidic, I.; Haegele, K.D.; Horning, M.G.; Stillwell, R.N. Atmospheric Pressure Ionization (API) MS. Solvent-Mediated Ionization of Samples Introduced in Solution and in a Liquid Chromatograph Effluent Stream. *J. Chromatogr. Sci.* **1974**, *12*, 725-729.

28. Carroll, D.I.; Dzidic, I.; Stillwell, R.N.; Horning, M.G.; Horning, E.C. Subpicogram Detection System for Gas Phase Analysis Based Upon Atmospheric Pressure Ionization (API) MS. *Anal. Chem.* **1974**, *46*, 706-704.

29. Blakley, C.R.; Carmody, J.J.; Vestal, M.L. A New Soft Ionization Technique for MS of Complex Molecules. *J. Am. Chem. Soc.* **1980**, *102*, 5931-5933.

30. Blakley, C.R.; Vestal, M.L. Thermospray Interface for Liquid Chromatography/Mass Spectrometry. *Anal. Chem.* **1983**, *55*, 750-754.

31. Vestal, M.L. Studies of Ionization Mechanisms Involved in Thermospray LC-MS. *Int. J. Mass Spectrom. Ion Phys.* **1983**, *46*, 193-196.

32. Wilkes, J.G.; Freeman, J.P.; Heinze, T.M.; Lay, J.O., Jr.; Vestal, M.L. AC Corona-Discharge Aerosol-Neutralization Device Adapted to Liquid Chromatography/Particle Beam/Mass Spectrometry. *Rapid Commun. Mass Spectrom.* **1995**, *9*, 138-142.

33. Vestal, M.L. High-Performance Liquid Chromatography-Mass Spectrometry. *Science* **1984**, *226*, 275-281.

34. Evans, C.A., Jr.; Hendricks, C.D. Electrohydrodynamic Ion Source for the MS of Liquids. *Rev. Sci. Inst.* **1972**, *43*, 1527-1530.

35. Simons, D.S.; Colby, B.N.; Evans, C.A., Jr. Electrohydrodynamic Ionization-MS. Ionization of Liquid Glycerol and Nonvolatile Organic Solutes. *Int. J. Mass Spectrom. Ion Phys.* **1974**, *15*, 291-302.

36. Zeleny, J. Instability of Electrified Liquid Surfaces. *Phys. Rev.* **1917**, *10*, 1-7.

37. Taylor, G.I. Disintegration of Water Drops in an Electric Field. *Proc. Royal Soc. London A* **1964**, *280*, 383-397.

38. Lord Rayleigh On the Equilibrium of Liquid Conducting Masses Charged With Electricity. *London, Edinburgh, Dublin Phil. Mag. J. Sci.* **1882**, *14*, 184-186.

39. Dülcks, T.; Röllgen, F.W. Ionization Conditions and Ion Formation in Electrohydrodynamic MS. *Int. J. Mass Spectrom. Ion Proc.* **1995**, *148*, 123-144.

40. Dülcks, T.; Röllgen, F.W. Ion Source for Electrohydrodynamic MS. *J. Mass Spectrom.* **1995**, *30*, 324-332.

41. Cook, K.D. Electrohydrodynamic MS. *Mass Spectrom. Rev.* **1986**, *5*, 467-519.

42. Dole, M.; Mack, L.L.; Hines, R.L.; Mobley, R.C.; Ferguson, L.D.; Alice, M.B. Molecular Beams of Macroions. *J. Chem. Phys.* **1968**, *49*, 2240-2249.

43. Dole, M.; Hines, R.L.; Mack, L.L.; Mobley, R.C.; Ferguson, L.D.; Alice, M.B. Gas Phase Macroions. *Macromolecules* **1968**, *1*, 96-97.

44. Gieniec, J.; Mack, L.L.; Nakamae, K.; Gupta, C.; Kumar, V.; Dole, M. ESI Mass Spectroscopy of Macromolecules: Application of an Ion-Drift Spectrometer. *Biomed. Mass Spectrom.* **1984**, *11*, 259-268.

45. Yamashita, M.; Fenn, J.B. ESI Ion Source. Another Variation on the Free-Jet Theme. *J. Phys. Chem.* **1984**, *88*, 4451-4459.

46. Yamashita, M.; Fenn, J.B. Negative Ion Production With the ESI Ion Source. *J. Phys. Chem.* **1984**, *88*, 4671-4675.

47. Whitehouse, C.M.; Robert, R.N.; Yamashita, M.; Fenn, J.B. ESI Interface for Liquid Chromatographs and Mass Spectrometers. *Anal. Chem.* **1985**, *57*, 675-679.

48. Bruins, A.P. Mass Spectrometry With Ion Sources Operating at Atmospheric Pressure. *Mass Spectrom. Rev.* **1991**, *10*, 53-77.

49. Kebarle, P.; Tang, L. From Ions in Solution to Ions in the Gas Phase - the Mechanism of ESI-MS. *Anal. Chem.* **1993**, *65*, 972A-986A.

50. Karas, M.; Bahr, U.; Dülcks, T. Nano-ESI-MS: Addressing Analytical Problems Beyond Routine. *Fresenius J. Anal. Chem.* **2001**, *366*, 669-676.

51. Anacleto, J.F.; Pleasance, S.; Boyd, R.K. Calibration of Ion Spray Mass Spectra Using Cluster Ions. *Org. Mass Spectrom.* **1992**, *27*, 660-666.

52. Hop, C.E.C.A. Generation of High Molecular Weight Cluster Ions by ESI; Implications for Mass Calibration. *J. Mass Spectrom.* **1996**, *31*, 1314-1316.

53. Kantrowitz, A.; Grey, J. High Intensity Source for the Molecular Beam. I. Theoretical. *Rev. Sci. Inst.* **1951**, *22*, 328-332.

54. Fenn, J.B. Mass Spectrometric Implications of High-Pressure Ion Sources. *Int. J. Mass Spectrom.* **2000**, *200*, 459-478.

55. Ikonomou, M.G.; Kebarle, P. A Heated ESI Source in Mass Spectrometry for Analytes From Aqueous Solutions. *J. Am. Soc. Mass Spectrom.* **1994**, *5*, 791-799.

56. Bruins, A.P.; Covey, T.R.; Henion, J.D. Ion Spray Interface for Combined Liquid Chromatography/Atmospheric Pressure Ionization-MS. *Anal. Chem.* **1987**, *59*, 2642-2646.

57. Covey, T.R.; Bruins, A.P.; Henion, J.D. Comparison of Thermospray and Ion Spray-MS in an Atmospheric Pressure Ion Source. *Org. Mass Spectrom.* **1988**, *23*, 178-186.

58. Covey, T.R.; Bonner, R.F.; Shushan, B.I.; Henion, J.D. The Determination of Protein, Oligonucleotide, and Peptide Molecular Weights by Ion-Spray-MS. *Rapid Commun. Mass Spectrom.* **1988**, *2*, 249-256.

59. Ikonomou, M.G.; Blades, A.T.; Kebarle, P. ESI - Ion Spray: a Comparison of Mechanisms and Performance. *Anal. Chem.* **1991**, *63*, 1989-1998.

60. Banks, J.F., Jr.; Quinn, J.P.; Whitehouse, C.M. LC/ESI-MS Determination of Proteins Using Conventional Liquid Chromatography and Ultrasonically Assisted ESI. *Anal. Chem.* **1994**, *66*, 3688-3695.

61. Banks, J.F., Jr.; Shen, S.; Whitehouse, C.M.; Fenn, J.B. Ultrasonically Assisted ESI for LC/MS Determination of Nucleosides From a Transfer RNA Digest. *Anal. Chem.* **1994**, *66*, 406-414.

62. Smith, R.D.; Barinaga, C.J.; Udseth, H.R. Improved ESI Interface for Capillary Zone

Electrophoresis-Mass Spectrometry. *Anal. Chem.* **1988**, *60*, 1948-1952.

63. Abian, J. The Coupling of Gas and Liquid Chromatography With Mass Spectrometry. *J. Mass Spectrom.* **1999**, *34*, 157-168.

64. Wilm, M.S.; Mann, M. Electrospray and Taylor-Cone Theory, Dole's Beam of Macromolecules at Last? *Int. J. Mass Spectrom. Ion Proc.* **1994**, *136*, 167-180.

65. Wilm, M.; Mann, M. Analytical Properties of the NanoESI Ion Source. *Anal. Chem.* **1996**, *68*, 1-8.

66. Wilm, M.; Shevshenko, A.; Houthaeve, T.; Breit, S.; Schweigerer, L.; Fotsis, T.; Mann, M. Femtomole Sequencing of Proteins From Polyacrylamide Gels by Nano-ESI-MS. *Nature* **1996**, *379*, 466-469.

67. Juraschek, R.; Dülcks, T.; Karas, M. NanoESI - More Than Just a Minimized-Flow ESI Source. *J. Am. Soc. Mass Spectrom.* **1999**, *10*, 300-308.

68. Smith, R.D.; Loo, J.A.; Barinaga, C.J.; Edmonds, C.G.; Udseth, H.R. Collisional Activation and Collision-Activated Dissociation of Large Multiply Charged Polypeptides and Proteins Produced by ESI. *J. Am. Soc. Mass Spectrom.* **1990**, *1*, 53-65.

69. Harvey, D.J. Collision-Induced Fragmentation of Underivatized *N*-Linked Carbohydrates Ionized by ESI. *J. Mass Spectrom.* **2000**, *35*, 1178-1190.

70. Schmidt, A.; Bahr, U.; Karas, M. Influence of Pressure in the First Pumping Stage on Analyte Desolvation and Fragmentation in Nano-ESI MS. *Anal. Chem.* **2001**, *71*, 6040-6046.

71. Jedrzejewski, P.T.; Lehmann, W.D. Detection of Modified Peptides in Enzymatic Digests by Capillary Liquid Chromatography/ESI-MS and a Programmable Skimmer CID Acquisition Routine. *Anal. Chem.* **1997**, *69*, 294-301.

72. Weinmann, W.; Stoertzel, M.; Vogt, S.; Svoboda, M.; Schreiber, A. Tuning Compounds for ESI/in-Source Collision-Induced Dissociation and Mass Spectra Library Searching. *J. Mass Spectrom.* **2001**, *36*, 1013-1023.

73. Huddleston, M.J.; Bean, M.F.; Carr, S.A. Collisional Fragmentation of Glycopeptides by ESI LC/MS and LC/MS/MS: Methods for Selective Detection of Glycopeptides in Protein Digests. *Anal. Chem.* **1993**, *65*, 877-884.

74. Chen, H.; Tabei, K.; Siegel, M.M. Biopolymer Sequencing Using a Triple Quadrupole Mass Spectrometer in the ESI Noz-zle-Skimmer/Precursor Ion MS/MS Mode. *J. Am. Soc. Mass Spectrom.* **2001**, *12*, 846-852.

75. Miao, X.-S.; Metcalfe, C.D. Determination of Carbamazepine and Its Metabolites in Aqueous Samples Using Liquid Chromatography-ESI Tandem-MS. *Anal. Chem.* **2003**, *75*, 3731-3738.

76. Gross, J.H.; Eckert, A.; Siebert, W. Negative-Ion ESI Mass Spectra of Carbon Dioxide-Protected N-Heterocyclic Anions. *J. Mass Spectrom.* **2002**, *37*.

77. de la Mora, J.F.; Van Berkel, G.J.; Enke, C.G.; Cole, R.B.; Martinez-Sanchez, M.; Fenn, J.B. Electrochemical Processes in ESI-MS. *J. Mass Spectrom.* **2000**, *35*, 939-952.

78. Van Berkel, G.J. Electrolytic Deposition of Metals on to the High-Voltage Contact in an ESI Emitter: Implications for Gas-Phase Ion Formation. *J. Mass Spectrom.* **2000**, *35*, 773-783.

79. Cole, R.B. Some Tenets Pertaining to ESI-MS. *J. Mass Spectrom.* **2000**, *35*, 763-772.

80. Gomez, A.; Tang, K. Charge and Fission of Droplets in Electrostatic Sprays. *Phys. Fluids* **1994**, *6*, 404-414.

81. Duft, D.; Achtzehn, T.; Müller, R.; Huber, B.A.; Leisner, T. Coulomb Fission. Rayleigh Jets From Levitated Microdroplets. *Nature* **2003**, *421*, 128.

82. Iribarne, J.V.; Thomson, B.A. On the Evaporation of Small Ions From Charged Droplets. *J. Chem. Phys.* **1976**, *64*, 2287-2294.

83. Thomson, B.A.; Iribarne, J.V. Field-Induced Ion Evaporation From Liquid Surfaces at Atmospheric Pressure. *J. Chem. Phys.* **1979**, *71*, 4451-4463.

84. Mack, L.L.; Kralik, P.; Rheude, A.; Dole, M. Molecular Beams of Macroions. II. *J. Chem. Phys.* **1970**, *52*, 4977-4986.

85. Labowsky, M.; Fenn, J.B.; de la Mora, J.F. A Continuum Model for Ion Evaporation From a Drop: Effect of Curvature and Charge on Ion Solvation Energy. *Anal. Chim. Acta* **2000**, *406*, 105-118.

86. Fenn, J.B. Ion Formation From Charged Droplets: Roles of Geometry, Energy, and Time. *J. Am. Soc. Mass Spectrom.* **1993**, *4*, 524-535.

87. Fenn, J.B.; Rosell, J.; Meng, C.K. In ESI, How Much Pull Does an Ion Need to Escape Its Droplet Prison? *J. Am. Soc. Mass Spectrom.* **1997**, *8*, 1147-1157.

88. Loo, J.A.; Edmonds, C.G.; Udseth, H.R.; Smith, R.D. Effect of Reducing Disulfide-

Containing Proteins on ESI Mass Spectra. *Anal. Chem.* **1990**, *62*, 693-698.

89. Chowdhury, S.K.; Katta, V.; Chait, B.T. Probing Conformational Changes in Proteins by-MS. *J. Am. Chem. Soc.* **1990**, *112*, 9012-9013.

90. Schmelzeisen-Redeker, G.; Bütfering, L.; Röllgen, F.W. Desolvation of Ions and Molecules in Thermospray-MS. *Int. J. Mass Spectrom. Ion Proc.* **1989**, *90*, 139-150.

91. Williams, E.R. Proton Transfer Reactivity of Large Multiply Charged Ions. *J. Mass Spectrom.* **1996**, *31*, 831-842.

92. Iavarone, A.T.; Jurchen, J.C.; Williams, E.R. Supercharged Protein and Peptide Ions Formed by ESI. *Anal. Chem.* **2001**, *73*, 1455-1460.

93. Iavarone, A.T.; Williams, E.R. Supercharging in ESI: Effects on Signal and Charge. *Int. J. Mass Spectrom.* **2002**, *219*, 63-72.

94. Iavarone, A.T.; Williams, E.R. Mechanism of Charging and Supercharging Molecules in ESI. *J. Am. Chem. Soc.* **2003**, *125*, 2319-2327.

95. Kebarle, P. A Brief Overview of the Present Status of the Mechanisms Involved in ESI-MS. *J. Mass Spectrom.* **2000**, *35*, 804-817.

96. Cody, R.B.; Tamura, J.; Musselman, B.D. ESI/Magnetic Sector Mass Spectrometry: Calibration, Resolution, and Accurate Mass Measurements. *Anal. Chem.* **1992**, *64*, 1561-1570.

97. Saf, R.; Mirtl, C.; Hummel, K. ESI-MS Using Potassium Iodide in Aprotic Organic Solvents for the Ion Formation by Cation Attachment. *Tetrahedron Lett.* **1994**, *35*, 6653-6656.

98. Chapman, J.R.; Gallagher, R.T.; Barton, E.C.; Curtis, J.M.; Derrick, P.J. Advantages of High-Resolution and High-Mass Range Magnetic-Sector Mass Spectrometry for ESI. *Org. Mass Spectrom.* **1992**, *27*, 195-203.

99. Cole, R.B.; Harrata, A.K. Solvent Effect on Analyte Charge State, Signal Intensity, and Stability in Negative Ion ESI-MS; Implications for the Mechanism of Negative Ion Formation. *J. Am. Soc. Mass Spectrom.* **1993**, *4*, 546-556.

100. Straub, R.F.; Voyksner, R.D. Negative Ion Formation in ESI-MS. *J. Am. Soc. Mass Spectrom.* **1993**, *4*, 578-587.

101. Mann, M.; Meng, C.K.; Fenn, J.B. Interpreting Mass Spectra of Multiply Charged Ions. *Anal. Chem.* **1989**, *61*, 1702-1708.

102. Labowsky, M.; Whitehouse, C.M.; Fenn, J.B. Three-Dimensional Deconvolution of Multiply Charged Spectra. *Rapid Commun. Mass Spectrom.* **1993**, *7*, 71-84.

103. Dobberstein, P.; Schroeder, E. Accurate Mass Determination of a High Molecular Weight Protein Using ESI With a Magnetic Sector Instrument. *Rapid Commun. Mass Spectrom.* **1993**, *7*, 861-864.

104. Haas, M.J. Fully Automated Exact Mass Measurements by High-Resolution ESI on a Sector Instrument. *Rapid Commun. Mass Spectrom.* **1999**, *13*, 381-383.

105. Hofstadler, S.A.; Griffey, R.H.; Pasa-Tolic, R.; Smith, R.D. The Use of a Stable Internal Mass Standard for Accurate Mass Measurements of Oligonucleotide Fragment Ions Using ESI Fourier Transform Ion Cyclotron Resonance-MS With Infrared Multiphoton Dissociation. *Rapid Commun. Mass Spectrom.* **1998**, *12*, 1400-1404.

106. Cooper, H.J.; Marshall, A.G. ESI Fourier Transform Mass Spectrometric Analysis of Wine. *J. Agric. Food Chem.* **2001**, *49*, 5710-5718.

107. Stenson, A.C.; Landing, W.M.; Marshall, A.G.; Cooper, W.T. Ionization and Fragmentation of Humic Substances in ESI Fourier Transform-Ion Cyclotron Resonance-MS. *Anal. Chem.* **2002**, *74*, 4397-4409.

108. Coles, J.; Guilhaus, M. Orthogonal Acceleration - a New Direction for Time-of-Flight-MS: Fast, Sensitive Mass Analysis for Continuous Ion Sources. *Trends Anal. Chem.* **1993**, *12*, 203-213.

109. Guilhaus, M. The Return of Time-of-Flight to Analytical-MS. *Adv. Mass Spectrom.* **1995**, *13*, 213-226.

110. Guilhaus, M.; Selby, D.; Mlynski, V. Orthogonal Acceleration Time-of-Flight-MS. *Mass Spectrom. Rev.* **2000**, *19*, 65-107.

111. Scalf, M.; Westphall, M.S.; Krause, J.; Kaufmann, S.L.; Smith, L.M. Controlling Charge States of Large Ions. *Science* **1999**, *283*, 194-197.

112. Scalf, M.; Westphall, M.S.; Smith, L.M. Charge Reduction ESI-MS. *Anal. Chem.* **2000**, *72*, 52-60.

113. Ebeling, D.D.; Westphall, M.S.; Scalf, M.; Smith, L.M. Corona Discharge in Charge Reduction ESI-MS. *Anal. Chem.* **2000**, *72*, 5158-5161.

114. Ebeling, D.D.; Scalf, M.; Westphall, M.S.; Smith, L.M. A Cylindrical Capacitor Ionization Source: Droplet Generation and Controlled Charge Reduction for MS. *Rapid Commun. Mass Spectrom.* **2001,** *15,* 401-405.

115. McLuckey, S.A.; Stephenson, J.L., Jr. Ion/Ion Chemistry of High-Mass Multiply Charged Ions. *Mass Spectrom. Rev.* **1998,** *17,* 369-407.

116. Herron, W.J.; Goeringer, D.E.; McLuckey, S.A. Product Ion Charge State Determination Via Ion/Ion Proton Transfer Reactions. *Anal. Chem.* **1996,** *68,* 257-262.

117. Zhu, J.; Cole, R.B. Formation and Decompositions of Chloride Adduct Ions, [M+Cl]⁻, in Negative Ion ESI-MS. *J. Am. Soc. Mass Spectrom.* **2000,** *11,* 932-941.

118. Choi, B.K.; Hercules, D.M.; Houalla, M. Characterization of Polyphosphates by ESI-MS. *Anal. Chem.* **2000,** *72,* 5087-5091.

119. Favaro, S.; Pandolfo, L.; Traldi, P. The Behavior of $[Pt(\eta^3\text{-Allyl})XP(C_6H_5)_3]$ Complexes in ESI Conditions Compared With Those Achieved by Other Ionization Methods. *Rapid Commun. Mass Spectrom.* **1997,** *11,* 1859-1866.

120. Feichtinger, D.; Plattner, D.A. Direct Proof for Mn^V-Oxo-Salen Complexes. *Angew. Chem. ,Int. Ed.* **1997,** *36,* 1718-1719.

121. Løver, T.; Bowmaker, G.A.; Henderson, W.; Cooney, R.P. ESI-MS of Some Cadmium Thiophenolate Complexes and of a Thiophenolate Capped CdS Cluster. *Chem. Commun.* **1996,** 683-685.

122. Reid, G.E.; O'Hair, R.A.J.; Styles, M.L.; McFadyen, W.D.; Simpson, R.J. Gas Phase Ion-Molecule Reactions in a Modified Ion Trap: H/D Exchange of Non-Covalent Complexes and Coordinatively Unsaturated Platinum Complexes. *Rapid Commun. Mass Spectrom.* **1998,** *12,* 1701-1708.

123. Volland, M.A.O.; Adlhart, C.; Kiener, C.A.; Chen, P.; Hofmann, P. Catalyst Screening by ESI Tandem-MS: Hofmann Carbenes for Olefin Metathesis. *Chem. - Eur. J.* **2001,** *7,* 4621-4632.

124. Jewett, B.N.; Ramaley, L.; Kwak, J.C.T. Atmospheric Pressure Ionization-MS Techniques for the Analysis of Alkyl Ethoxysulfate Mixtures. *J. Am. Soc. Mass Spectrom.* **1999,** *10,* 529-536.

125. Benomar, S.H.; Clench, M.R.; Allen, D.W. The Analysis of Alkylphenol Ethoxysulphonate Surfactants by HPLC, LC-ESI-MS and MALDI-MS. *Anal. Chim. Acta* **2001,** *445,* 255-267.

126. Eichhorn, P.; Knepper, T.P. ESI Mass Spectrometric Studies on the Amphoteric Surfactant Cocamidopropylbetaine. *J. Mass Spectrom.* **2001,** *36,* 677-684.

127. Levine, L.H.; Garland, J.L.; Johnson, J.V. HPLC/ESI-Quadrupole Ion Trap-MS for Characterization and Direct Quantification of Amphoteric and Nonionic Surfactants in Aqueous Samples. *Anal. Chem.* **2002,** *74,* 2064-2071.

128. Barco, M.; Planas, C.; Palacios, O.; Ventura, F.; Rivera, J.; Caixach, J. Simultaneous Quantitative Analysis of Anionic, Cationic, and Nonionic Surfactants in Water by ESI-MS With Flow Injection Analysis. *Anal. Chem.* **2003,** *75,* 5179-5136.

129. Little, D.P.; Chorush, R.A.; Speir, J.P.; Senko, M.W.; Kelleher, N.L.; McLafferty, F.W. Rapid Sequencing of Oligonucleotides by High-Resolution-MS. *J. Am. Chem. Soc.* **1994,** *116,* 4893-4897.

130. Limbach, P.A.; Crain, P.F.; McCloskey, J.A. Molecular Mass Measurement of Intact Ribonucleic Acids Via ESI Quadrupole MS. *J. Am. Soc. Mass Spectrom.* **1995,** *6,* 27-39.

131. Little, D.P.; Thannhauser, T.W.; McLafferty, F.W. Verification of 50- to 100-Mer DNA and RNA Sequences With High-Resolution-MS. *Proc. Natl. Acad. Sci. U. S. A.* **1995,** *92,* 2318-2322.

132. De Bellis, G.; Salani, G.; Battaglia, C.; Pietta, P.; Rosti, E.; Mauri, P. ESI-MS of Synthetic Oligonucleotides Using 2-Propanol and Spermidine. *Rapid Commun. Mass Spectrom.* **2000,** *14,* 243-249.

133. Greig, M.; Griffey, R.H. Utility of Organic Bases for Improved ESI-MS of Oligonucleotides. *Rapid Commun. Mass Spectrom.* **1995,** *9,* 97-102.

134. Pfenninger, A.; Karas, M.; Finke, B.; Stahl, B. Structural Analysis of Underivatized Neutral Human Milk Oligosaccharides in the Negative Ion Mode by Nano-ESI MS^n (Part 1: Methodology). *J. Am. Soc. Mass Spectrom.* **2002,** *13,* 1331-1340.

135. Pfenninger, A.; Karas, M.; Finke, B.; Stahl, B. Structural Analysis of Underivatized Neutral Human Milk Oligosaccharides in the Negative Ion Mode by Nano-ESI MS^n (Part 2: Application to

Isomeric Mixtures). *J. Am. Soc. Mass Spectrom.* **2002,** *13,* 1341-1348.

136. Weiskopf, A.S.; Vouros, P.; Harvey, D.J. ESI-Ion Trap-MS for Structural Analysis of Complex *N*-Linked Glycoprotein Oligosaccharides. *Anal. Chem.* **1998,** *70,* 4441-4447.

137. Metelmann, W.; Peter-Katalinic, J.; Müthing, J. Gangliosides From Human Granulocytes: a Nano-ESI QTOF-MS Fucosylation Study of Low Abundance Species in Complex Mixtures. *J. Am. Soc. Mass Spectrom.* **2001,** *12,* 964-973.

138. Tsuchiya, M. Atmospheric Pressure Ion Sources, Physico-Chemical and Analytical Applications. *Adv. Mass Spectrom.* **1995,** *13,* 333-346.

139. Carroll, D.I.; Dzidic, I.; Stillwell, R.N.; Haegele, K.D.; Horning, E.C. Atmospheric Pressure Ionization-MS. Corona Discharge Ion Source for Use in a Liquid Chromatograph-Mass Spectrometer-Computer Analytical System. *Anal. Chem.* **1975,** *47,* 2369-2373.

140. Dzidic, I.; Stillwell, R.N.; Carroll, D.I.; Horning, E.C. Comparison of Positive Ions Formed in Nickel-63 and Corona Discharge Ion Sources Using Nitrogen, Argon, Isobutane, Ammonia and Nitric Oxide As Reagents in Atmospheric Pressure Ionization-MS. *Anal. Chem.* **1976,** *48,* 1763-1768.

141. Mottram, H.; Woodbury, S.E.; Evershed, R.P. Identification of Triacylglycerol Positional Isomers Present in Vegetable Oils by High Performance Liquid Chromatography/Atmospheric Pressure Chemical Ionization-MS. *Rapid Commun. Mass Spectrom.* **1997,** *11,* 1240-1252.

142. Hayen, H.; Karst, U. Strategies for the Liquid Chromatographic-Mass Spectrometric Analysis of Non-Polar Compounds. *J. Chromatogr. A* **2003,** *1000,* 549-565.

143. Reemtsma, T. Liquid Chromatography-Mass Spectrometry and Strategies for Trace-Level Analysis of Polar Organic Pollutants. *J. Chromatogr. A* **2003,** *1000* , 477-501.

144. Robb, D.B.; Covey, T.R.; Bruins, A.P. Atmospheric Pressure Photoionization: an Ionization Method for Liquid Chromatography-Mass Spectrometry. *Anal. Chem.* **2000,** *72,* 3653-3659.

145. Keski-Hynnilä, H.; Kurkela, M.; Elovaara, E.; Antonio, L.; Magdalou, J.; Luukkanen, L.; Taskinen, J.; Koistiainen, R. Comparison of ESI, Atmospheric Pressure Chemical Ionization, and Atmospheric Pressure Photoionization in the Identification of Apomorphine, Dobutamine, and Entacapone Phase II Metabolites in Biological Samples. *Anal. Chem.* **2002,** *74,* 3449-3457.

146. Yang, C.; Henion, J.D. Atmospheric Pressure Photoionization Liquid Chromatographic-Mass Spectrometric Determination of Idoxifene and Its Metabolites in Human Plasma. *J. Chromatogr. A* **2002,** *970,* 155-165.

147. Kauppila, T.J.; Kuuranne, T.; Meurer, E.C.; Eberlin, M.N.; Kotiaho, T.; Koistiainen, R. Atmospheric Pressure Photoionization-MS. Ionization Mechanism and the Effect of Solvent on the Ionization of Naphthalenes. *Anal. Chem.* **2002,** *74,* 5470-5479.

148. Raffaeli, A.; Saba, A. Atmospheric Pressure Photoionization-MS. *Mass Spectrom. Rev.* **2003,** *22,* 318-331.

12 Hyphenated Methods

The analysis of complex mixtures particularly requires the combination of both separation techniques and mass spectrometry. [1-3] The first step in this direction was made by *gas chromatography-mass spectrometry coupling* (GC-MS), [4] and soon, GC-MS became a routine method. [5-7] The desire to realize a *liquid chromatography-mass spectrometry coupling* (LC-MS) [8,9] was the driving force for the development of API methods (Chap. 11). [10-13] Coupling of other liquid phase separation techniques to mass spectrometry followed: *capillary zone electrophoresis-mass spectrometry* (CZE-MS), [14-18] and *supercritical fluid chromatography-mass spectrometry* (SFC-MS). [19-23] Whatever the separation technique, it adds an additional dimension to the analytical measurement. The hyphen used to indicate the coupling of a separation technique to mass spectrometry led to the term *hyphenated methods*.

However, mass spectrometry itself offers two additional "degrees of freedom". One can either resolve the complexity of a sample by going to high or even ultrahigh mass resolution or one can employ tandem MS techniques to separate the fragmentation pattern of a single component from that of others in a mixture. [2,3] In practice, the coupling of separation techniques to mass spectrometry is often combined with advanced MS techniques to achieve the desired level of accuracy and reliability of analytical information. [1,7,24-27]

This chapter is about extending the range of samples that can be analyzed by mass spectrometry and about increasing the specificity of analytical information thereof. It briefly explains the basic concepts and methodologies such as handling of chromatograms, quantitation, as well as GC and LC interfaces.

12.1 General Properties of Chromatography-Mass Spectrometry Coupling

Chromatographic detectors such as the *flame ionization detector* (FID), frequently employed in GC, deliver a *chromatogram* that represents the mass flow eluting from the chromatographic column. Using a mass spectrometer instead, offers an added dimension of information, e.g., the mass spectra associated to any of the eluting components or other more selective means of target compound identification. Regardless of the separation method employed, there are some basic techniques in common to operate the mass spectrometer.

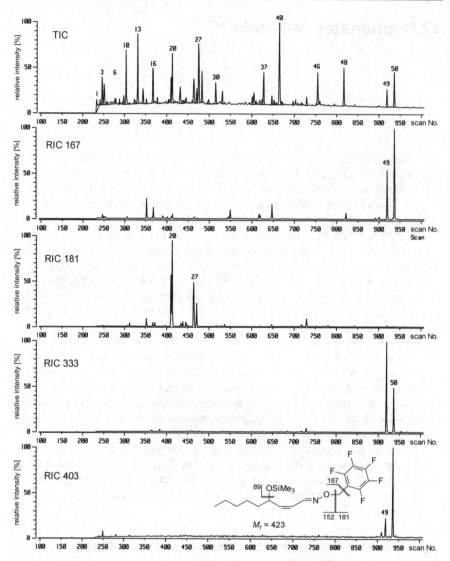

Fig. 12.1. TIC and RICs (*m/z* 167, 181, 333, and 403) obtained from an EC-GC-MS run from a plasma sample spiked with 4-HNE. [32] (The EC spectrum of the 4-HNE derivative is show in Fig. 12.2.)

Note: The abscissa of the TIC or RIC can either be plotted on the time scale, i.e., in units of seconds or minutes, or be labeled with scan numbers. Scan numbers are useful during data processing, while the time scale is better suited for comparison with other chromatography-mass spectrometry data.

12.1.1 Chromatograms and Spectra

A widespread mode of chromatography-mass spectrometry operation is to repetitively scan the mass analyzer over the m/z range of interest during the chromatographic run. [28-30] This generates a relationship between chromatogram and mass spectra of the eluting components. The conventional chromatogram is then replaced by the *total ion chromatogram* (TIC, Chap. 5.4.1). Knowledgeable handling of the TIC and suitable *reconstructed ion chromatograms* (RICs, Chap. 5.4.2) [30] presents the key to the effective assignment of chromatographic peaks to target compounds.

Example: 4-Hydroxynon-2-enal (4-HNE) is a major aldehydic product of lipid peroxidation (LPO), its products being indicators for oxidative stress. In order to introduce LPO products as biomarkers a GC-MS method for 4-HNE detection in clinical studies [31] was developed using a sample volume of 50 µl of plasma. For improved GC separation and subsequent mass spectral detection the aldehyde is converted into the pentafluorbenzyl-hydroxylimine and the hydroxy group is trimethylsilylated. [32] The TIC acquired in negative-ion electron capture mode (EC, Chap. 7.4) exhibits 50 chromatographic peaks (Fig. 12.1). Those related to the target compounds can easily be identified from suitable RICs. The choice of potentially useful m/z values for RICs is made from the EC mass spectrum of the pure 4-HNE derivative (below). In this case, $[M–HF]^{-\cdot}$, m/z 403, $[M–HOSiMe_3]^{-\cdot}$, m/z 333, and $[C_6F_5]^{-}$, m/z 167, are indicative, while $[CH_2C_6F_5]^{-}$, m/z 181, is not.

When a GC-MS experiment is performed in the repetitive scanning mode, each point of the TIC corresponds to a full mass spectrum. The time for the acquisition of a mass spectrum has to be shorter than the time to elute a component from the chromatographic column. In capillary GC-MS, this requires scan cycle times in the order of one second, i.e., a single chromatogram is represented by a set of about a thousand mass spectra. Nonetheless, the concentration of the eluting components still varies rapidly in time as compared to the time for a scan cycle. This affects the relative intensities of mass spectral peaks, i.e., high-mass ions are emphasized in an upwards mass scan at the onset of elution, but are underrepresented when elution fades out (Fig. 12.2). *Averaging* or *accumulation* of the scans contributing to a specific chromatographic peak compensates for that source of error. Additional *background subtraction* can substantially improve the signal-to-noise ratio (Chap. 5.2.3.) of the final spectrum. [29]

Note: High-quality mass spectra suitable for interpretation and/or data base search (Chap. 5.7) are only obtained from summing/averaging plus subsequent background subtraction. Particularly for components of low concentration the value of background subtraction cannot be overestimated.

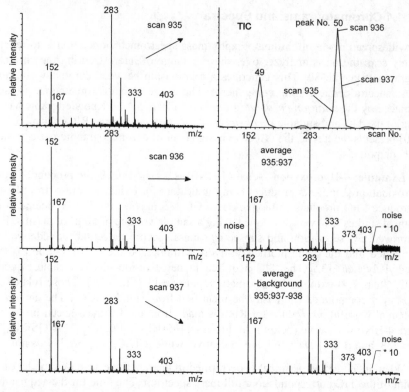

Fig. 12.2. Extraction of spectra from chromatographic peak No. 50 in the TIC. Left column: single scan spectra from scans 935, 936, and 937 show changing relative intensities; averaging scans 935:937 levels intensity but leaves noise; background subtraction reduces noise in addition. (For TIC and RICs see Fig. 12.1.)

12.1.2 Selected Ion Monitoring

The operation of magnetic sector (Chap. 4.3), linear quadrupole (Chap. 4.4), or quadrupole ion trap (Chap. 4.5) mass spectrometers in the repetitive scanning mode is useful for the identification of the components of a mixture. If *quantitation* is a major issue (below), *selected ion monitoring* (SIM) is preferably employed; the term *multiple ion detection* (MID) and some others are also in use. [33] In the SIM mode, the mass analyzer is operated in a way that it alternately acquires only the ionic masses of interest, i.e. it "jumps" from one *m/z* value to the next. [34-39] The information obtained from a SIM trace is equivalent to that from a RIC, but no mass spectra are recorded. Thus, the scan time spent on a diagnostically useless *m/z* range is almost reduced to zero, whereas the detector time for the ions of interest is increased by a factor of 10–100. [40] An analogous improvement in sensitivity (Chap. 5.2.3) is also observed.

As the monitored m/z values are selected to best represent the target compound, SIM exhibits high selectivity that can be further increased by *high resolution SIM* (HR-SIM) because this reduces isobaric interferences. [41-44] As HR-SIM requires precise and drift-free positioning on narrow peaks, one or several *lock masses* are generally employed although rarely explicitly mentioned. [44,45] The role of the lock mass is to serve as internal mass reference for accurate mass measurement. (Examples are given below.)

Note: Normally, three to ten m/z values are monitored 20–100 ms each in a cycle. Some settling time is needed for the mass analyzer after switching to the next value, e.g., 1–2 ms for pure electric scanning, 20–50 ms for a magnet scan.

12.1.3 Quantitation

Any ionization method exhibits compound-dependent ionization efficiencies (Chap. 2.4). Whether a specific compound is rather preferred or suppressed relative to another greatly depends on the ionization process employed. This makes a careful calibration of the instrument's response versus the sample concentration become prerequisite for reliable quantitation. [6,7]

12.1.3.1 Quantitation by External Standardization

External standardization is obtained by constructing a *calibration curve*, i.e., from plotting measured intensities versus rising concentration of the target compound. Calibration curves are generally linear over a wide range of concentrations. When concentration approaches the detection limit (Chap. 5.2.3) the graph deviates from

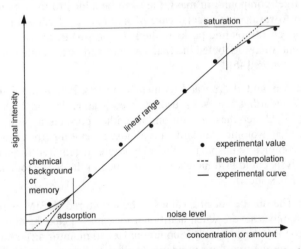

Fig. 12.3. General appearance of a calibration curve. The upper limit of the linear range is defined by saturation, the lower by memory and chemical background or adsorption. In addition, the noise level plays a role for the detection limit.

linearity, either towards underestimation or towards overestimation. Underestimation can be due to losses by adsorption, overestimation may either be due to "memory" from previous injections or result from chemical background. In the regime of high concentration, saturation of the detector or of the ion source cause an upper limit (Fig. 12.3). The preparation of a calibration curve requires repeated measurements which can be very time-consuming in case of a slow chromatographic separation. In addition, drifts in instrument sensitivity, e.g., due to ion source contamination, can deteriorate the quantitation result. Depending on the instrument's dynamic range (Chap. 4.2.5), the range of linear response is in the order of two to four orders of magnitude.

12.1.3.2 Quantitation by Internal Standardization

Internal standardization circumvents the effects of time-variant instrument response, but does not compensate for different ionization efficiencies of analyte and standard. For internal standardization, a compound exhibiting close similarity in terms of ionization efficiency and retention time is added to the sample at a known level of concentration, e.g., an isomer eluting closely to the analyte or a homologue may serve for that purpose. It is important to add the standard before any clean-up procedure in order not to alter the concentration of the analyte without affecting that of the standard. For reliable results, the relative concentration of analyte and standard should not differ by more than a factor of about ten.

12.1.3.3 Quantitation by Isotope Dilution

Identical or at least almost identical ionization efficiency (Chap. 2.4) for a pair of compounds is only given for isotopomers (Chap. 3.2.9). As these differ from the nonlabeled target compound in mass, they can be added to the mixture at known concentration to result in a special case of internal standardization. The relative intensity of the corresponding peaks in the RICs or SIM traces is then taken as the relative concentration of labeled internal standard and target compound; hence the term *isotope dilution*. [46]

Example: A potential drug and its metabolite in a liver sample are quantified by internal standardization with trideuterated standards for both compounds (Fig. 12.4). [25] Under these conditions it neither presents a problem that both analytes and their isotopic standards are almost co-eluting from the LC column, nor does the completely unspecific TIC play a role. If required for sensitivity reasons, this analysis could also have been performed in the SIM mode using the *m/z* values of the RICs shown.

Example: The number of *m/z* values to be monitored in SIM is limited. Such limitations are more severe when additional lock mass peaks have to be included in case of HR-SIM. Therefore, it is commonplace to monitor different sets of SIM traces during consecutive time windows leading to a sequence of different SIM setups during a single chromatographic separation. The quantitation of halogen-

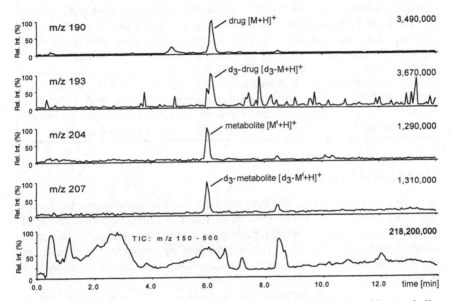

Fig. 12.4. RICs and TIC of an LC-ESI-MS quantification of a drug (M) and its metabolite (M') by isotope dilution. The numbers on the give the absolute intensity value for each trace. Reproduced from Ref. [25] by permission. © Elsevier Science, 2001.

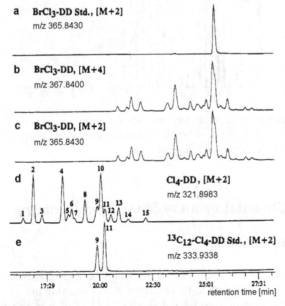

Fig. 12.5. Quantification of BrCl₃- and Cl₄-dibenzodioxins by HR-SIM. A $^{13}C_{12}$-labeled internal standard is added for the Cl₄-congener. Adapted from Ref. [43] by permission. © American Chemical Society, 1991.

ated dibenzo-*p*-dioxins in municipal waste incinerator fly ash at concentrations in the ppb to low-ppm range requires such a setup (Fig. 12.5). [43] Here, a combined approach of external standardization for the $BrCl_3$-species and internal standardization for the Cl_4-species has been realized.

Note: According to the enormous usage of isotope dilution, ^2H- (D) and ^{13}C-labeled standards [47] are commercially available for a wide range of applications. It is important not to select compounds with acidic hydrogens exchanged for deuterons. Other restrictions for internal standards apply analogously.

12.1.3.4 Retention Times of Isotopomers

Isotopomers exhibit slightly different retention times in chromatography. Deuterated compounds, for example, elute from chromatographic columns at slightly shorter retention times than their nonlabeled isotopomers. Normally, the difference is less than the peak width of the corresponding chromatographic peaks, but is still large enough to require their integration over separate time windows (Fig. 12.6).

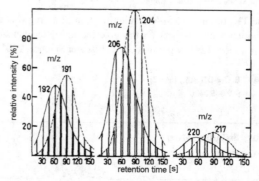

Fig. 12.6. Shorter retention times are observed for [D_7]glucose than for glucose eluting from a SE30 GC column as demonstrated for three pairs of peaks. Reproduced from Ref. [35] by permission. © American Chemical Society, 1966.

12.2 Gas Chromatography-Mass Spectrometry

12.2.1 GC-MS Interfaces

In early GC-MS with packed GC columns eluting several tens of milliliters per minute most of the flow had to be separated before entering the ion source to prevent the vacuum system from breakdown. [4,29,34] This was either effected by a simple split to divide the effluent in front of the inlet system by a factor of about 1 : 100 or by means of a more elaborate *separator*, the *jet separator* being the best known of those. [48,49] The advantage of separators over a simple split is the enrichment of the analyte relative to the carrier gas; in case of the jet separator this effect is based on a principle related to that of the nozzle-skimmer system in ESI (Chap. 11.2.1).

With the advent of capillary GC, [50-54] the need for separators and the concomitant risk of suppression of certain components vanished. Capillary columns are operated at flow rates in the order of 1 ml min^{-1}, and therefore can be directly interfaced to EI/CI ion sources. [48,49] Thus, a modern GC-MS interface basically consists of a heated (glass) line bridging the distance between GC oven and ion source. On the ion source block, an entrance port often opposite to the direct probe is reserved for that purpose (Chap. 5.2.1). The interface should be operated at the highest temperature employed in the actual GC separation or at the highest temperature the column can tolerate (200–300 °C). Keeping the transfer line at lower temperature causes condensation of eluting components to the end of the column.

In contrast to sample introduction via direct probe (Chap. 5.3.1), the components eluting from a GC capillary are quantitatively transferred into the ion source during a short time interval just sufficient to acquire about five mass spectra. Consequently, the partial pressure of the analyte is comparatively high during elution allowing sample amounts in the low nanogram range to be analyzed by capillary GC-MS.

12.2.2 Volatility and Derivatization

Gas chromatography requires a certain level of volatility and thermal robustness of the analyte. Both injection block of the chromatograph and interface region are always at high temperature even while the column oven is not. In order to adapt an analyte to these needs, *derivatization* is well established, [55,56] the most frequent derivatization procedures being silylation, acetylation, methylation, and fluoralkylation. As derivatization transforms XH groups into $XSiR_3$, $XC=OMe$, XMe, or $XCOCF_3$ groups, for example, polarity of the molecules largely decreases. This causes an improved volatility even though the molecular weight increases upon derivatization. Derivatization suppresses thermal decomposition, e.g., it protects alcohols from thermal dehydration. In particular fluoralkyl and fluoraryl groups are extremely useful to improve detection limits in EC-MS. [32,57]

12.2.3 Column Bleed

Rising temperature of the GC column not only assists transport of less volatile components, it also causes the slow release of the liquid phase from the inner wall of the capillary. As a result of slow thermal degradation, even chemically bonded liquid phases show such *column bleed* at elevated temperature. It is a characteristic of column bleed that it continuously rises as the temperature of the GC oven is raised and it falls again upon cooling of the system. Of course, the peaks from column bleed observed in the mass spectrum depend on the liquid phase of the GC capillary in use. In case of the frequently employed methyl-phenyl-siloxane liquid phases, abundant ions at m/z 73, 147, 207, 281, 355, 429, etc. are observed. Within one series the peaks are 74 u ($OSiMe_2$) distant. Fortunately, column bleed is easily recognized and can be removed by careful background subtraction. Similar background ions are also obtained from *septum bleed* and silicon grease. [6,58]

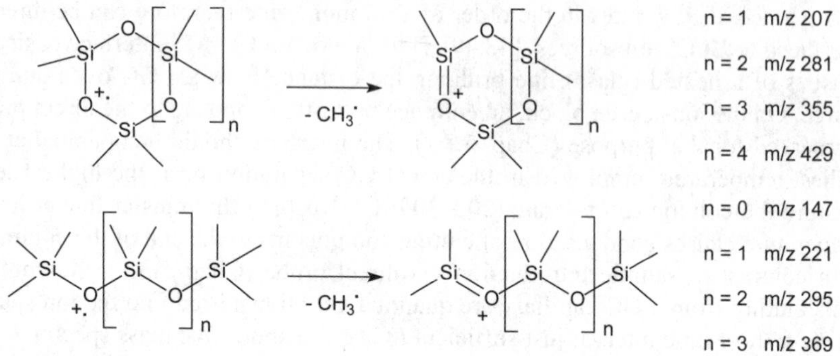

$n = 1$	m/z 207
$n = 2$	m/z 281
$n = 3$	m/z 355
$n = 4$	m/z 429
$n = 0$	m/z 147
$n = 1$	m/z 221
$n = 2$	m/z 295
$n = 3$	m/z 369

Scheme 12.1.

Example: In the partial TIC obtained by GC-EI-MS of an unknown mixture on a HP-5 column the chromatographic peak at 32.6 min is rather weak (Fig. 12.7). Scan 2045 extracted from its maximum yields a spectrum that is mainly due to background ions from column bleed as demonstrated by comparison to the average of scans 2068:2082. Finally, manual background subtraction ((2035:2052)-(2013:2020)) delivers a spectrum of reasonable quality, although some background signals are not completely erased.

12.2.4 Fast GC-MS

When a high sample throughput is of importance, *fast GC-MS* offers a time-saving concept for mixture analysis. [59-61] The GC separation can be accelerated by replacing standard size (20–60 m × 0.25–0.53 mm i.d.) capillary columns by short narrow-bore columns (2–5 m × 50 μm i.d.) and by applying sufficient pressure (8–10 bar) and rapid heating (50–200 °C min^{-1}). Thus, a conventional 30-min separation is compressed into a 3-min or even shorter time frame. However, in fast GC the half life of an eluting peak is too short for use with scanning quadrupole or magnetic sector analyzers. In fast GC-MS, oaTOF analyzers (Chap. 4.2.6) are typically employed because of their ability to acquire about a hundred spectra per second. Furthermore, the high duty cycle of oaTOFs offsets the difference in sensitivity between repetitive scanning and SIM analysis. At somewhat lower rates, advanced oaTOF systems even enable accurate mass determination at reasonable accuracy.

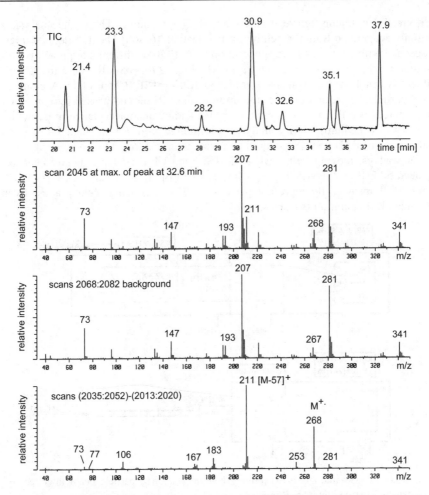

Fig. 12.7. Partial TIC and three EI spectra obtained by GC-MS of an unknown mixture on a 30-m HP-5 column. See text for discussion.

12.3 Liquid Chromatography-Mass Spectrometry

12.3.1 LC-MS Interfaces

Coupling of liquid chromatography to mass spectrometry has not only led to a wide variety of interfaces, but also initiated the development of new ionization methods. [8-13,62] In retrospect, the *moving belt interface* seems rather a curiosity than a LC-interface. The LC effluent is deposited onto a metal wire or belt which is heated thereafter to desolvate the sample. Then, the belt traverses a region of differential pumping before it enters an EI/CI ion source. There, the analyte is rapidly evaporated from the belt by further heating. [63-66] The moving belt interfaces LC with EI, CI, or FI [67] operation (Fig. 12.8). A different approach is presented by the *particle beam interface* (PBI, Fig. 12.8), [68-70] where the LC effluent is desolvated in a manner similar to API or TSP (Chap. 11.1). A cloud of fine particles is created upon evaporation of the solvent and passed into an EI/CI ion source. The PBI is comparatively robust and attained acceptance in particular for low- to medium-polarity analytes. *Continuous-flow FAB* (Chap. 9.4.5) is an adaptation of FAB/LSIMS to liquid chromatography at low flow rates.

Nowadays, moving belt, PBI, API, TSP, and CF-FAB have mostly been replaced by ESI [1,25], APCI, [23,71-73] and APPI [74,75] (Chap. 11). ESI, APCI, and APPI intrinsically represent perfect LC-MS interfacing technologies. Even LC-nanoESI operation is feasible. [18]

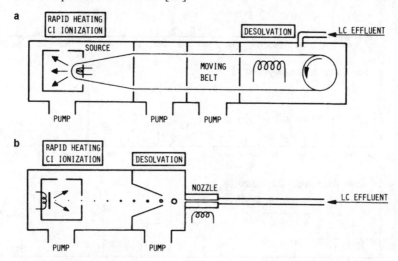

Fig. 12.8. LC-MS interfaces; (**a**) moving belt interface, (**b**) particle beam interface. By courtesy of Thermo Finnigan, Bremen.

Example: UV photodiode array (PDA) and ESI-TOF detection can be combined if the effluent is split or the PDA precedes the ESI interface. The detection methods complement each other in that their different sensitivities towards components of a mixture prevent substances from being overlooked. RICs help to differentiate a targeted compound – an unknown impurity in this case – from others and to identify eventually present isomers. Finally, accurate mass measurement helps in the identification of the unknown (Fig. 12.9). [25]

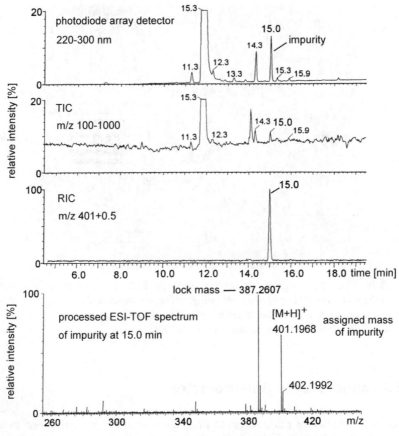

Fig. 12.9. Liquid chromatograms (from top) by photodiode array detection, TIC and RIC from ESI-MS and accurate mass measurement of one component. Adapted from Ref. [25] by permission. © Elsevier Science, 2001.

12.3.2 Multiplexed Electrospray Inlet Systems

Combinatorial chemistry and other high-throughput screening applications have forced the development of *multiplexed electrospray inlet systems* (MUX). Up to eight chromatographs can be connected in parallel to a single mass spectrometer by means of a MUX interface (Fig. 12.10). [76,77] Similar to fast GC-MS, this requires a sufficiently high speed of spectral acquisition and a high sensitivity of the instrument to compensate for the loss of sample by the alternating mode of operation. [78]

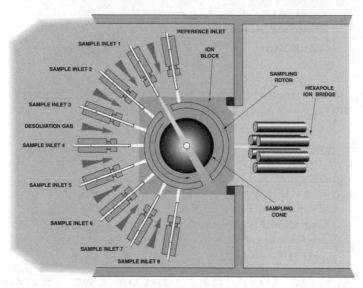

Fig. 12.10. Micromass z-spray with MUX technology. Eight ESI sprayers plus one reference sprayer are adjusted around the sampling orifice. However, only one sprayer has access to the orifice at a time, the others are blinded by a rotating aperture. By courtesy of Waters Corporation, MS Technologies, Manchester, UK.

12.3 Tandem Mass Spectrometry

In tandem MS, two or more stages of mass analysis are combined in one experiment. [79,80] Each stage provides an added dimension in terms of isolation, selectivity, or structural information to the analysis. Therefore, a tandem MS stage is equivalent to a chromatographic separation, provided the separation of isomers is not required. While chromatography distinguishes substances by their retention time, tandem MS isolates them by mass. [2,3,25] The principles of tandem MS have been discussed and some applications for structure elucidation and quantitation have already been shown (Table 12.1). However, the aspect of increased selectivity has not been addressed so far. [81]

The value of SIM in detecting or even quantifying a target trace compound is largely diminished if the compound of interest yields ions of *m/z* values where abundant background ions are also present. Techniques derived from *constant neutral loss* and *precursor ion scanning* are the key to added selectivity. Instead of simply monitoring a set of fragment ions as done by extracting RICs or acquiring SIM traces, these methods allow for the detection of a signal at a certain *m/z* only if the corresponding ion has been created from a precursor ion by a predefined process. This technique is known as *selected reaction monitoring* (SRM). SRM distinguishes the analyte ion from those of unspecific matrix compounds by classifying it due to a characteristic fragmentation process. [45,82-84]

Table 12.1. Tandem mass spectrometry referred to throughout the book

Chapter	Aspect of Tandem MS
2.12	CID and other activation techniques, NR-MS and ion-molecule reactions
4.2.7	tandem MS on TOF instruments
4.3.6	tandem MS with magnetic sector instruments; example: $B^2E = constant$ linked scan for caffeine quantitation using [D$_3$]caffeine internal standard
4.4.5	tandem MS with triple quadrupole analyzers
4.5.6	tandem MS with the quadrupole ion trap; example: LC-MS^4 on quadrupole ion trap to identify cyclic peptides
4.6.8	tandem MS with FT-ICR instruments
9.4.7	example: CID-FAB-MS/MS for peptide sequencing
10.5.4	example: structure elucidation of carbohydrates by PSD-MALDI
11.2.3	example: nanoESI peptide sequencing from a tryptic digest
11.5.5	example: structure elucidation of a nonasaccharide by nanoESI-CID-MS/MS on a Q-TOF hybrid instrument

Example: SIM is not sufficient for the LC-MS detection of 100 pg dextrometorphan (DEX) spiked into 1 ml of human plasma. The corresponding signal of the $[M+H]^+$ ion in the SIM trace at m/z 272 is barely detectable, whereas the SRM chromatogram obtained from the reaction $[M+H]^+ \rightarrow [M–C_8H_{15}N]^+$ shows a clean background and a signal-to-noise enhancement of more than 50-fold (Fig. 12.11). [25] Thus, replacing SIM by SRM offers not only improved detection limits due to increased selectivity; it can also serve to simplify clean-up procedures, to reduce sample amounts, or to speed up analysis by reducing LC time.

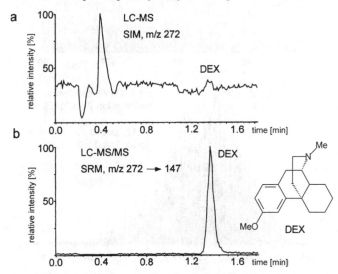

Fig. 12.11. The effect of SRM as compared to SIM in the detection of dextrometorphan. Reproduced from Ref. [25] by permission. © Elsevier Science, 2001.

> **Note:** SRM is preferably performed on triple quadrupole, quadrupole ion trap, and Q-TOF hybrid instruments due to their ease of setting up the experiment and to their speed of switching between channels if monitoring of multiple reactions plays a role.

12.4. Ultrahigh-Resolution Mass Spectrometry

High and in particular ultrahigh-resolution in combination with a soft ionization method such as ESI, MALDI, or FD presents another way to achieve the separation of the molecular species contained in a mixture. Given a sufficient level of resolution, isobaric ions are displayed separately in the range of their common nominal mass value (Chap. 3.3.2, 3.4).

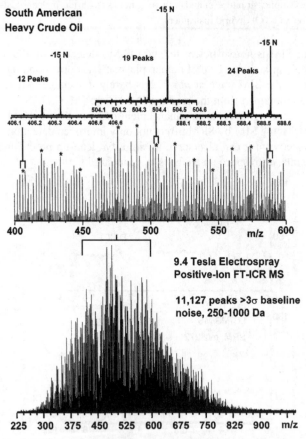

Fig. 12.12. Positive-ion ESI-FT-ICR spectrum of crude oil. The mass scale is successively expanded from the broadband spectrum into detail at three nominal mass values. Adapted from Ref. [86] by permission. © American Chemical Society, 2002.

Classically, high-resolution work is the domain of double-focusing magnetic sector instruments. More recently, TOF and to a certain degree triple quadrupole instruments are also capable of resolutions up to about 20,000. However, the rapid development of FT-ICR instruments has established those as the systems of choice if ultrahigh-resolution (>100,000) and highest mass accuracy (1 ppm) are required (Chap. 4.6).

The potential of ultrahigh-resolution mass spectrometry for the analysis of complex chemical mixtures is particularly illustrated by FT-ICR-MS which definitely sets a new standard. For example, ultrahigh-resolution was applied to separate several thousand components in crude oil, [85,86] fuels, [87,88] or explosion residues. [89]

Example: ESI selectively ionizes the basic compounds, i.e., only a small fraction of the entire chemical composition, in a sample of South American crude oil. Nevertheless, the positive-ion ESI-FT-ICR mass spectrum exhibits more than 11,100 resolved peaks, of which >75 % may be assigned to a unique elemental composition ($C_cH_hO_oN_nS_s$). Such a separation in mass is possible because the average mass resolution in the m/z 225–1000 broadband spectrum is approximately 350,000 (Fig. 12.12). This demonstrates the current upper limit for the number of chemically distinct components resolved and identified in a single step. [86]

Reference List

1. Williams, J.D.; Burinsky, D.J. Mass Spectrometric Analysis of Complex Mixtures Then and Now: the Impact of Linking Liquid Chromatography and Mass Spectrometry. *Int. J. Mass Spectrom.* **2001,** *212,* 111-133.

2. McLafferty, F.W. Tandem MS Analysis of Complex Biological Mixtures. *Int. J. Mass Spectrom.* **2001,** *212,* 81-87.

3. Kondrat, R.W. Mixture Analysis by Mass Spectrometry: Now's the Time. *Int. J. Mass Spectrom.* **2001,** *212,* 89-95.

4. Gohlke, R.S.; McLafferty, F.W. Early Gas Chromatography/Mass Spectrometry. *J. Am. Chem. Soc. Mass Spectrom.* **1993,** *4,* 367-371.

5. Message, G.M. *Practical Aspects of Gas Chromatography/Mass Spectrometry;* John Wiley & Sons: New York, 1984.

6. Hübschmann, H.-J. *Handbook of GC/MS: Fundamentals and Applications; Wiley-VCH:* Weinheim, 2001.

7. Budde, W.L. *Analytical Mass Spectrometry;* ACS and Oxford University Press: Washington, D.C. and Oxford, 2001.

8. *Current Practice of Liquid Chromatography-Mass Spectrometry;* Niessen, W.M.A.; Voyksner, R.D., editors; Elsevier: Amsterdam, 1998.

9. Ardrey, R.E. *Liquid Chromatography-Mass Spectrometry - An Introduction;* John Wiley & Sons: Chichester, 2003.

10. Horning, E.C.; Carroll, D.I.; Dzidic, I.; Haegele, K.D.; Horning, M.G.; Stillwell, R.N. Atmospheric Pressure Ionization (API) MS. Solvent-Mediated Ionization of Samples Introduced in Solution and in a Liquid Chromatograph Effluent Stream. *J. Chromatogr. Sci.* **1974,** *12,* 725-729.

11. Blakley, C.R.; Vestal, M.L. Thermospray Interface for LC-MS. *Anal. Chem.* **1983,** *55,* 750-754.

12. Vestal, M.L. High-Performance Liquid Chromatography-Mass Spectrometry. *Science* **1984,** *226,* 275-281.

13. Abian, J. The Coupling of Gas and Liquid Chromatography with Mass Spectrometry. *J. Mass Spectrom.* **1999,** *34,* 157-168.

14. Smith, R.D.; Barinaga, C.J.; Udseth, H.R. Improved Electrospray Ionization Interface for Capillary Zone Electrophoresis-Mass Spectrometry. *Anal. Chem.* **1988,** *60,* 1948-1952.

15. Schrader, W.; Linscheid, M. Styrene Oxide DNA Adducts: in Vitro Reaction and Sensitive Detection of Modified Oligonucleotides Using Capillary Zone Electrophoresis Interfaced to ESI-MS. *Archives of Toxicology* **1997**, *71*, 588-595.

16. Tanaka, Y.; Kishimoto, Y.; Otsuga, K.; Terabe, S. Strategy for Selecting Separation Solutions in CE-MS. *J. Chromatogr. A* **1998**, *817*, 49-57.

17. Siethoff, C.; Nigge, W.; Linscheid, M. Characterization of a CZE/ESI-MS Interface. *Anal. Chem.* **1998**, *70*, 1357-1361.

18. Hsieh, F.; Baronas, E.; Muir, C.; Martin, S.A. A Novel Nanospray CE/MS Interface. *Rapid Commun. Mass Spectrom.* **1999**, *13*, 67-72.

19. Smith, R.D.; Felix, W.D.; Fjeldsted, J.C.; Lee, M.L. Capillary Column Supercritical Fluid Chromatography Mass Spectrometry. *Anal. Chem.* **1982**, *54*, 1883-1885.

20. Arpino, P.J.; Haas, P. Recent Developments in SFC-MS Coupling. *J. Chromatogr. A* **1995**, *703*, 479-488.

21. Pinkston, J.D.; Chester, T.L. Guidelines for Successful SFC/MS. *Anal. Chem.* **1995**, *67*, 650A-656A.

22. Sjöberg, P.J.R.; Markides, K.E. New SFC Interface Probe for ESI and APCI-MS. *J. Chromatogr. A* **1997**, *785*, 101-110.

23. Combs, M.T.; Ashraf-Khorassani, M.; Taylor, L.T. Packed SFC-MS: A Review. *J. Chromatogr. A* **1998**, *785*, 85-100.

24. *Forensic Applications of Mass Spectrometry;* Yinon, J., editor; CRC Press: Boca Raton, 1994.

25. Hoke, S.H.II.; Morand, K.L.; Greis, K.D.; Baker, T.R.; Harbol, K.L.; Dobson, R.L.M. Transformations in Pharmaceutical Research and Development, Driven by Innovations in Multidimensional Mass Spectrometry-Based Technologies. *Int. J. Mass Spectrom.* **2001**, *212*, 135-196.

26. *Mass Spectrometry in Drug Discovery;* Rossi, D.T.; Sinz, M.W., editors; Marcel Dekker: New York, 2002.

27. *Modern Mass Spectrometry;* Schalley, C.A., editor; Springer: New York, 2003.

28. Hites, R.A.; Biemann, K. A Computer-Compatible Digital Data Acquisition System for Fast-Scanning, Single-Focusing Mass Spectrometers. *Anal. Chem.* **1967**, *39*, 965-970.

29. Hites, R.A.; Biemann, K. Mass Spectrometer-Computer System Particularly Suited for Gas Chromatography of Complex Mixtures. *Anal. Chem.* **1968**, *40*, 1217-1221.

30. Hites, R.A.; Biemann, K. Computer Evaluation of Continuously Scanned Mass Spectra of Gas Chromatographic Effluents. *Anal. Chem.* **1970**, *42*, 855-860.

31. Bertholf, R.L. Gas Chromatography and Mass Spectrometry in Clinical Chemistry, in *Encyclopedia of Analytical Chemistry*, Meyers, R.A., editor; John Wiley & Sons: Chichester, 2000; pp. 1314-1336.

32. Spies-Martin, D.; Sommerburg, O.; Langhans, C.-D.; Leichsenring, M. Measurement of 4-Hydroxynonenal in Small Volume Blood Plasma Samples: Modification of a GC-MS Method for Clinical Settings. *Journal of Chromatography B* **2002**, *774*, 231-239.

33. Price, P. Standard Definitions of Terms Relating to Mass Spectrometry. A Report From the Committee on Measurements and Standards of the Amercian Society for Mass Spectrometry. *J. Am. Soc. Mass Spectrom.* **1991**, *2*, 336-348.

34. Henneberg, D. Combination of Gas Chromatography and Mass Spectrometry for the Analysis of Organic Mixtures. *Zeitschrift für Analytische Chemie* **1961**, *183*, 12-23.

35. Sweeley, C.C.; Elliot, W.H.; Fries, I.; Ryhage, R. Mass Spectrometric Determination of Unresolved Components in Gas Chromatographic Effluents. *Anal. Chem.* **1966**, *38*, 1549-1553.

36. Crosby, N.T.; Foreman, J.K.; Palframan, J.F.; Sawyer, R. Estimation of Steam-Volatile N-Nitrosamines in Foods at the 136µg/Kg Level. *Nature* **1972**, *238*, 342-343.

37. Brooks, C.J.W.; Middleditch, B.S. Uses of Chloromethyldimethylsilyl Ethers As Derivatives for Combined GC-MS of Steroids. *Anal. Lett.* **1972**, *5*, 611-618.

38. Young, N.D.; Holland, J.F.; Gerber, J.N.; Sweeley, C.C. Selected Ion Monitoring for Multicomponent Analyses by Computer Control of Accelerating Voltage and Magnetic Field. *Anal. Chem.* **1975**, *47*, 2373-2376.

39. Tanchotikul, U.; Hsieh, T.C.Y. An Improved Method for Quantification of 2-Acetyl-1-Pyrroline, a "Popcorn"-Like Aroma, in Aromatic Rice by High-Resolution GC-MS/SIM. *J. Agric. Food Chem.* **1991**, *39*, 944-947.

40. Middleditch, B.S.; Desiderio, D.M. Comparison of SIM and Repetitive Scanning

During GC-MS. *Anal. Chem.* **1973**, *45*, 806-808.

41. Millington, D.S.; Buoy, M.E.; Brooks, G.; Harper, M.E.; Griffiths, K. Thin-Layer Chromatography and HR-SIM for the Analysis of C_{19} Steroids in Human Hyperplastic Prostate Tissue. *Biomed. Mass Spectrom.* **1975**, *2*, 219-224.

42. Thorne, G.C.; Gaskell, S.J.; Payne, P.A. Approaches to the Improvement of Quantitative Precision in SIM: High Resolution Applications. *Biomed. Mass Spectrom.* **1984**, *11*, 415-420.

43. Tong, H.Y.; Monson, S.J.; Gross, M.L.; Huang, L.Q. Monobromopolychlorodibenzo-*p*-Dioxins and Dibenzofurans in Municipal Waste Incinerator Fly Ash. *Anal. Chem.* **1991**, *63*, 2697-2705.

44. Shibata, A.; Yoshio, H.; Hayashi, T.; Otsuki, N. Determination of Phenylpyruvic Acid in Human Urine and Plasma by GC-NICI-MS. *Shitsuryo Bunseki* **1992**, *40*, 165-171.

45. Tondeur, Y.; Albro, P.W.; Hass, J.R.; Harvan, D.J.; Schroeder, J.L. Matrix Effect in Determination of 2,3,7,8-Tetrachlorodibenzodioxin by MS. *Anal. Chem.* **1984**, *56*, 1344-1347.

46. Heumann, K.G. Isotope Dilution Mass Spectrometry of Inorganic and Organic Substances. *Fresenius Z. Anal. Chem.* **1986**, *325*, 661-666.

47. Calder, A.G.; Garden, K.E.; Anderson, S.E.; Lobley, G.E. Quantitation of Blood and Plasma Amino Acids Using Isotope Dilution GC-EI-MS with U-^{13}C Amino Acids As Internal Standards. *Rapid Commun. Mass Spectrom.* **1999**, *13*, 2080-2083.

48. Gudzinowicz, B.J.; Gudzinowizc, M.J.; Martin, H.F. The GC-MS Interface, in *Fundamentals of integrated GC-MS Part III*, Marcel Dekker: New York, 1977; Chapter B, pp. 58-181.

49. McFadden, W.H. Interfacing Chromatography and Mass Spectrometry. *J. Chromatogr. Sci.* **1979**, *17*, 2-16.

50. Lindeman, L.P.; Annis, J.L. A Conventional Mass Spectrometer As a Detector for Gas Chromatography. *Anal. Chem.* **1960**, *32*, 1742-1749.

51. Gohlke, R.S. Time-of-Flight Mass Spectrometry: Application to Capillary-Column GC. *Anal. Chem.* **1962**, *34*, 1332-1333.

52. McFadden, W.H.; Teranishi, R.; Black, D.R.; Day, J.C. Use of Capillary Gas Chromatography With a Time-of-Flight Mass Spectrometer. *J. Food Sci.* **1963**, *28*, 316-319.

53. Dandeneau, R.D.; Zerenner, E.H. An Investigation of Glasses for Capillary Chromatography. *J. High Resol. Chromatogr. /Chromatogr. Commun.* **1979**, *2*, 351-356.

54. Dandeneau, R.D.; Zerenner, E.H. The Invention of the Fused-Silica Column: an Industrial Perspective. *LC-GC* **1990**, *8*, 908-912.

55. *Handbook of Derivates for Chromatography;* 1st ed.; Blau, G.; King, G.S., editors; Heyden & Son: London, 1977.

56. Halket, H.M.; Zaikin, V.G. Derivatization in Mass Spectrometry -1. Silylation. *Eur. Mass Spectrom.* **2003**, *9*, 1-21.

57. Aubert, C.; Rontani, J.-F. Perfluoroalkyl Ketones: Novel Derivatization Products for the Sensitive Determination of Fatty Acids by GC-MS in EI and NICI Modes. *Rapid Commun. Mass Spectrom.* **2000**, *14*, 960-966.

58. Spiteller, G. Contaminants in Mass Spectrometry. *Mass Spectrom. Rev.* **1982**, *1*, 29-62.

59. Leclerq, P.A.; Camers, C.A. High-Speed GC-MS. *Mass Spectrom. Rev.* **1998**, *17*, 37-49.

60. Prazen, B.J.; Bruckner, C.A.; Synovec, R.E.; Kowalski, B.R. Enhanced Chemical Analysis Using Parallel Column Gas Chromatography With Single-Detector TOF-MS and Chemometric Analysis. *Anal. Chem.* **1999**, *71*, 1093-1099.

61. Hirsch, R.; Ternes, T.A.; Bobeldijk, I.; Weck, R.A. Determination of Environmentally Relevant Compounds Using Fast GC/TOF-MS. *Chimia* **2001**, *55*, 19-22.

62. van der Greef, J.; Niessen, W.M.A.; Tjaden, U.R. Liquid Chromatography-Mass Spectrometry. The Need for a Multidimensional Approach. *J. Chromatogr.* **1989**, *474*, 5-19.

63. McFadden, W.H.; Schwartz, H.L.; Evans, S. Direct Analysis of Liquid Chromatographic Effluents. *J. Chromatogr.* **1976**, *122*, 389-396.

64. Karger, B.L.; Kirby, D.P.; Vouros, P.; Foltz, R.L.; Hidy, B. On-Line Reversed Phase LC-MS. *Anal. Chem.* **1979**, *51*, 2324-2328.

65. Millington, D.S.; Yorke, D.A.; Burns, P. A New Liquid Chromatography-Mass Spectrometry Interface. *Adv. Mass Spectrom.* **1980**, *8B*, 1819-1825.

66. Games, D.E.; Hirter, P.; Kuhnz, W.; Lewis, E.; Weerasinghe, N.C.A.; West-

wood, S.A. Studies of Combined LC-MS With a Moving-Belt Interface. *Journal of Chromatography* **1981**, *203*, 131-138.

67. Liang, Z.; Hsu, C.S. Molecular Speciation of Saturates by Online Liquid Chromatography-Field Ionization Mass Spectrometry. *Energy Fuels* **1998**, *12*, 637-643.

68. Willoughby, R.C.; Browner, R.F. Monodisperse Aerosol Generation Interface for Combining LC With MS. *Anal. Chem.* **1984**, *56*, 2625-2631.

69. Brauers, F.; von Bünau, G. Mass Spectrometry of Solutions: a New Simple Interface for the Direct Introduction of Liquid Samples. *Int. J. Mass Spectrom. Ion Proc.* **1990**, *99*, 249-262.

70. Winkler, P.C.; Perkins, D.D.; Williams, D.K.; Browner, R.F. Performance of an Improved Monodisperse Aerosol Generation Interface for LC-MS. *Anal. Chem.* **1988**, *60*, 489-493.

71. Chesnov, S.; Bigler, L.; Hesse, M. Detection and Characterization of Natural Polyamines by HPLC-APCI (ESI) MS. *Eur. Mass Spectrom.* **2002**, *8*, 1-16.

72. Hayen, H.; Karst, U. Strategies for the LC-MS Analysis of Non-Polar Compounds. *J. Chromatogr. A* **2003**, *1000*, 549-565.

73. Reemtsma, T. Liquid Chromatography-Mass Spectrometry and Strategies for Trace-Level Analysis of Polar Organic Pollutants. *J. Chromatogr. A* **2003**, *1000*, 477-501.

74. Robb, D.B.; Covey, T.R.; Bruins, A.P. Atmospheric Pressure Photoionization: an Ionization Method for LC-MS. *Anal. Chem.* **2000**, *72*, 3653-3659.

75. Raffaeli, A.; Saba, A. Atmospheric Pressure Photoionization Mass Spectrometry. *Mass Spectrom. Rev.* **2003**, *22*, 318-331.

76. Sage, A.B.; Giles, K. High Throughput Parallel LC/MS Analysis of Multiple Liquid Streams. *GIT Laboratory Journal* **2000**, *4*, 35-36.

77. Schrader, W.; Eipper, A.; Pugh, D.J.; Reetz, M.T. Second-Generation MS-Based High-Throughput Screening System for Enantioselective Catalysts and Biocatalysts. *Canadian J. Chem.* **2002**, *80*, 626-632.

78. Morrison, D.; Davies, A.E.; Watt, A.P. An Evaluation of a Four-Channel Multiplexed ESI Tandem MS for Higher Throughput Quantitative Analysis. *Anal. Chem.* **2002**, *74*, 1896-1902.

79. *Tandem Mass Spectrometry;* 1st ed.; McLafferty, F.W., editor; John Wiley & Sons: New York, 1983.

80. Busch, K.L.; Glish, G.L.; McLuckey, S.A. *Mass Spectrometry/Mass Spectrometry;* 1st ed.; Wiley VCH: New York, 1988.

81. Moritz, T.; Olsen, J.E. Comparison Between High-Resolution Selected Ion Monitoring, Selected Reaction Monitoring, and Four-Sector Tandem Mass Spectrometry in Quantitative Analysis of Gibberellins in Milligram Amounts of Plant Tissue. *Anal. Chem.* **1995**, *67*, 1711-1716.

82. Johnson, J.V.; Yost, R.A.; Faull, K.F. Tandem MS for the Trace Determination of Tryptolines in Crude Brain Extracts. *Anal. Chem.* **1984**, *56*, 1655-1661.

83. Dams, R.; Murphy, C.M.; Lambert, W.E.; Huestis, M.A. Urine Drug Testing for Opioids, Cocaine, and Metabolites by Direct Injection LC/Tandem MS. *Rapid Commun. Mass Spectrom.* **2003**, *17*, 1665-1670.

84. Miao, X.-S.; Metcalfe, C.D. Determination of Carbamazepine and Its Metabolites in Aqueous Samples Using LC-ESI Tandem MS. *Anal. Chem.* **2003**, *75*, 3731-3738.

85. Hughey, C.A.; Hendrickson, C.L.; Rodgers, R.P.; Marshall, A.G. Kendrick Mass Defect Spectrum: A Compact Visual Analysis for Ultrahigh-Resolution Broadband Mass Spectra. *Anal. Chem.* **2001**, *73*, 4676-4681.

86. Hughey, C.A.; Rodgers, R.P.; Marshall, A.G. Resolution of 11,000 Compositionally Distinct Components in a Single ESI-FT-ICR Mass Spectrum of Crude Oil. *Anal. Chem.* **2002**, *74*, 4145-4149.

87. Hsu, C.S.; Liang, Z.; Campana, J.E. Hydrocarbon Characterization by Ultrahigh Resolution FT-ICR-MS. *Anal. Chem.* **1994**, *66*, 850-855.

88. Hughey, C.A.; Hendrickson, C.L.; Rodgers, R.P.; Marshall, A.G. Elemental Composition Analysis of Processed and Unprocessed Diesel Fuel by ESI-FT-ICR-MS. *Energy Fuels* **2001**, *15*, 1186-1193.

89. Wu, Z.; Hendrickson, C.L.; Rodgers, R.P.; Marshall, A.G. Composition of Explosives by ESI-FT-ICR-MS. *Anal. Chem.* **2002**, *74*, 1879-1883.

Appendix

1 Isotopic Composition of the Elements

Table A.1 comprises the stable elements from hydrogen to bismuth with the radio-active elements technetium and promethium omitted. Natural variations in isotopic composition of some elements such as carbon or lead do not allow for more accurate values, a fact also reflected in the accuracy of their relative atomic mass. However, exact masses of the isotopes are not affected by varying abundances. The isotopic masses listed may differ up to some 10^{-6} u in other publications.

Table A.1. Isotopic mass, isotopic composition, and relative atomic mass [u] of non-radioactive elements. © IUPAC 2001.

Atomic Symbol	Name	Atomic No.	Mass No.	Isotopic Mass	Isotopic Comp.	Relative Atomic Mass
H	Hydrogen	1	1	1.007825	100	1.00795
			2	2.014101	0.0115	
He	Helium	2	3	3.016029	0.000137	4.002602
			4	4.002603	100	
Li	Lithium	3	6	6.015122	8.21	6.941
			7	7.016004	100	
Be	Beryllium	4	9	9.012182	100	9.012182
B	Boron	5	10	10.012937	24.8	10.812
			11	11.009306	100	
C	Carbon	6	12	12.000000	100	12.0108
			13	13.003355	1.08	
N	Nitrogen	7	14	14.003070	100	14.00675
			15	15.000109	0.369	
O	Oxygen	8	16	15.994915	100	15.9994
			17	16.999132	0.038	
			18	17.999116	0.205	
F	Fluorine	9	19	18.998403	100	18.998403
Ne	Neon	10	20	19.992402	100	20.1798
			21	20.993847	0.30	
			22	21.991386	10.22	
Na	Sodium	11	23	22.989769	100	22.989769
Mg	Magnesium	12	24	23.985042	100	24.3051
			25	24.985837	12.66	
			26	25.982593	13.94	
Al	Aluminium	13	27	26.981538	100	26.981538
Si	Silicon	14	28	27.976927	100	28.0855
			29	28.976495	5.0778	
			30	29.973770	3.3473	

P	Phosphorus	15	31	30.973762	100	30.973762
S	Sulfur	16	32	31.972071	100	32.067
			33	32.971459	0.80	
			34	33.967867	4.52	
			36	35.967081	0.02	
Cl	Chlorine	17	35	34.968853	100	35.4528
			37	36.965903	31.96	
Ar	Argon	18	36	35.967546	0.3379	39.948
			38	37.962776	0.0635	
			40	39.962383	100	
K	Potassium	19	39	38.963706	100	39.0983
			40	39.963999	0.0125	
			41	40.961826	7.2167	
Ca	Calcium	20	40	39.962591	100	40.078
			42	41.958618	0.667	
			43	42.958769	0.139	
			44	43.955481	2.152	
			46	45.953693	0.004	
			48	47.952534	0.193	
Sc	Scandium	21	45	44.955910	100	44.955910
Ti	Titanium	22	46	45.952629	11.19	47.867
			47	46.951764	10.09	
			48	47.947947	100	
			49	48.947871	7.34	
			50	49.944792	7.03	
V	Vanadium	23	50	49.947163	0.250	50.9415
			51	50.943964	100	
Cr	Chromium	24	50	49.946050	5.187	51.9962
			52	51.940512	100	
			53	52.940654	11.339	
			54	53.938885	2.823	
Mn	Manganese	25	55	54.938050	100	54.938050
Fe	Iron	26	54	53.939615	6.37	55.845
			56	55.934942	100	
			57	56.935399	2.309	
			58	57.933280	0.307	
Co	Cobalt	27	59	58.933200	100	58.933200
Ni	Nickel	28	58	57.935348	100	58.6934
			60	59.930791	38.5198	
			61	60.931060	1.6744	
			62	61.928349	5.3388	
			64	63.927970	1.3596	
Cu	Copper	29	63	62.929601	100	63.546
			65	64.927794	44.57	
Zn	Zinc	30	64	63.929147	100	65.39
			66	65.926037	57.37	

	Zn continued		67	66.927131	8.43	
			68	67.924848	38.56	
			70	69.925325	1.27	
Ga	Gallium	31	69	68.925581	100	69.723
			71	70.924705	66.367	
Ge	Germanium	32	70	69.924250	56.44	72.61
			72	71.922076	75.91	
			73	72.923459	21.31	
			74	73.921178	100	
			76	75.921403	20.98	
As	Arsenic	33	75	74.921596	100	74.921596
Se	Selenium	34	74	73.922477	1.79	78.96
			76	75.919214	18.89	
			77	76.919915	15.38	
			78	77.917310	47.91	
			80	79.916522	100	
			82	81.916700	17.60	
Br	Bromine	35	79	78.918338	100	79.904
			81	80.916291	97.28	
Kr	Krypton	36	78	77.920387	0.61	83.80
			80	79.916378	4.00	
			82	81.913485	20.32	
			83	82.914136	20.16	
			84	83.911507	100	
			86	85.910610	30.35	
Rb	Rubidium	37	85	84.911789	100	85.4678
			87	86.909183	38.56	
Sr	Strontium	38	84	83.913425	0.68	87.62
			86	85.909262	11.94	
			87	86.908879	8.48	
			88	87.905614	100	
Y	Yttrium	39	89	88.905848	100	88.905848
Zr	Zirconium	40	90	89.904704	100	91.224
			91	90.905645	21.81	
			92	91.905040	33.33	
			94	93.906316	33.78	
			96	95.908276	5.44	
Nb	Niobium	41	93	92.906378	100	92.906378
Mo	Molybdenum	42	92	91.906810	61.50	95.94
			94	93.905088	38.33	
			95	94.905841	65.98	
			96	95.904679	69.13	
			97	96.906021	39.58	
			98	97.905408	100	
			100	99.907478	39.91	
Ru	Ruthenium	44	96	95.907599	17.56	101.07
			98	97.905288	5.93	

	Ru continued		99	98.905939	40.44	
			100	99.904229	39.94	
			101	100.905582	54.07	
			102	101.904350	100	
			104	103.905430	59.02	
Rh	Rhodium	45	103	102.905504	100	102.905504
Pd	Palladium	46	102	101.905608	3.73	106.42
			104	103.904036	40.76	
			105	104.905084	81.71	
			106	105.903484	100	
			108	107.903894	96.82	
			110	109.905151	42.88	
Ag	Silver	47	107	106.905094	100	107.8682
			109	108.904756	92.90	
Cd	Cadmium	48	106	105.906459	4.35	112.412
			108	107.904184	3.10	
			110	109.903006	43.47	
			111	110.904182	44.55	
			112	111.902757	83.99	
			113	112.904401	42.53	
			114	113.903358	100	
			116	115.904755	26.07	
In	Indium	49	113	112.904061	4.48	114.818
			115	114.903879	100	
Sn	Tin	50	112	111.904822	2.98	118.711
			114	113.902782	2.03	
			115	114.903346	1.04	
			116	115.901744	44.63	
			117	116.902954	23.57	
			118	117.901606	74.34	
			119	118.903309	26.37	
			120	119.902197	100	
			122	121.903440	14.21	
			124	123.905275	17.77	
Sb	Antimony	51	121	120.903818	100	121.760
			123	122.904216	74.79	
Te	Tellurium	52	120	119.904021	0.26	127.60
			122	121.903047	7.48	
			123	122.904273	2.61	
			124	123.902819	13.91	
			125	124.904425	20.75	
			126	125.903306	55.28	
			128	127.904461	93.13	
			130	129.906223	100	
I	Iodine	53	127	126.904468	100	126.904468
Xe	Xenon	54	124	123.905896	0.33	131.29
			126	125.904270	0.33	
			128	127.903530	7.14	

	Xe continued		129	128.904779	98.33	
			130	129.903508	15.17	
			131	130.905082	78.77	
			132	131.904154	100	
			134	133.905395	38.82	
			136	135.907221	32.99	
Cs	Caesium	55	133	132.905447	100	132.905447
Ba	Barium	56	130	129.906311	0.148	137.328
			132	131.905056	0.141	
			134	133.904503	3.371	
			135	134.905683	9.194	
			136	135.904570	10.954	
			137	136.905821	15.666	
			138	137.905241	100	
La	Lanthanum	57	138	137.907107	0.090	138.9055
			139	138.906348	100	
Ce	Cerium	58	136	135.907145	0.209	140.116
			138	137.905991	0.284	
			140	139.905434	100	
			142	141.909240	12.565	
Pr	Praseodymium	59	141	140.907648	100	140.907648
Nd	Neodymium	60	142	141.907719	100	144.24
			143	142.909810	44.9	
			144	143.910083	87.5	
			145	144.912569	30.5	
			146	145.913112	63.2	
			148	147.916889	21.0	
			150	149.920887	20.6	
Sm	Samarium	62	144	143.911995	11.48	150.36
			147	146.914893	56.04	
			148	147.914818	42.02	
			149	148.917180	51.66	
			150	149.917271	27.59	
			152	151.919728	100	
			154	153.922205	85.05	
Eu	Europium	63	151	150.919846	91.61	151.964
			153	152.921226	100	
Gd	Gadolinium	64	152	151.919788	0.81	157.25
			154	153.920862	8.78	
			155	154.922619	59.58	
			156	155.922120	82.41	
			157	156.923957	63.00	
			158	157.924101	100	
			160	159.927051	88.00	
Tb	Terbium	65	159	158.925343	100	158.925343
Dy	Dysprosium	66	156	155.924279	0.21	162.50
			158	157.924405	0.35	

	Dy continued		160	159.925194	8.30	
			161	160.926930	67.10	
			162	161.926795	90.53	
			163	162.928728	88.36	
			164	163.929171	100	
Ho	Holmium	67	165	164.930319	100	164.930319
Er	Erbium	68	162	161.928775	0.42	167.26
			164	163.929197	4.79	
			166	165.930290	100	
			167	166.932045	68.22	
			168	167.932368	79.69	
			170	169.935460	44.42	
Tm	Thulium	69	169	168.934211	100	168.934211
Yb	Ytterbium	70	168	167.933894	0.41	173.04
			170	169.934759	9.55	
			171	170.936322	44.86	
			172	171.936378	68.58	
			173	172.938207	50.68	
			174	173.938858	100	
			176	175.942568	40.09	
Lu	Lutetium	71	175	174.940768	100	174.967
			176	175.942682	2.66	
Hf	Hafnium	72	174	173.940040	0.46	178.49
			176	175.941402	14.99	
			177	176.943220	53.02	
			178	177.943698	77.77	
			179	178.944815	38.83	
			180	179.946549	100	
Ta	Tantalum	73	180	179.947466	0.012	180.9479
			181	180.947996	100	
W	Tungsten	74	180	179.946707	0.40	183.84
			182	181.948206	86.49	
			183	182.950224	46.70	
			184	183.950933	100	
			186	185.954362	93.79	
Re	Rhenium	75	185	184.952956	59.74	186.207
			187	186.955751	100	
Os	Osmium	76	184	183.952491	0.05	190.23
			186	185.953838	3.90	
			187	186.955748	4.81	
			188	187.955836	32.47	
			189	188.958145	39.60	
			190	189.958445	64.39	
			192	191.961479	100	
Ir	Iridium	77	191	190.960591	59.49	192.217
			193	192.962924	100	
Pt	Platinum	78	190	189.959931	0.041	195.078

	Pt continued		192	191.961035	2.311	
			194	193.962664	97.443	
			195	194.964774	100	
			196	195.964935	74.610	
			198	197.967876	21.172	
Au	Gold	79	197	196.966552	100	196.966552
Hg	Mercury	80	196	195.965815	0.50	200.59
			198	197.966752	33.39	
			199	198.968262	56.50	
			200	199.968309	77.36	
			201	200.970285	44.14	
			202	201.970626	100	
			204	203.973476	23.00	
Tl	Thallium	81	203	202.972329	41.892	204.3833
			205	204.974412	100	
Pb	Lead	82	204	203.973029	2.7	207.2
			206	205.974449	46.0	
			207	206.975881	42.2	
			208	207.976636	100	
Bi	Bismuth	83	209	208.980383	100	208.980383
Th	Thorium*	90	232	232.038050	100	232.038050
U	Uranium*	92	234	234.040946	0.0055	238.0289
			235	235.043923	0.73	
			238	238.050783	100	

2 Carbon Isotopic Patterns

Read out the P_{M+1}/P_M ratio from a mass spectrum to calculate the approximate number of carbon atoms. Provided no other element contributing to M+1 is present, an M+1 intensity of 15 %, for example, indicates the presence of 14 carbons. (For the risk of overestimation due to autoprotonation cf. Chap. 7.2.1)

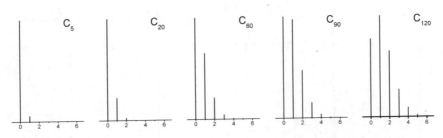

Fig. A.1. Calculated isotopic patterns for carbon. Note the steadily expanding width of the pattern as X+2, X+3, X+4,... become visible. At about C_{90} the X+1 peak reaches the same intensity as the X peak. At higher carbon number it becomes the base peak of the pattern.

Table A.2. Calculated isotopic distributions for carbon

Number of Carbons	X+1	X+2	X+3	X+4	X+5
1	1.1	0.00			
2	2.2	0.01			
3	3.3	0.04			
4	4.3	0.06			
5	5.4	0.10			
6	6.5	0.16			
7	7.6	0.23			
8	8.7	0.33			
9	9.7	0.42			
10	10.8	0.5			
12	13.0	0.8			
15	16.1	1.1			
20	21.6	2.2	0.1		
25	27.0	3.5	0.2		
30	32.3	5.0	0.5		
40	43.2	9.0	1.3	0.1	
50	54.1	14.5	2.5	0.2	0.1
60	65.0	20.6	4.2	0.6	0.2
90	97.2	46.8	14.9	3.5	0.6
120[a]	100.0	64.4	27.3	8.6	2.2

[a] The X peak has an abundance of 77.0 % in that case.

3 Silicon and Sulfur Isotopic Patterns

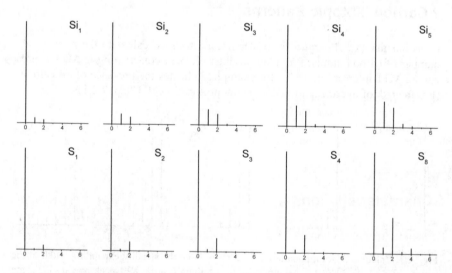

Fig. A.2. Isotopic patterns for silicon and sulfur. The peak at zero position corresponds to the monoisotopic ion at m/z X. The isotopic peaks are then located at m/z = X+1, 2, 3, ...

4 Chlorine and Bromine Isotopic Patterns

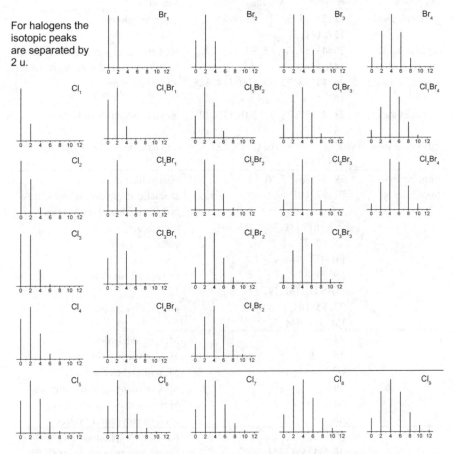

Fig. A.3. Calculated isotopic patterns for combinations of bromine and chlorine. The peak shown at zero position corresponds to the monoisotopic ion at *m/z* X. The isotopic peaks are then located at *m/z* = X+2, 4, 6, ... The numerical value of X is given by the mass number of the monoisotopic combination, e.g., 70 u for Cl_2.

5 Characteristic Ions

Remember that care should be taken when using tables of characteristic ions and neutral losses, because the values listet represent only a minor fraction of the fragmentations possible. There is a substantial risk of getting fixed on a certain fragment or structure too early.

Table A.3. Characteristic ion series and neutral losses

Ion Series	m/z and [M–X]⁺ Ions	Remarks
carbenium ions	15, 29, 43, 57, 71, 85, 99, 113, 127, 141, ...	any alkyl group
acylium ions	29, 43, 57, 71, 85, 99, 113, 127, 141, 155, ...	aliphatic aldehydes, ketones, carboxylic acids and their derivatives
immonium ions	30, 44, 58, 72, 86, 100, 114, 128, 142, 156, ...	aliphatic amines
oxonium ions	31, 45, 59, 73, 87, 101, 115, 129, 143, 157, ...	aliphatic alcohols and ethers
sulfonium ions	47, 61, 75, 89, 103, 117, 131, 145, 159, ...	aliphatic thiols and thioethers
from benzyl	39, 51, 65, 77, 91	phenylalkanes
from benzoyl	51, 77, 105	aromatic aldehydes, ketones, carboxylic acids and derivatives
	$[M-16]^+$, $[M-30]^+$, $[M-46]^{+\bullet}$	nitroarenes
	45, 60, 73, $[M-17]^+$, $[M-45]^+$	carboxylic acids
	59, 74, 87, $[M-31]^+$, $[M-59]^+$	methyl carboxylates
	73, 88, 101, $[M-45]^+$, $[M-73]^+$	ethyl carboxylates
	44	McL of aldehydes
	58	McL of methyl ketones
	60	McL of carboxylic acids
	59	McL of carboxylic acid amides
	74	McL of methyl carboxylates
	88	McL of ethyl carboxylates
	$[M-20]^{+\bullet}$	fluorine compounds
	35, $[M-35]^+$, $[M-36]^{+\bullet}$	chlorine compounds (plus Cl isotopic pattern)
	79, $[M-79]^+$, $[M-80]^{+\bullet}$	bromine compounds (plus Br isotopic pattern)
	127, $[M-127]^+$, $[M-128]^{+\bullet}$	iodine compounds
	$[M-15]^+$	loss of methyl
	$[M-17]^{+\bullet}$	loss of ammonia from amines,
	$[M-17]^+$	loss of OH˙ from (tert.) alcohols
	$[M-18]^{+\bullet}$	loss of water from alcohols
	$[M-27]^{+\bullet}$	loss of HCN from heterocycles or HNC from aromatic amines
	$[M-28]^{+\bullet}$	loss of CO, C_2H_4 or N_2
	$[M-44]^{+\bullet}$	loss of carbon dioxide
	$[M-91]^+$	loss of benzyl

6 Frequent Impurities

Table A.4. Recognition of frequent impurities

m/z	Source
18, 28, 32, 40, 44	residual air
149, 167, 279	phthalic acid esters (plasticizers)
149, 177, 222	diethyl phthalate (plasticizers)
73, 147, 207, 281, 355, 429	silicon grease or GC column bleed (Si_x isotopic pattern)
27, 29, 41, 43, 55, 57, 69, 71, 83, 85, 97, 99, 109, 111, 113, 125, 127, ..., up to m/z 500	hydrocarbons from grease or from suspensions in paraffin
32, 64, 96, 128, 160, 192, 224, 256	sulfur (S_x isotopic pattern)
51, 69, 119, 131, 169, 181, 219, 231, 243, 281, 317, 331, ...	background from PFK

Subject Index

A

α-cleavage
  of acetone 229
  of amines, ethers, and alcohols 235
  of halogenated hydrocarbons 243
  of ketones 232
  of thioethers 242
accumulation
  of spectra 477
accurate mass
  measurement 88
activation energy of the reverse reaction
  36
acylium ions 234
ADC *see* analog-to-digital converter
adiabatic ionization 19
AE *see* appearance energy
allylic bond cleavage
  isomerization prior to 255
  of alkenes 255
amu *see* atomic mass unit
analog-to-digital converter 124
analysis of complex mixtures 475
AP *see* appearance energy
APCI *see* atmospheric pressure
  chemical ionization
API *see* atmospheric pressure ionization
AP-MALDI *see* atmospheric pressure
  MALDI
appearance energy
  definition 23
  determination 48
appearance potential *see* appearance
  energy

APPI *see* atmospheric pressure
  photoionization
array detector 179
atmospheric pressure chemical
  ionization 465
atmospheric pressure ionization 441
  interface for AP-MALDI 431
atmospheric pressure ionization methods
  441
atmospheric pressure MALDI 431
atmospheric pressure photoionization
  467
atomic mass unit 72
atomic number 67
atomic weight *see* relative atomic mass
autoprotonation 333
averaging
  of spectra 477

B

background subtraction
  from spectra 477
bar graph
  representation of spectra 6
base peak
  definition 5
BAT *see* field desorption
benzyl/tropylium isomerization 251
benzylic bond cleavage 249
  phenylalkanes 249
BEqQ 173
Born-Oppenheimer approximation 18
breakdown graph 49

Printing: Mercedes-Druck, Berlin
Binding: Stein+Lehmann, Berlin